Advances in MEMS and Microfluidic Systems

Rajeev Kumar Singh
Department of Mechanical Engineering, Amity University, Noida, India

Rakesh Kumar Phanden
Department of Mechanical Engineering, Amity University, Noida, India

Basant Singh Sikarwar
Department of Mechanical Engineering, Amity University, Noida, India

J. Paulo Davim
Department of Mechanical Engineering, University of Aveiro, Portugal

IGI Global
PUBLISHER of TIMELY KNOWLEDGE

A volume in the Advances in Mechatronics and Mechanical Engineering (AMME) Book Series

Published in the United States of America by
IGI Global
Engineering Science Reference (an imprint of IGI Global)
701 E. Chocolate Avenue
Hershey PA, USA 17033
Tel: 717-533-8845
Fax: 717-533-8661
E-mail: cust@igi-global.com
Web site: http://www.igi-global.com

Library of Congress Cataloging-in-Publication Data

Names: Singh, Rajeev Kumar, 1969- editor.
Title: Advances in MEMS and microfluidic systems / edited by Rajeev Kumar Singh, Rakesh Kumar Phanden, Basant Singh Sikarwar, and J. Paulo Davim.
Description: Hershey, PA : Engineering Science Reference, [2023] | Includes bibliographical references and index. | Summary: "Advances in MEMS and Microfluidic Systems explores the emerging research and advances in MEMS device and microfluidic systems applications. It features in-depth chapters on microfluidic device design and fabrication as well as on the aspects of devices/systems, characterization, and comparative research findings. Covering topics such as biosensors, lab-on-a-chip, and microfluidic technology, this premier reference source is an indispensable resource for engineers, health professionals, students and educators of higher education, librarians, researchers, and academicians"-- Provided by publisher.
Identifiers: LCCN 2023000691 (print) | LCCN 2023000692 (ebook) | ISBN 9781668469521 (h/c) | ISBN 9781668469538 (s/c) | ISBN 9781668469545 (eISBN)
Subjects: LCSH: Microfluidics. | Microelectromechanical systems.
Classification: LCC TJ853.4.M53 A38 2023 (print) | LCC TJ853.4.M53 (ebook) | DDC 621.381--dc23/eng/20230126
LC record available at https://lccn.loc.gov/2023000691
LC ebook record available at https://lccn.loc.gov/2023000692

This book is published in the IGI Global book series Advances in Mechatronics and Mechanical Engineering (AMME) (ISSN: 2328-8205; eISSN: 2328-823X)

British Cataloguing in Publication Data
A Cataloguing in Publication record for this book is available from the British Library.

Advances in Mechatronics and Mechanical Engineering (AMME) Book Series

J. Paulo Davim
University of Aveiro, Portugal

ISSN:2328-8205
EISSN:2328-823X

Mission

With its aid in the creation of smartphones, cars, medical imaging devices, and manufacturing tools, the mechatronics engineering field is in high demand. Mechatronics aims to combine the principles of mechanical, computer, and electrical engineering together to bridge the gap of communication between the different disciplines.

The **Advances in Mechatronics and Mechanical Engineering (AMME) Book Series** provides innovative research and practical developments in the field of mechatronics and mechanical engineering. This series covers a wide variety of application areas in electrical engineering, mechanical engineering, computer and software engineering; essential for academics, practitioners, researchers, and industry leaders.

Coverage

- Manufacturing Methodologies
- Nanomaterials and nanomanufacturing
- Computer-Based Manufacturing
- Autonomous Systems
- Tribology and surface engineering
- Micro and nanomechanics
- Biologically Inspired Robotics
- Control Systems Modelling and Analysis
- Medical Robotics
- Design and Manufacture

IGI Global is currently accepting manuscripts for publication within this series. To submit a proposal for a volume in this series, please contact our Acquisition Editors at Acquisitions@igi-global.com or visit: http://www.igi-global.com/publish/.

The Advances in Mechatronics and Mechanical Engineering (AMME) Book Series (ISSN 2328-8205) is published by IGI Global, 701 E. Chocolate Avenue, Hershey, PA 17033-1240, USA, www.igi-global.com. This series is composed of titles available for purchase individually; each title is edited to be contextually exclusive from any other title within the series. For pricing and ordering information please visit http://www.igi-global.com/book-series/advances-mechatronics-mechanical-engineering/73808. Postmaster: Send all address changes to above address.

Titles in this Series

For a list of additional titles in this series, please visit: www.igi-global.com/book-series

Structural Mathematical Modeling Applications in Technological Machines and Transportation Vehicles
Andrey Eliseev (Irkutsk State Transport University, Russia)
Engineering Science Reference • ©2023 • 300pp • H/C (ISBN: 9781668472378) • US $225.00

Handbook of Research on Emerging Designs and Applications for Microwave and Millimeter Wave Circuits
Jamal Zbitou (LABTIC, ENSA of Tangier, University of Abdelmalek Essaadi, Morocco) Mostafa Hefnawi (Royal Military College of Canada, Canada) Fouad Aytouna (LABTIC, ENSA of Tetouan, University of Abdelmalek Essaadi, Morocco) and Ahmed El Oualkadi (LabTIC, ENSA of Tangier, University of Abdelmalek Essaadi, Morocco)
Engineering Science Reference • ©2023 • 542pp • H/C (ISBN: 9781668459553) • US $325.00

Trends, Paradigms, and Advances in Mechatronics Engineering
Mohamed Arezki Mellal (M'Hamed Bougara University, Algeria)
Engineering Science Reference • ©2023 • 299pp • H/C (ISBN: 9781668458877) • US $225.00

Handbook of Research on Aspects and Applications of Incompressible and Compressible Aerodynamics
Sathish K. Kumar (Nehru Institute of Engineering and Technology, India) and Naren Shankar Radhakrishnan (Vel Tech Rangarajan Dr. Sagunthala R&D Institute of Science and Technology, India)
Engineering Science Reference • ©2022 • 429pp • H/C (ISBN: 9781668442302) • US $335.00

Advances in Computational Approaches in Biomechanics
Pritam Pain (Haldia Institute of Technology, India) Sreerup Banerjee (Haldia Institute of Technology, India) and Goutam Kumar Bose (Haldia Institute of Technology, India)
Engineering Science Reference • ©2022 • 306pp • H/C (ISBN: 9781799890782) • US $260.00

Developing Charging Infrastructure and Technologies for Electric Vehicles
Mohammad Saad Alam (Aligarh Muslim University, Aligarh, India) Reji Kumar Pillai (India Smart Grid Forum, India) and N. Murugesan (India Smart Grid Forum, India)
Engineering Science Reference • ©2022 • 343pp • H/C (ISBN: 9781799868583) • US $240.00

Multidisciplinary Perspectives on Green Electromobility and Charging Stations
Yu Miao (East China University of Sciencer and Technology, China)
Engineering Science Reference • ©2022 • 305pp • H/C (ISBN: 9781799839972) • US $225.00

IGI Global
PUBLISHER of TIMELY KNOWLEDGE

701 East Chocolate Avenue, Hershey, PA 17033, USA
Tel: 717-533-8845 x100 • Fax: 717-533-8661
E-Mail: cust@igi-global.com • www.igi-global.com

Table of Contents

Detailed Table of Contents

Rajeev Kumar Singh, Department of Mechanical Engineering, Amity University, Noida, India
Rakesh Kumar Phanden, Department of Mechanical Engineering, Amity University, Noida, India
Adam Jacso, Budapest University of Technology and Economics, Hungary
Ankur Gupta, Indian Institute of Technology, Jodhpur, India

Microelectromechanical systems (MEMS) refer to systems with characteristic length ranging between 1 μm and 1 mm, fabricated by integrated circuits batch processing technologies and unite mechanical and electrical components. MEMS devices and systems have wide applications in multifarious medical and industrial applications with worldwide market of billions of dollars. Examples of MEMS devices are accelerometers for automobile airbags; micropumps for inkjet printing, electronic cooling, and environmental testing; infrared detectors, digital light processing chip for projection display, etc. Microfluidics refers to fluid flow at a small size scale that causes change in fluid behavior. Microfluidic devices/systems handle a small quantity (micro- or nanoliter) of fluids (liquid or gas). The major application for handling fluids in microfluidics relates to chemical and biomedical analyses. The benefit of application of microfluidics in chemical and biomedical analysis is that they provide a total solution from sample utilization to display of analytical results.

Dhirendra Patel, Department of Mechanical Engineering, Amity University, Noida, India

This chapter deals in detail with microfluidics and its fundamental concept. It gives a clear look at what actually is microfluid and lists down the details of the overview and development of the application of microfluidics. It lays down the properties of microfluidics and also describes how these properties help in making microfluidic devices, and how these devices are different from traditional ones. This chapter hereby illustrates the different challenges related to microfluidic device design and the possible solutions associated with it. It also deals in detail with the various types of microfluidic flow and discusses what is open microfluidic flow and continuous flow microfluid. This chapter has all the necessary content about the application of microfluid in different sectors and also inculcates the future scope of microfluidics. This chapter talks about how microfluidics has become an essential component of various devices that are integral to modern-day society. This chapter contains a discussion of a wide range of applications from drug delivery to DNA sequencing.

Microfluidic technologies have garnered interest due to their capacity to rapidly process samples and accurately control fluids in an assay. This short study seeks to introduce readers to the fascinating realm of microfluidics by introducing some fundamental theoretical principles and illustrative applications. Additionally, the physics of microfluidic devices, the worldwide market, and material types employed are discussed. In the end, the authors analyze promising developments and extrapolate lessons that will help microfluidic technologies advance in the future.

Micropumps play a crucial role in the microfluidic devices through their exact control and manipulation of fluid at the microscale. Advancements have been made in the creation of micropumps for applications ranging from chemical synthesis to biomedical applications. In this chapter an overview of the most recent developments in micropump technology and design, starting with understanding the working principles of micropumps like electrostatic, pneumatic, etc.as well as differences between Mechanical and Non-Mechanical pumps will be described. This chapter contains the recent developments in microfluidic channels, valves and actuators, as well as some general methods of fabrication of these micropumps (like soft-lithography, MEMS fabrication) along with examples. In the penultimate section, the chapter elaborates on the various applications of micropumps in environmental monitoring (such as particle counting, leak detection, etc.) and biomedical applications and point of care equipment. Finally, the chapter concludes about the future and markets of micropumps.

Microfluidics is a promising and upcoming technology for diagnostics and drug developments. With huge potential in bringing a revolution in biotechnology, there is going to be a big commercial demand for microfluidic systems in life science applications. The function of microfluidic system highly depends on the precise control of tiny fluid volumes by micropumps, and hence, design, fabrication, and integration of micropumps are the key components in microfluidic systems. Though the development of micropumps has been paid attention for the past two decades, it is still a great challenge in realizing a reliable and improved performance of a micropump. This chapter discusses a detailed study on various pumping mechanisms of micropumps reported with a perspective on the scope for manufacturing. The issues and challenges in commercialization will be discussed, and the possible solutions to overcome will also be highlighted.

Paper-based microfluidic devices have emerged as a promising technology for a variety of applications, including diagnostics, chemical synthesis, and environmental monitoring. These devices are inexpensive, portable, and easy to use, making them suitable for use in resource-limited settings. The chapter aims to highlight recent advances in the field of paper-based microfluidics, focusing on numerous fabrication techniques, materials, and applications. The authors also review the use of different types of paper and other cellulose-based materials as substrates, as well as the use of different types of adhesives and binders to fabricate the devices. Furthermore, they discuss various applications of paper-based microfluidic devices, including diagnostics, chemical synthesis, and environmental monitoring. Additionally, the chapter will discuss the potential of wireless communication and battery-free power sources integration with paper-based microfluidic devices to create fully autonomous devices that can be used in remote locations, along with current challenges faced and future outlook.

The μ-PADs have been acknowledged as an efficient analytical platform to monitor and detect various contaminants present in the environment. Paper-based devices have always been a choice raw material for several technologies as they aid in developing biodegradable, cost-effective, and readily available units with easily driven and changeable flow rates, and the ability to store reagents. In comparison to conventional analytical systems such as capillary electrophoresis, high-performance liquid chromatography, and gas and liquid mass spectroscopy, the use of μ-PADs brings more significant results in pesticide detection. This chapter discussed the strategies that are commonly used to detect pesticides which are double-catalytic amplification, UV-induced fluorescence spectroscopy, ratiometric fluorescence imaging (based on wavelength intensities to detect variation in the native environment), electrochemical sensing, electrospray ionization, and competitive-inhibiting reaction.

Preface

MEMS and microfabrication are emerging research fields due to their application in several diverse areas such as biology, chemistry, physical sciences, and medicine. Microfluidic devices typically handle the fluid in the range of picolitres to microliters. The advantages associated with these systems are low cost, specificity, low volume sampling, high sensitivity, controlled fluid handling and rapid readout. Lab-on-chip (LOC) devices are one of the emerging research areas of microfluidics which has revolutionized the field of biology and chemistry. LOCs are fabricated using microfabrication technologies in which entire chemical or biological laboratories are integrated in a single chip.

This book can be used as reference book for practicing engineers, academicians and researchers working in the field of MEMS and microfluidics or anyone who wish to gain knowledge of fabrication methods and working of microfluidic devices and MEMS and updates that has taken place in these fields.

This book has 13 chapters covering the basics as well as latest issues in microfluidic systems and MEMS such as microfluidic synthesis of nanoparticles, micropumps for lab-on-chip applications, fabrication of paper-based microfluidic devices, MEMS sensors in IOT integrated framework, microfluidics for rapid detection and quantization of microparticles, fabrication methods for Bio-MEMS, microfluidic technology for advanced drug delivery systems, lab-on-chip biosensors for COVID 19 detection and paper-based microfluidic devices for pesticide detection. The chapters provide systematic approach of devices fabrication and their characterization. Several issues such as leakage control, bonding, safe handling of pathogens etc. are addressed in-depth.

The focus of Chapter 1 is Microelectromechanical Systems (MEMS), which refers to systems with characteristic length ranging between 1 μm and 1 mm, fabricated by integrated circuits batch processing technologies and unite mechanical and electrical components. MEMS devices and systems has wide application in multifarious medical and industrial applications with worldwide market of billions of dollars. Examples of MEMS devices are accelerometers for automobile airbags; micropumps for inkjet printing, electronic cooling and environmental testing; infrared detectors, digital light processing chip for projection display etc. Microfluidics refers to fluid flow at small size scale that causes change in fluid behavior. Microfluidic devices/systems handle a small quantity (micro- or nanoliter) of fluids (liquid or gas). The major application for handling fluids in microfluidics relate to chemical and biomedical analyses. The benefit of application of microfluidics in chemical and biomedical analysis is that they provide a total solution from sample utilization to display of analytical results.

Chapter 2 deals in detail with microfluidics and its fundamental concept. It gives a clear look at what actually is microfluid and lists down the details of the overview and development of the application of microfluidics. It lays down the properties of microfluidics and also describes how these properties help in making microfluidic devices and how these devices are different from traditional ones. This chapter hereby illustrates the different challenges related to microfluidic device design and the possible solutions associated with it. It also deals in detail with the various types of microfluidic flow and discusses what is open microfluidic flow and continuous flow microfluid. This chapter has all the necessary content about the application of microfluid in different sectors and also inculcates the future scope of microfluidics. This chapter talks about how microfluidics has become an essential component of various devices that are integral to modern-day society. This chapter contains a discussion of a wide range of applications from drug delivery to DNA sequencing.

Microfluidic technologies have garnered interest due to their capacity to rapidly process samples and accurately control fluids in an assay. Chapter 3 seeks to introduce readers to the fascinating realm of microfluidics by introducing some fundamental theoretical principles and illustrative applications. Additionally, the physics of microfluidic devices, the worldwide market and material types employed are discussed. In the end, we analyze promising developments and extrapolate lessons that will help microfluidic technologies advance in the future.

Micropumps play a crucial role in the microfluidic devices through their exact control and manipulation of fluid at the microscale. Advancements have been made in the creation of micropumps for applications ranging, from chemical synthesis to biomedical applications. In this chapter an overview of the most recent developments in micropump technology and design, starting with understanding the working principles of micropumps like electrostatic, pneumatic, etc.as well as differences between Mechanical and Non-Mechanical pumps will be described. Chapter 4 contains the recent developments in microfluidic channels, valves and actuators, as well as some general methods of fabrication of these micropumps (like soft-lithography, MEMS fabrication) along with examples. In the penultimate section, the chapter elaborates on the various applications of micropumps in environmental monitoring (such as particle counting, leak detection, etc.) and biomedical applications and point of care equipment. Finally, the chapter concludes about the future and markets of micropumps.

Microfluidics is a promising and upcoming technology for diagnostics and drug developments. With a huge potential in bringing a revolution in biotechnology, there is going to be a big commercial demand for microfluidic systems in life science applications. The function of microfluidic system highly depends on the precise control of tiny fluid volumes by micropumps and hence design, fabrication and integration of micropumps are the key component in microfluidic systems. Though the development of micropumps has been paid attention for the past two decades, it is still a great challenge in realizing a reliable and improved performance of a micropump. Chapter 5 discusses a detailed study on various pumping mechanisms of micropumps reported till date with a perspective on the scope for manufacturing. The issues and challenges in commercialization are discussed in Chapter 5, and the possible solutions to overcome will also be highlighted.

Paper-based microfluidic devices have emerged as a promising technology for a variety of applications, including diagnostics, chemical synthesis, and environmental monitoring. These devices are inexpensive, portable, and easy to use, making them suitable for use in resource-limited settings. Chapter 6 aims to highlight recent advances in the field of paper-based microfluidics, focusing on numerous fabrication techniques, materials, and applications. We also review the use of different types of paper and other cellulose-based materials as substrates, as well as the use of different types of adhesives and binders to fabricate the devices. Furthermore, we discuss various applications of paper-based microfluidic devices, including diagnostics, chemical synthesis and environmental monitoring. Additionally, the chapter will discuss the potential of wireless communication and battery-free power sources integration with paper-based microfluidic devices to create fully autonomous devices which can be used in remote locations; along with current challenges faced and future outlook.

The μ-PADs have been acknowledged as an efficient analytical platform to monitor, and detect various contaminants present in the environment. Paper-based devices have always been a choice raw material for several technologies as they aid in developing biodegradable, cost-effective, and readily available units with easily driven and changeable flow rates, and the ability to store reagents. In comparison to conventional analytical systems such as capillary electrophoresis, high-performance liquid chromatography, and gas and liquid mass spectroscopy, the use of μ-PADs brings more significant results in pesticide detection. Chapter 7 discussed the strategies that are commonly used to detect pesticides which are: double-catalytic amplification, UV-induced fluorescence spectroscopy, ratiometric fluorescence imaging (based on wavelength intensities to detect variation in the native environment), electrochemical sensing, electrospray ionization and competitive-inhibiting reaction.

Chapter 8 highlights how future IoT-based hardware may greatly benefit from the potential of MEMS sensors. MEMS sensors are perfect for use in smaller spaces due to their small size, low cost, and low power requirements, which makes them suitable for IoT applications. These sensors have a high degree of sensitivity, sufficient dependability, and perfect accuracy, which ensures that your data is always accurate. MEMS technology is transforming how many things are produced. The high tolerances present in semiconductor process technology enable them to perform with excellent repeatability & astounding accuracy. The most current development in the electronics industry, nanotechnology, is used to design and create advanced sensors. These sensors offer a better degree of precision and accuracy than ever before because to nanotechnology. It has uses in every sphere of social and economic life and ranges in size from a few millimetres to a few nanometers. Consequently, MEMS sensors and IoT systems are growing in popularity as a result of their capacity to do effective and balanced work across a range of industries.

Microfluidic systems are promising alternatives for continuous synthesis of nanoparticles in controlled environments, overcoming challenges of conventional methods. Stable nanoparticles within a narrow range of size and shape distribution can be reliably produced by controlling geometrical configuration of microfluidic devices and optimizing various operational parameters. Chapter 9 provides an overview of fundamentals and advances in microfluidic synthesis of nanoparticles, categorizing devices into continuous and segmented flow microfluidic platforms. Strategies for functionalization of nanoparticles are explained, and their diverse applications in various fields are elaborated. Future prospects and underlying challenges are also examined for the commercialization of microfluidic devices for nanoparticle synthesis.

Rapid detection and quantization of microparticles whose size is comparable to biological entities provide a new avenue for rapid diagnosis and clinical applications. There are various methodologies have been already adopted for the detection and quantification of these micro entities like changes in permittivity, medium impedance, magnetic permeability, and other means. Recently, microfluidics plays an important role in the process of detection and diagnosis of these microparticles by using chip-based techniques. In this work, we have reviewed the necessity of microfluidic devices, various designs for rapid detection, processes of device fabrication, and range of applicability. Most recent research papers have been selected and studied. The outcomes of these papers are presented systematically in Chapter 10.

Discussed in Chapter 11, MEMS are the exotic relatives of semiconductors and integrated circuits (ICs), originally based on silicon wafer production processes but with the dimensions of space, flexion, and constantly varying output akin to analogue devices. Whereas BioMEMS refers to all interfaces and cross-sections of biological sciences and therapeutic disciplines with microsystems and nanotechnology. Commercial microfluidic devices rely heavily on material selection and production techniques. Molding, replication, casting, and bonding procedures are being developed that required for mass production with repeatability and functional dependability at a low cost, better design freedom, manufacturing ease which is critical in the medical disposable industry. Nowadays, BioMEMSs are the largest and most diversified applications of MEMS devices. Fabrication processes employed many steps process like wafer selection, lithography, etching and substrate binding. Furthermore, packaging for safety and biocompatibility is a major problem for the BioMEMS engineer.

Drug delivery is a procedure to transport drugs to an appropriate site for effective action, achieving the benefits of drugs. In the past few decades, nano-sized drug particles have attracted many researchers to develop an efficient drug delivery system. Microfluidic technology, for its ability to synthesize precise nanoparticles with uniform size, has emerged as a reliable technology for drug delivery. Specially, the microreactor has been established as a promising device due to its ability to provide intensive mixing in low-dimension microchannels having a high surface-to-volume ratio. It synthesizes drug particles with higher encapsulation efficiency and uniform particle size. The enhancement in the encapsulation efficiency results in the controlled release of drug particles at the appropriate site for effective medication. In addition, other applications of microreactors to synthesize various organic and inorganic nanoparticles are also discussed in Chapter 12.

The unexpected increase of the global SARS-CoV-2 or COVID-19 pandemic in early 2020 has required exceptional methods to accomplish instant detection and decrease the spread rate. Typical clinical inspective testing methods have been temperately actual in sufficient the growing demand for rapid recognition point-of-care methods to decelerate the additional remoteness. However, precise point-of-risk analysis of this evolving virus-related infection is superseding the essential for immediate standard functioning procedures and symptom supervision of COVID-19 will be the norm for the future generation. Chapter 13 mainly highlights the advance of lab-on-chip biosensing devices for the identification of COVID-19 infection. In this chapter we will study the roles of using lab-on-chip equipment in response to the present global COVID-19 pandemic, comprising their potential for rapid sample-to-answer processing times, economical, and facility of incorporation into a variety of healthcare backgrounds.

This book provides in-depth knowledge to new and experienced researcher working in the area of MEMS and microfluidics. The contribution to this book includes research work done in universities and research institutions in this area.

Rajeev Kumar Singh
Department of Mechanical Engineering, Amity University, India

Rakesh Kumar Phanden
Department of Mechanical Engineering, Amity University, India

Basant Singh Sikarwar
Department of Mechanical Engineering, Amity University, India

J. Paulo Davim
Department of Mechanical Engineering, University of Aveiro, Portugal

Acknowledgment

We extend our heartfelt gratitude to all those who have contributed to the creation of *Advances in MEMS and Microfluidic Systems*. We sincerely thank the authors for sharing their expertise and knowledge, enriching this book with valuable insights. Our appreciation also goes to the reviewers, publisher and institutions that supported this project. Additionally, we acknowledge the unwavering support of our families, friends, and colleagues. Their encouragement and understanding have been invaluable. Together, these collective efforts have made this book possible, and we are deeply grateful for the contributions of each individual involved in bringing it to fruition.

Introduction

It is with great pleasure that we introduce the edited book *Advances in MEMS and Microfluidic Systems*. In today's fast-paced technological landscape, microelectromechanical systems (MEMS) and microfluidic systems have emerged as transformative fields that drive innovation across various industries. This book serves as a comprehensive collection of the latest advancements and cutting-edge research in these fields, offering valuable insights into their significant impact on diverse applications.

As the world continues to evolve, MEMS and microfluidic systems have become increasingly crucial in addressing complex challenges in healthcare, biotechnology, environmental monitoring, energy systems, and beyond. This book brings together contributions from leading experts and researchers who have dedicated their efforts to pushing the boundaries of these technologies, enabling new possibilities and revolutionizing traditional approaches.

We would like to express our sincere appreciation to the contributors who have shared their expertise and insights, making this book a valuable resource for researchers, academics, students, and professionals in the field. Their dedication and passion for advancing MEMS and microfluidic systems have undoubtedly contributed to the progress and growth of these fields.

This book serves as a comprehensive reference for anyone interested in exploring the forefront of MEMS and microfluidic technologies. It is our hope that this book inspires and sparks new ideas, fostering a continued pursuit of knowledge and innovation in this ever-evolving domain.

Chapter 1
Introduction to MEMS and Microfluidics

Rajeev Kumar Singh
Department of Mechanical Engineering, Amity University, Noida, India

Rakesh Kumar Phanden
Department of Mechanical Engineering, Amity University, Noida, India

Adam Jacso
https://orcid.org/0000-0001-8247-2332
Budapest University of Technology and Economics, Hungary

Ankur Gupta
https://orcid.org/0000-0001-7855-2145
Indian Institute of Technology, Jodhpur, India

ABSTRACT

Microelectromechanical systems (MEMS) refer to systems with characteristic length ranging between 1 μm and 1 mm, fabricated by integrated circuits batch processing technologies and unite mechanical and electrical components. MEMS devices and systems have wide applications in multifarious medical and industrial applications with worldwide market of billions of dollars. Examples of MEMS devices are accelerometers for automobile airbags; micropumps for inkjet printing, electronic cooling, and environmental testing; infrared detectors, digital light processing chip for projection display, etc. Microfluidics refers to fluid flow at a small size scale that causes change in fluid behavior. Microfluidic devices/systems handle a small quantity (micro- or nanoliter) of fluids (liquid or gas). The major application for handling fluids in microfluidics relates to chemical and biomedical analyses. The benefit of application of microfluidics in chemical and biomedical analysis is that they provide a total solution from sample utilization to display of analytical results.

DOI: 10.4018/978-1-6684-6952-1.ch001

MICROELECTROMECHANICAL SYSTEMS (MEMS)

MEMS technology is employed to fabricate miniaturized systems or devices integrating electrical and mechanical components. These systems are created in batches using techniques borrowed from integrated circuit (IC) processing technology. These devices/systems are capable to sense, actuate and control at micro scale and consequently realize result at micro scale.

MEMS is an interdisciplinary field so design and manufacturing requires expertise from broad and multifaceted technical areas comprising electrical engineering, material science, chemical energy, electrical energy, mechanical engineering, optics, integrated circuit fabrication technology, fluid engineering and instrumentation & control. MEMS devices and systems covers substantial market range which is the indicator of perplexity of MEMS. These devices find applications in the field of communication, medical, defense, electronics, automobiles, semiconductor, aerospace, optics etc.

MEMS is the most propitious technology for the current century which has possibilities to transform consumer and industrial product by amalgamation of semiconductor-based microelectronics and microfabrication technology. MEMS devices and systems has capability to affect mode of living and daily life. Semiconductor microfabrication and MEMS are observed as first and second micromanufacturing revolution respectively.

The word MEMS was originated in the United States of America. The same system technology is also referred as Microsystem Technology and Micromachines in Europe and Japan respectively. In a MEMS chip, electronic components are fabricated using IC technology, micromechanical components fabrication requires forming of structures on substrate using micromachining technology. Selective removal of material or addition of structural layers on substrates to create mechanical or electromechanical components requires high aspect ratio micromachining and bulk & surface micromachining. Many MEMS devices are fabricated using silicon as substrate material due to its unique electrical and mechanical properties.

The general form of MEMS device is shown in Figure 1. The device may consist of microelectronic systems, microstructures, microactuators and microsensors. All these subsystems are integrated on the same silicon chip. The role of microsensor is to sense any appreciable change in the system's surrounding by estimating change in mechanical, electromagnetic, magnetic, chemical or thermal equilibrium and generate signal. These signals are processed by microelectronics and based on processing further input signal is generated for microactuators to create a suitable change in device hardware to accomplish the intended function. Microstructure forms the basis to form other miniature components. MEMS devices are miniature in size with microscopic components. MEMS technology was used to fabricate steam engines, motors, levers, pistons, gears etc., all at microscopic level. MEMS covers a wide engineering range. It means not all devices and systems are mechanical and made from silicon, but other systems and devices are fabricated without mechanical component and other diverse range of materials. MEMS represents devices and systems at microscale with complex subsystems having integrated microelectronics fabricated using batch fabrication techniques. Global MEMS market will cross USD 22.9 billion by the year 2027 as estimated by Markets and Markets (Markets and Markets, 2023).

Manufacturing techniques employed for MEMS fabrication include Electric Discharge Machining (EDM), Electrodeposition, Surface Silicon Micromachining, Lithography, Bulk Silicon Micromachining and LIGA (abbreviation for German word Lithographie Galvanoformung Abformung). Sensors and Actuators (collectively known as transducers) forms large portion of MEMS systems. The significant advantages of MEMS technology as compared to macroscopic electromechanical sensors are low noise

Figure 1. Schematic of MEMS and its subsystems

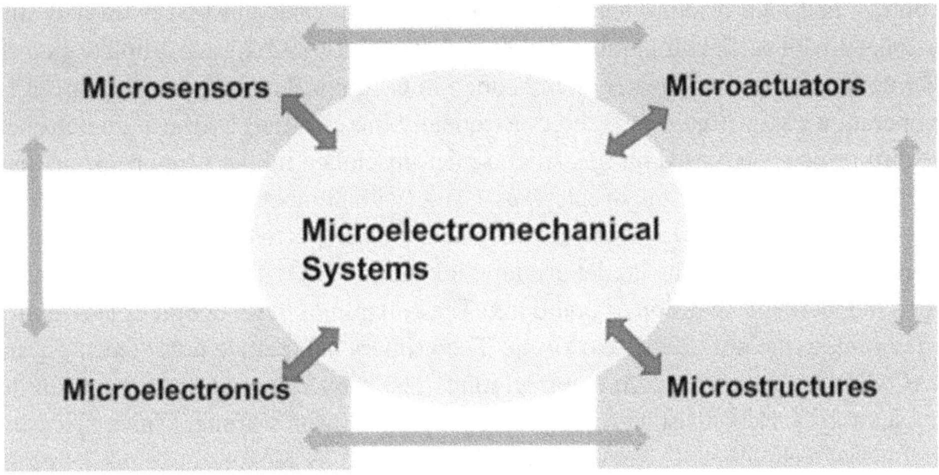

& cost and high sensitivity. Low cost of such systems results due to parallel batch fabrication and obviation of manual assembly steps.

History of MEMS Development

The MEMS fabrication techniques are borrowed from the integrated circuit (IC) technology. The invention of transistor in 1947 at the AT&T Bell Laboratories started revolution in computing systems. The number of transistors on Intel 4004 (1971), Intel 286 (1982), Pentium 3 (1999) chips were 2250, 120000 and 24 million respectively. Apple Inc., consumer electronics company, successfully fabricated M2 processor with 24 billion transistors in year 2022 (Apple, 2022). Gorden Moore, one of the cofounders of Intel corporation, observed in 1965 that number of transistors in IC doubles every 24 months (Schaller, 1997). There were worries about this predicted trend due to limitations imposed by engineering capabilities and fundamental physics, but engineers overcome the scientific barrier to manage the Moore's law continuing. The MEMS techniques were started from year 1960. Anisotropic silicon etching was invented to make 3D features in silicon substrate. The invention of piezoresistivity of single crystal and polycrystalline silicon for microsensors took place around year 1955. Harvey C. Nathanson, an electrical engineer at Westinghouse research labs in Pittsburgh PA invented resonant gate transistor (RGT) in year 1967. In this transistor, gate electrode was movable with respect to the substrate and not attached to the gate oxide. Electrostatic attractive forces were responsible to control the distance between the substrate and gate (Nathanson et al., 1967).

Based on piezoresistive transducer effect of silicon, several sensors for medical and automotive industry were developed. Later on, diaphragm type silicon micromachined pressure sensors were developed at IBM research laboratory. Piezoresistive sensors were embedded in thin silicon diaphragm allowing greater deformation under the application of pressure, hence greater sensitivity was achieved as compared with membrane type pressure sensors. Cost of individual pressure sensors were reduced by micromachining in batch so made them suitable for the medical industry. For normal working conditions, the micromachined pressure sensors are built using silicon substrate and for functioning in high temperature & harsh environment, non-silicon membrane materials e.g., diamond, ceramics etc. are used.

Ink jet printers use ink ejection nozzles realized by silicon micromachining technology. In order to get sharp contrast and high printing resolution, ink ejection nozzles must be extremely miniaturized and closely spaced with sufficient number. Printing is realized by small ink droplets ejected from array of ink jet nozzles. The small heater is embedded in each nozzle. High printing speed is achieved by rapid temperature rise of liquid ink (and consequently ink ejection) by these miniaturized heaters. Hewlett-Packard printers are based on this silicon micromachined ink jet printer nozzle discovered in year 1978. The average weight of ink droplet was 10 ng (with number of nozzles per cartridge as 300) and 10 pl (with resolution as 1000 dpi) in year 1995 and 2004 respectively. Canon pioneered in thermal bubble formation technology for its ink jet printers. Ink jet nozzles with equally small heaters generate bubbles due to the thermal expansion of liquid ink. The collapse of these bubbles eject ink from nozzle and draws more ink to the nozzle for next firing. The primary subtractive color dyes – cyan, magenta and yellow (CMY) are used to realize color printing. The printers based on thermal ink jet printing uses heat resistant dyes. Heat resistant dye is not required for Epson's printers since these are based on piezoelectric ink jet technology. Greater resolution is obtained by piezoelectric ink jet printers as ink dries quickly so spreading on paper is minimum. Today, ink jet technology is not only used for printing texts and images but also used for printing biological molecules (DNA microarray, stem cell printing), deposition of organic chemicals, elements for organic transistors, circuit boards, solar cells, delivering accurate quantity of drugs in pharmaceutical industry etc.

Surface micromachining started in year 1989 when an electrostatically driven micromotor with rotor diameter around 120 μm demonstrated by researchers. Later on, nanomotors with high power and torque were also demonstrated. Beyond year 1990, the Analog Devices and Texas Instruments successfully developed integrated inertial sensors and digital processing chip for automotive air bag system and projection display system respectively. Yearly market revenue of MEMS leading companies like NXP Semiconductors (smart switches for braking system, car audio amplifiers, vehicle network processors etc.), Hewlett-Packard (ink jet printer), Analog Devices (amplifiers, optical sensors, accelerometer, gyroscopes etc.) and Texas Instruments (microcontrollers, processors, sensors etc.) were US $ 11.06 billion, 20.13 billion, 7.318 billion and 183.44 billion respectively in year 2021.

The needs of miniaturized systems in display industry, automotive industry, communication industry, medical industry, gaming industry, cellular phone industry and personal electronics industry motivated the MEMS commercialization efforts. MEMS commercialization enhanced significantly in the last one and half decade due to the increase of new consumer products like electronics readers, handheld smart electronics, medical diagnostic products, cellular telephony, interactive electronics and gaming devices. Many new successful companies in MEMS field are MEMSIC (AMR magnetometer, thermal accelerometer), STMicroelectronics (acceleration sensors, e-compasses, microphones, humidity sensors, T-plus MEMS sensors), eInk (paper like display), SiTime (clock generators, kHZ and MHz MEMS resonators), TDK InvenSence (motion tracking device, CO_2 sensor, microphone etc.), MEDTRONIC (capsule endoscope), Abbott (CardioMEMS Heart Failure System) and Bruel and Kjaer (acoustic sensor). Many reputed large companies like Omron, General Electric, SONY, Honeywell, TRW Automotive, Qualcomm etc. are involved in MEMS products manufacturing. MEMS fabrication techniques are used to develop systems and devices in new areas including cell phone parts, wireless sensor networking, medical diagnosis and therapy, resonators, power and energy (e.g., micro fuel cells, solar cells, energy harvesters, smart grid management) etc.

Characteristics of MEMS

There are three characteristic distinction of MEMS devices viz., miniaturization, parallel fabrication and unification with microelectronics. Miniaturization leads to a greater number of devices per wafer and results reduction in cost per device. Cost of MEMS devices goes down with miniaturization. MEMS devices used in biomedical application (also called Bio-MEMS devices) causes less pain upon intrusion in patients due to less damage to sensory receptors because of miniaturization. Miniaturization results in greater sensitivity and fast response in devices. For example, time constant of ink droplet ejection from nozzle in ink jet printer is around 20 μs. With the help of scaling laws, the behavior of physical phenomena can be predicted for microscale system. Nanoelectromechanical Systems (NEMS) are electromechanical devices with characteristic length scale between 1 nm and 100 nm.

Unlike traditional machining technology, MEMS techniques are capable to fabricate miniaturized two- or three-dimensional features efficiently and reproducibly at low cost. Complex shapes at microscale such as cantilevers, high aspect ratio channels, through/blind holes of any geometry, membranes etc. can be fabricated with the help of photolithography (one of the MEMS techniques). It is impossible to fabricate some of these features at microscale with conventional manufacturing methods. It is a usual practice to make multiple copies of same device using MEMS techniques to lower the cost of individual unit. Uniform and finely defined features are generated across wafers and batches.

Some of the MEMS devices (e.g., digital light processor, accelerometer, ink jet printer head etc.) needs integration of microelectronic circuits to address functioning. For example, each mirror of digital light processor is controlled by CMOS logic circuit that is placed directly under it. This integration improves the quality of signal due to reduction in signal path length and noise. The integration of sensors and actuators with microelectronic systems at single wafer level is termed as monolithic integration. Devices are used in large and dense array micromechanical component integrated with electronic circuitry.

MEMS Devices

Transducer, consists of sensor and actuator, converts signals from one energy domain to some other energy domain. These energy domains can be classified as radiative, mechanical, electrical, magnetic, thermal and chemical domain. The domain and corresponding parameters are listed in Table 1.

The function of sensor is to detect change in environment surrounding it and generate corresponding signal while actuators produce mechanical motion and are hardware elements. Stimulus signal is converted into suitable parameter of energy domain by sensor to be detectable by electronic controllers

Table 1. The energy domains and corresponding parameters

Energy Domain	Common Parameters
Magnetic	Magnetic force, Lorentz force, electromagnetic force, magnetic field strength etc.
Thermal	Heat capacity, thermal conductivity, phase change, heat flux, temperature etc.
Chemical	pH, chemical concentration, reaction rate, DNA hybridization etc.
Mechanical	Pressure, speed, acoustic vibration, force, acceleration, viscosity, hardness, stress, strain etc.
Radiative	X-ray, UV radiation, IR radiation, color, plasma, electromagnetic waves etc.
Electrical	Resistance, current, voltage, capacitance, electrons, inductance, duty cycle, charge etc.

or computers. As an example, the temperature is stimulus for thermocouple which is converted into voltage (an electrical signal) that can be read by electronic controllers. It is not necessary that only two energy domains are involved in transducers but multiple domains may be involved to realize better performance. Each transduction pathway has its own merit based on cross-sensitivity, responsivity, sensitivity, temperature stability and cost. For a particular application, the transduction pathway is selected, taking account of ease in manufacturing, performance, robustness and cost.

Sensor performance can be evaluated based on sensitivity, accuracy, precision, resolution, noise, dynamic range, bandwidth, reliability and interference. Actuator performance is evaluated based on torque and force output capacity, range of motion, dynamic response speed, power consumption and energy efficiency, environmental stability and cross-sensitivity. Micro sensors and actuators have been successfully fabricated using MEMS technology (Liu, 2012).

Journals and Conferences

Main journals include the MEMS research are Journal of Micromechanics and Microengineering, Sensors and Actuators (A, B & C), Journal of Microelectromechanical Systems (JMEMS), Microsystem Technologies, Journal of Applied Physics (JAP), Micro & Nano Letters, Microelectronic Engineering, Nano-Micro Letters, Small and Micromachines. The leading conferences that cover the advances in MEMS and embody program sessions on microscale sensors and actuators are Micro Total Analysis Systems (µTAS), International Society for Optical Engineering (SPIE), International Conference on Solid-State Sensors and Actuators (Transducers), Solid-State Sensor and Actuator Workshop, International Conference on Micro Electro Mechanical Systems, International Conference on Micro and Nanotechnology for Power Generation and Energy Conversion Applications (PowerMEMS), International Conference Perspective technologies and methods in MEMS design and International Conference on MEMS (Maluf & Williams, 2004).

MICROFLUDICS

Fluidic processes are employed for mixing, separation, concentration and moving/stopping fluids in microfluidic systems used in biomedical applications. Pressure is most widely employed as driving force in fluid science. In microfluidics, viscous force is much larger than inertial force indicating low Reynold number flow. In micro flows, the high viscous shear stress owing to the large transverse velocity gradient is compensated by applied pressure. The applied pressure creates instantaneous effect in entire fluid in case of incompressible flows. This pressure gradient moves fluid along with immersed particles. Chip based systems uses peristalsis for fluid motion because rotary pump or positive displacement pump is difficult to manufacture and their integration to microfluidic systems is not easy. Surface tension can also be used as driving force in fluid motion at micro scale. Surface tension dominates at the reduced droplet size. Particle movement efficiency is enhanced in particle laden bulk fluid for discrete droplet flow. Surface electrical potential can be used to alter contact angle of liquid droplet with solid, a phenomenon known as electro-wetting. Varying contact angles will be produced by surface electrical potential gradient covering the droplet leading to unstable surface tension. This force moves the droplet in the direction of electrical potential gradient as determined by surface electrode pattern. Electrokinetic forces can be used to create relative motion between particles and bulk fluid as required in mixing and concentration. Electrokinetic effects apply forces on polarized or charged particles immersed in fluid to effect concentration in microfluidics. Dielectrophoresis can be used to manipulate particles in a fluid flow. Uncharged dielectric particles get polarized when they are kept in varying electric field and thus experiences a dielectrophoretic (DEP) force. Higher electric field gradient creates higher DEP force on the particles. The DEP force produced on the particles may be attractive or repulsive and depends on differences in complex permittivities of medium and particles. Particle separation is one of the applications of DEP force. The global microfluidics market would reach to USD 58.8 billion by the year 2026 (Market and Market, 2020a).

The systems that provide total solution from samples in the field of chemistry and biomedical sciences in microfluidics are known as lab-on-chip, biochip and micro total analysis systems (μTAS). Miniaturization of fluidic system results in less cost per analysis, enhancement of specificity and sensitivity, less contamination risk, reduction in sample/reagent consumption, increase of reliability and reduction of tedious operations. Initially, silicon was used for the fabrication of microfluidic devices but it offered difficulty in optical detection especially for chemical and biomedical analyses. So, glass and polymer had replaced silicon because these materials are optically transparent and having low cost. The polymers used for the fabrication of microfluidic devices were polystyrene (PS), polydimethyl siloxane (PDMS), polycarbonate (PC) and polymethyl methacrylate (PMMA). Among these polymers, PDMS is widely used for the fabrication of microfluidic devices in research labs. Soft lithography is one of the techniques to make devices and systems for PDMS. The reason of application of PDMS are biocompatibility, easy replication, low cost and optical transparency. Day to day applications of this technology includes blood glucose monitoring, pregnancy at-home-testing, COVID 19 testing, hepatitis A, B and C testing, human immunodeficiency virus (HIV) testing etc. Microfluidic chips (sole or with combination of other devices) are used for drug encapsulation, delivery & targeting, cell culture, disease diagnosis, cell analysis, nanoparticle preparation. Microneedles are another illustration of drug delivery based on microfluidics. In it, drugs are delivered through transdermal route accessing the microcirculation of the skin.

Interdependence Between MEMS and Microfluidic Systems

Early stage of microfluidics witnessed the development of micropumps, microvalves and microflow sensors in the late 1980s. Several new terms were coined for this new research discipline viz., "MEMS-fluidics", "microfluids" or "Bio-MEMS". With the advent of nanotechnology, "nanoflows" and "nanofluidics" gain popularity in fluidics research. For MEMS devices, the overall size is smaller than a millimeter while for microfluidic device, the length scale dimension is in microns that govern the flow behavior.

Silicon and other materials are used to fabricate microfluidic devices. Improved performance and novel effects are realized using scaling laws and continuum disruption in these devices. Regardless of the material and connected electronic systems, the fluid flow region has to be miniaturized. It is not necessary to miniaturize the entire microfluidic system. Microfluidic devices handle the fluid in microscopic quantity. The size of the fluid flow region is determined that causes changes in fluid behavior.

Microfluidics is a multidisciplinary field involving mechanical engineers, electrical engineers, biochemists, analytic chemists, chemical engineers etc. Mechanical and electrical engineers focus to reduce the device size using microfabrication technologies and device size is the criterion to define the microfluidics for them. Improved performance and new effects in the field of surface science is explored by biochemists, analytical chemists and chemical engineers using microfluidics. They are primarily interested to reduce the fluid quantity. So, microfluidics for them is the enabling technology to use

Figure 2. Comparison of common objects and microfluidic systems

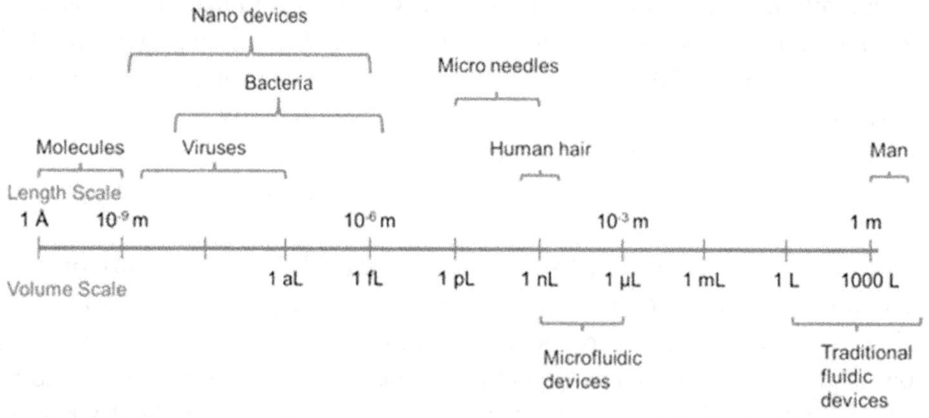

microscopic fluid quantities. Figure 1 shows the size of some of the common objects and microfluidic systems. Benefit of microscopic length scale in microfluidics lies in analyte concentration for sensor. Maximum sensitivity of sensor is dependent on analyte concentration. Sensor efficiency and analyte concentration can be related as

$$v = \frac{1}{N_A \eta_s A_i}$$

Where v represents sample volume, η_s represents sensor efficiency, N_A represents Avogadro number and A_i represents concentration of analyte. Value of η_s varies between 0 and 1. Sample volume kept inside microfluidic device is decided by the concentration of the intended analyte. Sample volume is decided such that it must contain target molecules. Sample volume without any target molecules would not be workable for detection purposes. Analyte concentration in case of human clinical chemistry assay lies in the range 10^{14} to 10^{21} copies/ml. Due to this high analyte concentration, sample volume in femtolitre range is sufficient for detection purposes. Immunoassay concentration range lies between 10^8 and 10^{18} copies/ml. In this case, sample volume on the order of nanoliters is sufficient for detection purposes. The concentration range for deoxyribonucleic acid (DNA) probe assay for bacteria, viruses and genomic molecules lies in the range from 10^2 to 10^7 copies/ml. This requires the sample volume on the order of ml.

Commercial Outlook

With recent advancement in the field of nanotechnology and biotechnology, the microfluidic systems have great commercial potential. These systems have capabilities and novel applications that cannot be realized by traditional systems. Commercial potential of microfluidic systems for real life applications boosted the involvement and interest of industry for microfluidic development and research. Because of the promising commercial potential, microfluidics today became a fast-growing segment of MEMS technology and facilitating technology for biotechnology and nanotechnology. Starting with miniaturized gas chromatograph on silicon wafer and currently devices to detect cancer cells in blood, a wide microfluidics market is now coming up (Nguyen & Wereley, 2006).

Global microfluidics market was estimated to USD 20.7 billion in year 2021 and is projected to reach USD 50.8 billion by year 2026 with a compounded annual growth rate of 23.2% during the predicted period. As per year 2020 statistics, the regions with descending order of global microfluidics markets were North America, Europe, the Asia Pacific and Latin America. The regions behind largest share of microfluidics market in North America were rising occurrences of chronic diseases, increased research funding for proteomics and genomics, enhanced funding for advancement of lab on chip systems, increasing requirement for POC testing, adequate reimbursement for medical devices, lab testing insurance coverage etc. (Market and Market, 2022b). Major international microfluidics companies are Micronit Microfluidics, Claros Diagnostics, Dolomite Microfluidics, Gyros AB, Sphere Fluidics Ltd., RainDance Technologies, Roche Diagnostics, Fluidigm Corporation, Advanced Liquid Logic, Inc., Cepheid, i-STAT Corp., Life Technologies Corporation, Agilent Technologies, Caliper Life Sciences etc. Their products find applications in human diseases detection, drug discovery, biomarker analyses, Immunoassays, genotyping and sequencing, molecular testing, microarray analysis, cell analysis etc. (Volpatti & Yetisen, 2014).

Scientific Outlook

Microfluidics is a multidisciplinary research field involving scientists and researchers from almost all science and engineering disciplines. It attained a rapid growth rate because of novel developments due to enhanced research funding and commercial potential. Novel enabling technologies of microfluidics were developed by mechanical and electrical engineers. MEMS fabrication techniques were borrowed from microelectronic manufacturing and microfluidics represent a part of MEMS technology. Novel fluid phenomena are possible at micro and nano scale in fluid mechanics. Conventional fluid mechanics deals with continuum-based hypotheses while microfluidic systems are dominated by a transitional regime between the molecular and continuum regimes. In microfluidics, computational models, analytical models and in situ micro instruments are used for fluid measurement. New effects are possible in microfluidics for chemists and life scientists. Novel chemical reactions & effects and use of micro instruments has opened the new possibilities in the field of bioengineering and chemistry. Major international conferences dedicated to microfluidics are "International Conference on Microfluidics", "International Conference on Microfluidics and Nanofluidics", "EMBL Conference – Microfluidics", "International Conference on Organ-on-Chip Technologies and Microfluidics (ICOCTM)", "International Conference on Fluid Dynamics and Microfluidics" etc. (Nguyen & Wereley, 2006).

REFERENCES

Apple. (2022). *Apple unveils M2, taking the breakthrough performance and capabilities of M1 even further*. Apple. https://www.apple.com/in/newsroom/2022/06/apple-unveils-m2-with-breakthrough-performance-and-capabilities/

Liu, C. (2012). *Foundations of MEMS* (2nd ed.). Pearson Education, Inc.

Maluf, N., & Williams, K. (2004). An Introduction to Microelectromechanical Systems Engineering. Artech House, Inc.

Markets and Markets. (2023). Microelectromechanical Systems (MEMS) Size. *Markets and Markets*. https://www.marketsandmarkets.com/Market-Reports/mems-market-13689179.html?gclid=EAIaIQobChMIrdXB9ve3_gIVhNCWCh3o2g70EAAYASAAEgIi9fD_BwE

Market and Market. (2020a). Microfluidics Market. *Market and Market.* https://www.marketsandmarkets.com/Market-Reports/microfluidics-market-1305.html

Market and Market. (2022b). *Microfluidics Market by Product (Devices, Components (Chip, Sensor, Pump, Valve)), Application (IVD (POC, Clinical, Veterinary), Research, Manufacturing, Therapeutics), End User (Hospital, Diagnostic Center, Academic Institutes) - Global Forecasts (2022 -)*. Markets and Markets Research Private Ltd. https://www.marketsandmarkets.com/Market-Reports/microfluidics-market-1305.html

Nathanson, H. C., Newell, W. E., Wickstrom, R. A., & Davis, J. R. (1967, March). The resonant gate transistor. *IEEE Transactions on Electron Devices*, *14*(3), 117–133. doi:10.1109/T-ED.1967.15912

Nguyen, N.-T., & Wereley, S. T. (2006). Fundamentals and Applications of Microfluidics. Artech House, Inc.

Schaller, R. R. (1997, June). Moore's law: Past, present and future. *IEEE Spectrum*, *34*(6), 52–59. doi:10.1109/6.591665

Volpatti, L. R., & Yetisen, A. K. (2014, July). Commercialization of microfluidic devices. *Trends in Biotechnology*, *32*(7), 347–350. doi:10.1016/j.tibtech.2014.04.010 PMID:24954000

Chapter 2
Exploring the Frontiers of Microfluidics:
Challenges and Future Prospects

Dhirendra Patel
https://orcid.org/0000-0003-3308-168X
Department of Mechanical Engineering, Amity University, Noida, India

ABSTRACT

This chapter deals in detail with microfluidics and its fundamental concept. It gives a clear look at what actually is microfluid and lists down the details of the overview and development of the application of microfluidics. It lays down the properties of microfluidics and also describes how these properties help in making microfluidic devices, and how these devices are different from traditional ones. This chapter hereby illustrates the different challenges related to microfluidic device design and the possible solutions associated with it. It also deals in detail with the various types of microfluidic flow and discusses what is open microfluidic flow and continuous flow microfluid. This chapter has all the necessary content about the application of microfluid in different sectors and also inculcates the future scope of microfluidics. This chapter talks about how microfluidics has become an essential component of various devices that are integral to modern-day society. This chapter contains a discussion of a wide range of applications from drug delivery to DNA sequencing.

INTRODUCTION TO MICROFLUIDICS

Fundamental Concepts in Micro Fluidics

Microfluidics The field of study known as microfluidics focuses on the handling, movement, and analysis of fluids at the micrometers size (Tabeling P, 2005).The field of micro fluidics is closely related to other fields such as nano-fluidics, macro-fluidics, and biophysics (Chokkalingam et al., 2010). Microfluidics is used to manufacture products, test biological fluids, and even extract DNA from cells. The field of microfluidics is expected to have a deep and lasting impact on the world and will become an essential

DOI: 10.4018/978-1-6684-6952-1.ch002

tool for scientists, researchers, and engineers. Microfluidics as a field is highly multidisciplinary (Kirby BJ, 2010). The application and caliber of this field ranges from pharmaceutical, healthcare, medical diagnostics, environment monitoring and control, chemical and process industries to space application and the list goes on (Karniadakis et al., 2005). As a mechanical engineering graduate, you can work on a variety of microfluidics related platforms. You can grab a research position/teaching position based on your research (Bruus H, 2007). May work on different medical sectors developing/analyzing different process/fabrication methods of microfluidics, you may work collaboratively with medical personnel's developing and searching for various methods of treating different diseases based on advances made with microfluidics, also you may work in the field of space exploration (f V, 2019; Shestopalo et al.,2004;Melin J et al.,2005).

Microfluidics is a field of technology, engineering, and technology that makes use of small fluid flows to manipulate and process substances on a sub-millimeter scale (Frisk T.et al., 2006; Berthier Jet al., 2016). In the simplest sense, microfluidics is the science of exercising control over small fluid flows in the sub-millimeter range (Pfohl T et al.,2003). Through the use of microfluidics, scientists and engineers have been able to perform a wide variety of functions and experiments, such as the generation of electrical currents, the separation of molecules, the delivery of bio-molecules. This wide variety of uses has led to the field becoming one of the most prominent subfields within modern biology and chemistry (D. Patel et al., 2023; Kaigala GV, 2012).

Microfluidics is the study of fluids at the micron level. It is the small-scale study of fluids under the influence of forces and pressures large enough to be observed in macroscopic systems, but small enough to be studied on a microscopic scale (Li C et al., 2013). The main implementation or use of microfluidics is within the fields of medicine, biology, and chemistry (Truckenmüller R, 2002). The ability to manipulate fluids at the micron scale has enabled the development of novel diagnostic and therapeutic technologies (Jeon JS et al., 2011; oung EW et al., 2011).

Microfluidics is the study of fluids at the microscale: small streams of liquid moving under their own power, with the ability to be manipulated and analyzed using advanced techniques (Bouaidat S et al., 2005). The first commercial applications of microfluidics were in the early 2000s to automate the separation of cells and DNA with the use of capillary electrophoresis, which had been used for decades to separate proteins (Ogawa M et al., 2015). Over the last decade, microfluidics has expanded to a variety of different applications, including drug delivery, cell encapsulation, and even the manufacture of complex products like micro-electromechanical systems (MEMS) (Konda A et al,.2021). More recently, the field has begun to merge with other fields, such as bioprinting and biofabrication (Chang H Cand, Yeo L 2009; Prakash M et al.,2007).

Microfluidics is the field of science that studies the movement and transport of fluids through very small channels and channels with very small cross sections (Tenje M st al., 2018). The main applications of microfluidics are in the field of medicine, where they are used to separate, concentrate, and manipulate fluids in the human body (Niu X et al., 2008). Microfluidics have also found a number of applications in other fields, such as biotechnology, where they are used to separate, purify, and manipulate fluids in the laboratory. Microfluidics to a certain extent has the caliber has to revolutionize the way that science is done (Zhang Yet al., 2017; Shilton Ret al., 2014).

Microfluidics is an in-depth study of fluids at the micrometer scale. The first commercial applications of microfluidics were in the early 2000s to automate the separation of cells and DNA using capillary electrophoresis, which had been used for decades to separate proteins. Over the last decade, microfluidics has expanded to multiple different applications, including drug delivery, cell encapsulation, and even the manufacture of complex products like microelectromechanical systems (MEMS) (Shemesh J et al., 2010; Berthier J et al.,2016). More recently, the field has begun to merge with other fields, such as bioprinting and bio fabrication, to create entirely new technologies (Li J et al., 2003).

Microfluidics is a technology that allows for the change/amendment of small volumes of liquid on a small scale. Microfluidics is often used in the field of bioengineering to separate, purify, and concentrate biological samples (Uram JD et al., 2006). Microfluidics has also found applications in the field of nanotechnology, where it has been used to build tiny electronic devices and sensors (Sen YH et al., 2009). The ability to build tiny electronic devices and sensors has led to a wide range of potential applications, such as the checking of the blood glucose level or the measurement of pollutants in the air (Lewpiriyawong N et al., 2011).

Microfluidic devices possess a very wide range of applications, from DNA sequencing to drug delivery (Gnyawali V et al., 2019). The ability to control fluid flow in tiny channels has enabled a host of new applications, but the small size and limited area of traditional microfluidic devices has limited their use to a handful of specialized applications (Xia Net et al.,2006).

The advent of high-throughput manufacturing and microfabrication technologies, however, has enabled the creation of a new class of micro fluidic devices: the nanofluidic device (Nguyen NIT; Wereley S 2006). These devices are smaller than microfluidic devices, but still offer the same level of control over fluid flow (Pawell R et al., 2013).

Microfluidic devices are a key enabling technology for many aspects of modern life. They are used in everything from blood testing to gene sequencing to drug delivery to generate tiny streams of liquid or gas, which can be manipulated and analyzed in ways that would be impossible using conventional devices (Barrett MP et al., 2017). One of the most common uses of microfluidic devices is in the field of DNA sequencing (Jing G et al.,2007). Microfluidic devices are particularly well-suited to DNA sequencing because they can be used to generate small quantities of fluid, which allows for the analysis of very small amounts of fluid at once (Chung BG et al.,2007; Hochstetter A., 2019).

Microfluidics is a field of science and engineering that allows the movement of small quantities of fluids at microscopic scales (Greener et al., 2012). These small-scale flows enable a wide range of applications, from simple droplet generation and measurement to complex fluidic filters and pumps (Greener et al.,2012). A key challenge in the field is to develop microfluidic devices that can perform complex tasks with high accuracy and resolution. One approach to this challenge is to consider the size and shape of microfluidics channels (Eduati F et al.,2018).

The behavior of a liquid in such microchannels differs dramatically from that of a macrofluidics in terms of surface effects and viscosity.Microfluidic device characteristics in terms of size (Liu Z et al.,2015; Liang LG et al.,2017).

All about The Scale of microfluidics

- Human hair diameter 25-100 μm
- Red blood cell diameter 6-8 μm
- Red blood cell thickness ~ 2 μm
- Cellular components (range 10-30 μm)
- Nucleus size ~ 3-10 μm
- Mitochondria size ~ 0.5-1.5 μm
- Virus diameter 10-400 nm
- DNA Alpha helix structure diameter is 2nm

Figure 1. Size characteristics of microfluidics devices

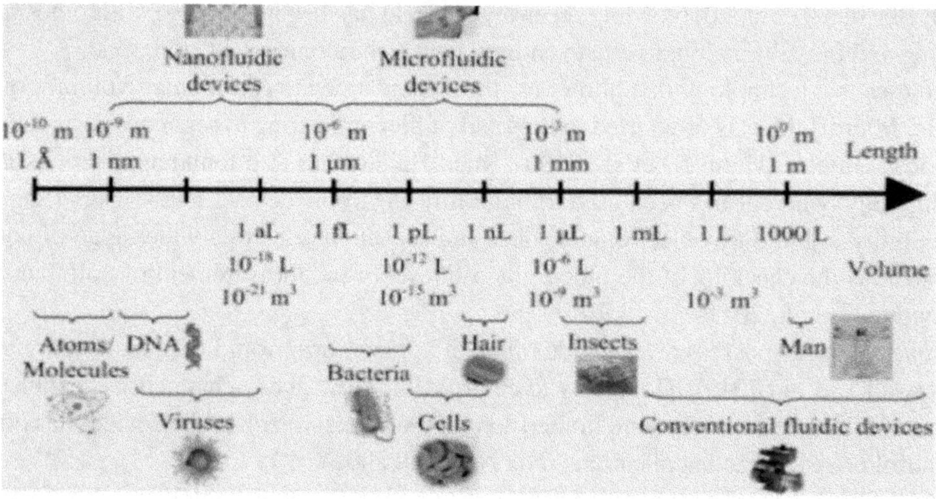

- Artificial nano-systems are in the range of 10-100 nm
- Atomic limit 0.1-0.5 nm

OVERVIEW AND DEVELOPMENT OF APPLICATIONS MICROFLUIDICS

Microfluidics is the field of engineering and science that deals with tiny fluid systems with diameters on the scale of microns (Karabacak NM et al., 2014). It is a key technology for the manufacture of nanotechnology and the future of medicine, biology and materials science (Hajji et al., 2020). The term "microfluidics" refers to the engineering and science of fluid systems with diameters on the scale of microns. Microfluidics also has an application of separation and manipulation tiny particles, such as cells, molecules and atoms, in a controlled environment (Harmon JB et al., 2020). This controllable environment can be used for a wide range of applications, including drug discovery and the study of materials, and is a key technology for the manufacture of nanotechnology (Trofimchu et al.,2020; Ko et al., 2021;Verma et al.,2021). The goal of microfluidics is to develop systems that can be used to manipulate and analyse fluids on the scale of microns (Mora MF et al.,edu 2015). This will enable the design of new nanomaterials, improved treatments for disease, and the future of medicine and biology (Zhu KY et al., 2012).

It can also be used to design and build devices to deliver drugs and other compounds to targeted places inside the human body (Hardouin J, et al., 2006). The ability to design and build devices that can deliver drugs to targeted places inside the human body is what has led to the greatest impact in the field of microfluidics (Ahmed T et al., 2010; P. Gravese et al.; 1993)

The technology is used to build devices that can be used in medicine, biology, chemistry and materials science. Microfluidics as a field has evolved rapidly in the past decade, with new technologies being developed that can manipulate fluids on a molecular scale (H. Bruus, 2008). The first microfluidic devices were used to separate tiny amounts of fluid, such as blood plasma, into individual components. As the field of microfluidics has developed, so have the uses for these devices (F.M. Whit, 2016).

In the future, microfluidics will be used to analyze and treat disease, create pharmaceutical drugs and materials, and much more. Microfluidics as a field began in the 1970s when scientists discovered that fluids could be forced through tiny channels (Gad-el-Hak M,2002).

This discovery led to the field of microfluidic engineering, which is the designing, fabrication, and testing of fluid systems on the micro-fluidics scale (G.M. Whitesides, 2016). Today, microfluidics is an interdisciplinary field that inculcates sources from various other fields, including engineering, physics, chemistry, biology, and medicine (IWA 2016; K. W., 2006; C. H. Ahn, J 2006 ; N. T. Nguyen et al.,2016).

Following applications of Microfluidic

2.1 microarray of DNA chips
2.2 Molecules of Biology
2.3 Evolutionary biologists physics
2.4 Activity within cells
2.5 the science of Biophysic
2.6 Optics
2.7 HPLC, or high-performance liquid chromatography, is a technique
2.8 Acoustical droplet ejection (ADE)
2.9 Fuel cell technology
2.10 Astrobiology
2.11 Food scientific research

PROPERTIES AND CHARACTERISTIC OF MICROFLUIDICS

Microfluidics is the field of engineering and science that deals with tiny fluid systems with diameters on the scale of microns (B.J. Kirby et al, 2010). It is a key technology for the manufacture of nanotechnology and the future of medicine, biology and materials science. The term "microfluidics" refers to the engineering and science of fluid systems with diameters on the scale of microns. Microfluidics can be used to separate and manipulate tiny particles, such as cells, molecules and atoms, in a controlled environment (P. Abgrall, A. M. Gue, 2017). This controllable environment can be used for a wide range of applications, including drug discovery and the study of materials, and is a key technology for the manufacture of nanotechnology. The goal of microfluidics is to develop systems that can be used to manipulate and analyze fluids on the scale of microns (W.Ehrfeld et al., 1998). This will enable the design of new nanomaterials, improved treatments for disease, and the future of medicine and biology (KERN Riwet, 2002) .This new technology could revolutionize the way we diagnose, prevent and treat disease. The evolution of microfluidics will also give us access to new microfluidics applications (K. Sugioka, et al.,2004; D. Braun, 2005; M. P. Cuchiara et al.,2010).

The history of microfluidics is an exciting story. A revolution in how we synthesize, separate and analyze small objects took place when, in the mid-twentieth century, researchers started to miniaturize devices for moving, measuring and manipulating fluids on the scale of microns (C. M. Hwang et al.,2010; M. Uygug et al.,2010; M. Heckel et al.,2004).

The Microfluidics is the engineering of Techniques for processing fluids with the properties of small volumes of fluid, such as blood, with unprecedented control (R.Truckenm€uller, et al., 2004). The fluid

Figure 2. Micro fluidics device design

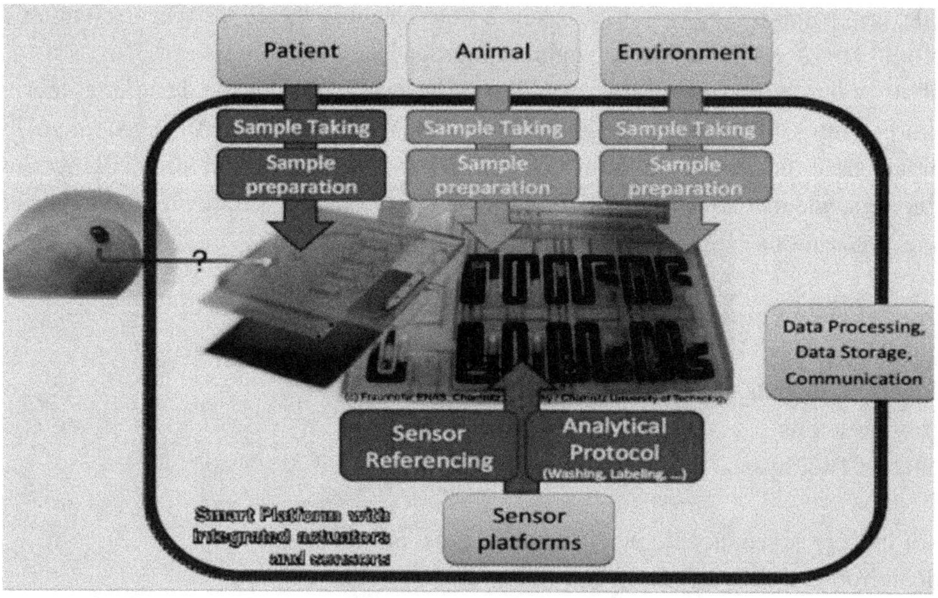

processing elements of the system, such as the micro-mixers, are fabricated in the lab and are similar in design to those used in many conventional laboratory procedures (U. M. Atti et al.2009).

Material science advancements are expected to lead to improved microfluidics devices that are fast, inexpensive and easy to build. One of our goals is to integrate microfluidics devices with inexpensive, low-volume manufacturing processes (P. S. Nune et al.,2010).

CHALLENGES OF MICROFLUIDICS DEVICE DESIGN

The implementation of discrete micro droplet manipulation and micro fluidic processes into a continuous flow and referred to as virtual micro fluidics (N. Petersen, et al, 2010). Electrolytic droplets have been actuated via surface anxiety is directly electrically controlled by a pair of obstructing planar electrodes turned into based on electro wetting-on-dielectric (J. M. Engel et al., 2005; M. Atkin et al., 2004). Shown the figure of 4.1 and 4.2 microfluidic device design and challenges of microfluidic

VARIOUS TYPES OF MICROFLUIDS FLOWS

Open Microfluids Flows

Over the years, the technology has been used to separate and purify biological fluids, such as blood, and to process and analyze tiny objects, such as cells and molecules (J. R.Andrews and B. Gerne 2004). The advent of microfluidics has revolutionized the way we perform many of the procedures in a conventional laboratory, such as the manufacture of vaccines and the DNA sequencing of genetic material(T. F. Hon et al.,2010; A. Imre et al.,2010). It has also opened the door to new applications, such as the delivery

Figure 3. Micro fluidics challenges

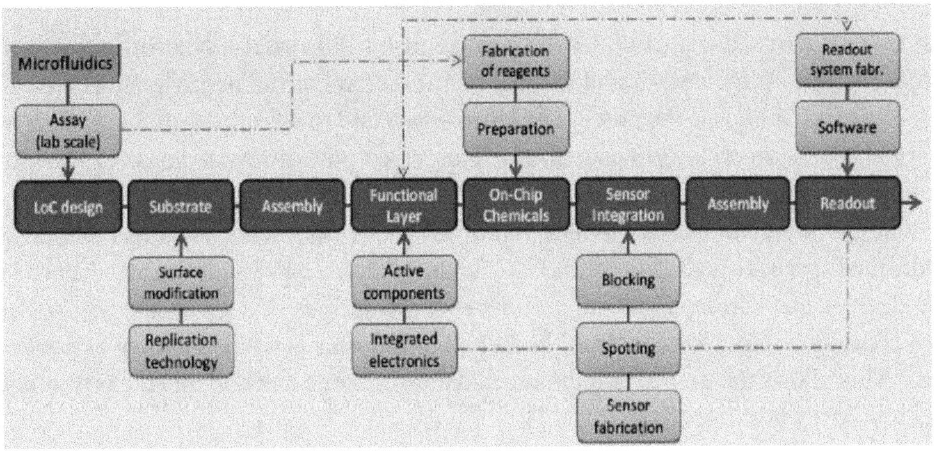

of drugs to specific cells in the body and the capture of viruses in the blood. Open microfluidics is the process of allowing a fluid to pass through the system without any physical barriers (L. Guan et al.,2009; W. Schrottet al.,2009; Y. Kakinuma et al.,2008; M. L. Huper et al.,2008; Unger.MA.et al.,2000; Wu, Het al.,2003; Zhang, Y. and Jiang, H.R. 2016).

Over the years, microfluidics has evolved into a diverse field of engineering and science, with applications in almost every field of science and engineering (Zhang, Y. and Ozdemir, P., 2009). The most common application of microfluidics is in the field of biochemistry, where it is used to separate and process tiny biological particles, such as cells, molecules and atoms (Park, S.et al., 2011). This is used to enrich and purify specific species of a given species, such as cells or proteins, and to separate and process large numbers of the same species at once (Diercks, A.H et al., 2009). Considering the capability of controlling the fluid flow at the micro scale will enable the design of new nanomaterials, improved treatments for disease, and the future of medicine and biology (Herr, A.E., et al.2007).

The ability to exercise control concerning the way fluids flow at the microscale will enable the design of new nanomaterials, improved treatments for disease, and the future of medicine and biology (Bhattacharyya, A. and Klapperich, C.M. 2007; Martinez, A.Wet al.,2010; Ellerbee, A.K et al.2009; Jeong, G.S et al.2010).

The behavior of liquids and the flows of liquids on surfaces have long captured the scientific imagination, as illustrated by the old adage that "oil and water don't mix." This adage arises from two different properties: oil is immiscible, and water is an electromagnetic fluid (D. Patel et al.2022; Yang and Z et al.,2001). It is only recently, however, that researchers have begun to look at the physics and chemistry of fluids at the micro scale in ways that lead to new applications (Lei, K.F et al.2008; Oh, K.W. and Ahn, C.H. 2006; Pollack, M.G.and Fair, R.B. 2000). One of these is the area of microfluidics. Microfluidics can be applied in creation of complex microfluidic devices that incorporate a variety of microfluidic functions, such as mixing, separation, concentration, and mixing and separation of multiple species at once (Wang,F. et al., 2009).

Continuous-Flow Microfluidics

Devices that carry micro size fluids under the surface of a solid surface. Microfluidics is a new field of science that focuses on the design and control of fluid flows at the micro scale (Patel, Dhiredra et al.,2022). Open micro fluidics is the process of allowing a fluid to pass through the system without any physical barriers (M. Squires, S.R. Quake 2005). This video will showcase some of the most exciting new applications being developed today (Lei, K.F. 2012; Lecault, V et al., 2012).

Devices that can separate and manipulate multiple types of particles, such as cells molecules and atoms in a confined space (Hung,P.J. et al. 2005; Chandrasekhar, S., 1981). It can also be used in creating microfluidic devices that can separate and process specific species of a given species, such as cells or proteins, in a continuous-flow format (Jeon, N. Let al.2002). This is what is known as continuous-flow microfluidics. This allows the user to manipulate and analyze cells in a contained environment in a way that was previously not possible (S. Kou 2013).

Systems that allow the continuous flow of fluids. This is particularly useful for the manipulation of small biological particles, such as cells, or the processing of small molecules, such as drugs (N.J. Hillson et al.2012; S.C.C. Shih et al.2015; H.S. Kim et al.2016; S.L. Sjostrom et al., 2017). It is also being used to create tiny bioreactors that can be used to grow small organisms, such as bacteria and yeast, on a small scale in a closed system (E.Eriksson et al.2010)] .This opens up the possibility for the mass production of small amounts of complex biological substances in a controlled environment (Yamada M, et al.2008).

Microfluids as Heat Transfer Fluids

Smaller fractional weights of micro particles, when suspended and dispersed steadily in fundamental fluid medium, provide significantly improve in the thermal and physical properties of fundamental fluids (Whitesides GM. Et al. 2002). This micro fluids technology, which is prioritized where thermal engineering and micro technology converge, has advanced significantly during the past ten years past(J et al. 2008). Micro fluids play a crucial the benefit of reaching the highest standard thermal characteristics with the smallest feasible particle fractional weights through uniform dispersion and stable suspension of micro particles in the base fluid medium (D. Patel et al.2021; D. Dumon et al.2020; Erickson D. et al.2004;van Deemter JJ et al.1995). Finding ways to improve thermal energy transmission in liquids is essential for achieving this goal. In the expanding field of micro fluids, a number of engineers and scientists have made groundbreaking research by examining surprising thermal properties of micro fluids and have proposed new, in the growing micro fluids area, have performed breakthrough research by accounting for unexpectable thermal properties of micro fluids and also proposed new mechanisms behind improved thermo physical and other properties heat transfer of micro fluids. New generation preparation of microfluidic method shown in figure 5.3 [50,55,60,85]

Microfluidics technology even provides a crucial component for the creation of small and efficient cooling systems. In the field of thermal science, microfluids are developed for their mysteriously improved thermal conductivities, which provide the idea of using microfluids as heat transfer fluids (Lin S et al.2014; A. K. Singh and D. Patel 2020). Figure 4.1 shows the preparation of microfluids, and Figure 5.2 shows a schematic representation of this heat transfer fluid (J.M. Chen et al.2007; N. Demierre et al.2008;Vasudev A et al.2012).

The enhanced thermal properties of microfluids offer a clearer understanding of an brief and objective innovation for the intensification of heat transfer, which is crucial to industrial sectors such as cooling,

Figure 4. Preparations of microfluidfluidic

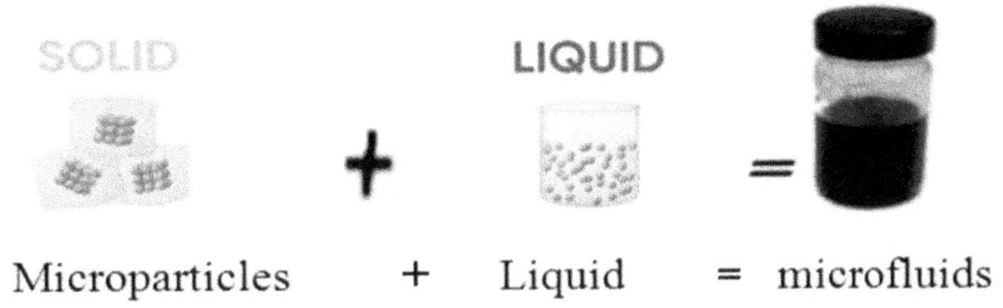

heating, and air conditioning as well as the production of power, transportation, and thermotherapy for the treatment of cancer (Ren K,et al. 2011).

ADVANTAGES OF MICROFLUIDIC SYSTEMS

Since it may be utilized to adjust the thermal properties of microfluidic systems as well as the stability of microparticle suspensions, particle size is the key physical parameter in microfluidic systems. So, with the dispersion of microparticles, microfluidic can flow freely through mini or micro channels (Friend J et al.2010; Balani, R.S et al.2021;Chin CD et al.2011;Focke M et al.2010; Cameron N et al.2006; Gomez FA et al. 2008; Yeo LY,et al.2011; Sackmann EK et al.2014; Heo H et al.2016; Peoples MC, Karnes HT 2008). The micro suspensions' increased heat conductivity is primarily attributable to improved convection between the surfaces of the microparticles and base liquid. Another advantage is that the microparticles are smaller in size, giving the impression that a base fluid molecule is suspended when the particles are scattered (Chikkaveeraiah BV, et al.2011; Chin CD et al.2011; Nahavandi S et al.2014; Pagaduan JV et al.2015; Sanjay ST,et al.2015; Tay A,et al.2016; Squires, T. M. and Quake S. R. 2005).

Figure 5. Schematic view of heat transfer microfluidic

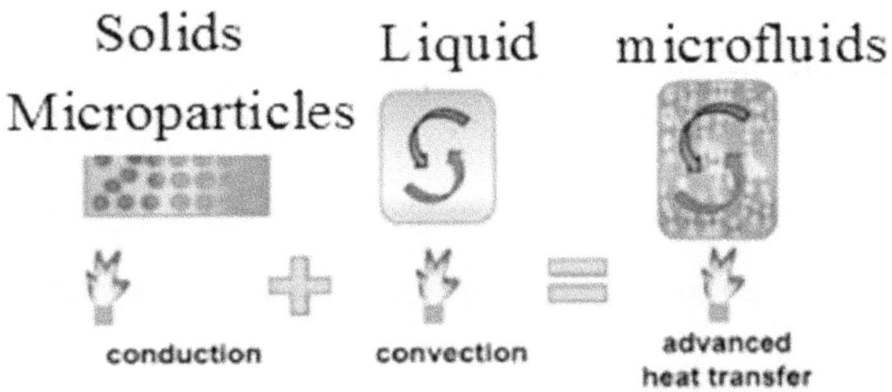

- Less expensive reagents are used, which reduces cost.
- Identification of molecules and separation with high resolution and sensitivity.
- Reduced analyzed of footprint.
- Increase in the net fluid's surface area and heat capacity.
- Increase in the thermal conductivity of the fluids.
- Intensified collision which leads to intense interaction among particles, the surface of flow passage, and base fluids.
- Reduction of particle clogging rather than conventional slurries.
- Smaller footprint of analytical and diagnostic equipment in comparison to large laboratory machines • Quicker analysis times and results
- Fluids flowing laminar or smoothly in small channels provide for better flow management.
- Better microscale control of experimental parameters and sample concentration

FUTURE SCOPE OF MICROFLUIDICS

Mechanical engineers in India will continue to have a future. The previous decades have a consistent pattern. The study of fluid and particle behavior at the micro scale, as well as the technology of creating procedures and tools that can precisely and successfully carry out such tasks, have all advanced the area of microfluidics. Additionally, this will continue to be true for many years to come. There are a load tonnes of businesses that are excellent for mechanical engineers; a few of them are as follows: NFL, NALCO, NHPC, NLC, Bosch group, BNPM, BPCL, Caterpillar, CELHPCL, Coal India Company Ltd, DRDO, Hyundai group, IOCL, IPR, Kirloskar group, L&T group, MDL, MECL, Mahendra & Mahendra group.

CONCLUSION

Controlling tiny quantities of liquids is the focus of the relatively new field of microfluidics. Surface tension and viscosity start to dominate when the channel dimensions get smaller. Low Reynolds numbers lead to more predictable fluid behavior and the dominance of the laminar flow regime. The liquid flow can then be generated or stopped using adjustments to the channel's dimensions or surface characteristics. The field of electromagnetic can be using to move or separate Fluids as well as ions, molecules, and cells. Discrete effects predominate and characterize the realm of nanofluidics when at least one dimension is below 100 nm. Additionally, surface to volume effects started to increase important as channel sizes get progressively smaller. In fact, surfaces have a big role.

REFERENCES

Abgrall, P., & Gue, A. M. (2007). Lab-on-chip technologies: Making a microfluidic network and coupling it into a complete microsystem—a review. *Journal of Micromechanics and Microengineering*, *17*(5), R15–R49. doi:10.1088/0960-1317/17/5/R01

Ahmed, T., Shimizu, T.S., Stocker, R. (2010). Microfluidics for bacterial chemotaxis. *Integrative Biology, 2*(11–12), 604–629 . . hdl:1721.1/66851. PMID 20967322 doi:10.1039/c0ib00049c

Andrews, J. R., & Gerner, B. (2004). Microfluidics, BioMEMS, and Medical Microsystems II. SPIE.

Atkin, M., Poetter, K., Cattrall, R., & Harvey, E. (2004). In P. Woias & I. Papautsky (Eds.), *Microfluidics, BioMEMS, and Medical Microsystems II* (Vol. 5345, pp. 138–146). Spie-Int Soc Optical Engineering. doi:10.1117/12.524321

Attia, U. M., Marson, S., & Alcock, J. R. (2009). Micro-injection moulding of polymer microfluidic devices. *Microfluidics and Nanofluidics*, *7*(1), 1–28. doi:10.100710404-009-0421-x

Balani, R. S., Patel, D., Barthwal, S., & Arun, J. (2021). Fabrication of Air Conditioning System Using the Engine Exhaust Gas. In P. K. Kapur, G. Singh, & S. Panwar (Eds.), *Advances in Interdisciplinary Research in Engineering and Business Management. Asset Analytics*. Springer., doi:10.1007/978-981-16-0037-1_29

Barrett, M. P., Cooper, J. M., Regnault, C., Holm, S. H., Beech, J. P., Tegenfeldt, J. O., & Hochstetter, A. (2017, October). Microfluidics-Based Approaches to the Isolation of African Trypanosomes. *Pathogens (Basel, Switzerland)*, *6*(4), 47. doi:10.3390/pathogens6040047 PMID:28981471

Berthier J, Brakke KA, Berthier E (2016-08-01). *Open Microfluidics*. Wiley. doi:10.1002/9781118720936

Berthier, J., Brakke, K. A., & Berthier, E. (2016). Open Microfluidics. John Wiley & Sons, Inc. . doi:10.1002/9781118720936

Bhattacharyya, A., & Klapperich, C. M. (2007). Design and testing of a disposable microfluidic chemiluminescent immunoassay for disease biomarkers in human serum samples. *Biomedical Microdevices*, *9*(2), 245–251. doi:10.100710544-006-9026-2 PMID:17165125

Bouaidat, S., Hansen, O., Bruus, H., Berendsen, C., Bau-Madsen, N.K., & Thomsen, P. (2005). Surface-directed capillary system; theory, experiments and applications. *Lab on a Chip, 5* (8), 827–836. . doi:10.1039/b502207j

Braun, D. (2005). *Polymer Synthesis: Theory and Practice—Fundamentals, Methods, Experiments* (4th ed.). Springer.

Cameron, N. S., Roberge, H., Veres, T., Jakeway, S. C., & Crabtree, H. J. (2006). High fidelity, high yield production of microfluidic devices by hot embossing lithography: Rheology and stiction. *Lab on a Chip*, *6*(7), 936–941. doi:10.1039/b600584e PMID:16804599

Chandrasekhar, S. (1981). *Hydrodynamic and Hydromagnetic Stability Dover.*

Chang, H. C., & Yeo, L. (2009). *Electrokinetically Driven Microfluidics and Nanofluidics*. Cambridge University Press.

Chen, J. M., Huang, P.-C., & Lin, M.-G. (2007). Analysis and experiment of capillary valves for microfluidics on a rotating disk. *Microfluidics and Nanofluidics*, *4*(5), 427–437. doi:10.100710404-007-0196-x

Chikkaveeraiah, B. V., Mani, V., Patel, V., Gutkind, J. S., & Rusling, J. F. (2011). Microfluidic electrochemical immunoarray for ultrasensitive detection of two cancer biomarker proteins in serum. *Biosensors & Bioelectronics*, *26*(11), 4477–4483. doi:10.1016/j.bios.2011.05.005 PMID:21632234

Chin, C. D., Laksanasopin, T., Cheung, Y. K., Steinmiller, D., Linder, V., Parsa, H., Wang, J., Moore, H., Rouse, R., Umviligihozo, G., Karita, E., Mwambarangwe, L., Braunstein, S. L., van de Wijgert, J., Sahabo, R., Justman, J. E., El-Sadr, W., & Sia, S. K. (2011). Microfluidics-based diagnostics of infectious diseases in the developing world. *Nature Medicine*, *17*(8), 1015–1019. doi:10.1038/nm.2408 PMID:21804541

Chin, C. D., Laksanasopin, T., Cheung, Y. K., Steinmiller, D., Linder, V., Parsa, H., Wang, J., Moore, H., Rouse, R., Umviligihozo, G., Karita, E., Mwambarangwe, L., Braunstein, S. L., van de Wijgert, J., Sahabo, R., Justman, J. E., El-Sadr, W., & Sia, S. K. (2011). Microfluidics-based diagnostics of infectious diseases in the developing world. *Nature Medicine*, *17*(8), 1015–1019. doi:10.1038/nm.2408 PMID:21804541

Chokkalingam, V., Weidenhof, B., Krämer, M., Maier, W. F., Herminghaus, S., & Seemann, R. (2010, July). Optimizeddroplet-basedmicrofluidicsschemeforol-gelreactions. *Lab on a Chip*, *10*(13), 17001705. doi:10.1039/b926976b PMID:20405061

Chung, B. G., Manbachi, A., Saadi, W., Lin, F., Jeon, N. L., & Khademhosseini, A. (2007). A gradient-generating microfluidic device for cell biology. *Journal of Visualized Experiments*, *7*(7), 271. doi:10.3791/271 PMID:18989442

Cuchiara, M. P., Allen, A. C. B., Chen, T. M., Miller, J. S., & West, J. L. (2010). Multilayer microfluidic PEGDA hydrogels. *Biomaterials*, *31*(21), 5491–5497. doi:10.1016/j.biomaterials.2010.03.031 PMID:20447685

Demierre, N., Braschler, T., Muller, R., & Renaud, P. (2008). Focusing and continuous separation of cells in a microfluidic device using lateral dielectrophoresis. *Sensors and Actuators. B, Chemical*, *132*(2), 388–396. doi:10.1016/j.snb.2007.09.078

Diercks, A. H., Ozinsky, A., Hansen, C. L., Spotts, J. M., Rodriguez, D. J., & Aderem, A. (2009). A microfluidic devicefor multiplexed protein detection in nano-liter volumes. *Analytical Biochemistry*, *386*(1), 30–35. doi:10.1016/j.ab.2008.12.012 PMID:19133224

Dumont-Fillon, D., Lamaison, D., & Chappel, E. (2020). Design and characterization of 3-stack MEMS-based passive flow regulators for implantable and ambulatory infusion pumps. *Journal of Microelectromechanical Systems*, *29*(2), 1–12. doi:10.1109/JMEMS.2019.2962107

Eduati F, Utharala R, Madhavan D, Neumann UP, Longerich T, Cramer T, et al. (June 2018). A microfluidics platform for combinatorial drug screening on cancer biopsies. *Nature Communications, 9* (1), 2434. . doi:10.1038/s41467-018-04919-w

Ehrfeld, W., Hessel, V., & Lehr, H. (1998). *Microsystem Technology in Chemistry and Life Science* (Vol. 194). Springer-Verlag. doi:10.1007/3-540-69544-3_10

Ellerbee, A. K., Phillips, S. T., Siegel, A. C., Mirica, K. A., Martinez, A. W., Striehl, P., Jain, N., Prentiss, M., & Whitesides, G. M. (2009). Quantifying colorimetric assays in paper-based microfluidic devices by measuring the transmission of light through paper. *Analytical Chemistry, 81*(20), 8447–8452. doi:10.1021/ac901307q PMID:19722495

Engel, J. M., Chen, J., Bullen, D., & Liu, C. (2005). *18th IEEE International Conference on Micro Electro Mechanical Systems, MEMS*. IEEE.

Erickson, D., & Li, D. (2004). Integrated microfluidic devices. *Analytica Chimica Acta, 507*(1), 11–26. doi:10.1016/j.aca.2003.09.019

Eriksson, E., Sott, K., Lundqvist, F., Sveningsson, M., Scrimgeour, J., & Hanstorp, D. (2010). *A microfluidic device for reversible environmental changes around single cells using optical tweezers for cell selection and positioning*. doi:10.1039/B913587A

Focke, M., Kosse, D., Müller, C., Reinecke, H., Zengerle, R., & von Stetten, F. (2010). Lab-on-a-foil: Microfluidics on thin and flexible films. *Lab on a Chip, 10*(11), 1365–1386. doi:10.1039/c001195a PMID:20369211

Friend, J., & Yeo, L. (2010). Fabrication of microfluidic devices using polydimethylsiloxane. *Biomicrofluidics, 4*(2), 026502-1, 026502–026505. doi:10.1063/1.3259624 PMID:20697575

Frisk, T., Rönnholm, D., van der Wijngaart, W., Stemme, G. (2006). A micromachined interface for airborne sample-to-liquid transfer and its application in a biosensor system. *Lab on a Chip, 6*(12), 1504 - 1509. . PMID 17203153. doi:10.1039/B612526N

Gad-el-Hak, M. (2002). Flow Physics. In M. Gad-el-Hak (Ed.), *The MEMS Handbook*. CRC Press.

Gnyawali, V., Strohm, E.M., Wang, J.Z., Tsai, S.S., Kolios, M.C. (2019). Simultaneous acoustic and photoacoustic microfluidic flow cytometry for label-free analysis. *Scientific Reports, 9*(1), 1585. . doi:10.1038/s41598-018-37771-5

Gomez, F. A. (Ed.). (2008). *Biological Applications of Microfluidics* (1st ed.). Wiley-Interscience.

Gravesen, P., Branebjerg, J., & Jensen, O. S. (1993). Microfluidics-a review. *Journal of Micromechanics and Microengineering, 3*(4), 168–182. doi:10.1088/0960-1317/3/4/002

Greener, J., Tumarkin, E., Debono, M., Dicks, A.P., Kumacheva, E. (2012). Education: a microfluidic platform for university-level analytical chemistry laboratories. *Lab on a Chip, 12*(4), 696–701. . doi:10.1039/c2lc20951a

Greener, J., Tumarkin, E., Debono, M., Kwan, C.H., Abolhasani, M., Guenther, A., Kumacheva, E. (2012). Development and applications of a microfluidic reactor with multiple analytical probes. *The Analyst, 137*(2) 444–450. . doi:10.1039/C1AN15940B

Guan, L., Peng, K. W., Yang, Y. L., Qiu, X. H., & Wang, C. (2009). *Nanotechnology*, 20. Institute of Physics.

H. Bruus, Theoretical Microfluidics. (2008). OUP.

Hajji, I., Serra, M., Geremie, L., Ferrante, I., Renault, R., Viovy, J.L., Descroix, S., & Ferraro, D. (2020). Droplet microfluidic platform for fast and continuous-flow RT-qPCR analysis devoted to cancer diagnosis application. *Sensors and Actuators B: Chemical, 303*, 127171. . S2CID 208705450 doi:10.1016/j.snb.2019.127171

Hardouin, J., Duchateau, M., Joubert-Caron, R., Caron, M. (2006). Usefulness of an integrated microfluidic device (HPLC-Chip-MS) to enhance confidence in protein identification by proteomics. *Rapid Communications in Mass Spectrometry, 20*(21), 3236244. . doi:10.1002/rcm.2725

Harmon, J. B., Gray, H. K., Young, C. C., & Schwab, K. J. (2020). Microfluidic droplet application for bacterial surveillance in fresh-cut produce wash waters. *PLoS One*, *15*(6), e0233239. doi:10.1371/journal.pone.0233239 PMID:32516315

Heckele, M., & Schomburg, W. K. (2004). Review on micro molding of thermoplastic polymers. *Journal of Micromechanics and Microengineering*, *14*(3), R1–R14. doi:10.1088/0960-1317/14/3/R01

Heo, H. Y., Chung, S., Kim, Y. T., Kim, D. H., & Seo, T. S. (2016). A valveless rotary microfluidic device for multiplex point mutation identification based on ligation-rolling circle amplification. *Biosensors & Bioelectronics*, *78*, 140–146. doi:10.1016/j.bios.2015.11.039 PMID:26606304

Herr, A. E., Hatch, A. V., Throckmorton, D. J., Tran, H. M., Brennan, J. S., Giannobile, W. V., & Singh, A. K. (2007). Microfluidic immunoassays as rapid saliva-based clinical diagnostics. *Proceedings of the National Academy of Sciences of the United States of America*, *104*(13), 5268–5273. doi:10.1073/pnas.0607254104 PMID:17374724

Hillson, N. J., Rosengarten, R. D., & Keasling, J. D. (2012). J.D. Keasling js DNA assembly design automation software. *ACS Synthetic Biology*, *1*(1), 14–21. doi:10.1021b2000116 PMID:23651006

Hochstetter, A. (2019, December). Presegmentation Procedure Generates Smooth-Sided Microfluidic Devices: Unlocking Multiangle Imaging for Everyone? *ACS Omega*, *4*(25), 20972–20977. doi:10.1021/acsomega.9b02139 PMID:31867488

Hong, T. F., Ju, W. J., Wu, M. C., Tai, C. H., Tsai, C. H., & Fu, L. M. (2010). Rapid prototyping of PMMA microfluidic chips utilizing a CO2 laser. *Microfluidics and Nanofluidics*, *9*(6), 1125–1133. doi:10.100710404-010-0633-0

Hung, P. J., Lee, P. J., Sabounchi, P., Aghdam, N., Lin, R., & Lee, L. P. (2005). A novel high aspect ratio microfluidic design to provide a stable and uniform microenvironment for cell growth in a high throughput mammalian cell culture array. *Lab on a Chip*, *5*(1), 44–48. doi:10.1039/b410743h PMID:15616739

Hupert, M. L., Guy, W. J., Llopis, S. D., Shadpour, H., Rani, S., Nikitopoulos, D. E., & Soper, S. A. (2007). Evaluation of micromilled metal mold masters for the replication of microchip electrophoresis devices. *Microfluidics and Nanofluidics*, *3*(1), 1–11. doi:10.100710404-006-0091-x

Hwang, C. M., Sim, W. Y., Lee, S. H., Foudeh, A. M., Bae, H., Lee, S.-H., & Khademhosseini, A. (2010). Benchtop fabrication of PDMS microstructures by an unconventional photolithographic method. *Biofabrication*, *2*(4), 045001. doi:10.1088/1758-5082/2/4/045001 PMID:21076185

Imre, A., Ocola, L. E., Rich, L., & Klingfus, J. (2010). Microelectronics and Nanometer Structures : Processing, Measurement, and Phenomena. *Journal of Vacuum Science & Technology. B: An Official Journal of the American Vacuum Society*, *28*, 304–309.

IWA23. (2016). *Interoperability of Microfluidic Devices— Guidelines for Pitch Spacing Dimensions and Initial Device Classification*. IWA 23.

Jeon, J. S., Chung, S., Kamm, R. D., & Charest, J. L. (2011, April). Hot embossing for fabrication of a microfluidic 3D cell culture platform. *Biomedical Microdevices*, *13*(2), 325–333. doi:10.100710544-010-9496-0 PMID:21113663

Jeon, N. L., Chiu, D. T., Wargo, C. J., Wu, H. K., Choi, I. S., Anderson, J. R., & Whitesides, G. M. (2002). *Biomedical Microdevices*, *4*(2), 117–121. doi:10.1023/A:1014683114796

Jeong, G. S., Chung, S., Kim, C. B., & Lee, S.-H. (2010). Applications of micromixing technology. *Analyst (London)*, *135*(3), 460–473. doi:10.1039/b921430e PMID:20174696

Jing, G., Polaczyk, A., Oerther, D. B., & Papautsky, I. (2007). Development of a microfluidic biosensor for detection of environmental mycobacteria. *Sensors and Actuators. B, Chemical*, *123*(1), 614–621. doi:10.1016/j.snb.2006.07.029

Kaigala, G. V., Lovchik, R. D., & Delamarche, E. (2012, November). Microfluidics in the "open space" for performing localized chemistry on biological interfaces. *Angewandte Chemie*, *51*(45), 11224–11240. doi:10.1002/anie.201201798 PMID:23111955

Kakinuma, Y., Yasuda, N., & Aoyama, T. (2008). Micromachining of Soft Polymer Material applying Cryogenic Cooling. *Journal of Advanced Mechanical Design, Systems and Manufacturing*, 2(4), 560 569. doi:10.1299/jamdsm.2.560

Karabacak, N. M., Spuhler, P. S., Fachin, F., Lim, E. J., Pai, V., Ozkumur, E., Martel, J. M., Kojic, N., Smith, K., Chen, P., Yang, J., Hwang, H., Morgan, B., Trautwein, J., Barber, T. A., Stott, S. L., Maheswaran, S., Kapur, R., Haber, D. A., & Toner, M. (2014, March). Microfluidic, marker-free isolation of circulating tumor cells from blood samples. *Nature Protocols*, *9*(3), 694–710. doi:10.1038/nprot.2014.044 PMID:24577360

Karniadakis, G. M., Beskok, A., & Aluru, N. (2005). *Microflows and Nanoflows. Springer Verlag. Bruus H (2007). Theoretical Microfluidics*. Oxford University Press.

KERN. (2002). *Riweta Material Selector*. KERN. www.kern.de.

Kim, H., Guzman, A., Thapa, H., Devarenne, T., & Han, A. (2016). A droplet-microfluidics platform for rapid microalgal growth and oil production analysis. *Biotechnolm Bioeng, 8,* 1691-1701.

Kirby, B. J. (2010). Micro- and Nanoscale Fluid Mechanics: Transport in Microfluidic Devices. Cambridge University Press.

Kirby, B. J. (2010). *Micro-and Nanoscale Fluid Mechanics: Transport in Microfluidic Devices*. Cambridge University Press.

Ko, C.-H., Liu, C.-C., Chen, K.-H., Sheu, F., Fu, L.-M., & Chen, S.-J. (2021). Microfluidic colorimetric analysis system for sodium benzoate detection in foods. *Food Chemistry, 345*, 128773. . doi:10.1016/j.foodchem.2020.128773

Konda, A., & Morin, S. A. (2017, June). Flow-directed synthesis of spatially variant arrays of branched zinc oxide mesostructures. *Nanoscale, 9*(24), 8393–8400. doi:10.1039/C7NR02655B PMID:28604901

Kou, S., Cheng, D., Sun, F., & Hsing, I. M. (2016). Microfluidics and microbial engineering. Lab on a Chip. doi:10.1039/C5LC01039J

Lecault, V., White, A. K., Singhal, A., & Hansen, C. L. (2012). Microfluidic single cell analysis: From promise to practice. *Current Opinion in Chemical Biology, 16*(3-4), 381–390. doi:10.1016/j.cbpa.2012.03.022 PMID:22525493

Lei, K. F. (2012). Microfluidic systems for diagnostic applications: A review. *JALA, 17*(5), 330–347. doi:10.1177/2211068212454853 PMID:22893635

Lei, K. F., & Li, W. J. (2008). A novel in-plane microfluidic mixer using vortex pumps for fluidic discretization. *Journal of the Association for Laboratory Automation, 13*(4), 227–236. doi:10.1016/j.jala.2008.03.008

Lewpiriyawong, N., Kandaswamy, K., Yang, C., Ivanov, V., & Stocker, R. (2011, December). Microfluidic characterization and continuous separation of cells and particles using conducting poly(dimethyl siloxane) electrode induced alternating current-dielectrophoresis. *Analytical Chemistry, 83*(24), 9579–9585. doi:10.1021/ac202137y PMID:22035423

Li, C., Boban, M., & Tuteja, A. (2017). Open-channel, water-in-oil emulsification in paper-based microfluidic devices. *Lab on a Chip, 17*(8), 1436–1441. . doi:10.1039/C7LC00114B

Li, J., Gershow, M., Stein, D., Brandin, E., & Golovchenko, J.A. (2003). DNA molecules and configurations in a solid-state nanopore microscope. *Nature Materials, 2*(9), 611–615. . doi:10.1038/nmat965

Liang, L.G., Kong, M.Q., Zhou, S., Sheng, Y.F., Wang, P., & Yu, T. (2017). An integrated double-filtration microfluidic device for isolation, enrichment and quantification of urinary extracellular vesicles for detection of bladder cancer. *ScientificReports, 7*(1), 46224. . doi:10.1038/srep46224

Lin, S.-L., Lin, T.-Y., & Fuh, M.-R. (2014). Microfluidic chip-based liquid chromatography coupled to mass spectrometry for determination of small molecules in bioanalytical applications: An update. *Electrophoresis, 35*(9), 1275–1284. doi:10.1002/elps.201300415 PMID:24165927

Liu Z, Lee Y, Jang JH, Li Y, Han X, Yokoi K, et al. (2015). Microfluidic cytometric analysis of cancer cell transportability and invasiveness. *Scientific Reports, 5*(1), 14272. . doi:10.1038/srep14272

Martinez, A. W., Phillips, S. T., Whitesides, G. M., & Carrilho, E. (2010). Diagnostics for the developing world: Microfluidic paper-based analytical devices. *Analytical Chemistry, 82*(1), 3–10. doi:10.1021/ac9013989 PMID:20000334

Melin, J., van der Wijngaart, W., & Stemme, G. (2005, June). Behaviour and design considerations for continuous flow closed-open-closed liquid microchannels. *Lab on a Chip, 5*(6), 682–686. doi:10.1039/b501781e PMID:15915262

Mora, M. F., Stockton, A. M., & Willis, P. A. (2015). Analysis of thiols by microchip capillary electrophoresis for in situ planetary investigations. In Microchip Capillary Electrophoresis Protocols. Methods in Molecular Biology (Vol. 1274, pp. 43–52). HumanaPress., , PMID 25673481. doi:10.1007/978-1-4939-2353-3_4

Nahavandi, S., Baratchi, S., Soffe, R., Tang, S.-Y., Nahavandi, S., Mitchell, A., & Khoshmanesh, K. (2014). Microfluidic platforms for biomarker analysis. *Lab on a Chip*, *14*(9), 1496–1514. doi:10.1039/C3LC51124C PMID:24663505

Nguyen, N. I. T., & Wereley, S. (2006). *Fundamentals and Applications of Microfluidics*. Artech House.

Nguyen, N. T., & Wu, Z. G. (2005). Micromixers—A review. *Journal of Micromechanics and Microengineering*, *15*(2), R1–R16. doi:10.1088/0960-1317/15/2/R01

Niu, X., Gulati, S., Edel, J. B., & deMello, A. J. (2008, November). Pillar-induced droplet merging in microfluidic circuits. *Lab on a Chip*, *8*(11), 1837–1841. doi:10.1039/b813325e PMID:18941682

Nunes, P. S., Ohlsson, P. D., Ordeig, O., & Kutter, J. P. (2010). Cyclic olefin polymers: Emerging materials for lab-on-a-chip applications. *Microfluidics and Nanofluidics*, *9*(2-3), 145–161. doi:10.100710404-010-0605-4

Ogawa, M., Higashi, K., & Miki, N. (2015). Development of hydrogel microtubes for microbe culture in open environment. *2015 37th Annual International Conference of the IEEE Engineering in Medicine and Biology Society (EMBC)*. (pp. 5896–5899). IEEE.

Oh, K. W., & Ahn, C. H. (2006). A review of microvalves. *Journal of Micromechanics and Microengineering*, *16*(5), R13–R39. doi:10.1088/0960-1317/16/5/R01

Oh, K. W., & Ahn, C. H. (2006). A review of microvalves. *Journal of Micromechanics and Microengineering*, *16*(5), R13–R39. doi:10.1088/0960-1317/16/5/R01

Oung, E.W., Berthier, E., Guckenberger, D.J., Sackmann, E., Lamers, C., Mcyvantsson, I. (2011). Rapid prototyping of arrayed microfluidic systems in polystyrene for cell-based assays. *Analytical Chemistry*, *83* (4), 1408–1417. . PMC 3052265. PMID 21261280. doi:10.1021/ac102897h

Pagaduan, J. V., Sahore, V., & Woolley, A. T. (2015). Applications of microfluidics and microchip electrophoresis for potential clinical biomarker analysis. *Analytical and Bioanalytical Chemistry*, *407*(23), 6911–6922. doi:10.100700216-015-8622-5 PMID:25855148

Park, S., Zhang, Y., Lin, S., Wang, T.-H., & Yang, S. (2011). Advances in microfluidic PCR for point-of-care infectious disease diagnostics. *Biotechnology Advances*, *29*(6), 830–839. doi:10.1016/j.biotechadv.2011.06.017 PMID:21741465

Patel, D., Mishra, A., & Nabeel, M. (2022). Heat Transfer Characteristics of Nanofluids of Silicon Oxides (Sio2) with Conventional Fluid, 2022 2nd International Conference on Innovative Practices in Technology and Management (ICIPTM), (pp. 420-423). IEEE. 10.1109/ICIPTM54933.2022.9754181

Patel, D., Sharma, A., & Mishra, A. (2021). Study of Convective Heat Transfer Characteristics of Nano Fluids in Circular Tube 2021 International Conference on Technological Advancements and Innovations (ICTAI), (pp. 264-267). IEEE., 10.1109/ICTAI53825.2021.9673432

Patel, D., Singh, R. P., Rajput, R. S., & Tiwari, P. (2022). *Thermophyscical Properties And Applications Nanofluids – On Review.* 2022 5th International Conference on Contemporary Computing and Informatics (IC3I), Uttar Pradesh, India. 10.1109/IC3I56241.2022.10072842

Pawell, R. S., Inglis, D. W., Barber, T. J., & Taylor, R. A. (2013). Manufacturing and wetting low-cost microfluidic cell separation devices. *Biomicrofluidics*, *7*(5), 56501. doi:10.1063/1.4821315 PMID:24404077

Peoples, M. C., & Karnes, H. T. (2008). Microfluidic immunoaffinity separations for bioanalysis. *Journal of Chromatography. B, Analytical Technologies in the Biomedical and Life Sciences*, *866*(1–2), 14–25. doi:10.1016/j.jchromb.2007.08.030 PMID:17869593

Petersen, N., Jensen, H., Hansen, S., Foss, S., Snakenborg, D., & Pedersen-Bjergaard, S. (2010). On-chip electro membrane extraction. *Microfluidics and Nanofluidics*, *9*(4-5), 881–888. doi:10.100710404-010-0603-6

Pfohl, T., Mugele, F., Seemann, R., & Herminghaus, S. (2003, December). Trends in microfluidics with complex fluids. *ChemPhysChem*, *4*(12), 1291–1298. doi:10.1002/cphc.200300847 PMID:14714376

Pollack, M.G., & Fair, R.B. (2000). Electrowetting-based actuation of liquid droplets for microfluidic applications. Appl. Phys. Lett., 77, 1725–1726.

Prakash, M., & Gershenfeld, N. (2007). Microfluidic bubble logic. *Science, 315* . doi:10.1126/science.1136907

Ren, K., Dai, W., Zhou, J., Su, J., & Wu, H. (2011). Whole-teflon microfluidic chips. *Proceedings of the National Academy of Sciences of the United States of America*, *108*(20), 8162–8166. doi:10.1073/pnas.1100356108 PMID:21536918

Sackmann, E. K., Fulton, A. L., & Beebe, D. J. (2014). The present and future role of microfluidics in biomedical research. *Nature*, *507*(7491), 181–189. doi:10.1038/nature13118 PMID:24622198

Sanjay, S. T., Fu, G., Dou, M., Xu, F., Liu, R., Qi, H., & Li, X. J. (2015). Biomarker detection for disease diagnosis using cost-effective microfluidic platforms. *Analyst (London)*, *140*(21), 7062–7081. doi:10.1039/C5AN00780A PMID:26171467

Schrott, W., Svoboda, M., Slouka, Z., & Snita, D. (2009). Metal electrodes in plastic microfluidic systems. *Microelectronic Engineering*, *86*(4-6), 1340–1342. doi:10.1016/j.mee.2009.01.001

Sen, Y.H., & Karnik, R. (2009). Investigating the translocation of lambda-DNA molecules through PDMS nanopores. *Analytical and Bioanalytical Chemistry, 394* (2), 437–446.. doi:10.1007/s00216-008-2529-3

Shemesh, J., Bransky, A., Khoury, M., & Levenberg, S. (2010). Advanced microfluidic droplet manipulation based on piezoelectric actuation. *Biomedical Microdevices, 12* (5), 907–914. . doi:10.1007/s10544-010-9445-y

Shestopalov, I., Tice, J.D., & Ismagilov, R.F. (2004). Multi-step synthesis of nanoparticles performed on millisecond time scale in a microfluidic droplet-based system. *Lab on a Chip, 4*(4), 316–321. doi: 15269797. doi:10.1039/b403378g.PMID

Shih, S. C. C., Goyal, G., Kim, P. W., Koutsoubelis, N., Keasling, J. D., Adams, P. D., Hillson, N. J., & Singh, A. K. (2015). A versatile microfluidic device for automating synthetic biology. *ACS Synthetic Biology*, *10*(10), 1151–1164. doi:10.1021/acssynbio.5b00062 PMID:26075958

Shilton, R. J., Travagliati, M., Beltram, F., & Cecchini, M. (2014, August). Nanoliter-droplet acoustic streaming via ultra high frequency surface acoustic waves. *Advanced Materials*, *26*(29), 4941–4946. doi:10.1002/adma.201400091 PMID:24677370

Shkolnikov, V. (2019). *Principles of Microfluidics*.

Singh, A. K., & Patel, D. (2020). Optimization of Air Flow over a Car by Wind Tunnel. *2020 International Conference on Computation, Automation and Knowledge Management (ICCAKM)*, (pp. 1-4). IEEE. 10.1109/ICCAKM46823.2020.9051457

Squires, M., & Quake Microfluidics, S. R. (2005). fluid physics at the nanoliter scale. *Reviews of Modern Physics*, *3*(3), 977–1026. doi:10.1103/RevModPhys.77.977

Squires, T. M., & Quake, S. R. (2005). Microfluidics: Fluid physics at the nanoliter scale. *Reviews of Modern Physics*, *77*(3), 977–1026. doi:10.1103/RevModPhys.77.977

Sugioka, K., Cheng, Y., Masuda, M., & Midorikawa, K. (2004). In P. R. Herman, J. Fieret, A. Pique, T. Okada, F. G. Bachmann, W. Hoving, K. Washio, X. Xu, J. J. Dubowski, D. B. Geohegan, & F. Trager (Eds.), *Photon Processing in Microelectronics and Photonics Iii* (Vol. 5339, pp. 205–213). Spie-Int Soc Optical Engineering. doi:10.1117/12.531630

Tabeling, P. (2005). *Introduction to Microfluidics*. Oxford University Press.

Tabeling, P. (2005). *Introduction to Microfluidics*. OUP.

Tay, A., Pavesi, A., Yazdi, S. R., Lim, C. T., & Warkiani, M. E. (2016). Advances in microfluidics in combating infectious diseases. *Biotechnology Advances*, *34*(4), 404–421. doi:10.1016/j.biotechadv.2016.02.002 PMID:26854743

Tenje, M., Fornell, A., Ohlin, M., & Nilsson, J. (2018, February). Particle Manipulation Methods in Droplet Microfluidics. *Analytical Chemistry*, *90*(3), 1434–1443. doi:10.1021/acs.analchem.7b01333 PMID:29188994

Trofimchuk, E., Nilghaz, A., Sun, S., & Lu, X. (2020). Determination of norfloxacin residues in foods by exploiting the coffee-ring effect and paper-based microfluidics device coupling with smartphone-based detection. *Journal of Food Science. 85* (3), 736–743. . doi:10.1111/1750-3841.15039

Truckenmüller, R., Rummler, Z., Schaller, T., & Schomburg, W. K. (2002). Low-cost thermoforming of micro fluidic analysis chips. *Journal of Micromechanics andMicroengineering. 12* (4), 375–379. . doi:10.1088/0960-1317/12/4/304

Unger, M. A., Chou, H. P., Thorsen, T., Scherer, A., & Quake, S. R. (2000). Monolithic microfabricated valves and pumps by multilayer soft lithography. *Science*, *288*(5463), 113–116. doi:10.1126cience.288.5463.113 PMID:10753110

Uram, J.D., Ke, K., Hunt, A.J., & Mayer, M. (2006). Label-free affinity assays by rapid detection of immune complexes in submicrometer pores. *Angewandte Chemie. 45* (14), 2281–2285. . hdl:2027.42/50668. PMID 16506296. doi:10.1002/anie.200502862

Uygun, M., Tasdelen, M. A., & Yagci, Y. (2010). Influence of Type of Initiation on Thiol-Ene "Click" Chemistry. *Macromolecular Chemistry and Physics, 211*(1), 103–110. doi:10.1002/macp.200900442

van Deemter, J. J., Zuiderweg, F. J., & Klinkenberg, A. (1995). Longitudinal diffusion and resistance to mass transfer as causes of nonideality in chromatography. *Chemical Engineering Science, 50*(24), 3869–3882. doi:10.1016/0009-2509(96)81813-6

Vasudev, A., Kaushik, A., Jones, K., & Bhansali, S. (2012). Prospects of low temperature co-fired ceramic (LTCC) based microfluidic systems for point-of-care biosensing and environmental sensing. *Microfluidics and Nanofluidics, 14*(3–4), 683–702.

Verma, K., Tarafdar, A., & Badgujar, P. C. (2021). Microfluidics assisted tragacanth gum based submicron curcumin suspension and its characterization. *LWT, 135*. . ISSN 0023-6438. S2CID 224875232 doi:10.1016/j.lwt.2020.110269

Wang, F. & Burns, M. (2009). A Performance of nanotechnology. *Microdevices, 11*, 1071–1080.

West, J., Becker, M., Tombrink, S., & Manz, A. (2008). J, Becker M, Tombrink S, Manz A. Micro total analysis systems: Latest achievements. *Analytical Chemistry, 80*(12), 4403–4419. doi:10.1021/ac800680j PMID:18498178

White, F. M. (2016). *Fluid Mechanics*. McGraw-Hill Education.

Whitesides, G. M. (2002). The origins and the future of microfluidics. Nature. 2006;442(7101):368–373. [66] Thorsen T, Maerkl SJ, Quake SR. Microfluidic large-scale integration. *Science, 298*(5593), 580–584. PMID:12351675

Whitesides, G. M. (2006). The origins and the future of microfluidics. *Nature, 442*(7101), 368–373. doi:10.1038/nature05058 PMID:16871203

Wu, H., Odom, T. W., Chiu, D. T., & Whitesides, G. M. (2003). Fabrication of complex three-dimensional microchannel systems in PDMS. *Journal of the American Chemical Society, 125*(2), 554–559. doi:10.1021/ja021045y PMID:12517171

Xia, N., Hunt, T.P., Mayers, B.T., Alsberg, E., Whitesides, G.M., Westervelt, R.M., Ingber, D.E. (2006). Combined microfluidic-micromagnetic separation of living cells in continuous flow. *Biomedical Microdevices, 8* (4), 299–308. . PMID 17003962. S2CID 14534776. doi:10.1007/s10544-006-0033-0

Yamada, M., Kobayashi, J., Yamato, M., Seki, M., & Okano, T. (2008). Millisecond treatment of cells using microfluidic devices via two-step carrier-medium exchange. [PubMed: 18432348]. *Lab on a Chip, 8*(5), 772–778. doi:10.1039/b718281c PMID:18432348

Yang, Z., Matsumoto, S., Goto, H., Matsumoto, M., & Maeda, R. (2001). Ultrasonic micromixer for microfluidic systems. *Sensors and Actuators. A, Physical, 93*(3), 266–272. doi:10.1016/S0924-4247(01)00654-9

Yeo, L. Y., Chang, H.-C., Chan, P. P. Y., & Friend, J. R. (2011). Microfluidic devices for bioapplications. *Small*, *7*(1), 12–48. doi:10.1002mll.201000946 PMID:21072867

Zhang, Y., & Jiang, H. R. (2016). A review on continuous-flow microfluidic PCR in droplets: Advances, challenges and future. *Analytica Chimica Acta*, *914*, 7–16. doi:10.1016/j.aca.2016.02.006 PMID:26965323

Zhang, Y. & Nguyen, N.T. (2017). Magnetic digital microfluidics - a review. *Lab on a Chip, 17*(6), 994–1008. . hdl:10072/344389. PMID 28220916. S2CID 5013542. doi:10.1039/C7LC00025A

Zhang, Y., & Ozdemir, P. (2009). Microfluidic DNA amplification – a review. *Analytica Chimica Acta*, *638*(2), 115–125. doi:10.1016/j.aca.2009.02.038 PMID:19327449

Zhu, K.Y., Leung, K.W., Ting, A.K., Wong, Z.C., Ng, W.Y., Choi, R.C. (March 2012). Microfluidic chip based nano liquid chromatography coupled to tandem mass spectrometry for the determination of abused drugs and metabolites in human hair. *Analytical and Bioanalytical Chemistry, 402* (9). . PMID 22281681. S2CID 7748546 doi:10.1007/s00216-012-5711-6

Chapter 3
Microfluidics: From Evolution and Applications to Potential Prospects and Challenges

Mohammad Irfan Hajam
https://orcid.org/0000-0002-3629-785X
National Institute of Technology, Srinagar, India

Mohammad Mohsin Khan
https://orcid.org/0000-0001-9465-5144
National Institute of Technology, Srinagar, India

ABSTRACT

Microfluidic technologies have garnered interest due to their capacity to rapidly process samples and accurately control fluids in an assay. This short study seeks to introduce readers to the fascinating realm of microfluidics by introducing some fundamental theoretical principles and illustrative applications. Additionally, the physics of microfluidic devices, the worldwide market, and material types employed are discussed. In the end, the authors analyze promising developments and extrapolate lessons that will help microfluidic technologies advance in the future.

INTRODUCTION AND BASICS OF MICROFLUIDICS

Microfluidics is science and technology used to manipulate fluid flow at sub millimetre level[1]. It involves the use of microchannels, microvalves, micromixers/separators and micropumps to integrate and automate the biochemical processes. High-tech revolutions in a variety of biomedical fields, including as diagnostics, single-cell analysis, the manufacturing of micro- and nanodevices, organ-on-chip platforms, and med-tech applications, have been sparked by advancements in the field of microfluidics. The rapid progress is driven by the synergistic collaboration of key nanomaterial developments and creative microfluidic capabilities in a variety of biomedical applications. In many organic, synthetic, and design

DOI: 10.4018/978-1-6684-6952-1.ch003

applications today, microfluidic devices play a crucial role. These applications use a variety of methods to build the necessary channels and emphasise measurements.

Over the past three decades, there has been a sustained research activity in the field of microfluidics owing to miniaturization (Figure 1). Even though the first microfluidic device was created in the 1970s, microfluidics did not become widely used until the early 1990s. For these years, the failure to incorporate microfluidic technology into basic and applied research has been blamed on the absence of novel, ground-breaking, and "out-of-the-box" applications. The mechanics, fluid dynamics, and thermodynamics that govern the operation of Lab on chip (LOC)systems are well understood; however, because of the condensed channel dimensions in microfluidics, the interplay of forces and their relative importance is completely different. As a result, a description of the scaling relations that exist between the macroscopic and microscopic systems is necessary.

Microfluidic technology is characterised by dimensions of the system used to manipulate fluids at the micrometre or sub-micrometre scale. At this length scale, the dominant fluid phenomena are quite different from those at the macroscale. At the macroscale, volumetric forces (such as gravity) are more dominant, but at the microscale, surface forces (such as surface tension and capillary forces) have the upper hand. These surface forces can be used for a variety of tasks such as forming monodisperse droplets in multiphase fluid streams, precisely patterning surfaces with user-defined substrates, filtering various analytes, and passively pumping fluids in microchannels. Despite these advances, the adoption of novel techniques has not matched the initial enthusiasm of the field.

Figure 1. Research articles per year

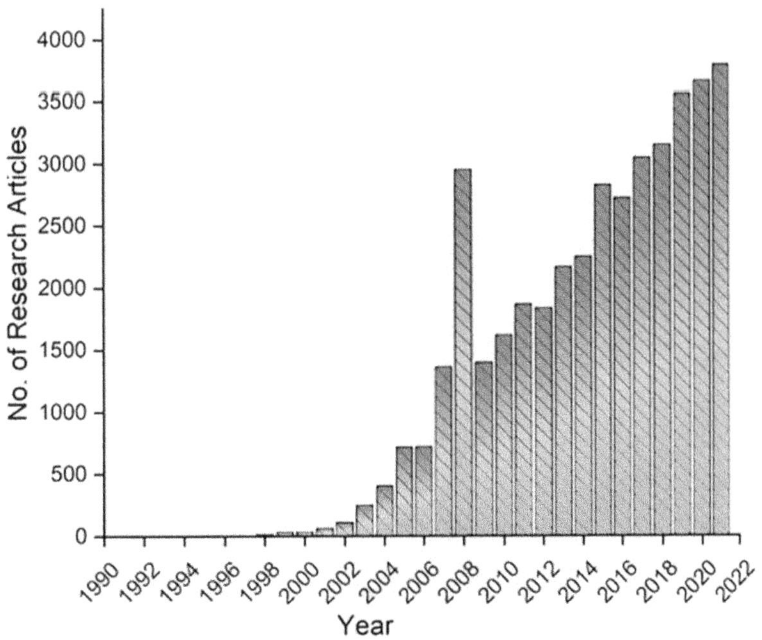

HISTORY OF MICROFLUIDICS

George McClelland Whitesides, field proponent, lists molecular biology, molecular analysis, microelectronics, and biodefense as the field's progenitors (Figure 2) (Whitesides, 2006). Despite the fact that the microfluidic technologies have been utilised in inkjet printers since the 1970's, it was in late 80's when the term was coined.

Microfluidics originated with the miniaturisation of analytical chemistry procedures for sample separation and analysis and this continues to be a primary emphasis of microfluidics today. The silicon-based chromatograph by Terry et al. (1979) is considered the first milestone in the field (Jerman and Terry, 1981). Following his publication, it became clear that microfluidics may help scientists create new, more effective techniques for analysing molecular data. Manz and other pioneers in the early 90's highlighted the microfluidic potential for tackling challenges in analytical methods notably phenomena of electrophoresis. They also coined the term "micro Total Analysis System" μTAS for short. Although the initial emphasis of these "lab-on-a-chip" devices was on analytical chemistry, microfluidics has subsequently spread into a wide range of fields, including chemistry and biology (Manz, Graber, and Widmer, 1990).

Figure 2. Parents of microfluidics

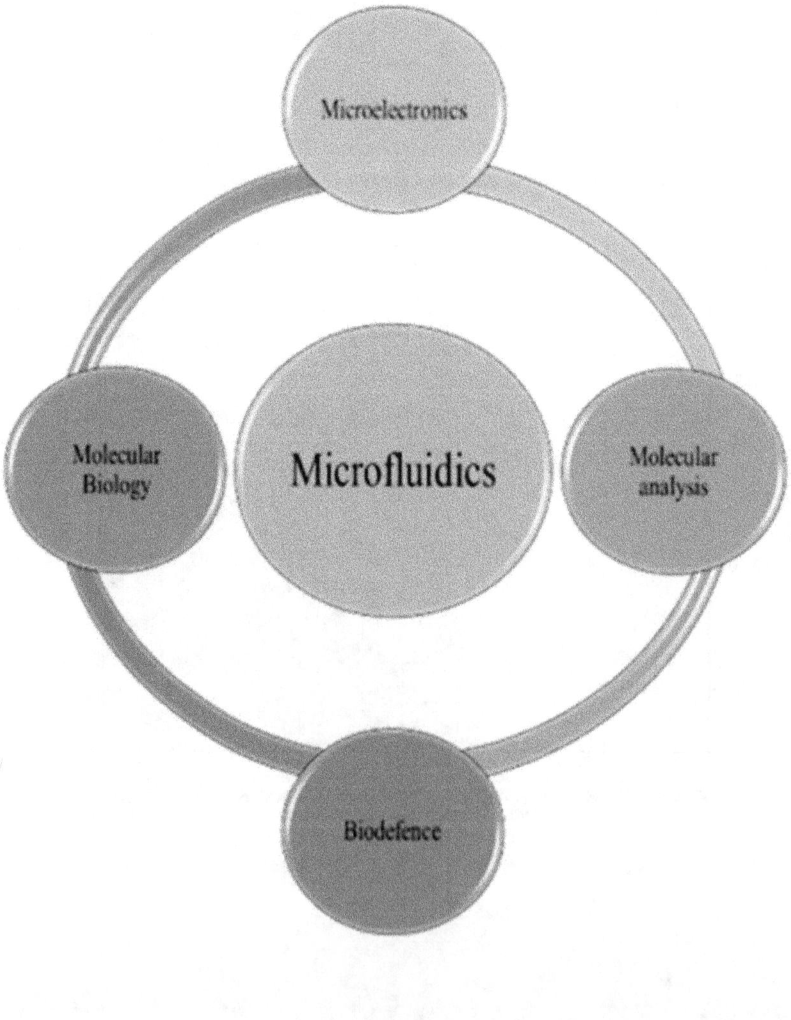

Figure 3. HGP

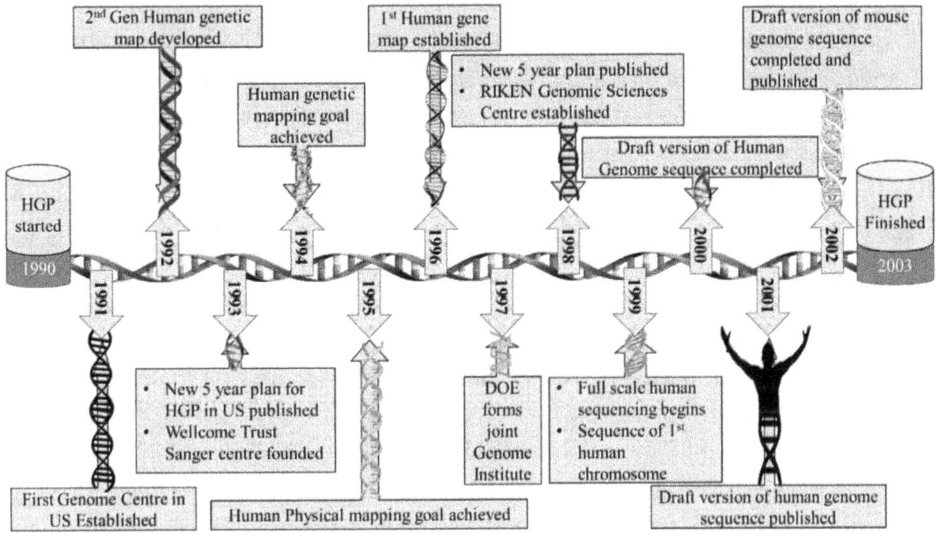

The human genome project (HGP) was another impute in the evolution of the field (Figure 3). Up to the 1990s Slab gel electrophoresis (SGE) was the prime method for DNA sequencing. It was possible to use slab gel electrophoresis for small-scale research applications, but this method would have made the HGP much more complicated since the preparation of the gels and the separation process were both arduous and time consuming. In response, researchers started applying the microfluidic principles to the development of a more robust technology for DNA sequencing. Following much research, Wooley et al. eventually were successful in fabricating glass-based microfluidic platforms that analysed far quicker and were easier to use (Woolley and Mathies, 1994).

In 1998, G. Whitesides introduced PDMS as a low-cost alternative to glass and silicon for rapid prototyping. Glass has limited gas permeability and silicon being too opaque makes them unsuitable for applied biology research. At the beginning of the 21st century, there was a major growth of technologies that were based on moulding micro-channels into polymers. Because of the decreased prices and length of time required to produce these devices, a significant number of research organisations were given the opportunity to participate in microfluidic research. One well-known example from this time period is the development of quake valves. Utilising the PDMS' elasticity, if enough pressure is applied on the wall of an adjacent channel by using flowing liquid, the main channel may be totally sealed off. Quake valves immediately surpassed all other types of microfluidic devices as a result of how well the field accepted this breakthrough (Lee, Bhattacharjee, and Folch, 2018).

Not only is the discipline of microfluidics well-established now, but new study fields based on its applications have also emerged. Terms such as Lab-on-a-chip, Organ-on-a-chip, µTAS may be regarded as their own domains with each one aiming to address various needs by creating unique implementations.

PHYSICS

It is crucial to understand the mechanics of microflows in order to fully comprehend the advantages of microfluidic devices. In the 1990s, research conducted at the University of Pennsylvania revealed a new type of flow behaviour in microflows that was not accounted for by continuum theory. This departure from canonical flows sparked interest in the study of "Low Reynold No." flows. Microflows for liquids exhibit granular flow behaviour, while gases exhibit rarefied flow behaviour with moving walls due to compressibility effects. Modifying boundary conditions is necessary for gas microflows to accurately account for compressibility effects, as the Navier-Stokes equation may yield inaccurate results. Fluid microflows require the use of hybrid atomistic-continuum (HAC) approaches instead of traditional atomistic models. Understanding fluid physics at the microscale is challenging due to the multitude of phenomena that occur simultaneously (Minnella, 2019). The microscale refers to the dimensions of channels and structures used in microfluidic devices, which typically range from a few micrometres to hundreds of micrometres. At this scale, several important phenomena become prominent, including surface tension, capillary forces, laminar flow, and viscous drag.

Surface tension is a dominant force at the microscale due to the high surface-to-volume ratio of microchannels. It causes liquid to behave differently compared to macroscale fluids. For example, droplets can be easily formed and manipulated in microchannels, and liquids can flow against gravity in narrow channels due to capillary action.

Capillary forces, which arise from the combination of surface tension and curvature, can be used to drive liquid flow in microchannels without the need for external pumps or pressure sources. This is known as passive pumping or capillary-driven flow, and it is a unique feature of microfluidics that allows for simple and cost-effective fluid manipulation.

Laminar flow is another key feature of microfluidics, where fluids flow in well-defined layers without significant mixing. This is in contrast to turbulent flow, which occurs at larger scales and is characterized by chaotic mixing. Laminar flow enables precise control of fluid transport and allows for precise manipulation of reagents and particles in microfluidic channels.

Viscous drag, which is the resistance of a fluid to flow due to its viscosity, becomes significant at the microscale. As the dimensions of microchannels decrease, the ratio of surface area to volume increases, leading to a higher viscous drag. This has important implications for the design and performance of microfluidic devices, as it affects flow rates, pressure drops, and mixing efficiency.

Table 1. Important dimensionless numbers

Dimensionless No.	Definition	Formula	Description
Reynolds No.	Inertial forces/Viscous force	$(\rho \text{ v d})/\mu$	• Ratio by which viscosity dominates momentum. • Predicts flow Regime
Capillary No.	Viscous force/Surface Tension force	$(\mu\text{v})/\gamma$	• Describes capillary flow
Peclet No.	advection/ diffusion	L v /D	• Ratio of advective to diffusive mass transport. • Mixing Phenomena
Weber No.	Dynamic pressure/Capillary Pressure	$\rho \text{ v}^2 \text{ l}/\gamma$	• Ratio of Dynamic pressure to Capillary Pressure.

ρ: density, v: average velocity, μ: dynamic viscosity, γ: Specific weight, L: characteristic length, and D: diffusion coefficient.

When discussing the motions of fluids, it is essential to think of them as a continuum. As a result, continuous counterparts of attributes like mass, pressure, and force should be used instead, namely density, pressure density, and force density, per unit volume. In reality, the system is separated into tiny fluid components, and the force for each component is assessed.

There is a wide variety of physical phenomena generated by microfluidic devices, all of which need to be weighed against one another to determine their relative importance. A system's place in fluidic parameter space is shown by dimensionless numbers that show the ratio of different events (Table 1).

Flow Laminarity

The Reynolds number, defined as the ratio of inertial to viscous forces, is the most significant of the dimensionless numbers in forced flows. It enables the determination of whether a flow is laminar or turbulent. Inertial forces result from changes in fluid state of motion, whereas viscous forces are caused by gradients in viscous stress.

In case of flow in a pipe, if $Re < 2000$ the flow is said to be laminar and if $Re > 3000$ the flow is said to be turbulent. In case of microfluidics the Re number is ~1 or even less that means in most of the cases the flow is laminar. Exploiting the advantage of laminar flow, the fluids can be regulated accurately, and molecules within the fluid may be readily managed to generate predictable gradients. This is made possible by the stable streamlines.

Diffusion

Diffusion is a physical process by which individual particles/molecules of a material move from a region of higher concentration to a region of lower concentration. At macroscale, mixing can typically be attributed to the vortices caused by the turbulence (Bragheri, Martinez Vazquez, and Osellame, 2016; Squires, Messinger, and Manalis, 2008). In microfluidics, due to the tiny length scale, the flow is usually in laminar regime. Depending on how fast the mixing needs to be, some applications, like microfluidic gradient generators (Poiseuille flow), use diffusion as the only mixing mechanism. Other applications, like microfluidic mixers, use both diffusion and chaotic mixing to quickly and thoroughly mix different components.

Surface Wettability

Surface wettability plays a critical role in microfluidics, which is the science and technology of manipulating and controlling fluids at the microscale. The wettability of microchannel surfaces affects the behaviour of fluids, including their flow, mixing, and interaction with solid surfaces, in microfluidic devices. The surface wettability in microfluidics is typically controlled by modifying the surface properties of microchannels to either promote or inhibit the spreading or spreading of fluids, depending on the desired application (Zhao, MacMinn, and Juanes, 2016).

Surface wettability is typically characterized by the contact angle, which is the angle formed between the liquid-vapor interface and the solid-liquid interface at the edge of a droplet or liquid film on a surface. The contact angle can be used to quantify the degree of wetting or spreading of a fluid on a surface. A high contact angle (greater than 90 degrees) indicates that the surface is hydrophobic, and the fluid tends to form droplets and bead up, while a low contact angle (less than 90 degrees) indicates that the surface is hydrophilic, and the fluid tends to spread and wet the surface.

In microfluidics, the control of surface wettability is crucial for various applications, including:

1. Flow control:

Surface wettability can be used to control the flow behaviour of fluids in microchannels. Hydrophobic surfaces can be used to promote fast and smooth flow of fluids, reducing the occurrence of flow disruptions such as air bubbles or droplets that can affect the performance of microfluidic devices. On the other hand, hydrophilic surfaces can be used to inhibit flow, for example, in microvalves or micropumps, by preventing the fluid from spreading or penetrating into certain regions of the microchannel.

2. Mixing and reaction control:

Surface wettability can be used to control the mixing and reactions of fluids in microchannels. By carefully designing the surface wettability of microchannels, it is possible to control the formation and transport of droplets, emulsions, or laminar flow streams, which can be utilized for various chemical or biological reactions, such as DNA amplification, enzyme reactions, or chemical synthesis.

3. Biofluidics and cell manipulation:

Surface wettability is crucial in microfluidic devices used for bio-fluidics and cell manipulation. The surface properties of microchannels can affect the behaviour of cells, such as their adhesion, migration, and proliferation. By controlling the surface wettability, it is possible to optimize the interactions between cells and microchannel surfaces, for example, to promote cell adhesion or to prevent cell fouling or aggregation.

4. Droplet-based microfluidics:

Surface wettability is critical in droplet-based microfluidics, where droplets of one fluid are manipulated in another immiscible fluid. By controlling the surface wettability of the microchannel surfaces, it is possible to control the formation, transport, and manipulation of droplets, which are widely used in applications such as droplet-based assays, drug screening, and bioassays (Dai et al., 2022; Cho, Zhao, and Kim, 2007; Huang et al., 2022; Umapathi et al., 2018; Demirci and Montesano, 2007).

5. Sample handling and transport:

Surface wettability can also affect the handling and transport of samples in microfluidic devices, such as sample loading, sample transport, and sample recovery. By optimizing the surface wettability of microchannels, it is possible to reduce sample loss, improve sample recovery, and enhance the overall performance of microfluidic devices.

MARKET

The global microfluidics market has been experiencing significant growth in recent years and is expected to continue expanding in the coming years. Microfluidic systems are required for various market segments in different ways. The biggest microfluidics market sector is in-vitro diagnostics (IVD) followed by therapeutics and pharmaceutical research. In 2020, IVD made up about 83.3% of the microfluidic market, which was worth about $15 billion. Grand View Research, a market research firm, projects that the size of the worldwide microfluidics market to increase from USD 17.9 billion in 2020 to USD 60.3

Figure 4. Expected microfluidics market

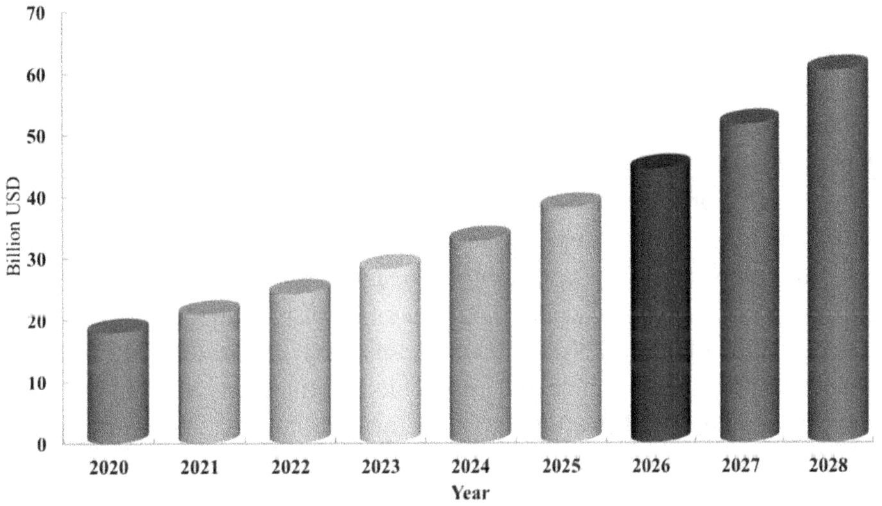

Figure 5. Material share of global microfluidics market

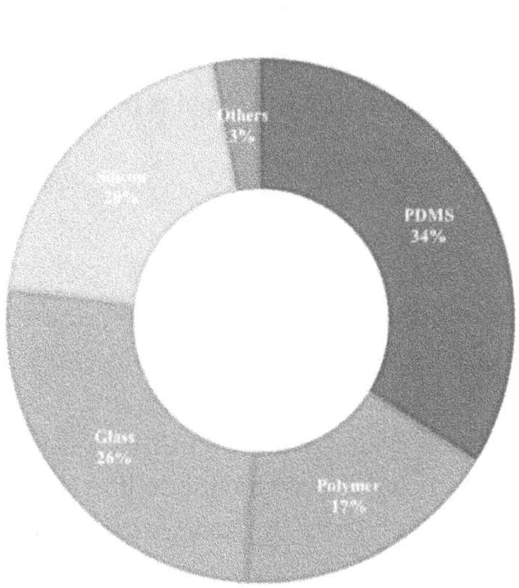

billion by 2028. The market is dominated by North America, which has 43.2% of the share. PDMS topped the microfluidics market in 2020 with 34% (USD 6.1 billion) of sales (Figure 4 and 5) (Grand View Research, 2021).

Several factors are driving the growth of the microfluidics market, including:

Increasing demand for point-of-care diagnostics: Point-of-care diagnostics, which enable rapid and accurate diagnosis at or near the site of patient care, have gained significant attention due to the need for timely and efficient healthcare delivery. Microfluidic-based diagnostic devices offer advantages such as miniaturization, portability, and the ability to handle small sample volumes, making them well-suited for point-of-care testing. The growing prevalence of chronic diseases, infectious diseases, and personalized medicine approaches are driving the demand for microfluidic-based point-of-care diagnostic devices.

Expanding applications in life sciences research: Microfluidics has become an essential tool in life sciences research, enabling precise manipulation of cells, biomolecules, and fluids at the microscale. Microfluidic devices are used for a wide range of applications, including genomics, proteomics, cell analysis, drug discovery, and drug delivery. The increasing focus on understanding complex biological processes, studying rare cells, and developing high-throughput screening methods has fuelled the demand for microfluidic platforms in life sciences research.

Advancements in microfluidic technologies: The field of microfluidics has been rapidly advancing with innovations in microfabrication techniques, materials, and device designs. These advancements have enabled the development of more sophisticated and functional microfluidic devices with improved performance, reliability, and scalability. For example, the integration of sensors, actuators, and automation technologies into microfluidic devices has expanded their capabilities and applications. Additionally, the increasing adoption of microfabrication methods, such as 3D printing and soft lithography, has made microfluidic devices more accessible and affordable, driving their adoption in various industries.

Growing investments and collaborations: The microfluidics market has been witnessing increased investments and collaborations among academic institutions, research organizations, and industry players. These collaborations are aimed at advancing microfluidic technologies, developing novel applications, and commercializing microfluidic-based products. The influx of funding and support for research and development in microfluidics has further accelerated the growth of the market.

Demand for improved healthcare outcomes and personalized medicine: The increasing focus on improving healthcare outcomes and advancing personalized medicine approaches has led to the adoption of microfluidics in areas such as liquid biopsy, single-cell analysis, and precision medicine. Microfluidic devices can provide precise and rapid analysis of small sample volumes, enabling early diagnosis, monitoring of disease progression, and tailoring of treatment plans based on individual patient characteristics.

MATERIALS

The commonly used materials for microfluidics include Rigid polymers; PDMS, inorganics like glass and silicon, paper, and mixtures of various materials, including those produced by 3D printing. Since most thermoplastics and PDMS are affordable and have seen much study, they have found widespread application. Paper microfluidics are inexpensive and can detect desired molecules visually. Many different 3D printing techniques, such as fused deposition modelling (FDM) and stereolithography (SLA), have been utilised successfully to create channels for transporting fluids.

Glass and Silicon

Given its optical transparency and potential compatibility with biological materials, glass microfluidics has found widespread use in biochemical investigation. However, since cleanroom setups with complex equipment are often needed, as well as appropriate safety gear for the application of necessary hazardous chemicals, the manufacture of glass microfluidic devices is costly. In addition, high temperature/pressure is often required for bonding etched devices to shape microfluidic features. Moreover, the incorporation of active fluid management into glass devices may be difficult, and low gas permeability of enclosed channels make them unsuitable for long-term cell growth (Ren, Zhou, and Wu, 2013; Wlodarczyk, Hand, and Maroto-Valer, 2019; Angell, Terry, and Barth, 1983; Zengerle et al., 1995; van Lintel, van De Pol, and Bouwstra, 1988).

For many years, silicon has been the go-to material for microfabrication thanks to its semiconducting characteristics. As a material for microfluidic devices, silicon is attractive because to its chemical resistance, design flexibility, and well-characterized surface modification capabilities based on the silanol group (-Si-OH). But because silicon has a high elastic modulus (150 GPa), it is hard to put in active parts like valves and pumps. Furthermore, since silicon devices are opaque to visible light, they are unsuitable for conventional fluorescence-based detection or direct fluid imaging.

It is possible to create fluidic platforms out of low-temperature co-fired ceramic, which is an aluminium oxide-based material with its own distinct surface chemistry and physical characteristics. The construction of intricate fluidic channels is made possible by the multilayer manufacturing method. Although microfluidic chips made from inorganic materials aren't the best choice for prototype research, they do have their uses. As a result, inorganic materials are preferable under extreme conditions such as high temperatures or pressures.

Thermoplastics

Thermoplastics have been employed in industry since the 1930s, and since then they have undergone extensive refinement, allowing for their mass manufacturing of high-quality items. Due to this, a large portion of microfluidics research geared toward practical applications has concentrated on building devices using rigid polymers as their main constituents. Development of microfluidic devices may become more challenging when they go from prototype to manufacturing since new technologies must be modified for mass production. Some researchers try to join in on the commercialization action, while others work to enhance fabrication techniques to do away with the need for prototyping at all.

FABRICATION TECHNIQUES

There are several fabrication techniques used in microfluidics to create microchannels, chambers, and other structures for fluid manipulation. Some of the common fabrication techniques include:

Soft lithography: Soft lithography is a widely used fabrication technique in microfluidics that involves creating microstructures using elastomeric materials, such as polydimethylsiloxane (PDMS). Soft lithography allows for the rapid prototyping of microfluidic devices with complex geometries, and it offers excellent flexibility in design and fabrication. Soft lithography typically involves replica moulding, where a master mould is created using photolithography or other techniques, and then the elastomeric material is cast against the master mould to create the microstructures.

Photolithography: Photolithography is a commonly used fabrication technique in microfluidics that uses photomasks to define patterns on a photosensitive material. It involves a series of steps, including applying a photoresist, exposing the resist to UV light through a photomask, developing the resist to remove the exposed areas, and then etching or depositing materials on the patterned substrate. Photolithography allows for high-resolution patterning of microstructures and is suitable for large-scale fabrication.

Laser machining: Laser machining is a versatile fabrication technique that uses lasers to ablate, etch, or cut materials to create microstructures. Laser machining can be used on a wide range of materials, including plastics, glass, metals, and ceramics, and it offers precise control over the dimensions and shapes of microstructures. Laser machining is particularly useful for creating microchannels, chambers, and other structures with complex geometries or for modifying the surface wettability of materials.

3D printing: 3D printing, also known as additive manufacturing, is a rapidly emerging fabrication technique in microfluidics that allows for the direct creation of three-dimensional microstructures. 3D printing offers high flexibility in design and fabrication, and it can create microstructures with complex geometries and internal features. Various 3D printing techniques, such as stereolithography, selective laser sintering, and fused filament fabrication, can be used in microfluidics to create microstructures, including microchannels, chambers, and connectors.

Micro-milling and micro-moulding: Micro-milling and micro-moulding are techniques used to create microstructures by subtractive and additive methods, respectively. Micro-milling involves using micro-scale milling tools to remove material and create microstructures, while micro-moulding involves using moulds to create microstructures by injecting or casting materials into the moulds. These techniques are particularly useful for creating microstructures in rigid materials, such as plastics, metals, and ceramics.

Bonding techniques: Bonding techniques are used to assemble microfluidic devices by joining different parts or layers together. Common bonding techniques in microfluidics include thermal bonding, solvent bonding, and adhesive bonding. These techniques involve applying heat, solvents, or adhesives to bond microstructures or layers of materials together, creating sealed microchannels and chambers.

Electroforming: Electroforming is a technique used to fabricate microstructures with high aspect ratios, such as microchannels with small diameters and large depths. It involves depositing a conductive material, typically metal, onto a patterned substrate using electroplating, and then separating the deposited structure from the substrate to create a free-standing microstructure. Electroforming is commonly used for fabricating microstructures made of metals, such as nickel, gold, or copper, and it allows for precise control over the dimensions and shapes of the microstructures.

Micro-embossing: Micro-embossing is a replication technique used to fabricate microstructures by pressing a heated material against a patterned mould. The material is softened by heat and takes the shape of the mould, creating microstructures when the material solidifies. Micro embossing is suitable for creating microstructures in thermoplastic materials, such as polystyrene, polycarbonate, or cyclic olefin copolymer (COC), and it offers high replication fidelity and cost-effectiveness for mass production.

Nanoimprint lithography: Nanoimprint lithography is a high-resolution patterning technique used to fabricate microstructures by imprinting a patterned mould onto a polymer material. The polymer material is softened by heat and pressure, allowing it to conform to the shape of the mould. After solidification, the mould is removed, leaving behind the desired microstructures. Nanoimprint lithography can achieve very high resolution down to the nanoscale and is suitable for creating microstructures with intricate patterns and features (Chou, Krauss, and Renstrom, 1996).

Microfluidic paper-based devices (μPADs): Microfluidic paper-based devices, also known as μPADs, are a low-cost and simple fabrication technique used to create microfluidic devices using porous paper as a substrate. μPADs are typically fabricated by patterning paper with hydrophobic barriers, such as wax or other hydrophobic materials, to create microchannels and chambers for fluid manipulation. μPADs are particularly useful for point-of-care diagnostics and other low-resource settings due to their simplicity, portability, and affordability.

Self-assembly techniques: Self-assembly techniques involve the spontaneous arrangement of micro- and nanoscale components into desired structures or patterns without external forces or energy inputs. Self-assembly techniques can be used to create microstructures with precise dimensions and orientations, and they are particularly useful for fabricating complex and hierarchical structures. Examples of self-assembly techniques used in microfluidics include self-assembly of nanoparticles, self-folding techniques, and self-organization of cells.

ADVANTAGES OF MICROFLUIDICS

Microfluidics offers several advantages over traditional macroscale analysis because of the smaller size of the individual components in a microfluidic circuit and the unexpected changes in the fluid's behaviour in microvolumes. On a single device, various components and processes may be integrated, resulting in a platform with a tiny footprint. Small dimensions also have the benefit of reducing the amount of sample and reagents, which reduces costs on reagents and waste. Microfluidic devices may be designed to meet the specific demands of different cell types and cell co-cultures can be executed on the same chip with tremendous flexibility. Also, it is possible to better imitate a cell's natural milieu using microfluidic cell culture (Yeo et al., 2011). The field of microfluidics has become more important in clinical diagnostics as a result of the fact that immunoassays and DNA analysis may be miniaturised via the use of microchannel-based methods or through microarrays. As a result of the high-throughput nature of microfluidic tests in the biological and chemical domains, microfluidics devices make it possible to describe heterogeneous systems with a better degree of accuracy (Dutse and Yusof, 2011; Swerdlow and Gesteland, 1990; Zhu et al., 2018).

For the monitoring of daily problems, such as in environmental and food analysis, as well as in forensic research, when only little quantities of material may be accessible, microfluidic devices may be used.

Custom chip designs, parallelism, automation, and direct connectivity to miniature downstream analytic platforms are all features of microfluidic devices that allow researchers to precisely control experimental settings.

Microfluidic systems employ the changed characteristics of liquids and gases at microscale. The amount of control provided by microfluidics has caused a shift in the methodology that scientists use to investigate how molecules interact with one another. This extremely effective instrument is finding applications in an increasing number of research domains, particularly in the field of bioassays.

Microfluidics offers improved capabilities for the complete integration and simplification of the biological process for the benefit of the end users. For personalised cancer therapy, the tumour's makeup and drug sensitivity are crucial factors. Utilising microfluidic technology, detection is more sensitive and throughput is enhanced, allowing for a customised cancer therapy that is tailored to the patient's needs (Hajii et al., 2020). As a result of the fact that single cells and even molecules may be modified experimentally in vitro, it is now possible to define the heterogeneity of a number of different factors in cell populations.

COVID19 has killed more than 6 million people and infected more than 662 million till date (World Health Organization, 2021), despite the fact that researchers have successfully replicated COVID19 in pigs (Weingartl et al., 2004; Chen, Yan, and Yang, 2005) hamsters (Chan et al., 2020), dogs (Shi et al., 2020), macaques (Bao et al., 2020; Rockx 2020), cats (Shi et al., 2020; Halfmann et al., 2020), minks (Oreshkova et al., 2020), and ferrets (Shi et al., 2020; Kim et al., 2020) etc none of these animal models accurately mimic the mortality and morbidity found in many human patients. As a result, numerous constraints associated with animal models continue to exist, potentially delaying or even preventing the development of vaccinations and treatments. Organ chip models can now or in future give chances to overcome many of those constraints and give fresh insight into the host response to infections and pathophysiology of diseases.

APPLICATIONS

Microfluidic technology has found diverse applications in various fields due to its advantages such as precise fluid handling, high throughput, automation, and portability. Here are some applications of microfluidics:

Biomedical Diagnostics: Microfluidics has revolutionized the field of diagnostics by enabling the development of portable, point-of-care diagnostic devices. These devices can quickly and accurately detect diseases such as infectious diseases, cancer, and genetic disorders using small volumes of samples. Microfluidic diagnostic devices are cost-effective, easy to use, and can provide rapid results, making them ideal for resource-limited settings and remote areas.

Drug Discovery and Development: Microfluidics has accelerated the drug discovery process by allowing high-throughput screening of drug candidates in a miniaturized format. Microfluidic devices can create controlled microenvironments that mimic human physiological conditions, allowing for more accurate testing of drug efficacy, toxicity, and pharmacokinetics. Microfluidic systems are also used for studying drug formulations, drug delivery, and personalized medicine.

Bioanalysis: Microfluidics has been widely used for bioanalysis, including DNA sequencing, protein analysis, and cell sorting. Microfluidic devices can manipulate and separate biomolecules with high precision, leading to faster and more efficient analysis. Microfluidic platforms have also been developed for single-cell analysis, including genomics, proteomics, and metabolomics, which have significant implications for cancer research, immunology, and stem cell research.

Environmental Monitoring: Microfluidic systems have been utilized for environmental monitoring and analysis, including water quality testing, air pollution monitoring, and soil analysis. Microfluidic devices can provide rapid and on-site analysis of various environmental parameters, such as pH, nutrients, heavy metals, and pathogens. These devices offer real-time monitoring and can aid in environmental management and pollution control.

Food and Agriculture: Microfluidics has been applied in the food and agriculture industry for various applications, such as food safety testing, crop disease monitoring, and animal health diagnostics. Microfluidic devices can detect contaminants, pathogens, and toxins in food and agricultural products rapidly and with high sensitivity. These devices can also facilitate precision agriculture by providing real-time data on soil conditions, crop health, and irrigation management.

Energy and Chemical Engineering: Microfluidics has been employed in the energy and chemical engineering fields for applications such as catalyst screening, fuel cell testing, and chemical synthesis. Microfluidic devices can create controlled reaction conditions, optimize reaction parameters, and reduce chemical waste. These devices also offer high-throughput screening of catalysts and reaction conditions, leading to accelerated process development and optimization.

Personalized Medicine: Microfluidics has the potential to revolutionize personalized medicine by enabling the development of miniaturized diagnostic and therapeutic devices that can be tailored to individual patients. Microfluidic devices can integrate sample preparation, diagnosis, and treatment in a single chip, allowing for rapid and targeted therapy. Personalized medicine based on microfluidics has the potential to improve patient outcomes, reduce healthcare costs, and enable precision healthcare.

Drug Delivery Systems: Microfluidics has been used to develop novel drug delivery systems that can precisely control the release of drugs with high spatiotemporal resolution. Microfluidic devices can create microscale drug carriers, such as liposomes, nanoparticles, and microcapsules, that can encapsulate and release drugs in a controlled manner. These drug delivery systems can improve the efficacy, safety, and patient compliance of drug therapies.

Organ-on-a-Chip: Also known as tissue-on-a-chip or organoid-on-a-chip, refers to microfluidic devices that are designed to mimic the structure and function of human organs or tissues. These devices typically consist of microscale channels, chambers, and compartments that are lined with living cells, allowing for the recreation of organ-level physiological functions in a controlled and customizable environment. Organ-on-a-chip platforms have gained significant attention in recent years due to their potential to revolutionize drug discovery, disease modelling, and personalized medicine. These devices offer several advantages over traditional in vitro cell culture and animal models. They can provide a more physiologically relevant microenvironment for cells, allowing for the recapitulation of complex tissue architecture, cell-cell interactions, and tissue-specific functions. Organ-on-a-chip platforms also allow for precise control of various parameters, such as fluid flow, oxygen levels, nutrient supply, and mechanical forces, which can closely mimic the in vivo conditions of human organs. Additionally, organ-on-a-chip platforms offer a high degree of customization, enabling the study of different organs or tissues in isolation or in interconnected multi-organ systems, which can provide insights into systemic interactions and organ-organ crosstalk.

Microscale Bio-fabrication: Microfluidics has been utilized for microscale bio fabrication, including 3D printing of living tissues and organs. Microfluidic devices can precisely control the deposition of cells, biomaterials, and growth factors to create complex 3D structures with high resolution. This has implications for tissue engineering, regenerative medicine, and transplantation.

Microscale Chemical Synthesis: Microfluidics has been used for microscale chemical synthesis, allowing for rapid and efficient synthesis of chemicals, pharmaceuticals, and other compounds. Microfluidic platforms can precisely control reaction conditions, reagent mixing, and reaction kinetics, leading to accelerated synthesis, reduced waste, and improved yields. Microscale chemical synthesis using microfluidics has the potential to revolutionize the field of chemistry and enable the synthesis of complex molecules with high precision.

Microscale Reaction Engineering: Microfluidics has been applied to reaction engineering at the microscale, enabling the study of reaction kinetics, transport phenomena, and reaction optimization. Microfluidic devices can create well-defined reaction conditions, such as temperature, pressure, and concentration gradients, allowing for precise control of reaction parameters. This has applications in areas such as catalysis, polymerization, and bioconversion, and can lead to more efficient and sustainable chemical processes.

Wearable and Implantable Devices: Microfluidics has been integrated into wearable and implantable devices for various applications, such as continuous monitoring of biomarkers, drug delivery, and diagnostic testing. Microfluidic devices can be miniaturized and integrated into wearable devices, patches, or implants, allowing for continuous monitoring of health parameters, drug administration, and disease management in real-time.

Microscale Robotics: Microfluidics has been used to develop microscale robots or "microbots" that can navigate, manipulate, and transport fluids at the microscale. These microbots can be remotely controlled or autonomous and can perform tasks such as sample collection, drug delivery, and cell manipulation. Microscale robotics using microfluidics has the potential to revolutionize fields such as medical robotics, environmental monitoring, and microscale assembly.

These are just a few examples of the wide-ranging applications of microfluidics. The field continues to advance rapidly, and microfluidic technology is expected to find even more diverse and innovative applications in the future.

CONCLUSION AND FUTURE SUGGESTIONS

The authors are of the opinion that there are no certain paths leading to the adoption of microfluidic technology in conventional biomedical research, and they believe that this is a significant question that is yet to be resolved. Biological research and diagnostics make use of microfluidics, yet the majority of research publications on the topic are still published in engineering journals. It may take some time for cutting-edge tools to fully replace or at least supplement more traditional approaches. Given the importance of this field to the advancement of technology, scientists need to come up with innovative and strategic approaches to study it. Further, the authors suggest that academicians and researchers from different domain should collaborate and hold conferences to foster this interdisciplinary research in this area.

As microfluidics continues to advance, there are several potential future directions and suggestions for further development and applications:

Miniaturization: Microfluidics has already enabled the miniaturization of complex fluidic systems into small, portable devices. In the future, there could be further advancements in miniaturization, allowing for even smaller and more integrated microfluidic devices. This could open up new possibilities for point-of-care diagnostics, wearable devices, and implantable devices.

Multiplexing and high-throughput screening: Microfluidic devices have the potential to enable high-throughput screening of multiple samples in parallel, leading to faster and more efficient drug discovery, diagnostics, and other applications. Future developments could focus on increasing the throughput and multiplexing capabilities of microfluidic devices, allowing for rapid analysis of large numbers of samples simultaneously.

Integration with other technologies: Microfluidics can be integrated with other technologies, such as biosensors, nanoparticles, and micro/nanofabrication techniques, to create more sophisticated and multifunctional microfluidic systems. Future developments could explore new ways to integrate microfluidics with other emerging technologies to create new functionalities and applications, such as on-chip sensing, drug delivery, and tissue engineering.

Automation and control: Automation and precise control are critical for the widespread adoption of microfluidics in various applications. Future developments could focus on the development of advanced automation and control methods, such as robotics, machine learning, and artificial intelligence, to enable more efficient and precise control of microfluidic devices, reducing human error and increasing reproducibility.

3D microfluidics: Three-dimensional (3D) microfluidic devices, which incorporate complex geometries and multi-layered structures, have the potential to offer enhanced functionalities and performance compared to traditional planar microfluidic devices. Future developments could focus on the fabrication and integration of 3D microfluidic structures, allowing for more sophisticated fluid manipulation, mixing, and reaction control.

Personalized medicine: Microfluidics has the potential to enable personalized medicine by allowing for the precise manipulation and analysis of small sample volumes, such as individual patient samples. Future developments could focus on the development of microfluidic devices and assays for personalized diagnostics, drug screening, and therapy monitoring, leading to more targeted and effective healthcare interventions.

Environmental and energy applications: Microfluidics can also find applications in environmental and energy-related fields, such as environmental monitoring, water quality analysis, and energy harvesting. Future developments could explore the use of microfluidics for environmental and energy applications, leading to more efficient and sustainable solutions for environmental monitoring and resource management.

Commercialization and accessibility: While microfluidics has shown great promise in various applications, there is still a need for further commercialization and accessibility of microfluidic devices and technologies. Future developments could focus on making microfluidic devices more affordable, user-friendly, and accessible to a broader range of users, leading to increased adoption and utilization in diverse fields.

REFERENCES

Angell, J. B., Terry, S. C., & Barth, P. W. (1983). Silicon Micromechanical Devices. *Scientific American, 248*(4), 44–55. doi:10.1038cientificamerican0483-44

Bao, L. (2020). Reinfection could not occur in SARS-CoV-2 infected rhesus macaques. bioRxiv 2020.03.13.990226

Bragheri, F., Martinez Vazquez, R. & Osellame, R. (2016). Microfluidics. *Three-Dimensional Microfabrication Using Two-Photon Polymerization: Fundamentals, Technology, and Applications,* 310–334. William Andrew. . doi:10.1016/B978-0-323-35321-2.00016-9

Chan, J. F. W. (2020). Simulation of the Clinical and Pathological Manifestations of Coronavirus Disease 2019 (COVID-19) in a Golden Syrian Hamster Model: Implications for Disease Pathogenesis and Transmissibility. *Clinical Infectious Diseases, 71,* 2428–2446. doi:10.1093/cid/ciaa644 PMID:32215622

Chen, W., Yan, M., Yang, L., Ding, B., He, B., Wang, Y., Liu, X., Liu, C., Zhu, H., You, B., Huang, S., Zhang, J., Mu, F., Xiang, Z., Feng, X., Wen, J., Fang, J., Yu, J., Yang, H., & Wang, J. (2005). Coronavirus Transmitted from Human to Pig. *Emerging Infectious Diseases, 11*(3), 2–4. doi:10.3201/eid1103.040824 PMID:15757562

Cho, S. K., Zhao, Y., & Kim, C. J. (2007). Concentration and binary separation of micro particles for droplet-based digital microfluidics. *Lab on a Chip, 7*(4), 490–498. doi:10.1039/b615665g PMID:17389966

Chou, S. Y., Krauss, P. R. & Renstrom, P. J. (1996). Imprint lithography with 25-nanometer resolution. *Science, 272*, 85–87.

Dai, R., Li, G., Xiao, L., Li, Y., Cui, Z., Jia, L., Zhou, M., Song, Y., Yang, Y., Cai, Y., Chen, C., & Yin, K. (2022). A droplet-driven micro-surfboard with dual gradients for programmable motion. *Chemical Engineering Journal, 446*, 136874. doi:10.1016/j.cej.2022.136874

Demirci, U., & Montesano, G. (2007). Single cell epitaxy by acoustic picolitre droplets. *Lab on a Chip, 7*(9), 1139–1145. doi:10.1039/b704965j PMID:17713612

Dutse, S. W., & Yusof, N. A. (2011). Microfluidics-based lab-on-chip systems in DNA-based biosensing: An overview. *Sensors (Basel), 11*(6), 5754–5768. doi:10.3390110605754 PMID:22163925

Hajji, I., Serra, M., Geremie, L., Ferrante, I., Renault, R., Viovy, J.-L., Descroix, S., & Ferraro, D. (2020). Droplet microfluidic platform for fast and continuous-flow RT-qPCR analysis devoted to cancer diagnosis application. *Sensors and Actuators. B, Chemical, 303,* 127171. doi:10.1016/j.snb.2019.127171

Halfmann, P. J., Hatta, M., Chiba, S., Maemura, T., Fan, S., Takeda, M., Kinoshita, N., Hattori, S., Sakai-Tagawa, Y., Iwatsuki-Horimoto, K., Imai, M., & Kawaoka, Y. (2020). Transmission of SARS-CoV-2 in Domestic Cats. *The New England Journal of Medicine, 383*(6), 592–594. doi:10.1056/NEJMc2013400 PMID:32402157

Huang, E., Huang, D., Wang, Y., Cai, D., Luo, Y., Zhong, Z., & Liu, D. (2022). Active droplet-array microfluidics-based chemiluminescence immunoassay for point-of-care detection of procalcitonin. *Biosensors & Bioelectronics, 195*, 113684. doi:10.1016/j.bios.2021.113684 PMID:34607116

il Kim, Y. (2020). Infection and Rapid Transmission of SARS-CoV-2 in Ferrets. *Cell Host & Microbe, 27*(5), 704–709.e2. doi:10.1016/j.chom.2020.03.023 PMID:32259477

Jerman, J. H., & Terry, S. C. (1981). A miniature gas chromatograph for atmospheric monitoring. *Environment International, 5*(2), 77–83. doi:10.1016/0160-4120(81)90126-4

Lee, Y. S., Bhattacharjee, N., & Folch, A. (2018). 3D-printed Quake-style microvalves and micropumps. *Lab on a Chip, 18*(8), 1207–1214. doi:10.1039/C8LC00001H PMID:29553156

Manz, A., Graber, N., & Widmer, H. M. (1990). Miniaturized total chemical analysis systems: A novel concept for chemical sensing. *Sensors and Actuators. B, Chemical, 1*(1-6), 244–248. doi:10.1016/0925-4005(90)80209-I

Minnella, W. (2019). Microfluidics applications: a short review - Elveflow. *Elveflow* https://www.elveflow.com/microfluidic-reviews/general-microfluidics/microfluidics-applications-a-short-review/

Oreshkova, N., Molenaar, R. J., Vreman, S., Harders, F., Oude Munnink, B. B., Hakze-van der Honing, R. W., Gerhards, N., Tolsma, P., Bouwstra, R., Sikkema, R. S., Tacken, M. G. J., de Rooij, M. M. T., Weesendorp, E., Engelsma, M. Y., Bruschke, C. J. M., Smit, L. A. M., Koopmans, M., van der Poel, W. H. M., & Stegeman, A. (2020). SARS-CoV-2 infection in farmed minks, the Netherlands, April and May 2020. *Eurosurveillance, 25*(23), 1. doi:10.2807/1560-7917.ES.2020.25.23.2001005 PMID:32553059

Ren, K., Zhou, J., & Wu, H. (2013). Materials for microfluidic chip fabrication. *Accounts of Chemical Research, 46*(11), 2396–2406. doi:10.1021/ar300314s PMID:24245999

Reseach, G. V. (2021). Injection Molded Plastics Market Size Report, 2021-2028. *Grand View Research.* https://www.grandviewresearch.com/industry-analysis/injection-molded-plastics-market

Rockx, B. (2020). Comparative pathogenesis of COVID-19, MERS, and SARS in a nonhuman primate model. *Science, 368*, 1012–1015.

Shi, J. (2020). Susceptibility of ferrets, cats, dogs, and other domesticated animals to SARS-coronavirus 2. *Science, 368*, 1016–1020.

Squires, T. M., Messinger, R. J. & Manalis, S. R. (2008). Making it stick: convection, reaction and diffusion in surface-based biosensors. *Nature Biotechnology, 26,* 417–426.

Swerdlow, H., & Gesteland, R. (1990). Capillary gel electrophoresis for rapid, high resolution DNA sequencing. *Nucleic Acids Research, 18*(6), 1415–1419. https://sci-hub.do/https://academic.oup.com/nar/article-abstract/18/6/1415/1325051. doi:10.1093/nar/18.6.1415 PMID:2326186

Terry, S. C., Jerman, J. H., & Angell, J. B. (1979). A gas chromatographic air analyzer fabricated on a silicon wafer. *IEEE Transactions on Electron Devices, 26*(12), 1880–1886. doi:10.1109/T-ED.1979.19791

Umapathi, U., Shin, P., Nakagaki, K., Leithinger, D. & Ishii, H. (2018). Programmable Droplets for Interaction. 1–1. . doi:10.1145/3170427.3186607

van Lintel, H. T. G., van De Pol, F. C. M., & Bouwstra, S. (1988). A piezoelectric micropump based on micromachining of silicon. *Sensors and Actuators, 15*(2), 153–167. doi:10.1016/0250-6874(88)87005-7

Weingart, H. M., & ... . (2004). Susceptibility of Pigs and Chickens to SARS Coronavirus. *Emerging Infectious Diseases*, *10*(2), 179–184. doi:10.3201/eid1002.030677 PMID:15030680

Whitesides, G. M. The origins and the future of microfluidics. Nature, 442. 368–373. (2006). doi:10.1038/nature05058

Wlodarczyk, K. L., Hand, D. P., & Maroto-Valer, M. M. (2019). Maskless, rapid manufacturing of glass microfluidic devices using a picosecond pulsed laser. *Scientific Reports*, *9*(1), 1–13. doi:10.103841598-019-56711-5 PMID:31882878

Woolley, A. T., & Mathies, R. A. (1994). Ultra-high-speed DNA fragment separations using microfabricated capillary array electrophoresis chips. *Proceedings of the National Academy of Sciences of the United States of America*, *91*(24), 11348–11352. doi:10.1073/pnas.91.24.11348 PMID:7972062

World Health Organization. (2021). WHO Coronavirus Disease (COVID-19) Dashboard With Vaccination Data. *World Health Organization*. https://covid19.who.int/%0Ahttps://covid19.who.int/%0Ahttps://covid19.who.int/region/searo/country/bd.

Yeo, L. Y., Chang, H. C., Chan, P. P. Y., & Friend, J. R. (2011). Microfluidic devices for bioapplications. *Small*, *7*(1), 12–48. doi:10.1002mll.201000946 PMID:21072867

Zengerle, R., Ulrich, J., Kluge, S., Richter, M., & Richter, A. (1995). A bidirectional silicon micropump. *Sensors and Actuators. A, Physical*, *50*(1-2), 81–86. doi:10.1016/0924-4247(96)80088-4

Zhao, B., MacMinn, C. W., & Juanes, R. (2016). Wettability control on multiphase flow in patterned microfluidics. *Proceedings of the National Academy of Sciences of the United States of America*, *113*(37), 10251–10256. doi:10.1073/pnas.1603387113 PMID:27559089

Zhu, S., Li, H., Yang, M., & Pang, S. W. (2018). Label-free detection of live cancer cells and DNA hybridization using 3D multilayered plasmonic biosensor. *Nanotechnology*, *29*(36), 365503. doi:10.1088/1361-6528/aac8fb PMID:29848789

Chapter 4
Advances in Micropumps for Microfluidic Systems:
Challenges and Opportunities – Recent Trends in Micropumps

Harsh Nigam
Birla Institute of Technology and Science Pilani, Goa, India

Jegatha Nambi Krishnan
Birla Institute of Technology and Science Pilani, Goa, India

ABSTRACT

Micropumps play a crucial role in the microfluidic devices through their exact control and manipulation of fluid at the microscale. Advancements have been made in the creation of micropumps for applications ranging from chemical synthesis to biomedical applications. In this chapter an overview of the most recent developments in micropump technology and design, starting with understanding the working principles of micropumps like electrostatic, pneumatic, etc.as well as differences between Mechanical and Non-Mechanical pumps will be described. This chapter contains the recent developments in microfluidic channels, valves and actuators, as well as some general methods of fabrication of these micropumps (like soft-lithography, MEMS fabrication) along with examples. In the penultimate section, the chapter elaborates on the various applications of micropumps in environmental monitoring (such as particle counting, leak detection, etc.) and biomedical applications and point of care equipment. Finally, the chapter concludes about the future and markets of micropumps.

INTRODUCTION

Chemical and biomedical engineering, biotechnology, and medical diagnostics are just a few of the domains where microfluidic systems and micropumps have found widespread application. Fluidic device miniaturization has made it possible to create cost-efficient, portable systems with excellent precision, sensitivity, and throughput (Barua et al., 2021). Researchers and scientists may now handle, analyze,

DOI: 10.4018/978-1-6684-6952-1.ch004

and process small amounts of fluids with little sample loss, little reagent usage (Dincer et al., 2017), and quick response times thanks to the use of microfluidic systems and micropumps. This article offers a thorough review of microfluidic systems and micropumps particularly, along with information on their principles of operation, design factors, fabrication methods, and applications. This chapter aims to provide a brief information for the readership interested in exploring the potential of microfluidic systems and micropumps for their projects and applications.

In this chapter, the concepts of Mechanical and Non-Mechanical micropumps, its working principles and the major differences (Mohith et al., 2019) along with some examples, would be discussed first along with some recent developments in the following field. Following that, design considerations and working principles of some types of commercially used micropumps, like Electrostatic Micropumps, Piezoelectric Micropumps (Haldkar et al., 2021) among others would be discussed. Recent developments in micropumps also come along with modifications and developments in actuators, valves (Zhang et al., 2007) and microfluidic channels. Now, these developments may not be achieved with just theoretical demonstrations, hence there is a need to know the various methods by which we can fabricate micropumps (Alvarez-Braña et al., 2021). General ways of various methods which could be used for the fabrication of micropump along with an example would be discussed (Aziz et al., 2019). Single micropumps are of great use, but better results could be accomplished by the combination of various micropumps, hence the concept of Pump Networks would also be explored. Keeping up with the smart world around us, insights on the integration of feedback control systems along with the micropumps and smart/ intelligent micropumps would be provided. Finally the various applications of the micropumps in environmental monitoring and biomedical applications (Wang et al., 2018) along with its future and market of micropumps would be elucidated (Bohr et al., 2019).

BRIEF OVERVIEW OF MECHANICAL AND NON-MECHANICAL MICROPUMPS

The exact manipulation and transportation of fluids at the microscale is made possible by micropumps, which are a crucial component of microfluidic systems. Based on the actuation principles used by the micropumps they can be divided into mechanical and non-mechanical groups. In this section, the fundamental distinctions between the two kinds of micropumps, their guiding principles will be described intensively.

Mechanical Micropumps

Mechanical micropumps utilize moving parts to generate flow, and they can be further categorized as either displacement pumps or dynamic pumps.

Displacement pumps (Mohith et al., 2019) work by displacing a fixed volume of fluid per cycle, such as a piston or diaphragm pump. These pumps can generate high pressures and they are commonly used for sample injection in microfluidic systems, as well as drug delivery devices and lab-on-a-chip systems.

Dynamic pumps work by generating flow through a moving impeller, such as a centrifugal or axial flow pump. These pumps can generate high flow rates and they are commonly used for microscale mixing and fluid circulation in microfluidic systems, as well as microscale cooling applications.

Figure 1. Schematic showing working principle of mechanical micropump
Source: Mohith et al. (2019)

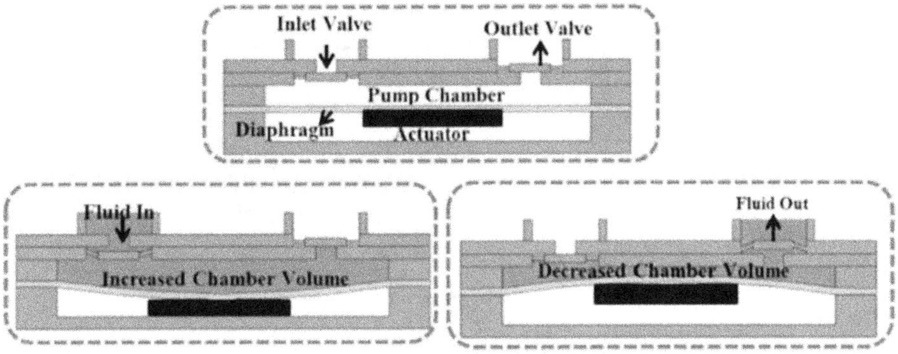

Examples of mechanical micropumps include peristaltic pumps, syringe pumps, and gear pumps. These pumps have been widely utilized in biomedical research, diagnostics, and drug discovery due to their reliability and ability to generate precise fluid flow.

Micropumps play a crucial role in the microfluidic devices through their exact control and manipulation of fluid at the microscale. Advancements have been made in the creation of micropumps for applications ranging from chemical synthesis to biomedical applications. In this chapter an overview of the most recent developments in micropump technology and design, starting with understanding the working principles of micropumps like electrostatic, pneumatic, etc. as well as differences between Mechanical and Non-Mechanical pumps will be described. This chapter contains the recent developments in microfluidic channels, valves and actuators, as well as some general methods of fabrication of these micropumps (like soft-lithography, MEMS fabrication) along with examples. In the penultimate section, the chapter elaborates on the various applications of micropumps in environmental monitoring (such as particle counting, leak detection, etc.) and biomedical applications and point of care equipment. Finally, the chapter concludes about the future and markets of micropumps.

Non-Mechanical Micropumps

Non-mechanical micropumps, on the other hand, do not have any moving parts, and they are typically powered by external stimuli such as electric fields, magnetic fields, or acoustic waves. Electrokinetic micropumps use an electric field to induce electroosmotic or electrophoretic motion, which creates fluid flow. By utilizing the interaction between the charged surfaces and the fluid, electroosmotic pumps, such as electroosmotic diode pumps, can produce flow. Dielectrophoretic pumps, for example, produce flow by taking advantage of the movement of charged particles in the fluid caused by an electric field. These pumps have been extensively used in applications for DNA sequencing, cell sorting, and microscale analysis. Magnetic fields are used by electromagnetic micropumps to create flow by causing magnetohydrodynamic motion in the fluid. The conventional construction of these pumps consists of a channel with a magnetic field applied perpendicular to the flow direction, creating a Lorentz force that propels fluid flow. These pumps have been utilized in microscale cooling applications and lab-on-a-chip systems.

Figure 2. Insulin micropump
Source: Debiotech

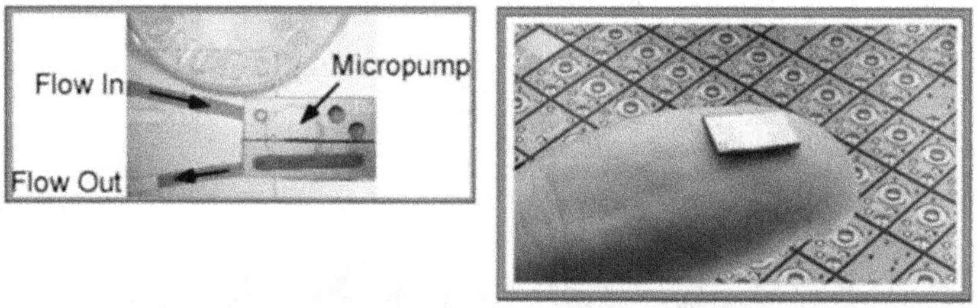

Insulin Micropump (Image courtesy of Debiotech, S. A., Switzerland)

Acoustic micropumps generate flow by utilizing acoustic waves to create pressure gradients in the fluid. These pumps can be used for precise liquid dispensing and mixing in microfluidic systems, as well as microscale particle manipulation and separation.

Current Technological Developments

More complicated and sophisticated micropumps have been created as a result of recent developments in microfabrication technologies. For instance, 3D printing has made it possible to create micropumps with intricate internal passageways (Amini et al., 2023). Additionally, developments in materials science have produced smart materials that can be used to create microactuators and valves, such as shape-memory alloys and electroactive polymers. Additionally, the development of lab-on-a-chip systems for point-of-care diagnostics and personalized medicine has been made possible by the integration of microfluidic systems with sensors and detectors. Microfluidic biosensors, for instance, can be used to quickly and accurately identify disease biomarkers in clinical samples, facilitating early diagnosis and therapy (Dincer et al., 2017).

Recent developments in mechanical micropumps also include the use of new materials such as graphene, carbon nanotubes, and nanocellulose, which offer high mechanical strength, high thermal conductivity, and biocompatibility. For example, researchers have developed a graphene-based micropump that can operate at high temperatures and pressures and is suitable for applications in harsh environments.

Non-mechanical micropumps are also being developed with new materials and techniques. For example, researchers have developed a microfluidic device that uses a thermoelectric effect to drive fluid flow. This device, which is based on the Seebeck effect, can generate fluid flows without any moving parts and has potential applications in microfluidic cooling and power generation.

For microfluidic devices, mechanical and non-mechanical micropumps each have certain benefits and drawbacks. Non-mechanical pumps have recently attracted interest because of their capacity to work without any moving components, enabling simpler and more reliable microfluidic control, whereas mechanical pumps have been frequently used due to their dependability and precision.

COMMERCIALIZED MICROPUMPS

Within the microfluidic world, physical laws often do tend to change their appearance. As an example, volumetric forces, such as weight or inertia [L^3, where L stands for a one dimension], often become negligible in comparison to surface force [L] which often tend to dominate fluidical behavior. With only a few exceptions, most of the micropumps rely on micro-actuation principles, which can reasonably be scaled up only to a certain size (Kim et al., 2018). Thus most of the pumps usually have the same mechanism for the input, output and motion of the fluid, but what differentiates them is the way in which actuation is achieved. Given below are some of the micropump categories which are mostly used in the commercial space.

Electrostatic Micropump

Electrostatic micropumps are a type of micropump that use electrostatic forces to generate fluid flow. These micropumps typically consist of two plates with one plate having a series of small channels and the other plate having a series of electrodes. When a voltage is applied to the electrodes, an electric field is generated, which interacts with the fluid in the channels and causes it to flow.

The electrostatic force applied on the electrostatic plates can be expressed by the equation

$F = dW/dX = 1/2\ (\varepsilon A\ V^2/X^2)$ Equation 1

(Where, F is electrostatic attraction force, W is energy stored, ε is dielectric constant, A is area of electrodes, X is electrode spacing and V is the applied voltage).

Electrostatic micropumps' capacity to generate high flow rates with little power usage is one of its key advantages. They are advantageous for applications that call for precise control of fluid flow since they can also generate very modest flow rates with high precision. Additionally, they have a relatively straightforward construction and are simple to incorporate into microfluidic systems. However, electrostatic micropumps do have significant disadvantages. One problem is that they may be sensitive to changes in the fluid's conductivity and viscosity, which may impair the pump's efficiency. Due to their narrow channel size, they may also be vulnerable to clogging or fouling, which can shorten their lifespan.

Electromagnetic Micropumps

Another type of micropump that uses magnetic fields to create fluid flow is called an electromagnetic micropump. The magnet and wire coil that make up these micropumps are often positioned close to one another. A magnetic field is created when an electrical current is run through the wire coil, interacting with the magnet to move it. The fluid flows in the microchannel of the micropump as a result of the magnetic movement.

One of the main benefits of electromagnetic micropumps is their ability to produce high flow rates with low power consumption. They are also capable of generating relatively high pressures, making them useful for applications that require the pumping of fluids against resistance. In addition, they are generally less prone to clogging and fouling compared to other micropump technologies. Though these pumps also do have some drawbacks, like they can be sensitive to fluctuations in the fluid parameters, such as viscosity, permittivity, and conductivity, which can alter the performance of the pump, just like

Figure 3. Proposed schematic illustration of electrostatic micropump
Source: Nisar (2008)

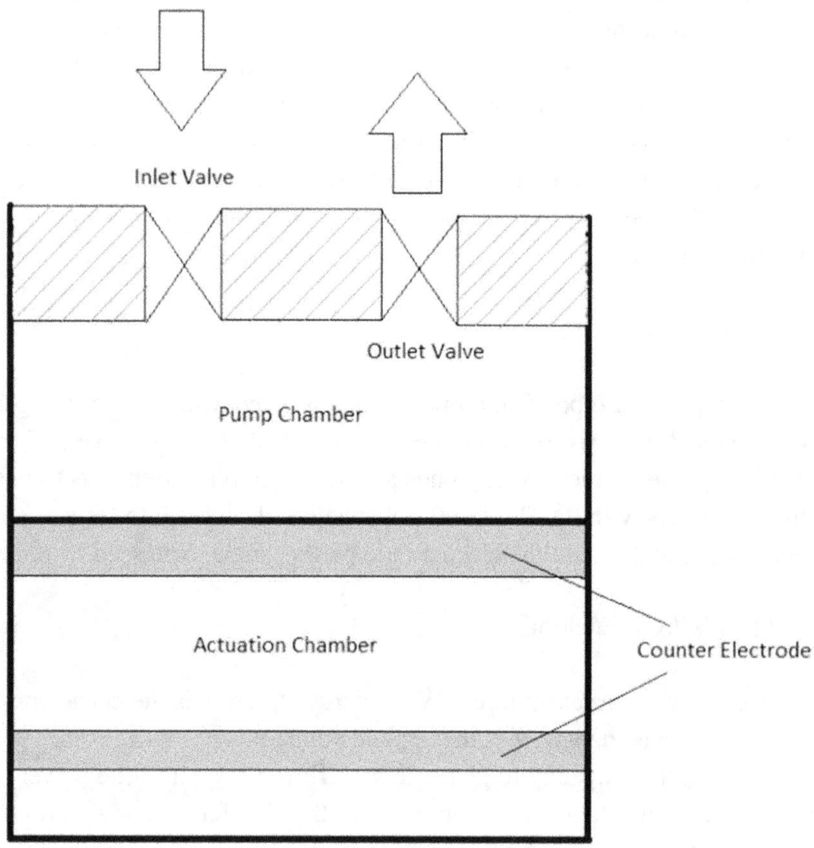

electrostatic pumps, which operate according to the same physics principles as electromagnetism. Another drawback is that they could produce heat while operating, which could be problematic for some applications in which speciality chemicals or fluids which can not be operated above a specific temperature are dealt with, for example, dialysis in which we can not have the bodily fluids being pumped above 37 °C.

Piezoelectric Micropumps

Piezoelectric micropumps are a particular kind of micropump that produce fluid flow using piezoelectric materials. Piezoelectric materials are usually materials which produce an electrical charge when it experiences a mechanical deformation and vice-versa. In these pumps, the same principle for actuation is used. A piezoelectric material deforms and generates pressure waves when a voltage is applied, which causes fluid to flow through the pump's microchannel (Haldkar et al., 2021).

The capacity to generate precise and adjustable fluid flow rates is one of the key features of Piezoelectric micropumps. The amount of voltage applied, which in turn helps us control the fluid flow can be precisely controlled. They are ideal for situations that call for pumping fluids against resistance since they can also generate reasonably high pressures. Furthermore, they are frequently rather easy to use and manufacture. Like other micropumps technology, piezoelectric micropumps are also prone to some drawbacks. One potential limitation is their relatively low flow rates compared to other micropump technologies. They can also be sensitive to variations in fluid properties, such as viscosity and temperature, which can affect their performance. Additionally, they may require a higher driving voltage compared to other micropumps.

Pneumatic Micropumps

Pneumatic micropumps, as the name suggests, produce fluid flow by using pressurized gas, often air or nitrogen, as its driving force. A chamber that is pressurized and a valve or diaphragm that regulates the flow of gas into and out of the chamber are the two main components of these pumps. A pressure difference caused by the gas being expelled from the chamber allows fluid to flow through the pump's microchannel.

The capacity to produce high flow rates with little power consumption is one of the key features of Pneumatic micropumps. They can be easily integrated into microfluidic systems and are also quite easy to construct and for application. Also, compared to other micropump technologies, pneumatic micropumps are less prone to clogging and fouling. Pneumatic micropumps have some disadvantages, such as the potential need for a bulky and, in some cases, expensive pressurized gas source. Furthermore, they might function differently depending on changes in temperature and gas pressure. Pneumatic micropumps may also vibrate and make noise while operating, which could be a problem in some applications. Though there exists a drawback for nearly all of the types of pumps mentioned above, these pumps have found use in a variety of applications, such as lab-on-a-chip devices, microscale mixing systems, drug delivery systems, microscale cooling systems, and microfluidic-based sensors. As research in this field continues, the performance and reliability of these pumps is going to increase eventually making them more useful for even a wider range of applications.

DESIGN CONSIDERATIONS OF MICROPUMPS

Brief Overview on the Utilization of Valves, Actuators, and Microfluidic Channels

One of the key areas of focus in recent years has been the development of new and improved valves, actuators, and microfluidic channels, which have led to significant improvements in the performance and versatility of micropumps.

Micropumps require valves because they control how much fluid flows through the system, making them an essential component. Precision, dependability, and speed have been highlighted in recent improvements in valve design. One method that has shown to be quite successful is the use of microvalves, which are tiny devices that may be put directly into microfluidic channels (Zhang et al., 2007). These valves offer highly accurate and repeatable fluid flow control. Different methods, such as pneumatic, electrostatic, or magnetic forces, can be used to operate them. Every day, researchers focus on the development of valveless micropumps.

Actuators are another critical component of micropumps, as they are responsible for converting energy into mechanical motion. Recent advances in actuator design have focused on improving their efficiency, response time, and durability. One promising approach that has emerged in recent years is the use of piezoelectric materials (Bußmann et al., 2021), which can generate motion in response to an applied electric field. These materials are particularly useful for driving micropumps, as they can generate high forces and displacements with low power consumption.

Recent developments in microfluidic channel design have emphasized enhancing their versatility and functionality. The utilization of multi-layer microfluidic channels, which can combine many components like valves, sensors, and mixers into a single integrated system, is one strategy that has proven particularly effective (Zhang et al., 2007). These channels can be made specifically to fit the requirements of various applications by using a number of fabrication methods, such as soft lithography or micro-electro-mechanical systems (MEMS). Overall, recent developments in micropump design have resulted in appreciable enhancements to their functionality and adaptability. The development of micropumps with greater accuracy, dependability, and functionality has been made possible by the use of new and enhanced valves, actuators, and microfluidic channels. This has created new opportunities for a variety of applications, including drug delivery, microscale chemical synthesis, and microfluidic-based biosensors. Future study in this area is likely to lead to the development of even more intriguing and cutting-edge micropump designs.

Nano/Micro Fabrication Techniques for the Construction of Micropumps

Micropumps are an essential component of microfluidic systems that play a critical role in manipulating and controlling the flow of fluids at the microscale. The ability to fabricate micropumps with precise dimensions, shapes, and functionalities is essential for the development of microfluidic systems with enhanced performance and functionality. Nano and microfabrication techniques offer a wide range of options for fabricating micropumps with high precision, complexity, and functionality. In this section, we will have a general discussion on the various nano and microfabrication techniques that can be used to fabricate micropumps.

Photolithography

Photolithography is a widely used technique in the field of microfabrication for the production of micropumps. It involves the use of light to pattern a photosensitive material, such as a photoresist, which can then be etched or used as a mold for further fabrication processes.

The fabrication of micropumps via photolithography entails a number of stages, such as photoresist application, substrate preparation, exposure to light, development, and etching. Prior to being coated with a photoresist material, a substrate material such as silicon, glass, or polymer must first be cleaned and prepared. A thin, even layer of photoresist is then spun onto the substrate surface. In order to create the required pattern of channels and chambers for the micropump, a mask with the desired pattern of channels and chambers is placed over the photoresist-coated substrate. Depending on whether a positive or negative photoresist was used, the exposed photoresist is subsequently developed using a chemical solution that either eliminates the exposed or unexposed areas of the photoresist layer. When a negative photoresist is exposed, the exposed portion becomes insoluble and is washed away, whereas when a positive photoresist is exposed, the exposed portion becomes soluble and is washed away. In most cases,

Figure 4. Schematic of general photolithography fabrication process
Source: Mohith et al. (2019)

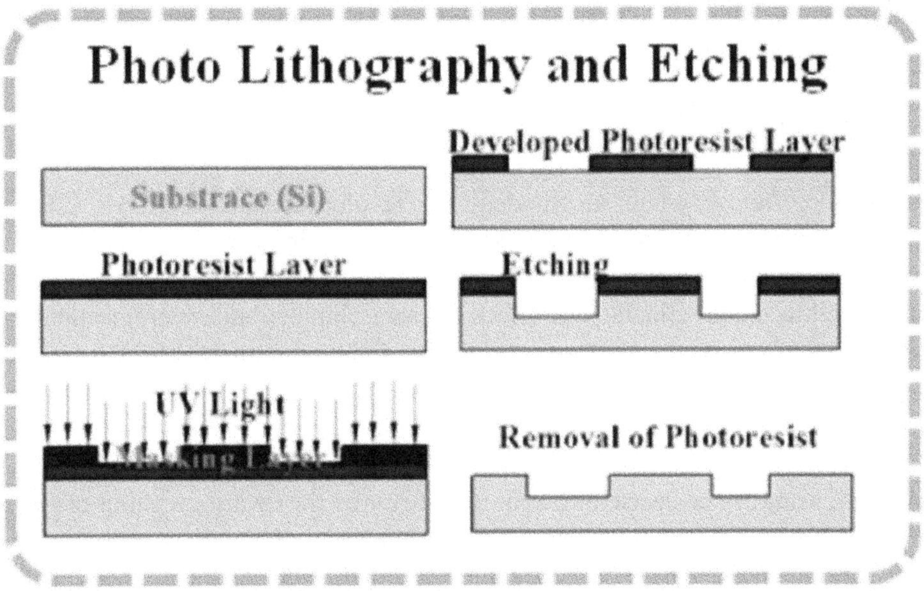

the complementary image is produced by using either a positive or negative photoresist. The substrate is covered with a patterned photoresist layer as a result of this procedure, which can later be erased or utilized as a mold for other fabrication steps (Mohith et al., 2019).

In summary, photolithography is a versatile and widely used technique for the fabrication of micropumps. Its high resolution and scalability makes it suitable for mass production, and its integration with other microfabrication techniques allows for the creation of complex and functional micropumps. However, the process can be time-consuming and requires specialized equipment and a clean room environment.

Soft Lithography

Soft lithography is a widely used technique in the field of microfabrication for the fabrication of microfluidic devices such as micropumps. It is a versatile and cost-effective technique that allows the creation of complex structures with high precision and accuracy. A master mold, which is a patterned substrate made using photolithography or other microfabrication methods, is often used in soft lithography. The patterned characteristics are then replicated using this mold on a soft elastomeric material, like polydimethylsiloxane. (PDMS) (Aziz et al., 2019). The final microfluidic device is made by bonding the PDMS duplicate to a substrate, commonly glass or silicon.

PDMS or other elastomeric materials are used to generate a copy after the master mold has been made. PDMS is normally poured onto the master mold after being combined with a curing agent. The PDMS duplicate is removed from the master mold after hardening, and any extra material is removed. Following that, an oxygen plasma treatment or other bonding procedures are used to attach the PDMS replica to a substrate. To encourage bonding with the PDMS, the substrate is frequently thinly coated with a hydrophobic substance like trimethylchlorosilane (TMCS).

In summary, soft lithography is an effective method for creating microfluidic devices like micropumps. Researchers working in the field of microfabrication frequently choose it because of its adaptability, low cost, and great precision. Although it's crucial to understand the limitations and carefully examine if it's appropriate for a given application.

Micro-Electro-Mechanical Systems (MEMS)

Micropumps and microfluidic systems have undergone a revolution thanks to MEMS (Microelectromechanical Systems) technology. MEMS-based micropumps are being used more frequently in a variety of applications, such as micro-reactors, lab-on-a-chip devices, and medication delivery systems (Barua et al., 2021). Several techniques, including surface micromachining, bulk micromachining, and LIGA (Matteucci et al., 2006), can be used to create MEMS-based micropumps. (LIGA is German acronym [contd.] for Lithographie, Galvanoformung, Abformung). Utilizing CAD (Computer-Aided Design) software, the micropump is first designed. Using photolithography, the design is then transferred to a silicon wafer, where a pattern is made on the wafer's surface using a photoresist layer. The patterned wafer is then etched using dry or wet etching techniques to create the required features of the micropump.

Surface micromachining involves the deposition of thin layers of materials, such as silicon dioxide, polysilicon, and metal, on the wafer's surface, followed by patterning and etching to form the required structures. Bulk micromachining (Mishra & Yadava, 2015) involves etching the silicon wafer from both sides to create the desired structures. LIGA involves the use of X-ray lithography to create high aspect ratio structures in thick photoresist layers, which are then transferred to metal or polymer materials using electroplating or casting.

MEMS-based micropumps offer several advantages, including high precision, accuracy, and repeatability. They can be integrated with other MEMS devices, such as sensors and actuators, to create complex microsystems. MEMS-based micropumps can also operate under a wide range of conditions, including high temperatures, pressures, and corrosive environments. They are also highly scalable, allowing for mass production of microdevices. MEMS-based micropumps do, however, have several shortcomings. The fabrication process can be difficult and drawn out, requiring specialized tools and knowledgeable workers. MEMS-based micropumps may cost more than other micropump fabrication methods as well. MEMS-based micropumps could also need extra parts, including power supplies and control circuitry, to work properly.

Electron Beam Lithography (EBL)

Micropumps have been made using the versatile nanofabrication method electron beam lithography (EBL), which has been applied to a large number of micro- and nanoscale devices. In EBL, a targeted electron beam is used to expose a resist material in a certain way before it is developed to create a pattern on a substrate. The patterned substrate is subsequently utilized in subsequent fabrication procedures as a mold or a mask.

The creation of a substrate, which could either be a silicon wafer, glass, or a polymer, is the first step in the EBL process. A resist, such as polymethyl methacrylate (PMMA), is applied to the substrate before it is subjected to the electron beam. A computer-controlled electron beam is directed to particular areas of the resist surface in order to create the desired pattern. In order to remove the exposed portions after exposure, the resist is developed (that is removed using a suitable solvent), leaving a patterned resist layer on the substrate. Following that, the patterned resist can be utilized as a mold or a mask for other manufacturing operations like material deposition or etching. EBL has several advantages over other nanofabrication techniques. One of the major advantages is that using EBL we can create complex patterns with high resolution and accuracy. EBL can achieve feature sizes down to a few nanometers, making it suitable for fabricating micro and nanoscale devices, such as micropumps. Additionally, EBL is a flexible technique that can be used to fabricate a wide range of materials, including metals, semiconductors, and polymers. However, EBL also has some limitations. One of the major limitations is the relatively slow fabrication speed, as the electron beam needs to be scanned over the entire surface to create the pattern, thus making EBL unsuitable for high-volume production. Another limitation is the high cost of the equipment which is needed for the process and the need for specialized expertise to operate it.

Laser Micromachining

A high-power laser is employed in the precise, non-contact production process known as laser micromachining to ablate or melt materials. It entails the selective removal of material from a substrate using a laser beam to produce microscale patterns and structures. The process of creating a micropump using laser micromachining involves several steps. First, a substrate material such as silicon, glass, or polymers must be chosen based on the specifications of the micropump design. The required micropump structure is then defined by patterning a layer of resist material onto the substrate using photolithography or other methods.

The substrate is then inserted into a laser micromachining system, which frequently includes a laser source, beam delivery system, and computer-controlled stage for accurate movement of the substrate. The resist-coated substrate is targeted by the laser beam, and the material is vaporized or ablated, as need be, to produce the micropump features. One of the key advantages of laser micromachining for micropump fabrication is its high precision and accuracy it can offer on a variety of materials, including metals, semiconductors, glass, and polymers, which can be utilized to create micropumps.

One of the main drawbacks of laser micromachining is that there is a large chance that the substrate material could be damaged by heat, especially when using high-power lasers. As a result, the substrate may deform or crack, endangering the micropump's ability to function. Another challenge is the decreased throughput of laser micromachining compared to other fabrication techniques like photolithography or soft lithography. Laser micromachining is typically a serial process, requiring the fabrication of each individual micropump one at a time. This can be time-consuming and costly for large-scale production.

In summary, laser micromachining is a potent and adaptable method that offers high precision and accuracy in the construction of complicated micropump designs. It is used to fabricate micropumps in microfluidic systems. While it has some drawbacks, such as potential heat damage and limited throughput, its unique capabilities make it a valuable tool in the field of microfluidics and micropump design.

Nanoimprint Lithography

Nanoimprint Lithography (NIL) is an extremely popular micro- and nano-fabrication technique which is used to produce high-resolution patterns and structures on surfaces. This technique is used in the production of micropumps for microfluidic systems. The basic concept behind nanoimprint lithography is the transfer of a patterned mold or template onto a substrate using a thin layer of resist material. The technique offers several advantages, including high resolution, low cost, and the ability to produce large-scale patterns. The creation of a mold or template is the first step in the nanoimprint lithography process. The mold, which is often constructed of a hard substance like silicon or quartz, is patterned using a different lithographic technique, such as electron beam lithography. To transfer the pattern from the mold to the substrate, a thin layer of resist material is spun onto the substrate to provide a uniform coating.

After the resist has been placed, the mold is brought into contact with the resist layer and pressed into it using a combination of heat and pressure. As a result of this step, a high-resolution pattern is produced on the substrate surface, which makes the resist material flow and fill the spaces between the mold features. Once the resist has been patterned, the substrate is subjected to a series of processing steps, including etching, deposition, and other techniques, to create the final micropump structure. In order to create high-resolution structures on a small scale, nanoimprint lithography is the best method. The resulting micropump features can be as small as a few nanometers.

One of the main benefits of nanoimprint lithography is its high resolution, which makes it possible to build microscopic structures with submicron accuracy. Sub-10 nm resolution is possible with nanoimprint lithography, which can create objects with enormous aspect ratios on the nanoscale. The process is also reasonably inexpensive when compared to other microfabrication techniques because it doesn't call for the use of pricey tools or materials. Additionally, the use of a mold enables the creation of large-scale patterns, enabling the rapid construction of intricate micropump systems.

The use of nanoimprint lithography does have some disadvantages, though. The inability to produce large-area designs with excellent uniformity is one of the key drawbacks, as the resist material may not always flow evenly across the substrate surface. Additionally, the technique may be limited by the quality of the mold or template used, as defects or variations in the mold pattern can result in defects in the final micropump structure.

Overall, nanoimprint lithography is a promising technique for fabricating micropumps with high resolution and low cost. While there are some challenges associated with the technique, advances in mold and template fabrication, as well as improvements in resist materials and processing techniques, are likely to further improve the reliability and accuracy of the technique in the future.

Example of a Fabrication Method to Construct a Micropump

This example (Aziz et al., 2019) presents a method for the fabrication of micropump using a combination of polylactic acid (PLA) and polydimethylsiloxane (PDMS) in 3D printing process. ANYCUBIC I3 Mega 3D printer (Evolve), Filament-PLA (Evolve), Fusion 360 (Autodesk) software, Ultimaker Cura 3.1.0, Sylgard 184 (Dow corning), Solenoid plunger CS-1, LCD and Electronic components (Standwell M. Sdn. Bhd.) & Arduino UNO (Mega) were the materials used for the fabrication of this specific micropump. The actuation module and a diffuser chip module make up the modular micropump design. The diffuser chip module is slid into the control module to carry out the micropumping. The membrane is flat in the free-pinch force condition (with the magnetic plunger in the upward position). The membrane is pinched when the piston is in the downward position, drawing fluid up from below the membrane. As a consequence of this cycle's repetitions, the outlet valve will continue to flow fluid. An ANYCUBIC I3 Mega 3D printer was used to create the micropump. The Ultimaker Cura 3.1.0 software was used for slicing and printing the 3D structures. Fusion 360 (Autodesk) was used to create the 3D structures. This fabrication method used polyactic acid (PLA), also referred to as 'bioplastic,' as the primary material to create a micropump. Due to its toughness, simplicity of fabrication (3D printing), biodegradability (similar to polypropylene), and affordability, PLA exhibits excellent mechanical properties. Three parts with dimensions in the region of mm range make up the developed micro-pump. The support body is at the top, the actuation module is in the centre with holes drilled for attachment to the support body, and the diffuser chip module is at the bottom. The separate membrane component will be fastened to the diffuser chip module's top. The membrane substance has a significant impact on the membrane's ideal deflection. Since PDMS is a flexible substance, an actuator's high stroke will cause a large amount of fluid to be pumped.

PUMP NETWORK

With the creation of micropumps and fluidic circuits for precise control of fluid flow at the microscale, the field of microfluidics has made major strides recently. These micropumps can be divided into mechanical pumps and non-mechanical pumps, each having specific benefits and drawbacks. Now different pumps can be connected, so that the drawbacks of one can be compensated using the other to get the most optimum network for the flow of our liquid. Thus, even larger advancements in fluid flow management and throughput have been made possible by the integration of numerous micropumps and the development of complex fluidic circuits.

The ability to simultaneously control the flow of several fluids is one of the key benefits of using numerous micropumps in a fluidic circuit. To accomplish this, distinct channels are made for each fluid, and the flow is pushed through each channel by a different micropump. By controlling the speed and direction of each micropump, it is possible to create a wide range of fluidic circuits that can be used for specific applications, and can be programmed easily to suit other specific applications. One example of the use of multiple micropumps in a fluidic circuit is the creation of microreactors for chemical synthesis. In this application, multiple micropumps are used to control the flow of reactants through the microreactor, allowing for precise control over reaction conditions and improving the yield and selectivity of the reaction. The ability to control the flow of multiple fluids in a microreactor also enables the creation of complex reaction networks, allowing for the synthesis of more complex molecules.

Another example of the use of multiple micropumps in a fluidic circuit is in the field of microfluidic cell culture. In this application, multiple micropumps are used to control the flow of different media through the culture chamber, allowing for the precise control of the nutrient supply to the cells. This has significant advantages over traditional cell culture techniques, where the nutrient supply can be limited and difficult to control. By using micropumps to control the flow of media, it is possible to create more physiologically relevant cell culture conditions, leading to improved cell viability and functionality.

In summary, the design of complex fluidic circuits and the employment of several micropumps have considerable advantages in terms of enhancing fluid flow control and throughput. With the continued development of microfabrication techniques, we can expect to see even greater improvements in the performance and functionality of microfluidic systems in the future.

INTEGRATION OF FEEDBACK CONTROL SYSTEMS

An important development in the field of microfluidics has been the integration of feedback control systems with micropumps. In microfluidic systems, the application of feedback control systems enhances accuracy, precision, and dependability by enabling real-time monitoring and manipulation of fluid flow rates, pressures, and volumes. Sensors are frequently used in the integration of feedback control systems to measure pertinent variables including flow rate, pressure, and temperature. The controller receives the measured data and compares it to a predetermined setpoint before adjusting the micropump to maintain the required flow rate, pressure, or volume. The technique used here is called closed-loop control.

The equipment used for integrating feedback control systems with micropumps can vary depending on the specific application. Generally, a sensor, a controller, and a micropump are required. The sensor could be a flowmeter, pressure sensor, or temperature sensor, while the controller can be a microprocessor or a programmable logic controller (PLC). The micropump can be either a mechanical or non-mechanical type, depending on the application requirements. It is necessary to employ mathematical equations to transform the measured data into useful information that the controller can use when integrating feedback control systems with micropumps. For instance, when using a flowmeter to detect fluid flow rates, the measured signal must be converted into the flow rate using the proper equation. Increasing throughput and improving fluid control are two major advantages of coupling feedback control systems with micropumps. Through the use of closed-loop control, fluid flow rates and pressures may be precisely and accurately regulated, improving fluid handling and lowering mistakes. Additionally, real-time fluid parameter monitoring and control allow for the identification and correction of any abnormalities or deviations from the setpoint, improving system stability and dependability.

Significant improvements in the coupling of feedback control systems with micropumps have been made in recent years. For instance, it is increasingly popular to employ microfluidic chips with built-in micropumps and feedback control systems. These devices allow for the creation of complex microfluidic systems with multiple channels and valves, enabling the manipulation and analysis of different fluids in parallel. Furthermore, the use of machine learning algorithms for closed-loop control has also shown promising results in improving the accuracy and precision of fluid control in microfluidic systems. To summarize, the integration of feedback control systems with micropumps has been a significant development in the field of microfluidics. The use of closed-loop control allows for real-time monitoring and regulation of fluid parameters, leading to increased accuracy, precision, and reliability in microfluidic systems. With continued advancements in this field, it is expected that the integration of feedback control systems with micropumps will become increasingly common and lead to further improvements in microfluidic system performance.

SMART/INTELLIGENT MICROPUMPS

In the realm of microfluidics, smart or intelligent micropumps are a relatively new invention that have the capacity to self-regulate and make real-time adjustments based on feedback from the system. The efficiency, accuracy, and dependability of microfluidic devices could be improved by these micropumps, which could revolutionise the industry. The functionality, uses, benefits, and drawbacks of intelligent/smart micropumps will be discussed in this section.

Working Principle

The working principle of smart/intelligent micropumps is based on feedback control systems. These micropumps are equipped with sensors that detect changes in the flow rate, pressure, or other relevant parameters, and send the information to a controller. The controller processes the data and sends a signal to the micropump to adjust its output accordingly. This process is continuous and occurs in real-time, allowing the micropump to respond rapidly to changes in the system.

Applications

There are numerous possible uses for smart/intelligent micropumps in microfluidic systems. The development of medication delivery systems is one of the most exciting applications. Patients might get accurate pharmaceutical doses from smart micropumps in real-time, with the dosage changing in response to the patient's demands. This would be especially helpful in cases like critical care where the patient's status is changing quickly. Lab-on-a-chip devices are a further potential application. Smart micropumps could be used to regulate the flow of fluids through various channels, giving experimenters greater exact control. This could be especially helpful in disciplines like chemistry and biology, where even slight variations in fluid flow can have a big impact on the results of an experiment.

Advantages

The advantages of smart/intelligent micropumps are numerous. By providing real-time feedback and control, these micropumps can improve the accuracy and precision of microfluidic devices. This can lead to more reliable results and improved efficiency in a wide range of applications.

Smart micropumps also have the potential to reduce the need for manual intervention in microfluidic systems. This can save time and reduce the risk of human error, particularly in situations where rapid adjustments are necessary.

Disadvantages

The higher power consumption of smart/intelligent micropumps is a possible drawback. In battery-powered devices or other applications where power consumption is a factor, additional energy requirements of the sensors and controllers can be problematic at certain instances.

To sum up, by offering real-time feedback and control, smart/intelligent micropumps have the potential to completely change the field of microfluidics. These micropumps have a wide range of potential applications, from drug delivery to lab-on-a-chip devices. While there are some potential drawbacks to this technology, the benefits of increased accuracy, precision, and reliability are significant. As research in this field continues, more advanced and sophisticated smart/intelligent micropumps in the future are foreseen.

APPLICATIONS OF MICROPUMPS

Environmental Monitoring

Environmental monitoring is essential for ensuring the safety and health of individuals and the environment. Among the most crucial aspects of environmental monitoring are particle counting, air quality monitoring, and leak detection. These three applications can be achieved through the use of micropumps and microfluidic systems, which offer accurate and reliable monitoring.

Particle Counting: Is essential in various applications, including industrial processes, healthcare, and environmental monitoring. Particles in the air or water can be efficiently and accurately detected with the use of micropumps and microfluidic systems. The procedure involves allowing or injecting the sample through a microfluidic channel, where the presence of particles is found by light scattering or absorption. Particle counters identify and count the particles while micropumps regulate the sample's flow through the microfluidic channels. This technique enables for immediate intervention in cases of hazardous particle levels and provides real-time monitoring of particle concentration.

Air Quality Monitoring: Is another critical application of micropumps and microfluidic systems. Poor air quality can lead to respiratory problems and other health complications, making air quality monitoring essential for public health. The process involves collecting samples of air and passing them through a microfluidic channel. The micropumps control the flow of the air samples, and sensors detect the concentration of pollutants such as carbon monoxide, ozone, and nitrogen oxides. This method provides real-time monitoring of air quality and allows for immediate action to be taken in case of hazardous pollutant levels.

Leak Detection: Is also crucial in environmental monitoring, particularly in industrial processes. Leaks can lead to environmental contamination, endangering human and animal health. Real-time leak detection is possible with the use of micropumps and microfluidic systems, enabling prompt action. A sample of the liquid is put via a microfluidic channel during the operation, and the flow rate is controlled by tiny pumps. Then, in case of a leak, sensors identify the existence or liquid and notify the operator. This approach provides a trustworthy and precise way to identify leaks and stop environmental contamination.

Particle counting, air quality monitoring, and leak detection are just a few of the numerous environmental monitoring applications for micropumps and microfluidic systems. These applications provide precise and dependable environmental monitoring techniques that enable prompt response to potentially dangerous situations. Because of their precision, dependability, and affordability, micropumps and microfluidic devices are projected to be used more in environmental monitoring.

Drug Delivery

Drug delivery is the process of administering a therapeutic agent into the body to achieve a desired therapeutic effect. Drug delivery has been revolutionized by micropumps and microfluidic devices because they allow for precise and regulated administration of medicinal substances to particular bodily regions. Compared to conventional medication delivery methods, the use of micropumps and microfluidic devices has a number of benefits, including enhanced patient compliance, better dose precision, and fewer adverse effects (Nisar et al., 2008).

Working Principles

Therapeutic substances are delivered using micropumps and microfluidic systems via intravenous, oral, and transdermal routes. The fundamental idea behind therapeutic agent delivery utilizing micropumps is to maintain a continuous and controlled flow inside the body. This is achieved by using various types of pumps such as peristaltic pumps, syringe pumps, and piezoelectric pumps.

Advantages

Micropumps and microfluidic systems offer several advantages in drug delivery, including:

- **Targeted Delivery:** Micropumps and microfluidic systems allow targeted delivery of drugs to specific parts of the body, minimizing side effects and increasing therapeutic efficacy.
- **Accurate Dosing:** Micropumps and microfluidic systems provide accurate dosing of therapeutic agents, eliminating errors associated with manual dosing.
- **Reduced Side Effects:** Micropumps and microfluidic systems provide a controlled and steady flow of therapeutic agents, reducing the likelihood of side effects associated with rapid changes in drug concentration.
- **Improved Patient Compliance:** Micropumps and microfluidic systems provide a convenient and easy-to-use method of drug delivery, improving patient compliance with medication regimens.

Disadvantages

While micropumps and microfluidic systems offer several advantages in drug delivery, they also have some disadvantages, including:

- **High Cost:** Micropumps and microfluidic systems can be expensive to manufacture, limiting their availability to certain patient populations.
- **Complexity:** The use of micropumps and microfluidic systems requires specialized training, which can limit their use in certain healthcare settings.
- **Risk of Infection:** The use of micropumps and microfluidic systems can increase the risk of infection due to the need for surgical implantation or use of catheters.

Applications

Micropumps and microfluidic systems have a wide range of applications in drug delivery, including:

- **Cancer Treatment:** Micropumps and microfluidic systems are used to deliver chemotherapy drugs directly to cancerous tumors, minimizing side effects and improving therapeutic efficacy.
- **Diabetes Management:** Micropumps and microfluidic systems are used to deliver insulin and other medications to manage diabetes, providing precise dosing and reducing the need for frequent injections (Yang et al., 2022).
- **Pain Management:** Micropumps and microfluidic systems are used to deliver pain medications directly to the site of pain, providing rapid and targeted relief.
- **Hormone Replacement Therapy:** Micropumps and microfluidic systems are used to deliver hormones for hormone replacement therapy, providing precise dosing and reducing the risk of side effects.

Micropumps and microfluidic systems have revolutionized drug delivery by providing controlled and targeted delivery of therapeutic agents to specific parts of the body. While they have some disad-

Figure 5. Schematic representation of drug delivery system
Source: Nisar et al., (2008)

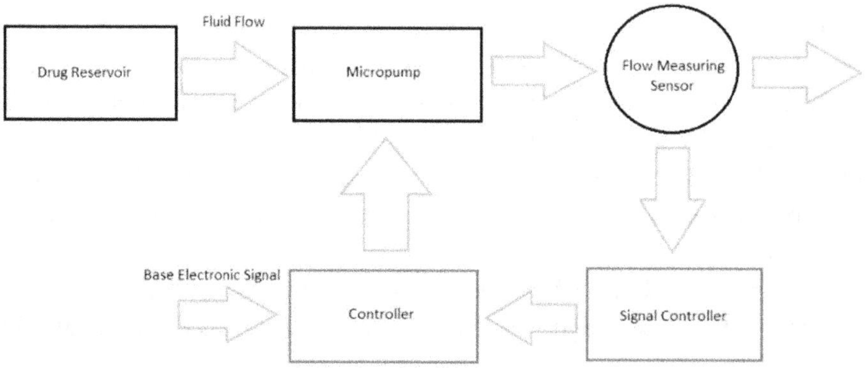

vantages, their many advantages make them a valuable tool in the field of drug delivery, with a wide range of applications in healthcare.

Point-of-Care Diagnostic Equipment (Lab on Chip) and Multiplexing in Biomarkers

There has been an increase in demand in recent years for portable, low-cost, and user-friendly point-of-care diagnostic tools. These tools are especially helpful in environments with limited resources and restricted access to conventional laboratory facilities. The utilization of micropumps and microfluidic systems, which enable precise and controlled fluid and sample manipulation, is one of the main elements of such diagnostic devices or tools.

One of the most promising applications of micropumps and microfluidic systems in point-of-care diagnostic equipment is in the area of multiplexing biomarkers (Dincer et al., 2017). Multiplexing refers to the simultaneous detection of multiple analytes in a single sample, and it has become an important tool for disease diagnosis and monitoring. Multiple biomarkers can be detected in a single experiment by using miniaturized immunoassay platforms made possible by micropumps and microfluidic systems. The immobilization of capture antibodies onto a microchannel surface, which can be accomplished via a variety of microfabrication processes, is the basis for the operation of such microfluidic systems. The target biomarker attaches to the immobilized antibody when a sample containing the biomarker is injected into the microchannel, creating a sandwich complex with a detecting antibody. This complex is then detected using various detection methods, such as fluorescence or electrochemical detection.

One of the advantages of using micropumps and microfluidic systems for multiplexing biomarkers is their ability to reduce the sample and reagent volumes required for each assay (Dincer et al., 2017). This reduces the cost of each assay and enables rapid screening of large numbers of samples. Additionally, the integration of many assays onto a single device is made possible by the use of micropumps and microfluidic systems, which lowers the price and complexity of the diagnostic platform even more.

Figure 6. Growth in market value and market share of mechanical and non-mechanical micropumps
Source: Research and Markets (n.d.)

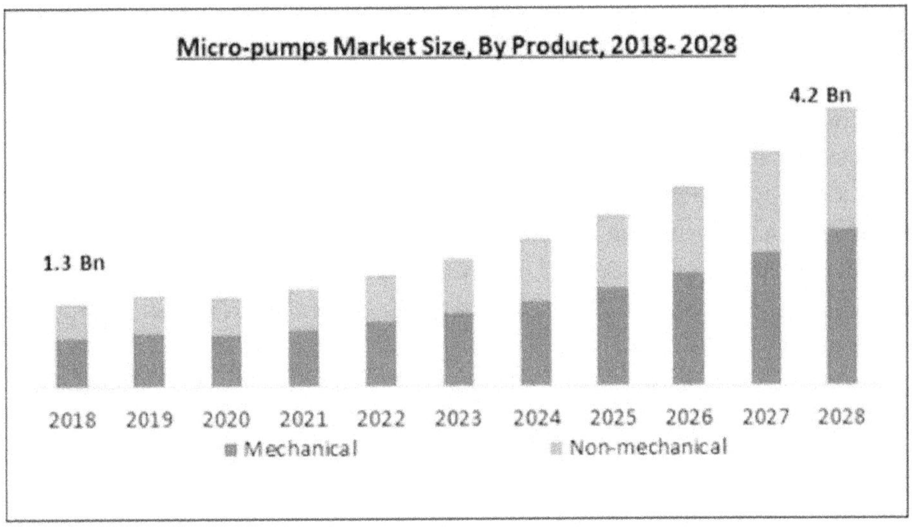

The use of micropumps and microfluidic systems for multiplexing biomarkers does have some restrictions. The necessity for precise and exact control of fluid flow and mixing in the microchannels is one of the major issues.

In spite of these challenges, point-of-care diagnostic equipment can be developed by multiplexing biomarkers using micropumps and microfluidic systems. Particularly in environments with limited resources, such gadgets have the potential to revolutionize the field of illness detection and surveillance. It is expected that as microfabrication methods advance, the creation of even more sophisticated microfluidic devices for multiplexing biomarkers, resulting in more accurate and dependable diagnostic platforms will be in progress.

FUTURE OF MICROPUMP'S MARKET

In the upcoming years, a number of sectors are anticipated to experience substantial changes as a result of micropumps. These tiny devices have already found use in industries including medicine delivery, point-of-care diagnostics, and environmental monitoring, among others, thanks to their capacity to pump and manipulate fluids at the microliter or even nanoliter scale. With rising demand for lab-on-a-chip systems, individualized medication, and point-of-care diagnostic tools, the future of micropumps appears promising. The global microfluidics market, which includes micropumps, is anticipated to reach $27.91 billion by 2025, expanding at a CAGR of 19.2% over the forecast period of 2020–2025, according to a report by MarketsandMarkets (n.d).

According to Global Market Insights, the global micropump market in different sectors would have an exponential growth.

Some of the key players in the micropump market include:

- **Bartels Mikrotechnik:** Bartels Mikrotechnik is a leading manufacturer of high-performance micropumps and microvalves, with a range of products designed for applications in areas such as medical technology, analysis, and process engineering.
- **Dolomite Microfluidics:** Dolomite Microfluidics specializes in the development and manufacture of microfluidic systems and components, including micropumps, microvalves, and chips, for a wide range of applications.
- **KNF Neuberger:** KNF Neuberger is a leading provider of micro and mini pumps and systems, with a focus on applications in medical and laboratory technology, environmental and process engineering, and instrumentation and analysis.
- **Microfluidica LLC:** Microfluidica LLC develops and manufactures microfluidic systems and devices, including micropumps, for use in areas such as medical diagnostics, drug discovery, and food and environmental testing.
- **Parker Hannifin Corporation:** Parker Hannifin Corporation is a global leader in motion and control technologies, with a range of products and systems designed for use in areas such as aerospace, industrial, and life sciences, including micropumps and microfluidic systems.

The growth of the micropump market is expected to be driven by the increasing demand for portable and point-of-care diagnostic devices, as well as the growing use of microfluidic systems in drug discovery and development. In the upcoming years, market expansion is also anticipated to be fueled by the growing use of micropumps in environmental monitoring and the rising need for personalized treatment. Overall, the future of micropumps appears promising due to rising demand across a variety of sectors and applications as well as ongoing R&D initiatives targeted at enhancing the functionality and potential of these tiny devices.

CONCLUSION

The exact control and manipulation of fluids at the microscale is made possible by the crucial role that micropumps play in microfluidic devices. Significant advancements have been made in the creation of micropumps for applications ranging from chemical synthesis to biomedical medications, over the last few decades and there are many more in the pipeline.

This chapter provides an overview of various kinds of micropumps, such as mechanical and non-mechanical micropumps along with the current technologies in place and the various forms of commercial micropumps widely used in the market. The different design concepts, manufacturing methods, and applications were articulated in this book chapter such as electrostatic, electromagnetic, piezoelectric and pneumatic to name a few. Following this, the most recent advances in micropump design were discussed including the utilization of microfluidic channels, valves, and actuators as well as the incorporation of micro/nanofabrication technologies such as soft lithography.

The requirement for greater throughput and improved fluid flow management is one of the main difficulties in micropump design. Therefore, the chapter has reviewed a number of solutions to this challenge, including the use of numerous pumps, the incorporation of feedback control systems, and the creation of intelligent micropumps that can detect and react to changes in the fluid environment.

In the last section of the chapter, various applications of micropumps in the fields such as environmental monitoring (such as Particle counting, Air quality monitoring and Leak detection), and disease detection and treatment (as in drug delivery systems and targeted delivery within implantable systems) were covered. Micropumps can be used efficiently in drug delivery systems as it does not require high flow rates and micropumps are precise in delivering small doses and have a flow independent of back pressure. In point-of-care diagnostic equipment, for instance, micropumps can be used to precisely control the amount and flow rate of reagents, enabling quick and exact detection of biomarkers. Biomarkers can be measured in cell culture supernatant, plasma, cerebrospinal fluid, bronchoalveolar lavage fluid depending on the type of prognosis. As some samples would only be available in small quantities, micropumps can be of great value in multiplexing biomarker analysis to gain as much information as possible.

Finally, this chapter concludes on the key market players present in the current market space of micropumps, such as Bartels Mikrotechnik. Apart from this, an exceptionally high growth rate at a CAGR of 19.2% has been observed that proves vast amount of possibilities and growth options available in this upcoming field.

ACKNOWLEDGMENT

I would like to thank BITS Pilani for providing Dr. Jegatha Nambi Krishnan with Interim Relief grant (1901220002) and all the sources available without which it would be impossible to write this chapter.

REFERENCES

Aishan, Y., Yalikun, Y., Shen, Y., Yuan, Y., Amaya, S., Okutaki, T., Osaki, A., Maeda, S., & Tanaka, Y. (2021). A chemical micropump actuated by self-oscillating polymer gel. *Sensors and Actuators B: Chemical, 337*. doi:10.1016/j.snb.2021.129769

Alvarez-Braña, Y., Etxebarria-Elezgarai, J., Ruiz de Larrinaga-Vicente, L., Benito-Lopez, F., & Basabe-Desmonts, L. (2021). Modular micropumps fabricated by 3D printed technologies for polymeric microfluidic device applications. *Sensors and Actuators B: Chemical, 342*. doi:10.1016/j.snb.2021.129991

Amini, A., Guijt, R. M., Themelis, T., De Vos, J., & Eeltink, S. (2023). Recent developments in digital light processing 3D-printing techniques for microfluidic analytical devices. *Journal of Chromatography A, 1692*. doi:10.1016/j.chroma.2023.463842

Augustine, S., Paul, L. J., & Paulo, D. (2021). Relevance of micromachining in microfluidics and biomedical devices. In Handbooks in Advanced Manufacturing, Advanced Machining and Finishing (pp. 339-373). Elsevier. doi:10.1016/B978-0-12-817452-4.00009-9

Aziz, A. A., Azmi, M. A. M., Nazri, M. N. A. M., Zainal, M. H. F., Khalfan, M. A., & Bahrain, A. K. (2019). Fabrication of micropump for microfluidics application. *Applied Physics of Condensed Matter*. doi:10.1063/1.5118148

Barua, R., Datta, S., Sengupta, P., Chowdhury, A. R., & Datta, P. (2021). Advances in MEMS micropumps and their emerging drug delivery and biomedical applications. In Advances and Challenges in Pharmaceutical Technology. Academic Press. doi:10.1016/B978-0-12-820043-8.00002-5

Bohr, A., Colombo, S., & Jensen, H. (2019). Future of microfluidics in research and in the market. In Micro and Nano Technologies, Microfluidics for Pharmaceutical Applications. William Andrew Publishing. doi:10.1016/B978-0-12-812659-2.00016-8

Bußmann, A. B., Durasiewicz, C. P., Sebastian, H. A. K., & Wald, C. K. (2021). Piezoelectric titanium based microfluidic pump and valves for implantable medical applications. *Sensors and Actuators A: Physical, 323*. doi:10.1016/j.sna.2021.112649

Chappel, E., & Dumont-Fillon, D. (2021). Micropumps for drug delivery. In *Developments in Biomedical Engineering and Bioelectronics. Drug Delivery Devices and Therapeutic Systems*. Academic Press. doi:10.1016/B978-0-12-819838-4.00015-8

DebiotechS. A. (n.d.). https://www.debiotech.com/

Dincer, C., Bruch, R., Kling, A., Dittrich, P. S., & Urban, G. A. (2017). Multiplexed Point-of-Care Testing – xPOCT. *Trends in Biotechnology, 35*(8). doi:10.1016/j.tibtech.2017.03.013

Haldkar, R. K., Khalatkar, A., Gupta, V. K., & Sheorey, T. (2021). New piezoelectric actuator design for enhance the micropump flow. *Materials Today: Proceedings, 44*(1), 776-781. doi:10.1016/j.matpr.2020.10.702

Hosseini, H. R., Nikookar, H., Yesiloz, G., Naseh, M., & Mohammadi, M. (2021). An overview on micropumps, micromixers, and their applications in bioprocess. In Advances in Bioenergy and Microfluidic Applications. Elsevier. doi:10.1016/B978-0-12-821601-9.00015-7

Jang, W. I., Choi, C. A., Jun, C. H., Kim, Y. T., & Esashi, M. (2004). Surface micromachined thermally driven micropump. *Sensors and Actuators A: Physical, 115*(1), 151-158. doi:10.1016/j.sna.2004.04.024

Kim, H., Hwang, H., Baek, S., & Kim, D. (2018). Design, fabrication and performance evaluation of a printed-circuit-board microfluidic electrolytic pump for lab-on-a-chip devices. *Sensors and Actuators A: Physical, 277*, 73-84. doi:10.1016/j.sna.2018.04.042

Lakhera, P., Chaudhary, V., Bhardwaj, B., Kumar, P., & Kumar, S. (2022). Development and recent advancement in microfluidics for point of care biosensor applications: A review. *Biosensors and Bioelectronics, 11*. doi:10.1016/j.biosx.2022.100218

Landari, H., Dussault, M.-A., Ruel, J., Begin-Drolet, A., & Miled, A. (2018). Biocompatible compact micropump with integrated unidirectional microvalves for low pressure microfluidic applications. *Sensors and Actuators A: Physical, 276*, 246-258. doi:10.1016/j.sna.2018.03.046

MarketsandMarkets. (n.d.). *About.* Markets and Markets. https://www.marketsandmarkets.com

Matteucci, M., Pérennès, F., Marmiroli, B., Miotti, P., Vaccari, L., Gosparini, A., Turchet, A., & Di Fabrizio, E. (2006). Compact micropumping system based on LIGA fabricated microparts. *Microelectronic Engineering, 83*(4–9). doi:10.1016/j.mee.2006.01.259

Mishra, S., & Yadava, V. (2015). Laser Beam MicroMachining (LBMM). A review. *Optics and Lasers in Engineering, 73*, 89-122. doi:10.1016/j.optlaseng.2015.03.017

Mohith, S., Navin Karanth, P., & Kulkarni, S. M. (2019). Recent trends in mechanical micropumps and their applications: A review. *Mechatronics, 60*, 34-55. doi:10.1016/j.mechatronics.2019.04.009

Nisar, A., Afzulpurkar, N., Mahaisavariya, B., & Tuantranont, A. (2008). MEMS-based micropumps in drug delivery and biomedical applications. *Sensors and Actuators B: Chemical, 130*(2), 917-942. doi:10.1016/j.snb.2007.10.064

Pourmand, A., Seyed, A. M. S., Ghavifekr, H. B., Aghdam, E. N., Dokmeci, M. R., Khademhosseini, A., & Zhang, Y. S. (2018). Fabrication of whole-thermoplastic normally closed microvalve, micro check valve, and micropump. *Sensors and Actuators B: Chemical, 262*, 625-636. doi:10.1016/j.snb.2017.12.132

Research and Markets. (n.d.). *Home.* Research and Markets. https://www.researchandmarkets.com

Shoji, S., Sato, H., & Zengerle, R. (2008). Liquid Micropumps. In Comprehensive Microsystems. Elsevier. doi:10.1016/B978-044452190-3.00041-0

Singh, S., Kumar, N., George, D., & Sen, A. K. (2015). Analytical modeling, simulations and experimental studies of a PZT actuated planar valveless PDMS micropump. *Sensors and Actuators A: Physical, 225*, 81-94. doi:10.1016/j.sna.2015.02.012

Wang, Y.-N. & Fu, L. -M. (2018). Micropumps and biomedical applications: A review. *Microelectronic Engineering, 195*, 121-138. doi:10.1016/j.mee.2018.04.008

Yang, Z., Dong, L., Wang, M., Liu, G., Li, X., & Li, Y. (2022). A wearable insulin delivery system based on a piezoelectric micropump. *Sensors and Actuators A: Physical, 347,* 113909. doi:10.1016/j.sna.2022.113909

Zhang, C., Xing, D., & Li, Y. (2007). Micropumps, microvalves, and micromixers within PCR microfluidic chips: Advances and trends. *Biotechnology Advances, 25*(5), 483-514. doi:10.1016/j.biotechadv.2007.05.003

Chapter 5
Micropumps for Lab-on-Chip Applications

K. Karthikeyan
M. Kumarasamy College of Engineering, India

L. Sujatha
Rajalakshmi Engineering College, India

ABSTRACT

Microfluidics is a promising and upcoming technology for diagnostics and drug developments. With huge potential in bringing a revolution in biotechnology, there is going to be a big commercial demand for microfluidic systems in life science applications. The function of microfluidic system highly depends on the precise control of tiny fluid volumes by micropumps, and hence, design, fabrication, and integration of micropumps are the key components in microfluidic systems. Though the development of micropumps has been paid attention for the past two decades, it is still a great challenge in realizing a reliable and improved performance of a micropump. This chapter discusses a detailed study on various pumping mechanisms of micropumps reported with a perspective on the scope for manufacturing. The issues and challenges in commercialization will be discussed, and the possible solutions to overcome will also be highlighted.

INTRODUCTION

Microfluidics is a multidisciplinary field which deals with manipulation of fluids at micro and nanoscale. Microfluidic systems can be used to develop portable and affordable diagnostic devices, such as microfluidic chips or lab-on-a-chip (LOC) devices, for applications such as disease diagnosis, drug screening, and environmental monitoring. Microfluidic systems are capable of handling small volumes of fluid with high precision, which is particularly useful for applications that require accurate dosing, mixing, and analysis. In general, a microfluidic system consists of a network of microchannels, reservoirs, inlet ports for injecting fluids, outlet port for removing fluids, micromixers, micropumps, microvalves and an integrated sensor system for analytical studies. Integration of these components makes miniaturization

DOI: 10.4018/978-1-6684-6952-1.ch005

Figure 1. Classification of micropumps

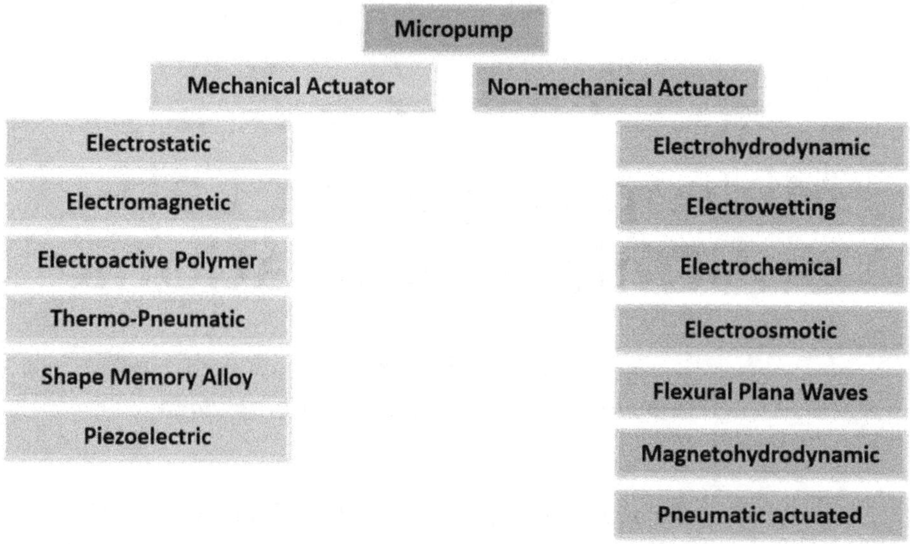

of diagnostic systems with multiple sensors known as Lab-On-Chip (LOC). Micropumps are required to pump the fluids between the reservoirs and the microfluidic chip, inside the microfluidic chip from one zone to another zone and then finally to send the fluid to the waste collecting reservoirs. There is a delay in realizing the commercialization of microfluidic systems due to lack of reliable microfluidic components especially micropumps and microvalves. These micropumps may be of mechanical or non-mechanical actuator type based on the principle of working and these will be discussed in detail in this chapter.

A micropump is a miniaturized pump that is capable of delivering small volumes of fluids typically in the picolitre or nanolitre range, with high precision and accuracy. Micropumps are being developed for use in a wide range of applications, including drug delivery, medical diagnosis, chemical analysis, and microfluidic systems (Karthikeyan et al. 2023). There are several types of micropumps, including mechanical, piezoelectric, electrokinetic, and thermopneumatic, each with its unique characteristics and advantages. Mechanical micropumps use moving parts to generate flow, while piezoelectric micropumps use mechanical vibrations to create pressure changes in the fluid. Electrokinetic micropumps employ the electric field to create fluid flow, and thermopneumatic micropumps use thermal expansion to generate pressure gradients. The choice of micropump type depends on the specific application and the desired performance criteria.

Overall, micropumps represent an exciting and rapidly developing technology with significant potential for a wide range of applications. Micropumps can be classified based on various factors, such as their pumping mechanism, fluid handling capabilities, and application areas. Here are a few common classification schemes for micropumps as shown in Figure 1. Micropumps are important in drug delivery uses because they allow for the exact movement of medicine from a reservoir to the target location. The micropumps are fabricated using Micro Electro Mechanical Systems (MEMS) technology and are highly suitable for wearable and implantable drug delivery systems for the management of chronic illnesses as well as the efficient discharge of insulin for diabetic patients.

Micropump Components

The components of a typical micropump depend on the specific design and pumping mechanism, but here are some common components that can be found in many micropump systems:

- **Pumping Mechanism:** The mechanism used to create the fluid flow, which can be based on electrokinetics, mechanical displacement, or other methods.
- **Pump Chamber:** The space where the fluid is stored and pumped.
- **Inlet and Outlet Ports:** The openings where the fluid enters and exits the pump chamber.
- **Valves:** Devices that control the direction of fluid flow and prevent backflow.
- **Actuators:** Components that generate the force needed to actuate the pumping mechanism, which can be based on electrical, magnetic, or other stimuli.
- **Power Supply:** The source of energy needed to operate the micropump, which can be based on batteries, external power sources, or energy harvesting.
- **Control Electronics:** The circuitry and software that regulates the pump operation, including timing, flow rate, and other parameters.

There are some factors to consider: micropumps may not have all of these components, or may have additional components depending on the specific design and application.

Design of Micropumps and Diaphragm Materials

The design of micropumps can vary depending on the specific application and pumping mechanism, but here are some general considerations and principles are often considered:

- **Pumping Mechanism:** The choice of pumping mechanism depends on the fluid properties, flow rate, pressure requirements, and other factors. Common pumping mechanisms for micropumps include electroosmosis, electrostatics, piezoelectric actuation, and diaphragm displacement.
- **Pump Chamber and Fluid Channels:** The design of the pump chamber and fluid channels should minimize dead volumes, turbulence, and flow resistance, while also ensuring proper sealing and alignment of the various components.
- **Valves:** The design of the valves should ensure reliable and fast operation, while also minimizing leakage and fluid shear stress.
- **Actuators and Power Supply:** The choice of actuator and power supply depends on the required force and speed of the pumping mechanism, as well as the size, weight, and power constraints of the micropump system.

As for the diaphragm materials, they should be selected based on their biocompatibility, chemical compatibility, mechanical properties, and durability. Common materials for diaphragms in micropumps include silicones, polyimides, and elastomers such as PDMS (polydimethylsiloxane). These materials have good flexibility, low hysteresis, and high fatigue resistance, which make them suitable for use in various micropump applications.

In a micropump, a diaphragm is positioned to separate the actuator and the fluid in the chamber. The flexible wall of the chamber, which is the diaphragm, comes into contact with the liquids inside.

Table 1. Materials used for the diaphragms of micropumps

Diaphragm Material	Reference
Polyethylene Terephthalate	Pabst et al. (2014)
Thermoplastic Polyurethane	Shaegh et al. (2015)
Polydimethylsiloxane	Kawnu et al. (2016)
Polyimide	Hamid et al. (2016)
Beryllium bronze	Wang et al. (2016)
Parylene	Johnson et al. (2016)
Elastomeric	Robertson et al. (2016)
Nafion/Pt	Shoji et al. (2016)
Soft Magnetorheological Elastomer	Ehsani et al. (2017)
Brass	Dong et al. (2017)
Earthwork muscle	Tanaka et al. (2017)
SU8	Le et al. (2017)
Silicon	Yazdi et al. (2019)

The diaphragm is held in place by clamps that attach it to the housing of the micropump, making the chamber and diaphragm the same diameter. On the opposite side of the diaphragm is where the actuator is located, which vibrates the diaphragm vertically. The vibration of the diaphragm in a micropump facilitates the transfer of fluid from the reservoir to the chamber and subsequently to the outlet, resulting in a pressure difference within the chamber. The selection of diaphragm material is a critical consideration in micropump design, and the intended application of the micropump must be considered when making this choice. The mechanical and physical characteristics of the diaphragm, such as flow rate and biosafety, are crucial factors to consider. Additionally, cost and abrasion resistance should also be considered (Asadi Dereshgi et al. 2018a and Asadi Dereshgi et al. 2018b). The materials used for the diaphragms of micropumps are presented in Table 1.

ACTUATOR TYPES AND OPERATION PARAMETERS

Actuators are in charge of the mechanical control and movement of a system, requiring an energy supply and control signal to be activated. The signal that governs the functioning of a system, whether it is pneumatic pressure or an electrical voltage or power of human, possesses a relatively low degree of energy. In most cased, micropumps are powered by sources of energy such as hydraulic fluid pressure, or electricity or pneumatic pressure. Based on their actuation techniques, which can be mechanical or non-mechanical, micropumps can be categorized into two main groups. Physical type devices are used to move microfluidic mass in mechanical micropumps, as shown in Table 2.

Non-mechanical micropumps utilize kinetic momentum to pump fluid, transforming non-mechanical energy into kinetic momentum. These non-mechanical pumps can be categorized into various types, such as electrical, magnetic, chemical, and so on. Table 3 lists studies that examine different non-mechanical actuators.

Table 2. Mechanical drivers of micropumps

Mechanical Actuator	Reference
Thermo-pneumatic	Chee et al. (2016)
Thermally expandable polymer	Tripathi et al. (2017)
Electroactive Polymer	Cheong et al. (2018)
Piezoelectric	Yildirim et al. (2018)
Shape memory alloy	Guerine et al. (2018)
Electromagnetic	Rusli et al. (2018)
Electrostatic	Uhlig et al. (2018)

Table 3. Studies on non-mechanical actuators of micropumps

Non-Mechanical Actuator	Reference
Evaporation	Nie et al. (2015)
Electrochemical	Uvarov et al. (2016)
Electrohydrodynamic	Russel et al. (2016)
Bubble formation	Uvaro et al. (2017)
Magnetohydrodynamic	Hasan et al. (2017)
Electroosmotic	Li et al. (2019)

Electromagnetic Actuators

Magnetic micropumps utilize electromagnetic principles to create fluid flow. These micropumps have dual permanent magnets placed on the bottom and top edges of the cavity, and the diaphragm of the

Figure 2. Schematic design of electromagnetic actuator
Source: Jumril Yunas et al. (2021)

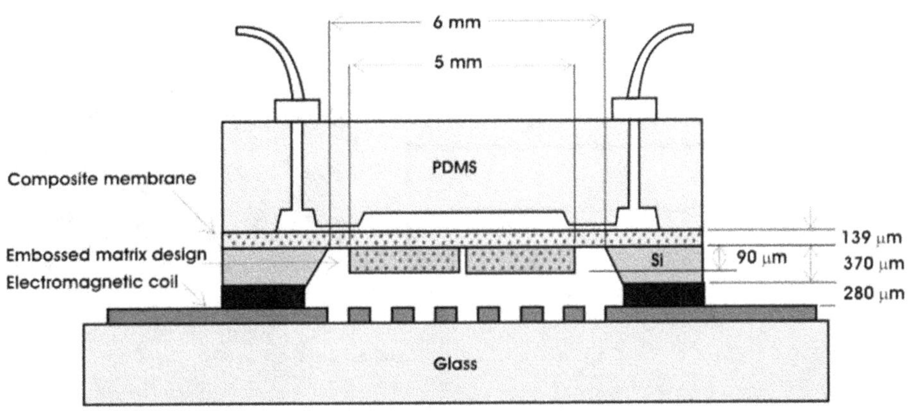

micropumps should be made of soft and elastic materials. The electromagnetic field enables efficient operation of pumping (Said et al. 2017).

The diaphragm of these micropumps vibrates due to the interaction of permanent magnets and the magnetic field of a micro-changeable coil (Sima et al. 2015). The capacity of the magnetic force to draw or repel causes the diaphragm to vibrate vertically when the current phase input is changed. Ultimately, fluid flow is generated due to the vibration of the diaphragm. Figure 2 shows the schematic design of electromagnetic actuator (Jumril Yunas et al. 2021).

Thermal Actuators

Thermo-pneumatic, shape memory alloy and thermally expandable polymer are the three primary components of thermally actuated micropumps. Thermo-pneumatic micropumps, in particular, are built with a cylinder, a diaphragm, and a microheater (Bonk et al., 2015; Benard et al., 1998; Chee et al., 2015).

The diaphragm is positioned between the microheater and chamber, with the pump chamber touching one side and the microheater touching the other. A vapour or liquid combination is heated in the space between the microheater and the diaphragm. When this fluid is heated, the diaphragm deforms, causing the pump chamber's pressure to fluctuate (Berg and Lammerink, 1998).

Bimetallic and shape memory alloy (SMA) are types of thermal actuations that feature unique alloys and possess a form memory effect. Vibrating these sensors requires a high-power, usually around 100 mW, but despite their high-power consumption, they can generate significant forces (Ali and Takahata, 2011; Fong et al., 2015). Meanwhile, in diaphragm micropumps, thermally expanded polymers serve as the actuators. These structures are made up of two components with different thermal expansion factors that are joined together. When subjected to temperature variations, the sensor moves vigorously but with less deflections (Spieth et al., 2012). Figure 3 shows the schematic design of thermal bubble micropump (Gang Guo et al., 2022).

Figure 3. Schematic design of the thermal bubble micropump
Source: Gang Guo et al. (2022)

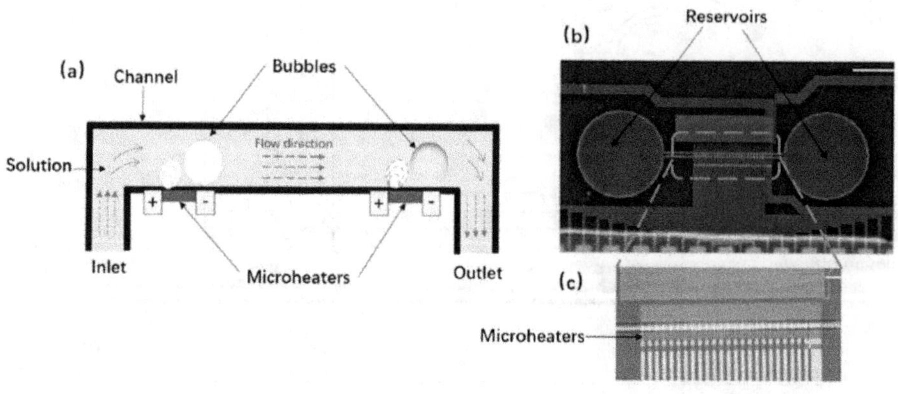

Figure 4. Working mechanism of bubble stimulated micro-pump
Source: Yang Qu et al. (2018)

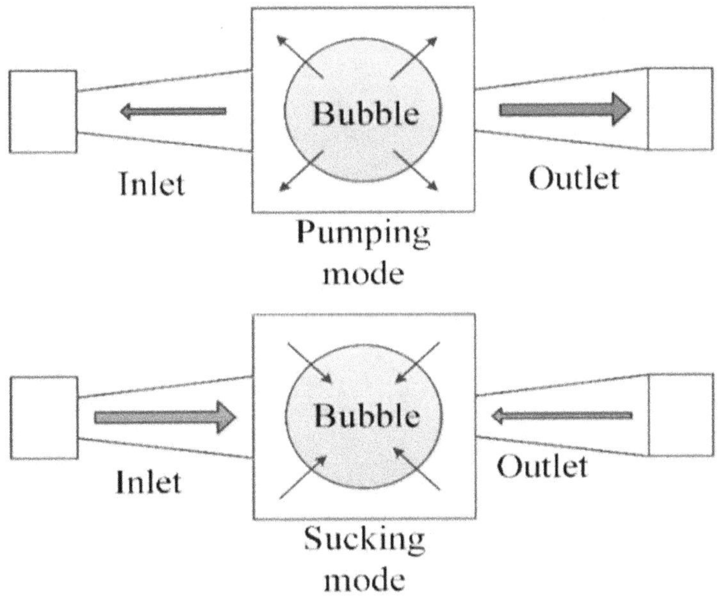

Bubble and Evaporation-Type Micropumps

Bubble-type and evaporation-type micropumps are two common types of micropumps that use different physical mechanisms to generate fluid flow. Bubble-type micropumps rely on the generation and movement of gas bubbles to create fluid flow. Typically, a gas such as hydrogen or oxygen is electrolyzed at an electrode surface to generate bubbles, which then move towards a fluid reservoir and push the fluid out of the pump. These pumps are simple and can generate high flow rates, but they are not suitable for applications requiring precise flow control.

Evaporation-type micropumps use the evaporation and condensation of a working fluid to generate fluid flow. Typically, the working fluid is heated at a specific location, causing it to evaporate and create a pressure gradient that drives the fluid towards a cooler region where it condenses. These pumps are more complex than bubble-type pumps but can be used for precise flow control and have the advantage of being able to generate bidirectional flow. Both types of micropumps have advantages and disadvantages and can be used in various microfluidic applications. The choice of micropump will depend on factors such as the required flow rate, pressure, and precision, as well as the specific application requirements. Figure 4 shows the working mechanism of bubble stimulated micro-pump (Yang Qu et al., 2018).

Flexural Planar Waves Actuators

Flexural planar wave actuators (FPWAs) are a type of micropump that use the flexural deformation of a thin plate to create fluid flow. The thin plate is typically made of piezoelectric material, which is capable

Figure 5. Schematic design of flexural planar waves actuators
Source: Ersin Sayar et al. (2013)

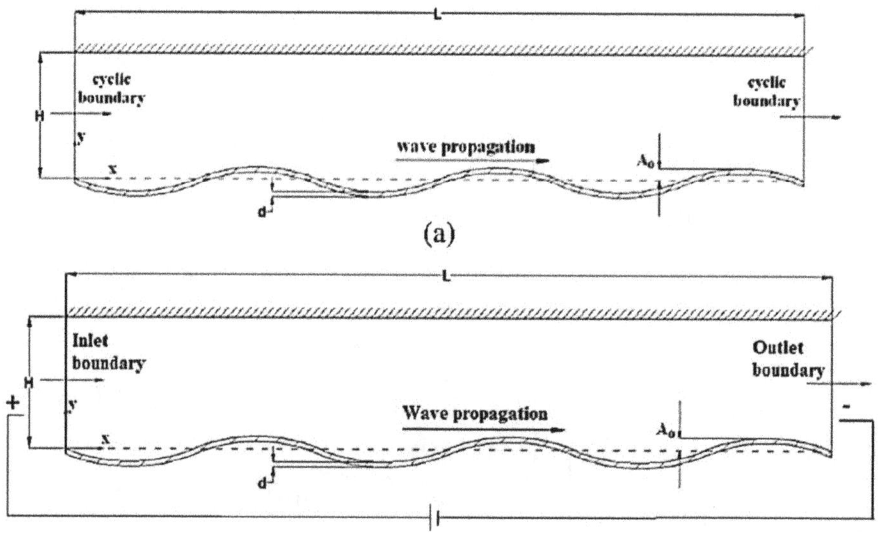

of generating mechanical deformation in response to an applied electrical field. When an AC electrical field is applied to the piezoelectric plate, it generates flexural planar waves that propagate through the plate and create a pressure gradient in the surrounding fluid. This pressure gradient drives fluid flow in the direction perpendicular to the wave propagation direction. Figure shows the schematic design of flexural planar waves actuators.

Figure 5 shows the schematic design of flexural planar waves actuators. FPWAs have several advantages as micropumps, including high efficiency, low power consumption, and the ability to generate high flow rates with low vibration and noise. They can also be used for both pumping and mixing applications, and can be easily integrated with other microfluidic components. However, FPWAs have some drawbacks, such as the difficulty of regulating the flow rate and the susceptibility of pumping efficacy to external variables such as temperature and humidity (Ersin Sayar et al., 2013) FPWAs are a potential micropump technology that could be used in a variety of applications such as lab-on-a-chip devices, medicine delivery systems, and microscale chemical reactions.

Piezoelectric Actuators

Piezoelectric actuators are a common type of micropump that use the piezoelectric effect to generate mechanical deformation in a thin plate or membrane. The deformation generates a pressure gradient that drives fluid flow through microchannels. Piezoelectric micropumps can produce high flow rates and pressures while using little electricity, which is one of their many benefits. Additionally, they are lightweight, compact, and simple to combine with other microfluidic parts. Based on their mode of operation, piezoelectric micropumps can be further divided into several kinds, including valve-based pumps, diaphragm-based and peristaltic. In terms of accuracy, pressure, and flow rate, each variety has particular benefits and restrictions. One of the major challenges of piezoelectric micropumps is the

Figure 6. Piezoelectric micropump
Source: Rakesh Kumar Haldkar et al. (2021)

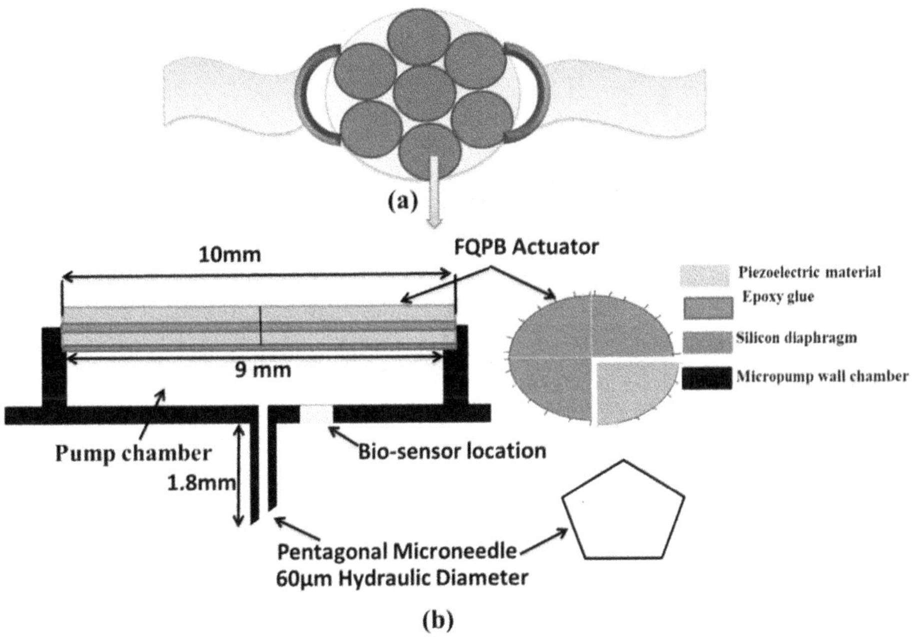

design and optimization of the pumping performance. The geometric design, material properties, and operating conditions all affect the pumping performance and should be carefully enhanced to achieve the desired pressure and flow rate.

Piezoelectric micropumps remain widely used and versatile technology for microfluidic applications, including drug delivery, lab-on-a-chip devices, and microscale chemical reactors. Several substances and crystals exhibit a piezoelectric effect, which allows mechanical energy to be converted into electrical energy.

When a piezoelectric substance is exposed to pressure, the direct piezoelectric effect occurs. When an electrical voltage is given to a substance, it produces a reverse piezoelectric effect. Deflection of the piezoelectric element can be affected by both the magnitude and direction of the applied voltage. Figure 6 shows the four-quarter piezoelectric bimorph actuator (Rakesh Kumar Haldkar et al., 2021).

A piezoelectric material undergoes polarization reversed when it is subjected to inverse pressure. Further, by deviating the electrical field applied to the material, the direction of the oscillation can be altered. Micrometer and nanometer-sized deflections are possible with piezoelectric materials. Piezoelectric materials are typically used as actuators in modern micropumps, as they are suitable from both a size and performance perspective (Nguyen et al., 2002). It is possible to generate suction and discharge using reverse piezoelectrics in these micropumps. When an alternating current voltage is introduced, piezoelectric oscillations occur in a horizontal plane. Despite the fact that the margins of the piezoelectric material are restrained on the diaphragm and the piezoelectric actuator and diaphragm are obliquely connected to the pumping membrane.

As a consequence of the horizontal deformation, the piezoelectric substance vibrates vertically. In addition to the mechanical characteristics and physical characteristics of the actuator, the frequency,

Figure 7. Structure of the magnetohydrodynamic (MHD) micropump
Source: Xuyan Zhou et al. (2017)

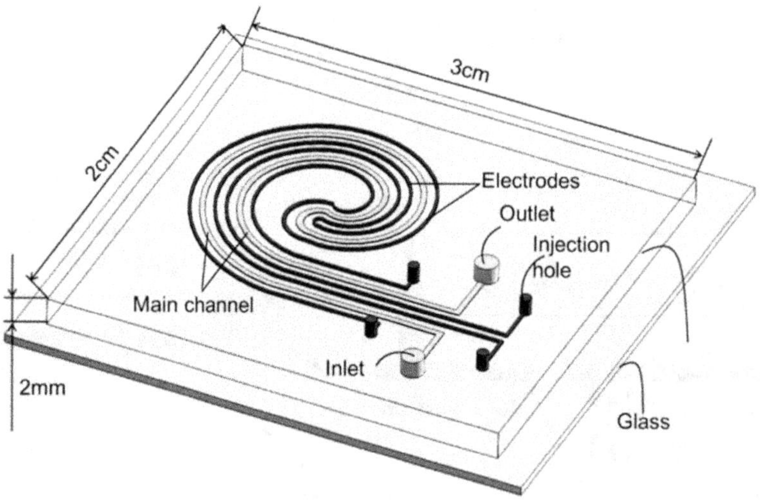

voltage, border conditions, and pressure of fluid affect the resonance rates and mode shapes of plate-type piezoelectric actuators. An electric potential applied to a piezoelectric actuator produces transverse vibrations parallel to its surface normal through the reverse piezoelectric effect. A piezoelectric micropump produces an oscillating flow rate through periodic changes in fluid volume caused by the diaphragm's transverse vibrations (Karthikeyan et al., 2021, 2022; Yildiz et al., 2019).

Magnetohydrodynamic Actuators

Magnetohydrodynamic (MHD) actuators are a type of micropump that use the interaction between an electric field and a magnetic field to generate fluid flow. MHD micropumps consist of a microchannel filled with a conducting fluid, and two perpendicular magnetic fields are applied to the channel. The fluid is then subjected to an electric field, which combines with the magnetic fields to produce a Lorentz force that forces fluid flow in the desired direction. Figure 7 shows the schematic design of the magnetohydrodynamic for the liquid-metal based micropump using glass - polydimethylsiloxane microfluidic device (Xuyan Zhou et al., 2017).

MHD micropumps have several compensations, including low power consumption, high efficiency, and the ability to generate high flow rates without using any movable components. Additionally, they are lightweight, compact, and simple to combine with other microfluidic parts. However, MHD micropumps also have some limitations, including the difficulty of controlling the flow rate and the sensitivity of the pumping performance to the magnetic field strength and direction. Overall, MHD micropumps are a promising technology for microfluidic applications, including lab-on-a-chip devices, drug delivery systems, and microscale chemical reactors. However, further research is needed to optimize their performance and overcome some of their limitations.

An area of study known as magnetohydrodynamics examines the magnetohydrodynamic systems magnetic characteristics. The study of MHD, the interactions of ferroelectric particles with electrical currents and electromagnetic fields. There are many ways in which fluid can be pumped in non-mechanical pumps, but MHD is one of the most significant. Lorentz force drives MHD micropumps during pumping. The Lorentz force, according to Duwairi and Abdullah (2006), is caused by the interaction of the magnetic and electric fields.

Electrowetting and Electrochemical Micropumps

Electrowetting and electrochemical micropumps are two different types of micropumps that utilize electrical forces to manipulate fluid at the microscale. Electrowetting micropumps use an electric field to change the wetting properties of a surface, which in turn alters the fluid flow. The electric field is typically applied to a small electrode that is placed near a droplet of fluid on a hydrophobic surface. When the electric field is turned on, the droplet of fluid is attracted to the electrode and the surface becomes more hydrophilic, causing the droplet to spread out and flow. The path and velocity of fluid movement can be altered by selectively activating and deactivating the electric field.

Electrochemical micropumps, on the other hand, use electrical current to drive electrochemical reactions that generate gas bubbles, which in turn generate fluid flow. The current is applied to electrodes that are placed in the fluid, causing a chemical reaction to occur that produces gas bubbles. The gas bubbles then expand and flow through the microchannel, creating a net flow of fluid. Both electrowetting and electrochemical micropumps have several advantages, including the ability to generate high flow rates with low power consumption, and the ability to be easily integrated with other microfluidic components. However, these micropumps can also be sensitive to changes in the operating conditions, and the pumping performance can be affected by factors such as temperature, humidity, and surface roughness. Overall, electrowetting and electrochemical micropumps are promising technologies for microfluidic applications, including lab-on-a-chip devices, drug delivery systems, and microscale chemical reactors (Karthikeyan et al. 2023). Further research is needed to optimize their performance and overcome some of their limitations. Figure 8 shows the electrochemical micropump (Ilia V. Uvarov et al., 2021).

Figure 8. Electrochemical micropump
Source: Ilia V. Uvarov et al. (2021)

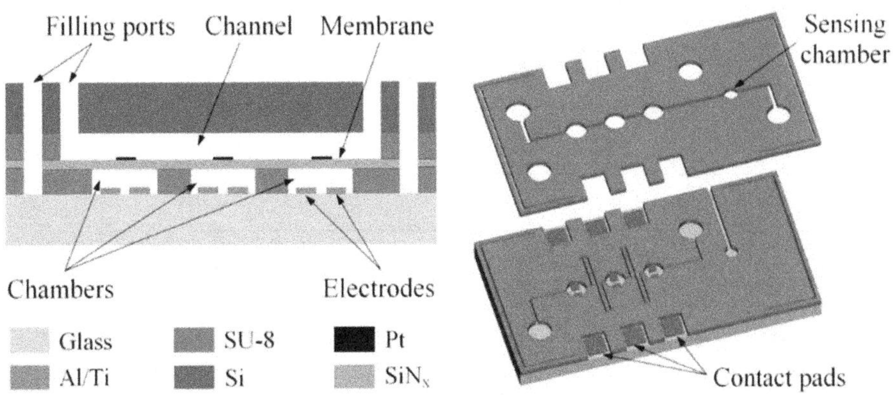

An electrowetting (EW) micropump can increase the wetness of a surface by altering its hydrophobicity. EW micropumps utilize dielectric layers and apply voltage to them to drive fluid movement because surface tension reduces the interfacial energy between liquids and solids (Javadi et al., 2013). Through the application of a DC voltage to containing two platinum electrodes with saltwater solution, electrochemical micropumps produce electricity by causing gas to be produced between the platinum electrodes. A force for fluid flow is ultimately generated by the electrochemical bubbles created (Kjeang et al., 2009).

Electroosmotic Actuators

Electroosmotic actuators micropumps are devices that use the electroosmotic effect to pump fluids at a microscale level. Electroosmotic micropumps have many potential applications in microfluidic systems, including drug delivery, chemical analysis, and biological assays. They are particularly useful in situations where precise control of fluid flow and manipulation is required, such as in microfluidic systems for diagnostic and LOC devices applications. Figure 9 shows the schematic design of electroosmotic flow micropump (Yuki Okamoto et al., 2020). Electroosmotic actuators micropumps typically consist of a microchannel, two electrodes, and a power source. A power source is used to spread an electric field across the conduit, which has electrodes placed at either end. The electroosmotic action causes the fluid to be propelled through the conduit as it moves through it.

Figure 9. Schematic design of electroosmotic flow micropump (EOF: electro osmotic flow; HV: high voltage; CLK: clock signal; NCLK: negative clock signal; V_{DD}: supply voltage; V_{th}: threshold voltage of MOS transistor; V_{out}: output voltage)
Source: Yuki Okamoto et al. (2020)

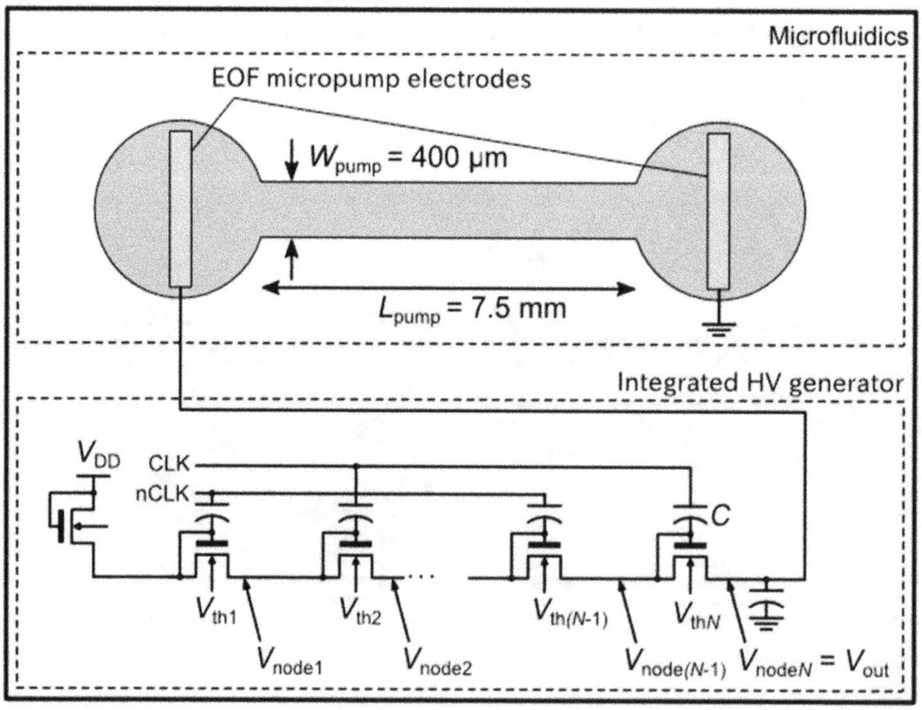

These micropumps have a number of benefits, including compact size, low power usage, and exact flow rate. They are also easily integrated into microfluidic systems and can handle a wide range of substances, including biological data. Electroosmotic actuator micropumps do, however, have some drawbacks. For instance, because of the tiny size of the microchannels, they may be susceptible to variations in the electrical conductivity of the fluid being circulated. Overall, electroosmotic actuators micropumps are a promising technology for microfluidics applications and are the subject of ongoing research and development (Karthikeyan et al., 2017; Karthikeyan et al., 2018a, 2018b). In electroosmotic pumps, fluid is moved through a conduit with a charged wall, which is subjected to an external electric field (Chen and Santiago, 2002). An electric field is produced through a tube constructed of silica or glass with electrodes for electroosmotic micropumps (Lin et al., 2013).

Electrostatic and Electroactive Polymer Actuators

Electrostatic and electroactive polymer actuators are a type of micropump that use electric fields to generate mechanical motion. These actuators are made from a class of materials known as smart materials, which can change their shape or size in response to an electrical stimulus. An electrostatic micropump is a type of microfluidic pump that uses an electrostatic force to generate fluid flow. The pump consists of two parallel plates with an insulating layer in between, and a microchannel that connects the inlet and outlet ports. The liquid transfers towards the exit from the inlet when a voltage is given to the plates, creating an electrostatic charge. The electrostatic force is created by the attraction between opposite charges on the plates and the fluid. As a result, the electrostatic micropump is highly efficient and consumes very little power (Ghazali et al., 2017).

Additionally, since there are no moving parts involved, the pump is highly reliable and has a long lifespan. One of the difficulties in building electrostatic micropumps is optimising device factors such as fluid properties, applied voltage and plate spacing. Researchers have used numerical simulations and experimental techniques to optimize the design parameters and to study the flow characteristics of the pump. Lab-on-a-chip systems, Drug distribution, and microfluidic cooling systems are just a few of the uses for electrostatic micropumps. They are especially helpful in applications that require minimal power usage, compact size, and dependability. Figure 10 depicts the conceptual structure of a micropump test structure chip, which consists of a three-caste silicon wafer mound sandwich (Sebastian Uhlig et al., 2018).

Electroactive polymer (EAP) actuators are a type of soft actuator that can undergo large deformations when stimulated by an electric field. These actuators have been widely investigated for their potential use in micropumps, which are devices that can transport fluids at the microscale. One promising approach for developing EAP-based micropumps involves using a type of EAP called dielectric elastomers. When a voltage is applied to a dielectric elastomer, it undergoes a large deformation, which can be harnessed to actuate a micropump. The flow rate and orientation of the micropump can be regulated by using voltage waveforms and various electrode geometries (Nam and Ahn, 2012).

Several challenges remain in the development of EAP-based micropumps, including improving their reliability, robustness, and integration with microfluidic systems. Nevertheless, the potential advantages of using EAPs for micropumps, including their low weight, low power consumption, and biocompatibility makes them an appealing tool for a wide range of applications in microfluidics, in delivery of drug, LOC devices, and microfluidic cooling. (Karthikeyan et al., 2018c; Khawwaf et al., 2017).

Figure 10. Conceptual structure of a micropump test structure chip consists of a three- caste silicon wafer mount
Source: Sebastian Uhlig et al. (2018)

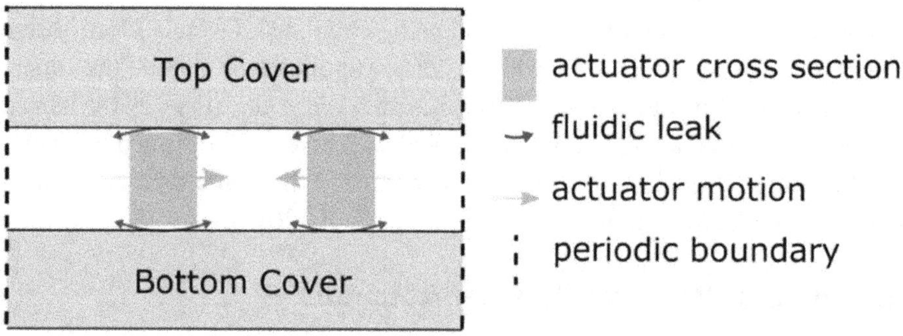

Electrohydrodynamic Actuators

Electrohydrodynamic (EHD) actuators are mechanisms that use electric fields to transform electrical energy into fluid motion. They have been widely used in microfluidic systems, including micropumps, due to their high efficiency, low power consumption, and rapid response time. EHD actuators for micropumps typically consist of two parallel electrodes separated by a dielectric layer, with a fluidic channel located on top of the electrodes. The fluid feels an electrically induced body force that pushes it in that direction when an electric field is present. One of the key advantages of EHD actuators for micropumps is their ability to pump fluids without the need for moving parts, which can lead to issues such as clogging, wear, and maintenance.

Additionally, capable of producing extremely low flow rates, EHD micropumps are simple to combine with other microfluidic parts. However, optimizing the device settings to achieve the desired flow rate and effectiveness is one of the difficulties in creating EHD actuators for micropumps. Researchers have used numerical simulations and experimental techniques to optimize the design parameters, such as electrode geometry, applied voltage, and fluid properties, for EHD micropumps. Overall, EHD actuators for micropumps are a hopeful technology for microfluidic systems because they have the potential for miniaturization and integration with other microfluidic parts, as well as high efficacy and low power usage (Ze-bing Mao et al., 2022).

Figure 11 shows the schematic structure of electrohydrodynamic micropump, (a) Electrodes (b) Single EHD pumps (c) stacked EHD pumps. The study of fluid dynamics including electrical charges is known as electrohydrodynamic. The fluid is energized by charging the electrodes and generating an electric field. The fluid then moves as a result of the electric field's strength (Sato et al., 2017). A liquid with a low viscosity and a high dielectric constant is required to attain high flow rates in EHD micropumps (Russel et al., 2016).

Figure 11. Schematic structure of electrohydrodynamic pumps
Source: Ze-bing Mao et al. (2022)

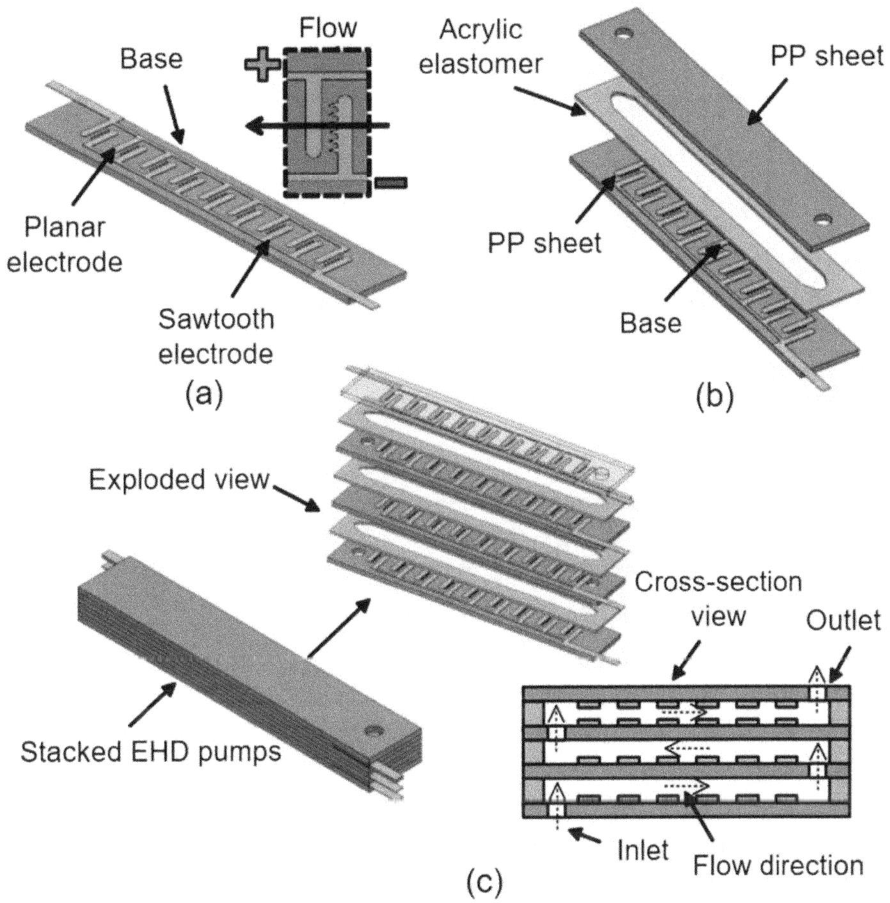

Shape Memory Alloy

Shape Memory Alloys (SMAs) are a class of materials that, in response to a specific stimulus like heat or an electrical current, have the remarkable capacity to regain their initial form. SMAs were employed in the creation of micropumps, among other technical uses. SMAs can be incorporated into micropumps as thin filament or film. The SMA contracts when a current is applied, causing it to go through a phase transition. This mechanical force can be used to activate a micropump. The high actuation force and quick reaction time of SMAs, which enable the creation of high flow rates and the exact regulation of fluid movement, are among the benefits of using them in micropumps. SMAs are also highly durable and have a long lifespan, making them ideal for use in microfluidic systems. However, one of the challenges in using SMAs in micropumps is the optimization of the device parameters, such as the choice of SMA material, the design of the actuator, and the control of the applied current. Researchers have used numerical simulations and experimental techniques to optimize the design parameters for SMA-based micropumps.

Figure 12. Schematic structure of SMA micropump
Source: Zainal et al. (2015)

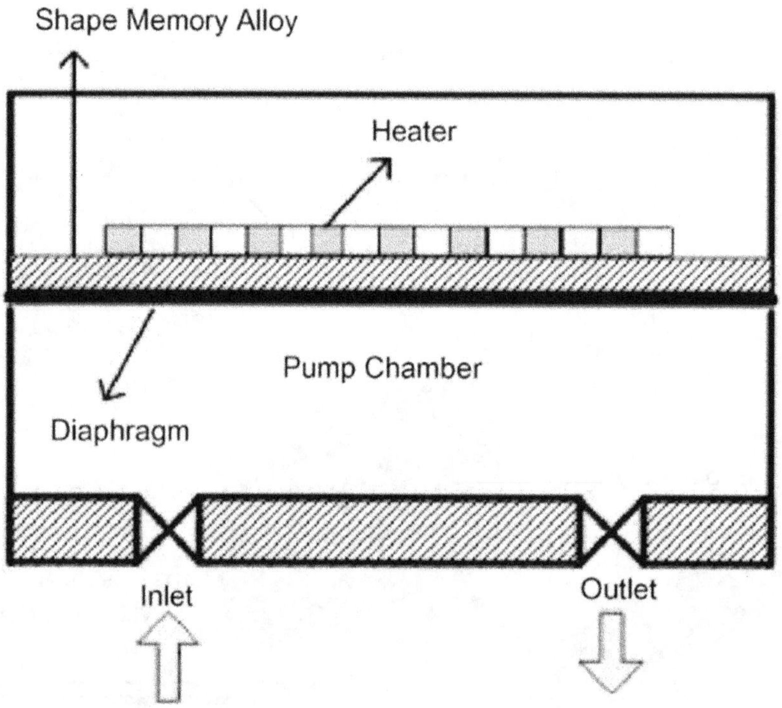

Figure 12 shows the schematic structure of SMA micropump. SMAs are a promising material for the development of micropumps, offering high actuation force, fast response time, and durability. The use of SMAs in micropumps has the potential to improve the performance and reliability of microfluidic systems, and could lead to the development of new and innovative microfluidic applications (Zainal et al., 2015).

Pneumatic Actuated Micropump

A pneumatic actuated micropump is a device that uses air pressure to generate fluid flow. It usually consists of a piston or diaphragm that oscillates inside of a chamber, producing a vacuum or pressure that propels fluid movement through the pump. An elastic piston or flexible diaphragm is housed in a chamber in a pneumatically driven micropump's fundamental construction. The chamber is connected to an inlet and outlet port, with a one-way valve placed between the inlet and chamber, and another one-way valve placed between the chamber and outlet. When compressed air is introduced into the chamber, the diaphragm or piston moves forward, creating a vacuum and drawing fluid into the chamber through the inlet valve. When the air pressure is released, the diaphragm or piston moves backward, generating pressure and forcing the fluid out of the chamber through the outlet valve.

One of the advantages of pneumatic actuated micropumps is their ability to generate high flow rates with low power consumption. They can also be designed to produce pulsatile flow, which is useful in

Figure 13. Fabrication process flow of pneumatic micropump
Source: Xuling Liu et al. (2020)

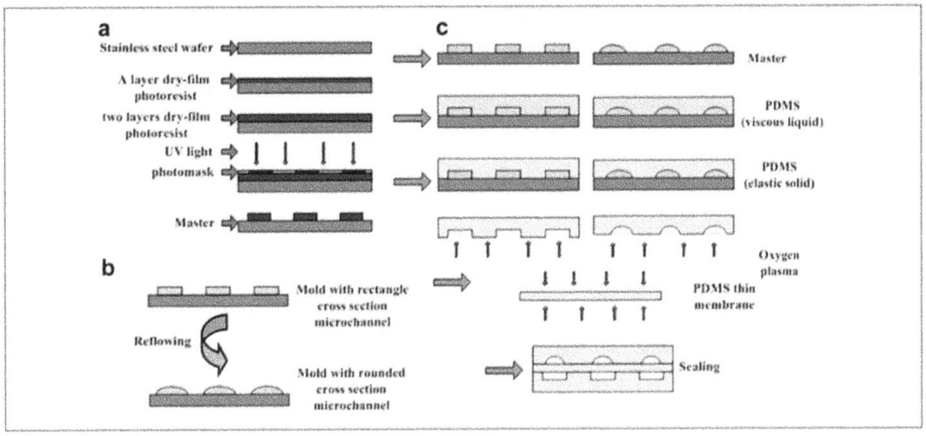

certain medical applications. Additionally, because they do not require any electrical power to operate, they can be used in environments where electrical interference or electromagnetic radiation is a concern (M. S. Xavier et al., 2022).

There are several different types of pneumatic actuated micropumps available, including diaphragm pumps, piston pumps, and peristaltic pumps. Diaphragm pumps use a flexible membrane to generate the flow, while piston pumps use a sliding piston to create pressure. Peristaltic pumps use a series of rollers to compress and move fluid through a flexible tube. The choice of micropump type will depend on the specific application requirements, such as flow rate, pressure, and accuracy. Some micropumps may also require additional components, such as pressure sensors or flow controllers, to achieve optimal performance. Figure 12 shows the fabrication process flow of pneumatic micropump (Yang J et al., 2021).

These types of micropumps are commonly used in applications that require precise control of fluid flow rates, such as in medical devices, lab-on-a-chip systems, and microfluidics. They are also popular in applications where small size, low power consumption, and low noise are important, such as in portable analytical instruments and environmental monitoring systems. However, pneumatic actuated micropumps can be sensitive to changes in air pressure, humidity, and temperature. They may also require precise control of the air pressure and flow rate to achieve optimal performance. Overall, pneumatic actuated micropumps are a versatile and reliable option for generating fluid flow in miniature applications (Karthikeyan et al., 2018d, 2018e; Davis et al., 2018). These micropumps have a wide range of applications in various fields, including medical, analytical chemistry, and industrial automation. Here are some specific examples of the application of pneumatic actuated micropumps:

Pneumatic actuated micropumps are commonly used in medical devices, such as insulin pumps and drug delivery systems. These pumps can precisely deliver small volumes of fluids or medications to a patient with great accuracy (Ankur Gupta et al., 2019). In microfluidic systems for chemical research, such as liquid chromatography and mass spectrometry, pneumatically actuated micropumps are employed. For an exact chemical study, they can carefully and reliably regulate the passage of gases and liquids through these devices (Atsushi Takashima et al., 2010). Pneumatic actuated micropumps are also used in environmental monitoring applications, such as air and water quality monitoring. They can be used to sample and analyze small volumes of fluids or gases in the field, which is useful for monitoring environmental pollution levels (Hamid et al., 2016). Pneumatic actuated micropumps are used in industrial automation applications, such as in microfluidic systems for controlling fluid flow and mixing. They are also used in pneumatic control systems for controlling the movement of robotic arms and other automation equipment (Xuling Liu et al., 2021). Overall, pneumatic actuated micropumps have a wide range of applications in various fields, where precise and accurate control of fluid or gas flow is essential.

CONCLUSION

Micropumps have emerged as a promising technology with a wide range of applications in various fields, including medical, environmental, and industrial. They offer several advantages over traditional pumps, such as their compact size, low power consumption, high precision, and suitability for use in microfluidic systems. The ongoing development of micropumps is expected to bring further improvements, such as enhanced efficiency, greater reliability, and expanded functionality, thereby enabling new applications and opening up new avenues for research. With the increasing demand for compact and efficient fluid handling systems, the outlook for micropumps looks bright, and they are likely to find increased use in diverse areas in the future. The limited access for microfabrication facilities and the issues in the interface of microfluidic components are the major barrier in developing the products for commercialization for several diagnosis. Centralized Multi-user Microfluidic fabrication facilities similar to Multi-user MEMS Processes (MUMPs) foundries may help in bridging the gap between the biotechnologists and microfluidic engineering and hence to develop new products into market in a short time. Microfluidic platform will bring a very big revolution in medical field with the potential advancement in integration of micropumps, microvalves and biosensors into the microfluidic chip in the forthcoming years.

ACKNOWLEDGMENT

The authors acknowledge the Department of Science & Technology (DST) for the financial support of the project (File No. DST/TDT/TDP-03/2019 (c) dated 12.03.2020).

REFERENCES

Ali & Takahata. (2011). Wireless microfluidic control with integrated shape-memory-alloy actuators operated by field frequency modulation. *J Micromech Microeng, 21*(7).

Asadi Dereshgi, H., & Yildiz, M. Z. (2018a). Investigation of electromechanical factors effecting piezoelectric actuator for valveless micropump characteristics. *J Eng Sci Technol, 13*, 2843–2856.

Asadi Dereshgi, H., & Yildiz, M. Z. (2018b). A novel micropump design, Investigation of the voltage effect on the net flow rate. *Sakarya Univ J Sci, 22*(4), 1152–1156. doi:10.16984aufenbilder.388658

Benard, W., Kahn, H., Heuer, A., & Huff, M. (1998). Thin-film shape memory alloy actuated micropumps. *Journal of Microelectromechanical Systems, 7*(2), 245–251. doi:10.1109/84.679390

Berg, A. V. D., & Lammerink, T. S. J. (1998). Micro total analysis systems: Microfluidic aspects, integration concept and applications in Microsystem Technology in Chemistry and Life Science. *Topics in Current Chemistry, 194*, 21–49. doi:10.1007/3-540-69544-3_2

Bonk, S., Stubbe, M., Buehler, S., Tautorat, C., Baumann, W., Klinkenberg, E.-D., & Gimsa, J. (2015). Design and characterization of a sensorized microfluidic cell-culture system with electrothermal micropumps and sensors for cell adhesion, oxygen, and pH on a glass chip. *Biosensors (Basel), 5*(3), 513–536. doi:10.3390/bios5030513 PMID:26263849

Chee, P. S., Minjal, M. N., Lcow, P. L., & Ali, M. S. M. (2015). Wireless powered thermo pneumatic micropump using frequency-controlled heater. *Sensors and Actuators. A, Physical, 233*, 1–8. doi:10.1016/j.sna.2015.06.017

Chee, P. S., Nafea, M., Leow, P. L., & Ali, M. S. M. (2016). Thermal analysis of wirelessly powered thermo-pneumatic micropump based on planar LC circuit. *Journal of Mechanical Science and Technology, 30*(6), 2659–2665. doi:10.100712206-016-0527-5

Chen, C. H., & Santiago, J. (2002). A Planar Electroosmotic Micropump. *Journal of Microelectromechanical Systems, 11*(6), 672–683. doi:10.1109/JMEMS.2002.805055

Cheong, H. R., Lai, K. C., & Chee, P. S. (2018). A wireless powered electroactive polymer using magnetic resonant coupling. *IOP Conf Ser Mater Sci Eng, 409*, 1-6. 10.1088/1757-899X/409/1/012002

Davis. (2018). Pneumatic Actuators. *Actuators, 7*, 62.

Dong, J. S., Liu, R. G., Liu, W. S., Chen, Q. Q., Yang, Y., Wu, Y., Yang, Z. G., & Lin, B. S. (2017). Design of a piezoelectric pump with dual vibrators. *Sensors and Actuators. A, Physical, 257*, 165–172. doi:10.1016/j.sna.2017.02.001

Duwairi, H., & Abdullah, M. (2006). Thermal and flow analysis of a magneto-hydrodynamic micropump. *Microsystem Technologies, 13*(1), 33–39. doi:10.100700542-006-0258-0

Ehsani & Nejat. (2017). Conceptual design and performance analysis of a novel flexible-valve micropump using magnetofluid–solid interaction. *Smart Mater Struct, 26*.

Fong, J., Xiao, Z., & Takahata, K. (2015). Wireless implantable chip with integrated nitinol-based pump for radio-controlled local drug delivery. *Lab on a Chip*, *15*(4), 1050–1058. doi:10.1039/C4LC01290A PMID:25473933

Ghazali, F. A. M., Mah, C. K., & Abuzaiter, A. (2017). Soft dielectric elastomer actuator micropump. *Sensors and Actuators. A, Physical*, *263*, 276–284. doi:10.1016/j.sna.2017.06.018

Guerine, A., Merzouki, T., Hami, A. E., & Zineb, T. B. (2018). Uncertainty analysis of an actuator for a shape memory alloy micro-pump with uncertain parameters. *Advances in Engineering Software*, *122*, 22–30. doi:10.1016/j.advengsoft.2018.02.011

Guo, G., Wu, X., Liu, D., Liao, L., Zhang, D., Zhang, Y., Mao, T., He, Y., Huang, P., & Wang, W. (2021). A Self-Regulated Microfluidic Device with Thermal Bubble Micropumps. *Micromachines, 13.*

Gupta, A., & Pal, P. (2019). 8 - Micro-electro-mechanical system–based drug delivery devices, In Woodhead Publishing Series in Electronic and Optical Materials. Bioelectronics and Medical Devices, 183-210.

Haldkar, R.K., Gupta, V.K., Sheorey, T., & Parinov, I.A., (2021). Design, Modeling, and Analysis of Piezoelectric-Actuated Device for Blood Sampling. *Appl. Sci., 11.*

Hamid, N. A., Majlis, B. Y., Yunas, J., Syafeeza, A. R., Wong, Y. C., & Ibrahim, M. (2016). A stack bonded thermopneumatic micro-pump utilizing polyimide-based actuator membrane for biomedical applications. *Microsystem Technologies*, *23*(9), 4037–4043. doi:10.100700542-016-2951-y

Hasan, M. I., Ali, A. J. F., & Tufah, R. S. (2017). Numerical study of the effect of channel geometry on the performance of Magnetohydrodynamic micro pump. *Eng Sci Technol Int J*, *20*(3), 982–989. doi:10.1016/j.jestch.2017.01.008

Javadi, A., Habibi, M., & Taheri, F.S. (2013). Effect of wetting on capillary pumping in microchannels. *Sci Rep., 3.*

Johnson & Borkholder. (2016). Towards an Implantable, Low Flow Micropump That Uses No Power in the Blocked-Flow State. *Micromachines, 7.*

Karthikeyan, K. (2023). Numerical simulation and parameter optimization of micromixer device using fuzzy logic technique. *RSC Advances*, *13*(7), 4504–4522. doi:10.1039/D2RA07992E PMID:36760289

Karthikeyan, K., Kishor, A., Karthikeyan, P., & Karthikeyan, S. (2022). Design and Numerical Simulation of RF Antenna for Bio-Medical Application. *International Conference on Applied Artificial Intelligence and Computing (ICAAIC)*, 1222-1227.

Karthikeyan, K., & Sujatha, L. (2017) Design and fabrication of microfluidic device for mercury ions detection in water. *International Conference on Nextgen Electronic Technologies: Silicon to Software (ICNETS2)*, 218-221. 10.1109/ICNETS2.2017.8067934

Karthikeyan, K., & Sujatha, L. (2018a). Detection of Mercury Ion in a Microfluidic Device Using Colorimetric Method. *International Journal of Advances in Science, Engineering and Technology*, *6*(2), 67–70.

Karthikeyan, K., & Sujatha, L. (2018c). Fluorometric Sensor for Mercury Ion Detection in a Fluidic MEMS Device. *IEEE Sensors Journal*, *18*(13), 5225–5231. doi:10.1109/JSEN.2018.2840331

Karthikeyan, K., & Sujatha, L. (2018d). Study of Permissible Flow Rate and Mixing Efficiency of the Micromixer Devices. *International Journal of Chemical Reactor Engineering, 17*(1).

Karthikeyan, K., Sujatha, L., & Sudharsan, N. M. (2018e). Numerical Modeling and Parametric Optimization of Micromixer for Low Diffusivity Fluids. *International Journal of Chemical Reactor Engineering, 16*(3).

Karthikeyan, K., Sujatha, L., & Sundar, R. (2021). A Low-Cost Fabrication and Numerical Simulation of a MEMS Acoustic Transducer Using Polyimide Membrane on FR4 Substrate. *Journal of Electronic Materials, 50*(11), 6489–6503. doi:10.100711664-021-09184-0

Karthikeyan, K., Sujatha, L., Sundar, R., & Sharma, S. K. (2018b). Dimension Tolerances in Fabrication of Polymer Microfluidic Devices. *Journal of Semiconductor Technology and Science-IEIE, 18*(2), 262–269. doi:10.5573/JSTS.2018.18.2.262

Kawun, P., Leahy, S., & Lai, Y. (2016). A thin PDMS nozzle/diffuser micropump for biomedical applications. *Sensors and Actuators. A, Physical, 249*, 149–154. doi:10.1016/j.sna.2016.08.032

Khawwaf, J., Zheng, J., Lu, R., Al-Ghanimi, A., Kazem, B. I., & Man, Z. (2017). Robust tracking control of an IPMC actuator using nonsingular terminal sliding mode. *Smart Materials and Structures, 26*(9), 095042. doi:10.1088/1361-665X/aa7d69

Kim, H., Kim, K., & Lee, S.J. (2016). Compact and Thermosensitive Natureinspired Micropump. *Scientific Reports, 6*.

Kjeang, E., Djilali, N., & Sinton, D. (2009). Microfluidic fuel cells, a review. *Journal of Power Sources, 186*(2), 353–369. doi:10.1016/j.jpowsour.2008.10.011

Le, S., & Hegab, H. (2017). Investigation of a multistage micro gas compressor cascaded in series for increase pressure rise. *Sensors and Actuators. A, Physical, 256*, 66–76. doi:10.1016/j.sna.2017.01.016

Li, L., Wang, X., Pu, Q., & Liu, S. (2019). Advancement of electroosmotic pump in microflow analysis: A Review. *Analytica Chimica Acta, 1060*, 1–16. doi:10.1016/j.aca.2019.02.004 PMID:30902323

Lin, S.-C., Lu, J.-C., Sung, Y.-L., Lin, C.-T., & Tung, Y.-C. (2013). A low sample volume particle separation device with electrokinetic pumping based on circular travelling-wave electroosmosis. *Lab on a Chip, 13*(15), 3082–3089. doi:10.1039/c3lc50343g PMID:23753015

Liu, B., Sun, J., Li, D., Zhe, J., & Oh, K. W. (2016). A high flow rate thermal bubble-driven micropump with induction heating. *Microfluidics and Nanofluidics, 20*(11), 155. doi:10.100710404-016-1822-2

Liu, X., & Li, S. (2020). Fabrication of a Three-Layer PDMS Pneumatic Microfluidic Chip for Micro Liquid Sample Operation. *SLAS Technology, 25*(2), 151–161. doi:10.1177/2472630319870126 PMID:31425005

Liu, X., Zhang, D., Liu, J., Wang, L., & Li, S. (2021). RGB Color Model Analysis for a Simple Structured Polydimethylsiloxane Pneumatic Micromixer. *SLAS Technology, 26*(5), 510–518. doi:10.1177/24726303211008868 PMID:34384310

Mao, Z., Asai, Y., Wiranata, A., Kong, D., & Man, J. (2022). Eccentric actuator driven by stacked electrohydrodynamic pumps. *Journal of Zhejiang University. Science A*, *23*(4), 329–334. doi:10.1631/jzus.A2100468

Meng, A. H., Nguyen, N.-T., & White, R. M. (2000). Focused flow micropump using ultrasonic flexural plate waves. *Biomedical Microdevices*, *2*(3), 169–174. doi:10.1023/A:1009905926740

Nam, D. N. C., & Ahn, K. K. (2012). Design of an IPMC diaphragm for micropump application. *Sensors and Actuators. A, Physical*, *187*, 174–182. doi:10.1016/j.sna.2012.08.027

Nguyen, N.-T., Huang, X., & Chuan, T. K. (2002). MEMS-micropumps; A Review. *Journal of Fluids Engineering*, *124*(2), 384–392. doi:10.1115/1.1459075

Nie, C., Frijns, A.J.H., Mandamparambil, R., & Toonder, J.M.J.D. (2015). A microfluidic device based on an evaporation-driven micropump. *Biomed Microdevices, 17*(47).

Okamoto, Y., Ryoson, H., Fujimoto, K., Ohba, T., & Mita, Y. (2020). On-Chip CMOS-MEMS-Based Electroosmotic Flow Micropump Integrated With High-Voltage Generator. *Journal of Microelectromechanical Systems*, *29*(1), 1–10. doi:10.1109/JMEMS.2019.2953290

Pabst, O., Ho¨lzer, S., Beckert, E., Perelaer, J., Schubert, U. S., Eberhardt, R., & Tünnermann, A. (2014). Inkjet printed micropump actuator based on piezoelectric polymers, Device performance and morphology studies. *Organic Electronics*, *15*(11), 3306–3315. doi:10.1016/j.orgel.2014.09.007

Qu, Y., Zhou, J., & Wu, W. (2018). Theoretical and Experimental Research on Bubble Actuated Micro-Pumps. *Micromachines*, *9*(225), 1–11. doi:10.3390/mi9050225 PMID:30424158

Robertson, J. M., Rodriguez, R. X., Holmes, L. R. Jr, Mather, P. T., & Wetzel, E. D. (2016). Thermally driven microfluidic pumping via reversible shape memory polymers. *Smart Materials and Structures*, *25*(8), 085043. doi:10.1088/0964-1726/25/8/085043

Rusli, M., Chee, P. S., Arsat, R., Lau, K. X., & Leow, P. L. (2018). Electromagnetic actuation dual-chamber bidirectional flow micropump. *Sensors and Actuators. A, Physical*, *282*, 17–27. doi:10.1016/j.sna.2018.08.047

Russel, M., Selvaganapathy, P., & Ching, C. (2016). Ion drag electrohydrodynamic (EHD) micro-pumps under a pulsed voltage. *Journal of Electrostatics*, *82*, 48–54. doi:10.1016/j.elstat.2016.05.003

Said, M.M., Yunas, J., & Bais, B. (2017). The design, fabrication, and testing of an electromagnetic micropump with a matrix-patterned magnetic polymer composite actuator membrane. *Micromachines, 9*(13).

Sato, T., Yamanishi, Y., Cacucciolo, V., Kuwajima, Y., Shigemune, H., Cianchetti, M., Laschi, C., & Maeda, S. (2017). Electrohydrodynamic conduction pump with asymmetrical electrode structures in the microchannels. *Chemistry Letters*, *46*(7), 950–952. doi:10.1246/cl.170217

Sayar, E., & Farouk, B. (2013). Electroosmotic Augmentation in Flexural Plate Wave Micropumps. *Journal of Microelectromechanical Systems*, *22*(2), 372–385. doi:10.1109/JMEMS.2012.2226930

Shaegh, S. A. M., Wang, Z., Ng, S. H., Wu, R., Nguyen, H. T., Chan, L. C. Z., Toh, A. G. G., & Wang, Z. (2015). Plug-and-play microvalve and micropump for rapid integration with microfluidic chips. *Microfluidics and Nanofluidics, 19*(3), 557–564. doi:10.100710404-015-1582-4

Shoji, E. (2016). Fabrication of a diaphragm micropump system utilizing the ionomer-based polymer actuator. *Sensors and Actuators. B, Chemical, 237*, 660–665. doi:10.1016/j.snb.2016.06.153

Sima, A.H., Salari, A., & Shafii, M.B. (2015). Low-cost reciprocating electromagnetic-based micropump for high-flow rate applications. *J Micro/Nanolithogr MEMS MOEMS, 14.*

Spieth, S., Schumacher, A., & Kallenbach, C. (2012). The Neuro Medicator a micropump integrated with silicon microprobes for drug delivery in neural research. *J Micromech Microeng, 22.*

Takashima, A., Kojima, K., & Suzuki, H. (2010). Autonomous Microfluidic Control by Chemically Actuated Micropumps and Its Application to Chemical Analyses. *Analytical Chemistry, 82*(16), 6870–6876. doi:10.1021/ac1009657 PMID:20669895

Tanaka, Y., Noguchi, Y., Yalikun, Y., & Kamamichi, N. (2017). Earthworm muscle driven bio-micropump. *Sensors and Actuators. B, Chemical, 242*, 1186–1192. doi:10.1016/j.snb.2016.09.123

Tripathi, D., Sharma, A., & Be'g, O. A. (2017). Electrothermal transport of nanofluids via peristaltic pumping in a finite micro-channel, effects of Joule heating and Helmholtz-Smoluchowski velocity. *International Journal of Heat and Mass Transfer, 111*, 138–149. doi:10.1016/j.ijheatmasstransfer.2017.03.089

Uhlig, S., Gaudet, M., & Langa, S. (2018). Electrostatically driven inplane silicon micropump for modular configuration. *Micromachines, 9.*

Uhlig, Gaudet, Langa, Schimmanz, Conrad, Kaiser, & Schenk. (2018). Electrostatically Driven In-Plane Silicon Micropump for Modular Configuration. *Micromachines, 9*(4).

Uvarov, I.V., Lemekhov, S.S., & Melenev, A.E. (2016). A simple electrochemical micropump, design and fabrication. *J Phys Conf Ser, 741.*

Uvarov, I. V., Lemekhov, S. S., Melenev, A. E., & Svetovoy, V. B. (2017). Exploding microbubbles driving a simple electrochemical micropump. *Journal of Micromechanics and Microengineering, 27*(10), 105009. doi:10.1088/1361-6439/aa8914

Uvarov, I. V., Shlepakov, P. S., Melenev, A. E., Ma, K., Svetovoy, V. B., & Krijnen, G. J. M. (2021). A Peristaltic Micropump Based on the Fast-Electrochemical Actuator, Design, Fabrication, and Preliminary Testing. *Actuators, 10*(3), 62. doi:10.3390/act10030062

Wang, J., Liu, Y., Shen, Y., Chen, S., & Yang, Z. (2016). A resonant piezoelectric diaphragm pump transferring gas with compact structure. *Micromachines, 7*(12), 219. doi:10.3390/mi7120219 PMID:30404390

Xavier, M. S., & Charbel, D. (2022). Soft Pneumatic Actuators, A Review of Design, Fabrication, Modeling, Sensing, Control and Applications. *IEEE Access : Practical Innovations, Open Solutions, 10*, 59442–59485. doi:10.1109/ACCESS.2022.3179589

Yang, J., & Li, X. (2021). Research and development of a novel pneumaticelectric hybrid actuator. *Journal of Mechanical Science and Technology, 35*(6).

Yazdi, S. A. F. F., Corigliano, A., & Ardito, R. (2019). 3-D design and simulation of a piezoelectric micropump. *Micromachines*, *10*(4), 259. doi:10.3390/mi10040259 PMID:31003481

Yildirim, Y. A., Toprak, A., & Tigli, O. (2018). Piezoelectric membrane actuators for micropump applications using PVDF-TrFE. *Journal of Microelectromechanical Systems*, *27*(1), 86–94. doi:10.1109/JMEMS.2017.2773090

Yildiz, M. Z., & Asadi Dereshgi, H. (2019). Design of PZT micropumps for biomedical applications, glaucoma treatment. *Journal of Engineering Research*, *7*, 226–241.

Zainal, M. A., Sahlan, S., & Mohamed, S. M. A. (2015). Micromachined Shape-Memory-Alloy Microactuators and Their Application in Biomedical Devices. *Micromachines*, *7*(7), 879–901. doi:10.3390/mi6070879

Zhang, Z., Kan, J., Cheng, G., Wang, H., & Jiang, Y. (2013). A piezoelectric micropump with an integrated sensor based on space-division multiplexing. *Sensors and Actuators. A, Physical*, *203*, 29–36. doi:10.1016/j.sna.2013.08.027

Zhou, X., Gao, M., & Gui, L. (2017). A Liquid-Metal Based Spiral Magnetohydrodynamic Micropump. *Micromachines*, *8*(365), 1–10. PMID:30400555

Chapter 6
Advances in Paper-Based Microfluidics:
From Fabrication to Applications

Ashish Mathur
Centre for Interdisciplinary Research and Innovation, University of Petroleum and Energy Studies, India

Anu Bharti
Centre for Interdisciplinary Research and Innovation, University of Petroleum and Energy Studies, India

Souradeep Roy
Centre for Interdisciplinary Research and Innovation, University of Petroleum and Energy Studies, India

Shikha Wadhwa
Department of Chemistry, University of Petroleum and Energy Studies, India

ABSTRACT

Paper-based microfluidic devices have emerged as a promising technology for a variety of applications, including diagnostics, chemical synthesis, and environmental monitoring. These devices are inexpensive, portable, and easy to use, making them suitable for use in resource-limited settings. The chapter aims to highlight recent advances in the field of paper-based microfluidics, focusing on numerous fabrication techniques, materials, and applications. The authors also review the use of different types of paper and other cellulose-based materials as substrates, as well as the use of different types of adhesives and binders to fabricate the devices. Furthermore, they discuss various applications of paper-based microfluidic devices, including diagnostics, chemical synthesis, and environmental monitoring. Additionally, the chapter will discuss the potential of wireless communication and battery-free power sources integration with paper-based microfluidic devices to create fully autonomous devices that can be used in remote locations, along with current challenges faced and future outlook.

DOI: 10.4018/978-1-6684-6952-1.ch006

INTRODUCTION TO PAPER MICROFLUIDICS

Microfluidics is defined as the science of controlling fluid flow in extremely low quantities, that is within the range of few μL – nL, through chambers or channels having dimensions in the micron – sub micron scale. The behavior of fluids at these dimensions are governed by the principles which are markedly different from the macroscopic scenario. The liquid flow through microscale channels have been known to be Newtonian and incompressible (Kirby, 2010). In such approximated cases, the flow pattern can be modelled using the Navier-Stokes equation as shown in equations 1 and 2 (Lammertyn et al., 2006):

$$\nabla.\mathbf{v} = 0 \tag{1}$$

$$\acute{A}(\nabla.\mathbf{v})\,\mathbf{v} = -\nabla p + \cdot\, \nabla^2 \mathbf{v} \tag{2}$$

It is seen that the velocity gradient (∇v) and pressure gradient (∇p) are the key parameters which needs to be computed, and the optimization of which is vital for developing efficient microfluidic platforms/devices.

Paper-based microfluidics has drawn considerable attention in the scientific community over the past decade, and is defined as the use of paper substrates in microfluidic devices (Anushka et al., 2022). Research in the field of paper-based microfluidics was first initiated by Martinez and co-workers, demonstrating the simultaneous detection of glucose and protein on a patterned cellulose sheet which utilized sample volume as less as ~ 5 μL (Martinez et al., 2007). One of the key advantages of paper substrates, as compared to conventional ones namely glass, silicon and polymers, is its flexibility, lightweight, abundant availability and biocompatibility which allows for the development of cost-effective and portable as well as wireless healthcare and environment monitoring modules (Shen et al., 2020). These devices are then termed as Microfluidic Paper-based Analytical Devices (μPADs). Fluid flow in paper microfluidics occurs without the requirement of external perturbation, and is governed primarily by capillary action. The widely known Lateral Flow Immunoassays (LFAs) such as pregnancy test strips and the recently commercialized home-based COVID-19 monitoring electrodes, the latter developed by Mylab Discovery Solutions Pvt. Ltd. India, work on the principles of capillary-driven flow. There are normally two opposing forces which governs the interaction between paper substrates and the liquid which is in contact with the former. These are the cohesive and adhesive forces. When the liquid is brought in contact with paper, there exists an intermolecular force between the fluid molecules at liquid–air interface termed as cohesion as well as between the solid-liquid interface known as adhesion. The adhesive force is results in the spreading of the liquid over the porous surface, while the cohesive force (surface tension) leads to reduction in the liquid contact area. Therefore, the liquid flows only when the adhesive force is greater than the cohesive counterpart. Moreover, the wicking process depends on numerous geometrical properties of the paper such as paper structure, pore size, permeability as well as on the physical properties of the liquid. The liquid transport can generally be categorized into two groups. These are the wet-out and fully wetted flow processes. In the former, the liquid front simply remains in contact with the porous paper surface, which then can be modelled by the Lucas-Washburn equation as per equation-3 (Mathur & Roy, 2022).

$$x(t) = \sqrt{\frac{\sigma tr(\cos\theta)}{2\eta}} \tag{3}$$

where,
x(t) = distance traversed by liquid under capillary pressure.

σ = surface tension at liquid-air interface
t = time required for traversal of liquid
r = average pore radius of paper substrate
θ = contact angle at liquid-paper interface.
η = viscosity of the liquid.

In the second case, liquid transport takes place over the porous paper substrates which can be modelled using Darcy's law as shown in equation-4 (Bhattacharya et al., 2019).

$$v = -\frac{k}{\eta}\nabla P \tag{4}$$

where,

v = velocity of liquid flow.
k = permeability of paper material.
∇P = pressure gradient

Figure 1. Schematic highlighting the advantages of paper microfluidics and few key application sectors

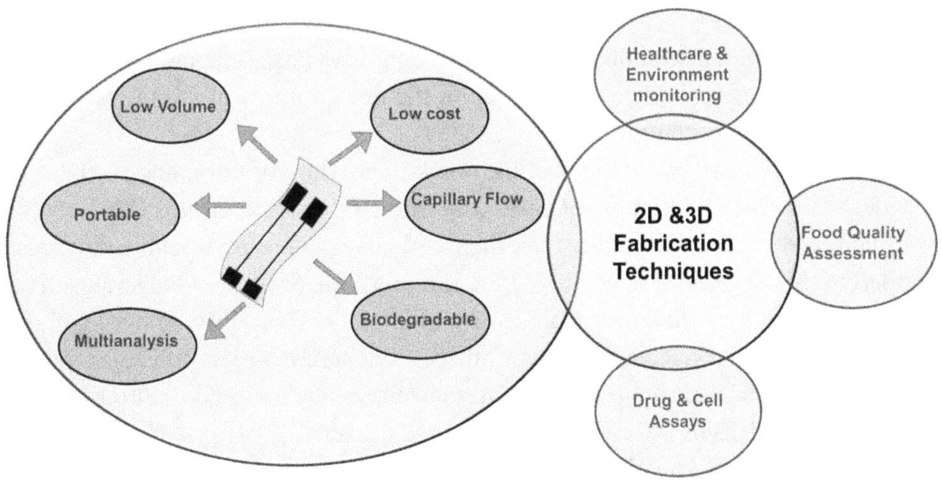

These parameters can be optimized for specific types of paper substrates, the choice of which depends on the application of interest. Figure 1 highlights the advantages of paper-based microfluidic substrates as well as few areas of application.

The following sections will briefly highlight some of the commonly used paper substrates for the development of μPADs, choice of adhesives & binders, few commonly implemented fabrication techniques followed by a detailed focus on various applications – ranging from biomedical to environmental analysis.

TYPES OF PAPER SUBSTRATES IN MICROFLUIDICS

The choice of suitable paper substrates is crucial for specific applications in paper-based microfluidics. Each type of paper has unique properties which affects the liquid flow rate and contact angle depending on the pore dimensions, which further has a direct influence on the capillary pressure gradient ∇P as indicated in equation 4.

Cellulose paper is the most commonly used substrate in paper microfluidics, and the resulting μPADs are termed as Lab-on-Paper devices (Dukle et al., 2022). It is a cost-effective and easily available material that has unique properties suitable for microfluidic applications. Cellulose paper is a highly porous (2 – 25 μm, depending on grade) material that enables fluid transport by capillary action, thereby negating the requirement of external flow pumps. The cellulose networks behave as micro-channels which enables retainment of suitable materials and also allowing liquid flow from over the paper substrate (Das et al., 2022). Another key property of cellulose-based papers is their high specific surface area aided by a high degree of hydrophilicity. Such phenomena help in efficient immobilization of biomolecules like enzymes, DNA, aptamers and antibodies/antigens for the selective recognition of suitable biomarkers, microbial activity and environmental contaminants to name a few (Anushka et al., 2022).

A new class of cellulose named nanocellulose have recently emerged. The advantages of cellulose sheets can be extrapolated further when these chains are bundled together. The resulting high degree of ordered regions isolated as nanoparticles are known as nanocellulose. These materials exhibit exotic properties such as excellent chemical inertness, stiffness, low thermal expansion, high dimensional stability and diverse possibilities of modifying its surface chemistry which are deemed extremely suitable for microfluidic applications. Although a detailed description on nanocellulose is beyond the scope of this chapter, a comprehensive analysis of its surface chemistry and morphology can be found elsewhere (Trache et al., 2020).

Apart from the conventional cellulose matrix, nitrocellulose-based papers have also been widely employed in paper-based microfluidics – especially in the development of LFAs. A typical schematic of LFA-based strip is shown in Figure 2.

Nitrocellulose is synthesized by treating cellulose with a mixture of nitric and sulfuric acids, which changes the widely available OH groups (of cellulose) into NO_3 or nitro groups (McKeen, 2016). This makes them comparatively less hydrophilic than their cellulose counterpart, which otherwise may lead to rapid degradation due to excessive wetting. In this regard, nitrocellulose substrates are considered comparatively high durability with appreciable hydrophilicity and flexibility. The role of nitrocellulose membranes/substrates in paper microfluidics is primarily to capture target molecules via immobilized bioreceptors. Apart from this, these porous substrates have been widely used in ultrafiltration, osmosis, electrophoresis films as well as in scaffolds in cell culture (A. Li et al., 2013). Overall, the use of nitrocellulose paper in microfluidics has revolutionized the field of diagnostics and made it possible to develop

Figure 2. Schematic of the structure and constituents of a lateral flow assay strip

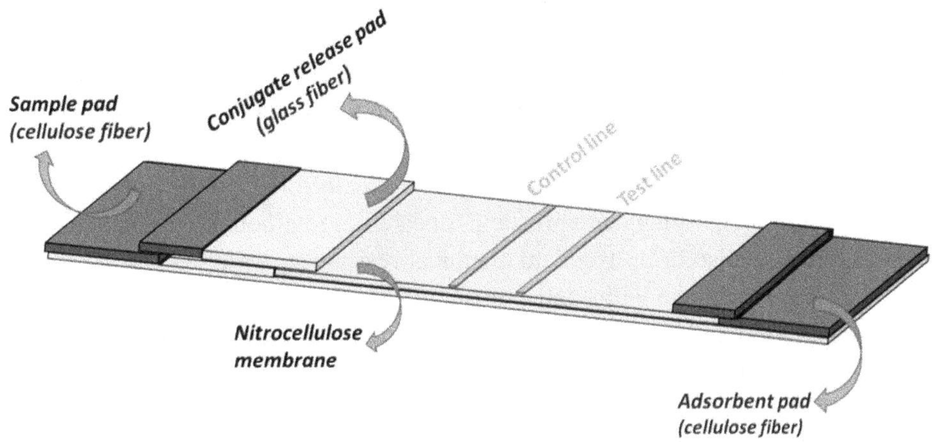

low cost, portable and user-friendly devices that can be used in a variety of settings. Its versatility and unique properties make it a valuable material for developing new microfluidic technologies.

Apart from the choice of paper, the selection of adhesives and binders also play a crucial role in the fabrication of efficient, stable and robust microfluidic devices. The next section will briefly highlight on this matter.

TYPES OF BINDERS/ADHESIVES

Another important consideration for paper-based microfluidics is the use of appropriate materials and adhesives. The choice of paper, adhesives, and other materials can affect the mechanical properties, chemical compatibility, and biocompatibility of the devices. Many researchers are interested to improve the designing and fabrication steps involved in paper microfluidics. The types of paper and fabrication technique employed for designing these devices have great impact on the device performance and have been studied vastly (Cate et al., 2015; Nishat et al., 2021; Sinha et al., 2022) Moreover, the types of materials used for making channels & reservoirs is also a key component in the paper based microfluidic devices and the materials are called as adhesives and binders. Paper used in the fabrication of microfluidic platforms/devices is a relatively fragile material and can get deformed easily during handling. Therefore, use of Adhesives can stabilize and reinforce the paper, helping to prevent it from tearing, wrinkling or collapsing during use. Adhesives/binders can improve the wettability of the paper, allowing for better capillary flow and easier fluid handling. They help to overcome the limitations of paper as a material and enable the development of simple, low-cost, and disposable microfluidic devices for a range of applications. Here are some commonly used adhesives or linkers in paper-based microfluidic devices:

1. **Epoxy:** Epoxy is a two-part adhesive that is commonly used in paper-based microfluidic devices. It is more complex to use than other adhesives, but it forms a strong bond and can withstand high pressures and temperatures. The epoxy binder is typically applied by spraying or brushing a solu-

tion of the epoxy onto the paper. By providing a stable, water-resistant surface and improving the wettability of the paper, the epoxy allows for reliable and robust fluid handling in these simple and cost-effective devices.

2. **Polyurethane:** These binders offer a range of advantages that make them an excellent choice for this application. One of the primary benefits of using polyurethane binders is their excellent adhesion properties. They are able to create strong bonds between different types of paper substrates, making them an ideal choice for developing multi-layered microfluidics devices. This results in robust and reliable devices that can withstand repeated use without delaminating or falling apart. Polyurethane binders also offer excellent chemical resistance, which is crucial for microfluidics applications. They are able to withstand exposure to a wide range of solvents and chemicals without degrading or breaking down, which ensures the integrity and functionality of the microfluidic device. In addition, polyurethane binders are highly flexible and can conform to a variety of shapes and sizes. This allows for the development of more complex and intricate microfluidics designs, which can be tailored to specific applications and requirements.
3. **Silanes:** They have been widely used as coupling agents and surface modifiers due to their ability to form strong covalent bonds with many different surfaces. Silanes have been used as binders for a variety of different biomolecules, including enzymes, antibodies, and DNA. They have also been used to immobilize nanoparticles and other materials onto paper substrates. One advantage of using silanes as binders is that they can form strong and stable covalent bonds, which can help prevent the detachment of the target material from the paper substrate.
4. **Polystyrene:** Polystyrene is used as a binder in paper-based microfluidic devices to improve their mechanical properties, such as stiffness and durability. When added to the paper substrate, polystyrene can increase the hydrophobicity of the surface, which can aid in controlling fluid flow. It can also act as a sealant, preventing the leakage of fluids from the device. However, the use of polystyrene as a binder in paper-based microfluidic devices can have environmental concerns, as polystyrene is not biodegradable and can persist in the environment for hundreds of years.

The choice of adhesive or linker depends on the application and the materials being used in the microfluidic device. It is important to consider factors such as compatibility with the fluids being used, ease of use, cost, and durability when selecting an adhesive or linker for paper-based microfluidic devices.

FABRICATION TECHNIQUES FOR MPADS

The fabrication of devices based on paper microfluidics can be achieved using numerous techniques. Few of them are briefly highlighted below:

Laser Cutting/Engraving: This is a widely used technique for the precision fabrication of paper-based microfluidic devices. A laser is used to cut channels and other structures into paper. The laser can be used to cut through the entire thickness of the paper or to create shallow cuts on the surface of the paper. This technique is precise, fast, and can be used to create complex geometries.

Stamping: This technique involves the use of a stamp or a template to create channels and other structures in paper. The stamp is pressed against the paper, and the channels are created by the indentations left by the stamp. This technique is simple, fast, and can be used to create large arrays of devices.

Table 1. Few advantages and disadvantages of commonly implemented fabrication strategies in paper microfluidics

Sl. No	Fabrication Techniques	Advantages	Disadvantages
1	Laser cutting	Precise and fast, can create complex geometries	Expensive, can create heat damage to the paper
2	Stamping	Simple and inexpensive, can create large arrays of devices	Not as precise as laser cutting, limited resolution
3	Inkjet printing	Fast, allows for integration of sensing technologies	Limited resolution, shallow channels
4	Microfabrication	Precise, versatile, and fast, can create complex geometries	Expensive, not as easy to scale up
5	3D printing	Can create complex geometries, allows for easy scalability	Limited resolution, may not be as robust as other materials
6	Thermal bonding	Strong and stable bonds, can create large arrays of devices	Requires specialized equipment, not as precise as other techniques
7	Electrospinning	Can create complex geometries, high surface area	Requires specialized equipment, not as precise as other techniques

Inkjet Printing: This technique involves the use of an inkjet printer to create channels and other structures in paper. The printer is used to deposit droplets of a liquid, such as a hydrogel, onto the paper to create channels. This technique is fast, can be used to create large arrays of devices, and allows for the integration of sensing and detection technologies.

Microfabrication: This technique involves the use of traditional microfabrication techniques, such as photolithography, to create channels and other structures in paper. This technique is precise, fast and can be used to create complex geometries, but it requires specialized equipment and is more expensive than other techniques.

3D Printing: This technique involves the use of a 3D printer to create channels and other structures in paper. The printer is used to deposit droplets of a liquid, such as a hydrogel, onto the paper to create channels. This technique is fast, can be used to create large arrays of devices, and allows for the integration of sensing and detection technologies.

Each of the mentioned fabrication techniques provide unique advantages, and hence, the choice of adopting a suitable strategy solely depends on the desired application. Table 1 highlights the advantages and disadvantages of few commonly used fabrication techniques in the field of paper microfluidics.

APPLICATIONS OF PAPER MICROFLUIDICS

Paper based microfluidic devices have become popular due to their deployment in a plethora of applications which includes diagnostic testing, environmental monitoring, food safety and chemical synthesis. This is primarily due to their advantages such as simplicity of fabrication, portability, affordability and requirement of less sample volume (discussed in Table 1). These devices are a highly promising and

Table 2. Few widely implemented applications of paper-based microfluidic devices

Applications	Description
Diagnostics	Paper microfluidic devices have been implemented to develop early detection methods for various pathogens such as bacteria and viruses, as well as other disease biomarkers such as lactate, creatinine and glucose.
Environmental monitoring	Paper based devices can be used to detect contaminants in water, soil, or air. These devices can be used to monitor water quality, detect heavy metals and pesticides, and can be used for air quality monitoring.
Chemical synthesis	They have been used for chemical synthesis by performing efficient mixing, separation, and reaction of small volumes of fluids. These devices can be used for synthesizing small molecules, peptides, and for performing chemical reactions in small volumes.
Cell culture	Paper microfluidic devices have been used for cell culture, providing a simple and effective way to culture cells in small volumes. These devices can be used for a variety of applications such as stem cell research, drug development, and tissue engineering.
DNA and PCR analysis	These can be used to perform DNA analysis, such as PCR and gel electrophoresis, in small volumes. These devices can be used for various purposes such as forensic analysis, disease detection, and genetic research.
Drug delivery	They are found to be used for controlled drug delivery, by precisely controlling the flow of drugs in small volumes. These devices can be used for a variety of applications such as cancer therapy, pain management, and wound healing.
Lab-on-a-chip	A vast array of applications related to lab-on-a-chip can be addressed for paper microfluidics such as PCR, DNA analysis, protein analysis, and cell counting. These devices can provide a simple and cost-effective way to perform these assays in small volumes.
Point-of-care testing	Paper based microfluidic devices also shows potential in point-of-care testing, providing a simple and portable way to perform diagnostic assays in resource-limited settings. This can improve access to healthcare in low-income and rural areas, as well as in disaster and military settings.

reliable platform that has demonstrated huge potential in the development of point-of-care diagnostics particularly in resource-limited locations and developing nations.

The integration of different sensing and detection technologies with paper-based microfluidics, such as electrochemical, optical, colorimetric and piezoelectric sensors, can enhance the performance of the devices and expand their capabilities (Fu & Wang, 2018). Most studies conducted so far have utilized well-established sensing mechanisms like colorimetric and electroanalytical sensing in paper-microfluidic devices. While μPADs have proven useful, achieving low limits of detection with good selectivity has been a challenge in colorimetric detection. However, by the coupling μPADs with electrochemical transduction, otherwise termed as Microfluidic Paper-based Electroanalytical Devices (μPEDs), it is possible to achieve highly sensitive and selective detection at a low cost.

Electroanalytical detection involves the utilization of electrodes or probes to interact with chemical species and obtain electrical output in terms of voltage or current, which enables the precise evaluation of various target molecules. The process can initiate a chemical reaction by providing external electrical field across the electrode system dipped in an electrolytic cell. The former consists of three different electrodes. The working electrode is the surface of interest which is coated with suitable nano-biohybrids for the detection of suitable target species. This is accomplished by redox reactions (electro-catalysis) or binding events (DNA based and immunosensing), all of which occurs at the working electrode. The counter electrode serves primarily to complete the overall electric/electrochemical circuit. These electrodes possess high surface area and are made of inert materials (most commonly platinum), so that the electric current can essentially flow through the sink without reacting with chemical species. Meanwhile, the reference electrode functions as a support which is responsible for maintain the potential of working electrode as well as provide IR compensation, which would otherwise result in misinterpretations. The silver-silver chloride (Ag/AgCl) wire is the commonly used reference electrode. Hence, integration of paper microfluidics in electrochemical setup has led to the development of affordable, portable, and easy-to-use detection tool with improved sensitivity by selecting appropriate potential range and electrode materials (Fu & Wang, 2018). Various methods can be used to fabricate microfluidic channels on cellulose paper, as mentioned earlier. Similarly, electrodes can be created on paper using several methods such as screen-printing, wax printing, direct pencil/conductive pen writing and spray coating to name a few. Despite the availability of multiple options, the screen-printing strategy is the most commonly deployed technique for fabrication of paper electrodes.

Moreover, receptor molecules are also important component in the development of any type of biosensor for disease monitoring. These molecules have the ability to specifically bind to target of interest, and can be used to detect their presence in a sample. Commonly used receptor molecules for biosensor development include antibodies, nucleic acids/aptamers, and enzymes. Antibodies are widely used due to their high specificity and selectivity, while aptamers are a newer class of receptor molecules that are generated through an in vitro selection process and have the potential to offer improved sensitivity and stability. Enzymes can also serve as receptor molecules in biosensors, and their activity can be measured to indicate the presence of a disease biomarker. The selection of the appropriate receptor molecule is critical for the development of a sensitive and specific biosensor for disease monitoring. Nowadays, nanomaterials have become increasingly important in the development of paper-based microfluidic sensors, as they can be conveniently coated or printed onto paper. One of the key advantages of using nanoparticles is their ability to add conductive material, which can significantly enhance the electrode's chemical functionality and increase its overall surface area. There has been extensive research exploring the potential of various nanomaterials, including metal nanoparticles, metal oxides, graphene and graphene oxide, quantum dots, hybrid materials, and metal-organic frameworks, each of which has brought significant advances in µPEDs design (Guo et al., 2019; Sudhakara Prasad et al., 2020). Subsequently, the detailed applications of the recently reported µPEDs in clinical diagnosis and environmental contexts are discussed in this section.

Clinical Diagnosis

Although clinical diagnostic tests are essential for patient care, they often rely on expensive and complicated instrument-based methods, limiting their availability to centralized medical centers and hospitals. This poses a significant challenge for making these technically sophisticated tests accessible to patients. However, paper-based electrochemical microfluidic devices, which has enormous potential, can overcome this barrier and revolutionize the delivery of diagnosis to patients. Researchers have been motivated to explore and harness the potential of µPEDs in various clinically relevant applications. The versatility and convenience of µPEDs have made them an attractive option for healthcare providers, researchers, and patients alike, and they hold great promise for improving healthcare outcomes and reducing healthcare costs. µPEDs are designed to fulfill the "REASSURED" criteria, which prioritizes attributes such as easy sample collection, cost-effectiveness along with high degree of sensitivity, user-friendliness, precision, speed, durability, and adaptability to function as point-of-care settings with limited resources and equipment (Gong & Sinton, 2017). They have been utilized for detecting metabolites, cancer biomarkers, pathogens such as bacteria and viruses as shown in Figure 3.

Figure 3. Various healthcare applications of microfluidic paper-based electroanalytical devices in clinical diagnosis

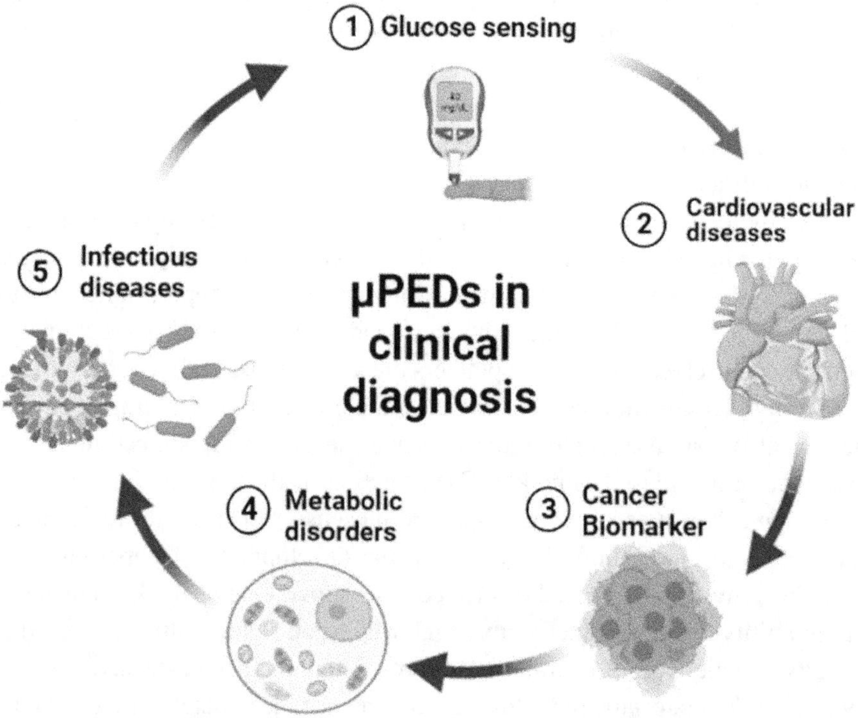

μPEDs for Metabolite Testing

Microfluidic paper-based detection methods have the potential to revolutionize clinical diagnosis for a wide range of medical conditions, including metabolic disorders. Metabolic disorders are conditions that result from abnormalities in the body's metabolic processes, leading to the accumulation or deficiency of certain metabolites. These disorders can cause a variety of symptoms and can be life-threatening if not diagnosed and treated promptly. Currently, metabolic testing is primarily performed in centralized laboratories using complex and expensive equipment. This process can take several days or even weeks, delaying the diagnosis and treatment of patients. μPEDs offer a solution to this problem by providing a portable and cost-effective alternative for metabolite testing. For example, μPEDs have been used to analyse lactate, glucose and uric acid in various biological samples, including blood, urine, and sweat. These devices have also been used to diagnose metabolic disorders such as diabetes, gout, and lactic acidosis. Over the past few years, significant efforts have been made towards the advancement of microfluidic devices for disease diagnosis, specifically that employ electrochemical detection methods (given in Table 2).

Among chronic conditions, diabetes results from inadequate insulin production, leading to long-term effects on the body. Fortunately, the advent of commercially available point-of-care devices, such as glucometer kits, has transformed the lives of many people with diabetes by enabling them to conveniently and regularly check their blood glucose levels. As a result, these devices have been widely adopted by individuals seeking a more accessible and hassle-free way to manage their diabetes (Zhu et al., 2018).

The initial μPED was designed to detect glucose, uric acid, and lactate, which are routinely evaluated metabolites in blood serum. The device consists of three separate wells, each containing a Prussian Blue (PB)-modified Screen printed carbon electrode (SPCE) and reagents that were specifically targeted to detect each metabolite. This multiplexed design allowed the device to wick samples from the inlet and accurately detect the activity of oxidase enzymes for each species (Dungchai et al., 2009).

Chaiyo et al. (2018) have developed a μPED that can non-enzymatically detect glucose in honey, white wine, and human serum. Their approach involved modifying a screen-printed carbon electrode with cobalt phthalocyanine, graphene, and an ionic liquid (CoPc/G/IL/SPCE) and the resulting electrode on μPED demonstrated good electrocatalytic activity for glucose. Cinti and colleagues have demonstrated the potential of filter paper as a versatile container for reactions (Cinti et al., 2018). By producing Prussian Blue Nanoparticles (PBNPs) on filter paper, they were able to develop a reagentless electrochemical method for glucose monitoring. The device was based on glucose oxidase and could detect glucose concentrations up to 25 mmol/L. This innovative approach holds significant promise for developing inexpensive and dispenseable glucose sensors for various applications. Monitoring the levels of ions, such as sodium, calcium, and potassium, in blood and urine is crucial for maintaining overall health. However, detecting ions of alkali and alkaline metals presents a challenge due to the need for selectivity to differentiate between them. To address this issue, an electrochemical system has been developed using wax-patterning technique. This system utilizes an ion-selective membranes to ensure selectivity and can simultaneously detect sodium ions, calcium ions, potassium ions and chloride ions in blood sample (Jin et al., 2018).

During pregnancy, the placenta secretes a hormone called human chorionic gonadotropin (HCG). HCG is important for the maintenance of the corpus luteum, which in turn produces progesterone to help support the early stages of pregnancy. HCG is often used as a marker for pregnancy, as it can be detected in the blood and urine of pregnant women. Cao et al. (2017) presented a highly sensitive immunological assay for detecting HCG on a paper-based microfluidic device. Detecting 17β-estradiol (E2) is crucial due to its vital contribution to the growth of female reproductive tissues and its significance as a clinical biomarker. Ming et al. (2019) demonstrated the detection of 17β-estradiol (E2) on μPEDs using a folding aptasensor platform and electrochemical detection.

According to a recent publication by Mathur and co-workers, a paper-based device has been developed for the electroanalytical sensing of glucose and L-tyrosine (Mathur et al., 2021). To modify the working electrodes for glucose and L-tyrosine testing, α-MnO_2-GQD/GOx and α-MnO_2-GQD/tyrosinase hybrids were used. The system worked effectively for a linear dynamic range of 50-800 mg/dL and 1-500 μM for glucose and L-tyrosine, respectively. The detection method also exhibited a detection limit of ~58 mg/dL for glucose and ~0.3 μM for L-tyrosine and exhibited a shelf life of ~ 1 month. This device's ability to detect L-tyrosine levels is especially significant as it can aid in identifying diabetic foot ulcers.

To quantify the biomarkers for oxidative and nitrative stress, specifically 8-OHdG and 3-NT, a highly precise μPED was designed using molecular imprinted polymers (MIPs) modified electrodes (Nontawong et al. (2022). The device was capable of analyzing urine and serum samples within a detection range of 0.05 μM-500.0 μM for 8-OHdG and 0.01 mM-500.0 μM for 3-NT. The device was also designed with a 3D sample well built by hydrophobic barriers created using a surface coating material i.e. penguard enamel. To fabricate the reference and working electrodes, combination of 8-OHdG-MIP/graphene-ink and Ag/AgCl ink were screen printed, respectively. Another working and counter electrode were screen printed using a graphene-based ink blended with 3NT-MIP.

Despite their potential, μPEDs for metabolite testing are still in the early stages of development. Further research is needed to optimize their sensitivity, specificity, and accuracy for clinical use. μPEDs have the potential to revolutionize metabolite testing in clinical diagnosis by providing a portable, cost-effective, and easy-to-use alternative to centralized laboratory testing. However, further research and regulatory approval are needed to realize their full potential.

μPEDs for Cancer Biomarkers

Cancer is a widespread and life-threatening disease that can affect any part of the body. Cancer cells divide uncontrollably, leading to tumor formation that can spread to other parts of the body through a process called metastasis. During this process, a circulating tumor cell breaks off from the original tumor and forms a secondary tumor in a different location through the bloodstream. Metastasis is responsible for most cancer-related deaths, emphasizing the urgent need for effective treatments against this disease. Traditional cancer diagnosis and treatment methods involve exposure to high doses of radiation or chemotherapy, which can cause many side effects. Due to the increased demand for point-of-care analysis and health monitoring, researchers have been exploring the use of μPEDs in therapeutic applications. The timely and accurate detection and measurement of cancer biomarkers play a crucial role in disease diagnosis and monitoring treatment progress.

The cost-effective, highly sensitive, and rapid testing have sparked significant interest in utilizing μPEDs for immunoassays aimed at detecting macromolecules. Zang et al. (2012) were the first to develop electrochemical immunoassays on paper-based platforms for the multiplexed testing of protein based cancer markers. Antibodies were attached to chitosan/CNT coated cellulose fibers immobilized on SPCE in order to capture carcinoma antigen 125 (CA125), alpha-fetoprotein (AFP), carcinoembryonic antigen (CEA) and carcinoma antigen 199 (CA199). The HRP-tagged antibodies were then employed for the labelling of the captured proteins. The experiment employed an overall incubation time of 4 minutes to produce sufficient 2,2'-diaminoazobenzene, and the compound was then analyzed at -0.57 V using DPV. In standard solutions, the limits of detection were 6.0 mU/mL, 0.01 ng/mL, 5.0 pg/mL and 8.0 mU/mL for CA125, AFP, CEA and CA199, respectively. In combination with nanoparticles, several alternative detection techniques or tags have been proposed to enhance electrochemical immunoassay detection. Qi and coworkers developed a movable valve microfluidic paper-based electroanalytical device (Bio-MIP-μPEDs) with in-situ MIPs to enable antibody-free biomarker analysis (Qi et al., 2019). The CEA detection range of the device was reported to be 1-500 ng/mL, with detection limit of 0.32 ng/mL. In addition to antibodies, aptamers have also been investigated for their potential use in detecting cancer biomarkers (Cunningham et al., 2014).

The level of glutathione within mammalian cells is of great significance. Abnormal concentration of blood glutathione have been found to be associated with many disorders, such as Alzheimer's, Parkinson's, diabetes, and different types of cancer. Additionally, a convenient method for detecting glutathione in blood using a disposable strip has been developed by Tomei et al. (2019). The detection process involves the thiol-disulfide exchange reaction between pre-loaded cystamine and glutathione, which is released through blood lysis. This reaction results in the production of cysteamine, a compound with excellent electrocatalytic properties due to the presence of prussian blue in the working electrode. Wei's group, fabricated a highly sensitive electrochemical biosensor using gold nanoparticles (AuNPs), reduced graphene oxide (rGO), and thionine (THI) nano composites as working electrodes (Wei et al., 2018). Zheng and colleagues developed a porous AuNP-modified paper working electrode (Au-PWE) with

Figure 4. A schematic representation of μPEDs for detection of MCF-7 cells in breast cancer

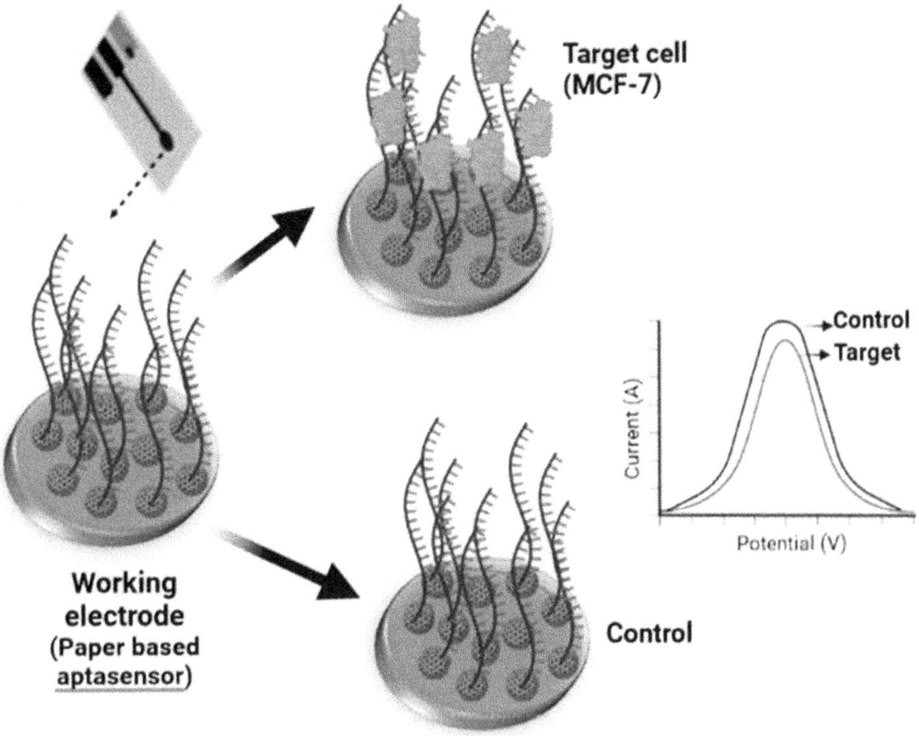

all-round conductivity and numerous active sites for biological ligand attachment (Zheng et al., 2018). The platform was successfully employed for the enzyme-free sensing of two cancer related biomarkers, i.e. CEA, and prostate-specific antigen (PSA).

Khoshroo et al. (2022) designed a paper-based microfluidic device that utilizes aptasensing for circulating tumor cells, which is critical for early detection and monitoring of breast cancers. The device was constructed using commercially available wax paper and contained three screen-printed electrodes, as shown in Figure 4. The team used a laboratory-made carbon ink, consisting of cellulose acetate and graphite, for screen-printing. They modified the reference electrode with silver and iron chloride to create an Ag/AgCl electrode, and the working electrode with gold nanoparticles. Furthermore, they immobilized an aptamer on the surface of the working electrode, which has a strong affinity for mucin 1 proteins. The device achieved a high sensitivity, detecting circulating tumor cells at concentrations as low as 7 cells/mL in serum samples, with a linear detection range of 20-106 cells/mL.

μPEDs for Virus and Bacteria

Till now, human population has faced numerous pandemics caused by a variety of viruses and bacteria that have posed significant threats to the life. Some notable examples include the Black Death, Spanish flu, Antonine Plague, and the ongoing coronavirus disease pandemic. To combat these invisible enemies, numerous research groups are dedicated to developing and enhancing methods of detection that can quickly and effectively diagnose patients with viral infections. Through their tireless efforts, these groups are working towards protecting public health and saving lives.

Kaushik and colleagues constructed an electrochemical immunosensing platform for the quick and accurate detection of Ebola virus at POC settings (Kaushik et al., 2016). The platform showed detection limit of about pmol/L concentration within 40 minutes, and was found to be highly selective for Ebola virus. In a study, Li and Liu designed a novel paper based device, which employed zinc oxide nanowires for the detection of rabbit IgG and the immunodeficiency virus p24 antigen (X. Li & Liu, 2016). The entire detection process could be completed in under 25 minutes. The biosensor achieved a LOD of 60

Table 3. Some of the clinically relevant diagnostic μPEDs for metabolite testing, cancer biomarker and virus as well as bacteria detection

Application Field	Target Analyte	Type of Sample	Detection Limit	Detection Range	Reference
Metabolite testing	Glucose	Urine	2.0 mM	2.0-20.0 mM	(Z. Li et al., 2015)
	Dopamine	Serum	0.37 μM	1–100 μM	(Rattanarat et al., 2012)
	17β-estradiol (E2)	Serum	5 pg/Ml	10 pg/mL-500 ng/mL	(Ming et al., 2019)
	17β-estradiol (E2)	Serum	10 pg/mL	0.01-100 ng/mL	(Wang et al., 2018)
	[a]HCG	Serum	0.36 mIU/mL	1.0mIU/mL-100.0 IU/mL	(Cao et al., 2017)
	5-hydroxy-l-tryptophan	Pharmaceutical sample	7.6 μM	25-1000 μM	(Canhete de Moraes et al., 2022)
Cancer biomarkers	Glutathione	Blood	60 μM	0.25-10 mM	(Tomei et al., 2019)
	[b]CEA and [c]PSA	Serum	0.002 ng/mL 0.001 ng/mL	0.005-100 ng/mL 0.002-40 ng/mL	(Zheng et al., 2018)
	[c]PSA	Serum	10 pg/mL	0.05-200 ng/mL	(Wei et al., 2018)
	[b]CEA and [d]NSE	Serum	2 pg/mL 10 pg/mL	0.01–500 ng/mL 0.05–500 ng/mL	(Wang et al., 2019)
	miRNA-155 and miRNA-21	-	25.7 nM	33.8 nM–135.3 nM	(Torul et al., 2021)
Virus and bacteria	[e]HIV p24	Serum	60 fg/mL	1 fg/mL-1 mg/mL	(X. Li & Liu, 2016)
	[e]HIV	Serum	0.13 nmol/L	10 nmol/L-1 μmol/L	(Lu et al., 2021)
	Hepatitis B and Hepatitis C	Serum	18.2 pg/mL 1.19 pg/mL	0.1–250 ng/mL 0.001–250 ng/mL	(Boonkaew et al., 2021)
	Herpes	Urine	97 copies/mL	Two ranges 113-10^3 and 3 × 10^5-10^6 copies/mL	(Narang et al., 2018)

[a]human chorionic gonadotropin;[b]carcinoembryonic antigen; [c]prostate-specific antigen; [d]neuron-specific enolase; [e]human immunodeficiency virus

fg/mL for rabbit IgG in phosphate buffered saline, and an LOD of 300 fg/mL for the p24 antigen in human serum using the μPED detection platform.

Saadati and colleagues developed a flexible μPED using a DNA modifier for detecting Haemophilus influenzae, a bacterium that causes respiratory tract infections (Saadati et al., 2021). Influenza, a highly contagious virus, has three distinct strains, and its potential to cause severe symptoms in patients underscores the need for rapid and precise detection methods. To meet this demand, researchers have developed several microfluidic systems that can scan for multiple strains of influenza simultaneously. These cutting-edge technologies enable early diagnosis and prompt treatment, offering hope for improved outcomes in patients affected by this viral infection. The device's electrodes were hand-drawn on photographic paper using Ag and Au@Ag core-shell/GQDs ink. The working electrode surface was immobilized with a DNA probe that had an affinity for cDNA (H. influenzae). The device demonstrated high sensitivity and selectivity when tested with a cDNA sample, with a linear detection range (100.0 ZM-0.1 pM) and a detection limit of 100.0 ZM.

In 2019, the outbreak of SARS-CoV-2 (severe acute respiratory syndrome coronavirus 2) gave rise to a widespread worldwide pandemic. In response to the urgent need for rapid and sensitive diagnostic tools, Yakoh and co-workers created a novel label-free paper-supported electroanalytical immunosensor for the detection of SARS-CoV-2 (Yakoh et al., 2021). This sensor was found to be highly effective in detecting SARS-CoV-2 antibodies in real serum samples from infected patients. This label-free approach has the potential to significantly enhance the accuracy and sensitivity of SARS-CoV-2 diagnosis, making it a promising tool in the ongoing battle against this global health crisis. Various other ePADs were also developed for the detection of infectious diseases like herpes (Narang et al., 2018), HIV (Lu et al., 2021), hepatitis B and C viruses (Boonkaew et al., 2021), and sars-cov-19 detection (Chaibun et al., 2021; Fabiani et al., 2021).

With the advancement of technology in microfluidic devices in recent years, some of the clinically relevant diagnostic tools have been extensively explored are given in Table 3, paving the way for exciting new possibilities in the field of electrochemical biosensing.

Environmental Monitoring

Environmental monitoring is a critical task in today's world, where pollution levels are increasing at an alarming rate. It is essential to monitor air, water, and soil quality regularly to ensure the health and well-being of living organisms. Microfluidics is a rapidly evolving field that has the potential to revolutionize the way we monitor the environment (Meredith et al., 2016). For various environmental applications, paper-based analytical devices based on electrochemical detection have surfaced as a low-cost, portable, and disposable platforms as shown in the Figure 5.

Several studies have demonstrated the use of μPADs for the detection of various environmental contaminants. Here are some examples to demonstrate the environmental applications of paper based microfluidic devices:

Water Quality Monitoring

Nowadays, water pollution is a serious global concern that affects numerous water bodies, such as rivers, oceans, lakes, and aquifers. This contamination primarily arises from human activities, involving various chemicals and pathogens. It has become critical to detect water contaminants due to the main

Figure 5. A schematic of a paper-based electrode for environmental monitoring

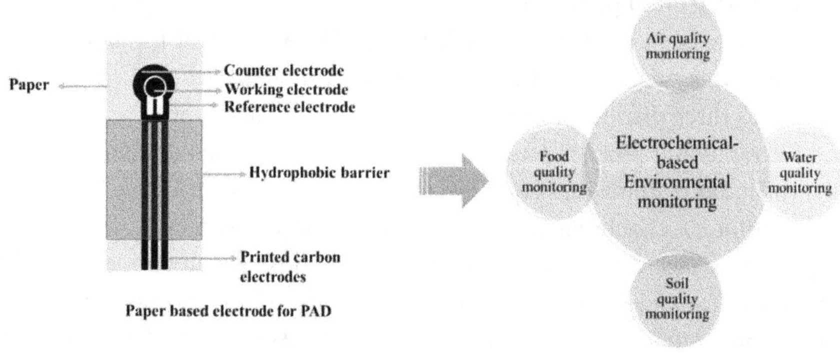

cause of sickness throughout the globe. To address this issue, many µPED platforms have been proposed for analyzing river water. These devices typically target analytes like heavy metals (mercury, lead, zinc, cobalt, chromium, copper, cadmium, silver and others. µPEDs have also been developed for the detection of various waterborne pathogens, including bacteria, viruses, and parasites. A review by Hara and Singh features the recent advances in the progress of µPEDs for water quality monitoring, involving the detection of heavy metals, pesticides, and pathogens (Hara & Singh, 2021). Although, colorimetric and electrochemical analytical techniques have been widely explored for water quality monitoring based on paper-based microfluidics. Electro-analytical detection technique offers qualitative as well as quantitative information within nanomolar range, and enables quick response.

When plastic bottles are transported improperly under sunlight, they can release Bisphenol A into drinking water, leading to contaminated water that poses health risks to both humans and the environment. Therefore, it is crucial to detect Bisphenol A in drinking water quickly and on-site to safeguard public health. Dhouha et al. (2020) developed a µPEDs which incorporates the electrochemical cell printed on filter paper along with the chemical reagents for detection in the cellulose fiber network. This method offers a reagent-free analysis using square wave voltammetry technique (SWV) for the detection of Bisphenol-A in river water. The sensor can detect Bisphenol A with a limit of detection of 0.03 µM and has two linear ranges of 0.1-0.9 µM and 1 µM-50 µM, using square wave voltammetry as the electrochemical technique. The sensor was tested on river and drinking water samples and showed satisfactory recovery values, making it a suitable tool for screening analyses in water samples. Mohan and coworkers developed a new µPED that can detect minute levels of picric acid (Mohan et al., 2021). The device was created by inkjet printing electrodes on a paper substrate using an inexpensive, highly viscous carbon-based ink. To modify the reference and working electrodes, silver/silver chloride paste and ZnO nanoparticles were used, respectively. The detection method can detect picric acid concentrations between 4.0 and 60.0 µM, having LOD of 4.04 µM which is below the recommended safe concentration i.e. 8.0 µM. The µPED was also utilized for the analysis of lake and tap water samples. Several other µPEDs have been developed for detecting contaminants in water (Colozza et al., 2021; Pokpas et al., 2022; Silva et al., 2022).

Air Quality Monitoring

Conventional air quality monitoring techniques are expensive, inaccessible and uses complex instruments which makes them cumbersome to use. Airborne pathogens such as gases and particulate matter have been detected by μPEDs. Channels are patterned on a piece of paper using wax printing or other techniques to develop μPED devices. These channels are designed to execute different chemical reactions or analyses, for a variety of applications. For air quality monitoring, the fabricated channels on paper can be used to recognize the presence of certain pollutants. These devices have demonstrated potential for the recognition of numerous air pollutants, which includes particulate matter, volatile organic compounds (VOCs), and nitrogen oxides. μPEDs demonstrate ample benefits for air quality monitoring. Park and colleagues recently reported a paper-based analytical method, having peptide-modified working electrode for the rapid and label free testing of *Bacillus subtilis* spores (Park et al., 2021). The analytical setup was constructed by printing wax hydrophobic barriers on the paper substrate, followed by stencil-printing carbon paste working/counter electrodes and Ag/AgCl reference electrodes. With excellent sensitivity and stability, the device could detect spores with a concentration as low as 6.9×10^2 CFU/mL in just 30 minutes. Furthermore, using a sampler, the team successfully tested the device on air samples, indicating that it could be customized for immediate and continuous monitoring of airborne spores. Apart from being low-cost & portable, they are easy to handle for applications and generate minimal waste, thus serving as ideal devices for remote or low-resource settings. With further research and advancement, μPEDs possess immense potential to transform air quality monitoring and enhance public health.

Soil Analysis

Massive industrialization and urbanization caused substantial soil pollution in various regions of the world. It degrades the soil fertility and impairs the nutrient and water retention capacity of soil. Therefore, soil quality monitoring is crucial for safeguarding the wellbeing of agricultural land and the vegetation grown on it. μPEDs have been demonstrated to serve for the analysis of soil samples for several environmental pollutants including heavy metals and pesticides. These devices for soil quality monitoring comprises of a paper strip which is pre-treated with specific reagents to detect the presence of specific soil contaminants. A small soil sample is added to one end of the strip folded into a channel. The sample is then mixed with a buffer solution, this mixture flows through the channel on the strip via capillary action. Numerous detection techniques have been designed for the testing of common soil contaminants, such as chlorate, explosive residues, ammonia, copper, phosphate, heavy metals, soil-transmitted helminth, and tetrabromobisphenol A by using electrochemical methods. Cioffi et al. (2021) reported an innovative electrochemical method to monitor organophosphorus pesticide levels. Unlike conventional plastic-based strips, this sensor was made with the help of regular copy paper that provided numerous benefits. The electrodes were screen-printed onto the paper, making it possible to print and alter them with biohybrid nano-sized probes. The device was also disposable, reducing waste disposal. The team tested the sensor on soil and fruit vegetable samples, performing in situ quantification of organophosphorus pesticides. This portable sensor had detection limit of 1.3 ng/mL, proving its efficacy as a tool to monitor pesticide concentrations in agricultural soils. μPEDs demonstrate a cost effective and user friendly solution for monitoring soil quality. It could be used in agricultural settings to quickly and efficiently assess soil quality, enabling farmers to take appropriate measures to improve the health of their land and crops.

Food Safety Analysis

Paper based microfluidic devices have appeared as a favorable tool for food safety analysis due to their low cost, portability, ease of use, and capability to detect many contaminants (Kuswandi et al., 2022; Nilghaz et al., 2021; Weng & Neethirajan, 2017). Some examples on the application of paper-based microfluidics for food safety analysis are:

1. **Detection of Pathogens:** Paper-based microfluidic devices are used for the analysis of pathogens such as Salmonella, E. coli, and Listeria in food samples.
2. **Detection of Toxins:** These devices can also be used for the detection of toxins like aflatoxins and mycotoxins in food samples.
3. **Detection of Allergens:** Paper-based microfluidics can also be used for detection of allergens such as peanuts and soy in food samples.
4. **Monitoring of Food Quality:** They can be used to monitor the quality of food by detecting changes in pH, glucose, and other parameters.

Although paper-based microfluidic devices offer many advantages, there are limitations to their use in certain applications when compared to traditional microfluidic devices. For example, it can be difficult to collect products or recover samples from paper substrate since the material gets absorbed into the paper fibers. Another disadvantage is that the inherent variability of paper can introduce undefined errors into applications of paper devices. Even two adjacent areas on the same piece of paper can have dramatically different structures and layouts of paper fibers. These alterations can have impact on the wicking rates, accuracy of hydrophobic barriers printed into the paper substrate, and hence the outcomes of paper-based assays. Another limitation is the short shelf life of the chemical compounds that are deposited on the paper based substrates. For e.g. enzymes gets denatured rapidly when immobilized onto the paper-based devices, which can limit the accuracy and reliability of assays. Although recent research has shown promising results regarding the shelf life of enzymes on paper, further studies are necessary to improve their longevity and stability on these devices. Ensuring specificity is crucial to avoid false positives, especially when testing complex samples from diverse environments. The presence of matrix components can interfere with the detection of specific analytes, making it necessary to develop methods to eliminate or minimize these interferences. One of the primary causes of imprecision in many PEDs is the limited and unpredictable dissolution of dry reagents deposited on the paper substrate. There are several research areas that can advance the field of paper-based microfluidics. One direction is to develop assays with detection limits comparable to conventional laboratory-based methods.

WIRELESS COMMUNICATION AND BATTERY-FREE POWER SOURCES

Additionally, there has been a growing interest in the integration of wireless communication and battery-free power sources with paper-based microfluidic devices, to create fully portable and autonomous diagnostic devices. This can greatly improve access to healthcare in low-income and rural areas, as well as in disaster and military settings. Wireless communication can be integrated into paper-based microfluidic devices by using wireless chips or sensors. These devices can be connected to a mobile phone or a computer, allowing for real-time data transmission and remote monitoring of the device. Battery-free

Table 4. Advantages of wireless communication and battery free power sources

Advantages	Description
Fully portable and autonomous diagnostic devices	The integration of wireless communication and battery-free power sources eliminates the need for external power sources and data transmission devices, allowing for the use of the device in remote locations.
Real-time monitoring	Wireless communication enables real-time data transmission, allowing for the monitoring of various parameters such as pH, temperature, and chemical concentrations in real-time.
Remote control	Wireless communication can enable remote control of the device's operation, such as the flow rate of fluids, the temperature of the device, and the timing of the device's operation.
Data sharing and analysis	Connecting the device to the cloud via wireless communication enables the sharing of data and information with other devices and with remote servers, which can be useful for data analysis and remote monitoring.
Cost-effective	Battery-free power sources eliminates the need for replaceable batteries, reducing the overall cost of the device.

power sources can also be integrated into paper-based microfluidic devices. This can be done by using energy harvesting technologies such as solar cells, piezoelectric generators, or thermoelectric generators. These technologies can convert ambient energy into electrical energy, which can be used to power the device. This eliminates the need for batteries, making the device more portable and convenient to use. The integration of wireless communication and battery-free power sources can also enable the development of smart paper-based microfluidic devices, which can communicate with other devices and share data through the internet of things (IoT).

The integration of wireless communication and battery-free power sources with paper-based microfluidic devices can also enable the development of real-time monitoring systems, remote control systems, and cloud-connected devices. These advancements can greatly enhance the accessibility and affordability of diagnostic tests, and can also have a wide range of other applications in various fields such as environmental monitoring, industrial processes and more. Various advantages of these devices are discussed in Table 4.

While integrating wireless communication and battery-free power sources with paper-based microfluidic devices can greatly enhance their capabilities and accessibility, it also poses several challenges that need to be addressed.

1. One of the main challenges is power management. The integration of wireless communication and battery-free power sources requires efficient power management to ensure that the device has enough power to operate and transmit data. This can be challenging, as the amount of power available from energy harvesting sources can be limited and unpredictable.
2. Another challenge is data security. The integration of wireless communication can open the device to potential security threats, such as hacking and data breaches. Ensuring the security of the device and the data transmitted is essential to protect sensitive information and to maintain the integrity of the device.
3. The device's reliability and robustness is also a challenge in wireless devices. Paper-based microfluidic devices are delicate and can be easily damaged, which can make it difficult to ensure the device's reliability and robustness. This is particularly important for applications in resource-limited

settings, where the device may be used in harsh conditions and may not have access to maintenance and repair services.

CONCLUSION

Paper-based microfluidics is an emerging field with a lot of potential for innovation and development. Researchers are developing new techniques for fabricating paper-based microfluidic devices, improving their performance, and expanding their capabilities. The use of paper as a substrate allows for simple and low-cost fabrication methods, and paper's biocompatibility and portability make it an attractive option for a wide range of applications. The field is multidisciplinary, with contributions from various fields such as chemistry, biology, engineering, and materials science. The use of different types of paper and other cellulose-based materials, adhesives and binders, surface modifications, and multi-layer devices are some of the areas where research is on-going and many new developments are expected in the future. However, further research is needed to address the limitations of paper-based microfluidics and to improve the performance of the devices. Furthermore, paper-based microfluidic devices have a wide range of applications including diagnostics, chemical synthesis, food safety and environmental monitoring etc. These devices are inexpensive, portable, and easy to use, making them suitable for use in resource-limited settings. The recent advances in the field of paper-based microfluidics have focused on fabrication techniques, materials, and applications. The integration of wireless communication and battery-free power sources has the potential to create fully portable and autonomous diagnostic devices, which can be used in remote locations. This can greatly enhance the accessibility and affordability of diagnostic tests for people living in rural and under-served areas. The integration can also enable the development of real-time monitoring systems, remote control systems, and cloud-connected devices. It also poses several challenges such as power management, data security, device's reliability and robustness, and cost.

Overall, the field of paper-based microfluidics is a rapidly growing area of research, with many ongoing developments and potential applications. The continued research and development efforts are expected to address the challenges and limitations of paper-based microfluidic devices and to further enhance their capabilities and accessibility.

CHALLENGES AND FUTURE OUTLOOK

Although there are various existing challenges in implementing paper microfluidics, the future outlook for this field is very promising, with many potential advancements and improvements in fabrication techniques, materials, and applications. One of the main areas of focus for future research is the development of new fabrication techniques for paper-based microfluidic devices. This includes the use of advanced manufacturing techniques such as 3D printing and microfabrication, which can enable the creation of more complex and sophisticated devices. Additionally, research is also being conducted on the use of new materials and adhesives, which can improve the biocompatibility and chemical compatibility of paper-based microfluidic devices. Another area of focus is the development of new applications for paper-based microfluidic devices. This includes the development of new diagnostic tests for a wide range of diseases, as well as the development of new chemical synthesis and environmental monitoring applications. Researchers are also working on developing new materials and adhesives for use in these devices, which can improve their biocompatibility and chemical compatibility. Wireless communication and battery-free power sources integration will be essential for the future of paper-based microfluidic devices, as it will enable them to be more portable and autonomous and can be used in remote locations. This can greatly enhance the accessibility and affordability of diagnostic tests for people living in rural and under-served areas. Additionally, it will enable the development of real-time monitoring systems, remote control systems, and cloud-connected devices.

One of the main challenges is power management in wireless communication in which battery-free power sources requires efficient power management to ensure that the device has enough power to operate and transmit data. This can be challenging, as the amount of power available from energy harvesting sources can be limited and unpredictable. Another challenge is data security. The integration of wireless communication can open the device to potential security threats, such as hacking and data breaches. Ensuring the security of the device and the data transmitted is essential to protect sensitive information and to maintain the integrity of the device. Another challenge is the device's reliability and robustness. Paper-based microfluidic devices are delicate and can be easily damaged, which can make it difficult to ensure the device's reliability and robustness. This is particularly important for applications in resource-limited settings, where the device may be used in harsh conditions and may not have access to maintenance and repair services.

Additionally, the integration of artificial intelligence and machine learning algorithms with paper-based microfluidic devices will be a promising area of research. This can enable the devices to analyze data and make decisions on their own, without the need for human intervention.

REFERENCES

Anushka, B., Bandopadhyay, A., & Das, P. K. (2022). Paper based microfluidic devices: A review of fabrication techniques and applications. *The European Physical Journal. Special Topics*. Advance online publication. doi:10.1140/epjs11734-022-00727-y PMID:36532608

Bhattacharya, S., Kumar, S., & Agarwal, A. (2019). *Paper Microfluidics- Theory and Applications*. Academic Press.

Boonkaew, S., Yakoh, A., Chuaypen, N., Tangkijvanich, P., Rengpipat, S., & Chailapakul, O. (2021). An automated fast-flow/delayed paper-based platform for the simultaneous electrochemical detection of hepatitis B virus and hepatitis C virus core antigen. *Biosensors & Bioelectronics*, *193*, 113543. doi:10.1016/j.bios.2021.113543 PMID:34416431

Canhete de Moraes, N., Marques Petroni, J., de Lima, F., Souza Ferreira, V., & Gabriel Lucca, B. (2022). Paper-based electrochemical platform modified with graphene nanoribbons: A new and affordable approach for analysis of 5-hydroxy-l-tryptophan. *Microchemical Journal*, *183*, 108030. doi:10.1016/j.microc.2022.108030

Cao, L., Fang, C., Zeng, R., Zhao, X., Jiang, Y., & Chen, Z. (2017). Paper-based microfluidic devices for electrochemical immunofiltration analysis of human chorionic gonadotropin. *Biosensors & Bioelectronics*, *92*, 87–94. doi:10.1016/j.bios.2017.02.002 PMID:28189070

Cate, D. M., Adkins, J. A., Mettakoonpitak, J., & Henry, C. S. (2015). Recent Developments in Paper-Based Microfluidic Devices. *Analytical Chemistry*, *87*(1), 19–41. doi:10.1021/ac503968p PMID:25375292

Chaibun, T., Puenpa, J., Ngamdee, T., Boonapatcharoen, N., Athamanolap, P., O'Mullane, A. P., Vongpunsawad, S., Poovorawan, Y., Lee, S. Y., & Lertanantawong, B. (2021). Rapid electrochemical detection of coronavirus SARS-CoV-2. *Nature Communications*, *12*(1), 802. doi:10.103841467-021-21121-7 PMID:33547323

Chaiyo, S., Mehmeti, E., Siangproh, W., Hoang, T. L., Nguyen, H. P., Chailapakul, O., & Kalcher, K. (2018). Non-enzymatic electrochemical detection of glucose with a disposable paper-based sensor using a cobalt phthalocyanine-ionic liquid-graphene composite. *Biosensors & Bioelectronics*, *102*, 113–120. doi:10.1016/j.bios.2017.11.015 PMID:29128713

Cinti, S., Cusenza, R., Moscone, D., & Arduini, F. (2018). Paper-based synthesis of Prussian Blue Nanoparticles for the development of whole blood glucose electrochemical biosensor. *Talanta*, *187*, 59–64. doi:10.1016/j.talanta.2018.05.015 PMID:29853066

Cioffi, A., Mancini, M., Gioia, V., & Cinti, S. (2021). Office Paper-Based Electrochemical Strips for Organophosphorus Pesticide Monitoring in Agricultural Soil. *Environmental Science & Technology*, *55*(13), 8859–8865. doi:10.1021/acs.est.1c01931 PMID:34165948

Colozza, N., Tazzioli, S., Sassolini, A., Agosta, L., di Monte, M. G., Hermansson, K., & Arduini, F. (2021). Vertical-Flow Paper Sensor for On-Site and Prompt Evaluation of Chloride Contamination in Concrete Structures. *Analytical Chemistry*, *93*(43), 14369–14374. doi:10.1021/acs.analchem.1c03363 PMID:34669396

Cunningham, J. C., Brenes, N. J., & Crooks, R. M. (2014). Paper electrochemical device for detection of DNA and thrombin by target-induced conformational switching. *Analytical Chemistry*, *86*(12), 6166–6170. doi:10.1021/ac501438y PMID:24871788

Das, S., Gagandeep, & Bhatia, R. (2022). Paper-based microfluidic devices: Fabrication, detection, and significant applications in various fields. *Reviews in Analytical Chemistry*, *41*(1), 112–136. doi:10.1515/revac-2022-0037

Dhouha, J., Marcoccio, E., Moscone, D., Dridi, C., & Arduini, F. (2020). Highly sensitive paper-based electrochemical sensor for reagent free detection of bisphenol A. *Talanta*, *216*, 120924. doi:10.1016/j.talanta.2020.120924 PMID:32456933

Dukle, A., Nathanael, A. J., Panchapakesan, B., & Oh, T. H. (2022). Role of Paper-Based Sensors in Fight against Cancer for the Developing World. *Biosensors (Basel)*, *12*(9), 737. doi:10.3390/bios12090737 PMID:36140122

Dungchai, W., Chailapakul, O., & Henry, C. S. (2009). Electrochemical Detection for Paper-Based Microfluidics. *Analytical Chemistry*, *81*(14), 5821–5826. doi:10.1021/ac9007573 PMID:19485415

Fabiani, L., Saroglia, M., Galatà, G., De Santis, R., Fillo, S., Luca, V., Faggioni, G., D'Amore, N., Regalbuto, E., Salvatori, P., Terova, G., Moscone, D., Lista, F., & Arduini, F. (2021). Magnetic beads combined with carbon black-based screen-printed electrodes for COVID-19: A reliable and miniaturized electrochemical immunosensor for SARS-CoV-2 detection in saliva. *Biosensors & Bioelectronics*, *171*, 112686. doi:10.1016/j.bios.2020.112686 PMID:33086175

Fu, L. M., & Wang, Y. N. (2018). Detection Methods and Applications of Microfluidic Paper-Based Analytical Devices. *Trends in Analytical Chemistry*, *107*, 107. doi:10.1016/j.trac.2018.08.018

Gong, M. M., & Sinton, D. (2017). Turning the Page: Advancing Paper-Based Microfluidics for Broad Diagnostic Application. *Chemical Reviews*, *117*(12), 8447–8480. doi:10.1021/acs.chemrev.7b00024 PMID:28627178

Guo, X., Zong, L., Jiao, Y., Han, Y., Zhang, X., Xu, J., Li, L., Zhang, C. W., Liu, Z., Ju, Q., Liu, J., Xu, Z., Yu, H. D., & Huang, W. (2019). Signal-Enhanced Detection of Multiplexed Cardiac Biomarkers by a Paper-Based Fluorogenic Immunodevice Integrated with Zinc Oxide Nanowires. *Analytical Chemistry*, *91*(14), 9300–9307. doi:10.1021/acs.analchem.9b02557 PMID:31241314

Hara, T. O., & Singh, B. (2021). Electrochemical Biosensors for Detection of Pesticides and Heavy Metal Toxicants in Water: Recent Trends and Progress. *ACS ES&T Water*, *1*(3), 462–478. doi:10.1021/acsestwater.0c00125

Jin, J. H., Kim, J. H., Lee, S. K., Choi, S. J., Park, C. W., & Min, N. K. (2018). A Fully Integrated Paper-Microfluidic Electrochemical Device for Simultaneous Analysis of Physiologic Blood Ions. *Sensors (Basel)*, *18*(1), 104. doi:10.339018010104 PMID:29301270

Kaushik, A., Tiwari, S., Dev Jayant, R., Marty, A., & Nair, M. (2016). Towards detection and diagnosis of Ebola virus disease at point-of-care. *Biosensors & Bioelectronics*, *75*, 254–272. doi:10.1016/j.bios.2015.08.040 PMID:26319169

Khoshroo, A., Fattahi, A., & Hosseinzadeh, L. (2022). Development of paper-based aptasensor for circulating tumor cells detection in the breast cancer. *Journal of Electroanalytical Chemistry (Lausanne, Switzerland)*, *910*, 116182. doi:10.1016/j.jelechem.2022.116182

Kirby, B. J. (2010). *Micro- and Nanoscale Fluid Mechanics: Transport in Microfluidic Devices*. Cambridge University Press.

Kuswandi, B., Hidayat, M. A., & Noviana, E. (2022). Paper-Based Electrochemical Biosensors for Food Safety Analysis. *Biosensors (Basel)*, *12*(12), 1088. doi:10.3390/bios12121088 PMID:36551055

Lammertyn, J., Verboven, P., Veraverbeke, E. A., Vermeir, S., Irudayaraj, J., & Nicolaï, B. M. (2006). Analysis of fluid flow and reaction kinetics in a flow injection analysis biosensor. *Sensors and Actuators. B, Chemical*, *114*(2), 728–736. doi:10.1016/j.snb.2005.06.050

Li, A., Wang, Y., Deng, L., Zhao, X., Yan, Q., Cai, Y., Lin, J., Bai, Y., Liu, S., & Zhang, Y. (2013). Use of nitrocellulose membranes as a scaffold in cell culture. *Cytotechnology*, *65*(1), 71–81. doi:10.100710616-012-9458-x PMID:22717658

Li, X., & Liu, X. (2016). A Microfluidic Paper-Based Origami Nanobiosensor for Label-Free, Ultrasensitive Immunoassays. *Advanced Healthcare Materials*, *5*(11), 1326–1335. doi:10.1002/adhm.201501038 PMID:27122227

Li, Z., Li, F., Hu, J., Wee, W. H., Han, Y. L., Pingguan-Murphy, B., Lu, T. J., & Xu, F. (2015). Direct writing electrodes using a ball pen for paper-based point-of-care testing. *Analyst (London)*, *140*(16), 5526–5535. doi:10.1039/C5AN00620A PMID:26079757

Lu, Q., Su, T., Shang, Z., Jin, D., Shu, Y., Xu, Q., & Hu, X. (2021). Flexible paper-based Ni-MOF composite/AuNPs/CNTs film electrode for HIV DNA detection. *Biosensors & Bioelectronics*, *184*, 113229. doi:10.1016/j.bios.2021.113229 PMID:33894427

Martinez, A. W., Phillips, S. T., Butte, M. J., & Whitesides, G. M. (2007). Patterned paper as a platform for inexpensive, low-volume, portable bioassays. *Angewandte Chemie International Edition*, *46*(8), 1318–1320. doi:10.1002/anie.200603817 PMID:17211899

Mathur, A., Nayak, H. C., Rajput, S., Roy, S., Nagabooshanam, S., Wadhwa, S., & Kumar, R. (2021). An Enzymatic Multiplexed Impedimetric Sensor Based on α-MnO_2/GQD Nano-Composite for the Detection of Diabetes and Diabetic Foot Ulcer Using Micro-Fluidic Platform. *Chemosensors (Basel, Switzerland)*, *9*(12), 339. doi:10.3390/chemosensors9120339

Mathur, A., & Roy, S. (2022). 10-Microfluidics and lab-on-a-chip. In G. Maruccio & J. Narang (Eds.), *Electrochemical Sensors* (pp. 261–287). Woodhead Publishing. doi:10.1016/B978-0-12-823148-7.00010-6

McKeen, L. W. (2016). *Permeability Properties of Plastics and Elastomers* (4th ed.).

Meredith, N. A., Quinn, C., Cate, D. M., Reilly, T. H. III, Volckens, J., & Henry, C. S. (2016). Paper-based analytical devices for environmental analysis. *Analyst (London)*, *141*(6), 1874–1887. doi:10.1039/C5AN02572A PMID:26901771

Ming, T., Wang, Y., Luo, J., Liu, J., Sun, S., Xing, Y., Xiao, G., Jin, H., & Cai, X. (2019). Folding Paper-Based Aptasensor Platform Coated with Novel Nanoassemblies for Instant and Highly Sensitive Detection of 17β-Estradiol. *ACS Sensors, 4*(12), 3186–3194. doi:10.1021/acssensors.9b01633 PMID:31775503

Mohan, J. M., Amreen, K., Kulkarni, M. B., Javed, A., Dubey, S. K., & Goel, S. (2021). Optimized ink jetted paper device for electroanalytical detection of picric acid. *Colloids and Surfaces. B, Biointerfaces, 208*, 112056. doi:10.1016/j.colsurfb.2021.112056 PMID:34425529

Narang, J., Singhal, C., Mathur, A., Sharma, S., Singla, V., & Pundir, C. S. (2018). Portable bioactive paper based genosensor incorporated with Zn-Ag nanoblooms for herpes detection at the point-of-care. *International Journal of Biological Macromolecules, 107*(Pt B), 2559-2565.

Nilghaz, A., Mousavi, S. M., Li, M., Tian, J., Cao, R., & Wang, X. (2021). Paper-based microfluidics for food safety and quality analysis. *Trends in Food Science & Technology, 118*, 273–284. doi:10.1016/j.tifs.2021.08.029

Nishat, S., Jafry, A., Martinez, A., & Awan, F. (2021). Paper-based microfluidics: Simplified fabrication and assay methods. *Sensors and Actuators. B, Chemical, 336*, 129681. doi:10.1016/j.snb.2021.129681

Nontawong, N., Ngaosri, P., Chunta, S., Jarujamrus, P., Nacapricha, D., Lieberzeit, P. A., & Amatatongchai, M. (2022). Smart sensor for assessment of oxidative/nitrative stress biomarkers using a dual-imprinted electrochemical paper-based analytical device. *Analytica Chimica Acta, 1191*, 339363. doi:10.1016/j.aca.2021.339363 PMID:35033235

Park, C., Lee, J., Lee, D., & Jang, J. (2021). Paper-based electrochemical peptide sensor for label-free and rapid detection of airborne Bacillus anthracis simulant spores. *Sensors and Actuators. B, Chemical, 355*, 131321. doi:10.1016/j.snb.2021.131321

Pokpas, K., Jahed, N., Bezuidenhout, P., Smith, S., Land, K., & Iwuoha, E. (2022). Nickel contamination analysis at cost-effective silver printed paper-based electrodes based on carbon black dimethylglyoxime ink as electrode modifier: Original scientific paper. *Journal of Electrochemical Science and Engineering, 12*(1), 153–164. doi:10.5599/jese.1173

Qi, J., Li, B., Zhou, N., Wang, X., Deng, D., Luo, L., & Chen, L. (2019). The strategy of antibody-free biomarker analysis by in-situ synthesized molecularly imprinted polymers on movable valve paper-based device. *Biosensors & Bioelectronics, 142*, 111533. doi:10.1016/j.bios.2019.111533 PMID:31377573

Rattanarat, P., Dungchai, W., Siangproh, W., Chailapakul, O., & Henry, C. S. (2012). Sodium dodecyl sulfate-modified electrochemical paper-based analytical device for determination of dopamine levels in biological samples. *Analytica Chimica Acta, 744*, 1–7. doi:10.1016/j.aca.2012.07.003 PMID:22935367

Saadati, A., Kholafazad Kordasht, H., Ehsani, M., Hasanzadeh, M., Seidi, F., & Shadjou, N. (2021). An innovative flexible and portable DNA based biodevice towards sensitive identification of Haemophilus influenzae bacterial genome: A new platform for the rapid and low cost recognition of pathogenic bacteria using point of care (POC) analysis. *Microchemical Journal, 169*, 106610. doi:10.1016/j.microc.2021.106610

Shen, L.-L., Zhang, G.-R., & Etzold, B. J. M. (2020). Paper-Based Microfluidics for Electrochemical Applications. *ChemElectroChem, 7*(1), 10–30. doi:10.1002/celc.201901495 PMID:32025468

Silva, M. K. L., Sousa, G. S., Simoes, R. P., & Cesarino, I. (2022). Fabrication of paper-based analytical devices using a PLA 3D-printed stencil for electrochemical determination of chloroquine and escitalopram. *Journal of Solid State Electrochemistry*, *26*(2), 581–586. doi:10.100710008-021-05075-w PMID:34751209

Sinha, A., Basu, M., & Chandna, P. (2022). Paper based microfluidics: A forecast toward the most affordable and rapid point-of-care devices. *Progress in Molecular Biology and Translational Science*, *186*(1), 109–158. doi:10.1016/bs.pmbts.2021.07.010 PMID:35033281

Sudhakara Prasad, K., Cao, X., Gao, N., Jin, Q., Sanjay, S. T., Henao-Pabon, G., & Li, X. (2020). A Low-Cost Nanomaterial-based Electrochemical Immunosensor on Paper for High-Sensitivity Early Detection of Pancreatic Cancer. *Sensors and Actuators. B, Chemical*, 305. PMID:32863588

Tomei, M. R., Cinti, S., Interino, N., Manovella, V., Moscone, D., & Arduini, F. (2019). Paper-based electroanalytical strip for user-friendly blood glutathione detection. *Sensors and Actuators. B, Chemical*, *294*, 294. doi:10.1016/j.snb.2019.02.082

Torul, H., Yarali, E., Eksin, E., Ganguly, A., Benson, J., Tamer, U., Papakonstantinou, P., & Erdem, A. (2021). Paper-Based Electrochemical Biosensors for Voltammetric Detection of miRNA Biomarkers Using Reduced Graphene Oxide or MoS_2 Nanosheets Decorated with Gold Nanoparticle Electrodes. *Biosensors (Basel)*, *11*(7), 236. doi:10.3390/bios11070236 PMID:34356708

Trache, D., Tarchoun, A. F., Derradji, M., Hamidon, T. S., Masruchin, N., Brosse, N., & Hussin, M. H. (2020). Nanocellulose: From Fundamentals to Advanced Applications. *Frontiers in Chemistry*, *8*, 392. doi:10.3389/fchem.2020.00392 PMID:32435633

Wang, Y., Luo, J., Liu, J., Li, X., Kong, Z., Jin, H., & Cai, X. (2018). Electrochemical integrated paper-based immunosensor modified with multi-walled carbon nanotubes nanocomposites for point-of-care testing of 17β-estradiol. *Biosensors & Bioelectronics*, *107*, 47–53. doi:10.1016/j.bios.2018.02.012 PMID:29428366

Wang, Y., Luo, J., Liu, J., Sun, S., Xiong, Y., Ma, Y., Yan, S., Yang, Y., Yin, H., & Cai, X. (2019). Label-free microfluidic paper-based electrochemical aptasensor for ultrasensitive and simultaneous multiplexed detection of cancer biomarkers. *Biosensors & Bioelectronics*, *136*, 84–90. doi:10.1016/j.bios.2019.04.032 PMID:31039491

Wei, B., Mao, K., Liu, N., Zhang, M., & Yang, Z. (2018). Graphene nanocomposites modified electrochemical aptamer sensor for rapid and highly sensitive detection of prostate specific antigen. *Biosensors & Bioelectronics*, *121*, 41–46. doi:10.1016/j.bios.2018.08.067 PMID:30196046

Weng, X., & Neethirajan, S. (2017). *Paper based microfluidic aptasensor for food safety*. Academic Press.

Yakoh, A., Pimpitak, U., Rengpipat, S., Hirankarn, N., Chailapakul, O., & Chaiyo, S. (2021). Paper-based electrochemical biosensor for diagnosing COVID-19: Detection of SARS-CoV-2 antibodies and antigen. *Biosensors & Bioelectronics*, *176*, 112912. doi:10.1016/j.bios.2020.112912 PMID:33358057

Zang, D., Ge, L., Yan, M., Song, X., & Yu, J. (2012). Electrochemical immunoassay on a 3D microfluidic paper-based device. *Chemical Communications*, *48*(39), 4683–4685. doi:10.1039/c2cc16958d PMID:22301713

Zheng, X., Li, L., Cui, K., Zhang, Y., Zhang, L., Ge, S., & Yu, J. (2018). Ultrasensitive Enzyme-free Biosensor by Coupling Cyclodextrin Functionalized Au Nanoparticles and High-Performance Au-Paper Electrode. *ACS Applied Materials & Interfaces*, *10*(4), 3333–3340. doi:10.1021/acsami.7b17037 PMID:29318878

Zhu, X., Sarwar, M., Zhu, J.-J., Zhang, C., Kaushik, A., & Li, C.-Z. (2018). Using a glucose meter to quantitatively detect disease biomarkers through a universal nanozyme integrated lateral fluidic sensing platform. *Biosensors & Bioelectronics*, 126. PMID:30544082

Chapter 7
Paper-Based Microfluidic Devices for Pesticide Detection

Hardeep Kaur
Department of Biotechnology, University Institute of Biotechnology, India

Shinar Athwal
Department of Biotechnology, University Institute of Biotechnology, India

ABSTRACT

The µ-PADs have been acknowledged as an efficient analytical platform to monitor and detect various contaminants present in the environment. Paper-based devices have always been a choice raw material for several technologies as they aid in developing biodegradable, cost-effective, and readily available units with easily driven and changeable flow rates, and the ability to store reagents. In comparison to conventional analytical systems such as capillary electrophoresis, high-performance liquid chromatography, and gas and liquid mass spectroscopy, the use of µ-PADs brings more significant results in pesticide detection. This chapter discussed the strategies that are commonly used to detect pesticides which are double-catalytic amplification, UV-induced fluorescence spectroscopy, ratiometric fluorescence imaging (based on wavelength intensities to detect variation in the native environment), electrochemical sensing, electrospray ionization, and competitive-inhibiting reaction.

INTRODUCTION

Microfluidic devices are the analytical platform that processes small proportions of fluids (unit), utilizing channels having dimensions ranging from 1-100 micrometers. Microfluidic devices have emerged from the combination of principles and technologies from various already existing areas, such as physics, and chemistry. biology, microelectronics, material science, and fluid dynamics (Niculescu et al., 2021). Since the microfluidic devices cover various domains so they can preferably be used for precise, sensitive detection of the desired target molecule in the provided sample. These microfluidic units are commonly made up of polymers (Hoehr et al., 2019), glass (Hwang et al., 2019), inorganic entities (Kane et al., 2020), silicon (Akita et al., 2020), ceramics (Liang et al., 2021), and paper (Walgama et al., 2021). Of all

DOI: 10.4018/978-1-6684-6952-1.ch007

these, paper-based microfluidic devices (μ-PADs) offer a green and sustainable approach to developing economical, biodegradable, non-toxic, and easily available analytical platforms to sense analyte (Tian et al., 2018). Compared to typical analytical systems such as capillary electrophoresis, high-performance liquid chromatography, and gas and liquid mass spectroscopy, the utilization of μ-PADs brings forth remarkable results in target detection. The capillary action in the μ-PADs allows smooth movement of sample solution through the channels along with tunable and optimized flow rate to accurately embed the particles of the sample over the paper for detection purposes (Dabbagh et al., 2021). For over a decade, μ-PADs have been used to detect analytes in several fields, including biomedical treatments (Chakraborty, Kumar, and Sen, 2018; Lim et al, 2019) environmental monitoring (Kasoju et al., 2020), and industrial operations (Nak et al., 2020). One such use of μ-PADs has been acknowledged to screen pesticides in the environment.

Pesticides have been used to eliminate unwanted weeds and other living organisms from agricultural fields to promote the growth and development of crops. However, farmers have been extensively using pesticides for the early and fast growth of crops which has ultimately led to noxious effects on humans and the environment (Richardson et al., 2019). The presence of pesticides in food materials can have a carcinogenic impact on human health while pesticides when consumed by fishes, birds, and other living organisms, can affect the food chain and in severe conditions even death (Goswami and Manna, 2021). In the past few years, the use of pesticides has been limited but still use of pesticides is prevalent in the agricultural sector which raises concern. To this μ-PADs offer a reliable platform to detect pesticides in the samples. μ-PAD units allow timely, on-site, and easily readable screening of pesticides. The nerve agents such as organophosphates and carbamate are the commonly found pesticides in food samples responsible for having noxious effects (Md Meftaul et al., 2020). The strategies utilized to sense pesticides are the DCA technique, UV-induced fluorescence, ratiometric fluorescence, electrochemical platform, electrospray ionization, and competitive inhibition. Some μ-PADs account for enzyme inhibition processes involving AchE and BChE inhibition which produces reliable signals (Khalil et al., 2018). Coupling of μ-PADs with plastic, SERS (C. Wang et al., 2021), and DLLM (Hui et al., 2019) can increase capillarity, the intensity of detected signals, and sensitivity, respectively. The introduction of smartphone analysis with μ-PADs can increase the ease of detection and consequently its employability (Alizadeh, Salimi, and Hallaj, 2018). Moreover, quantum dots hybridized μ-PADs depicted efficacy during the detection process (Kaur et al., 2021). However, a few limitations in capillary action, flow rate, sensitivity, and organic solvents can be encountered while working with paper-based analytical devices. The use of pesticides should be limited as per the guidelines but still, one or the other form of pesticide can be traced in the environment. The adverse effects of these pesticides can be limited if detected timely and with precision. So using μ-PADs accurate and early screening of pesticides is possible along with the enhancement of a sustainable approach.

STRATEGIES TO DETECT PESTICIDES

The usually used strategies to detect pesticides are double-catalytic amplification (involving two current amplification entities), UV-induced fluorescence spectroscopy (based on the absorbance of the particular pesticide), ratiometric fluorescence imaging (based on wavelength intensities to detect variation in the native environment), electrochemical sensing (involves an electrochemical reaction between target and electrolyte), electrospray ionization (electrically assisted ion transfer for the analysis of target) and competitive-inhibiting reaction (involving interrupted chemical pathway).

Double Catalytic Amplification

Dual catalytic amplification (DCA) involves two current amplification entities which significantly increases the detection efficiency of the sensor platforms. DCA technology has already been used in various fields as a sensor material to detect biomarkers (F. Li et al., 2020), glycoproteins (W. Yang et al., 2018), cancerous cells (C. Liu et al., 2021), etc. So the involvement of DCA technology in pesticide detection can open the way for precise sensing systems. A novel DCA unit involving silver-terephthalate metal-organic frameworks and Fe_3O_4 nanozyme-assisted carbon dots (Fe_3O_4/C-dots@Ag-MOFs) was used by Li et al. in the year 2021 as current amplification moieties for pesticide detection (Figure 1) (S. Li et al., 2021). The novel amplification unit specifically detected parathion-methyl an insecticide from the given sample. Hydrothermal synthesis followed by augmentation of molecularly imprinted polymer (PM as template element) can produce a stable amplification unit. The working of the catalytic unit was dependent on the PM entrapment in to the reaction region by MIP, which initiated a reduction-current stimulation at -0.53 V within a three-electrode platform present in a chip. Simultaneously catalytic interaction by Ag-MOFs and Fe_2O_3/C-dots considerably increased the signals. The dual catalytic platform can depict a LOD of about $1.16 \times 10-11$ mol L−1, >5.0% of standard deviation, and a recovery rate ranging between 82.7% and 109%. The combination between MIP and DCA remarkably enhances the selectivity and sensitivity of the chip for the potential detection of pesticides in a sample (S. Li et al., 2021). The catalytic reaction of the involved entities elevates signals thereby improvising the detection of the target molecule.

Figure 1. Graphical representation of DCA platform depicting Fe_3O_4/C-dots@Ag-MOFs, channels for samples, electrochemical workstation with three-electrodes, catalytic and reduction reactions
Source: S. Li et al. (2021)

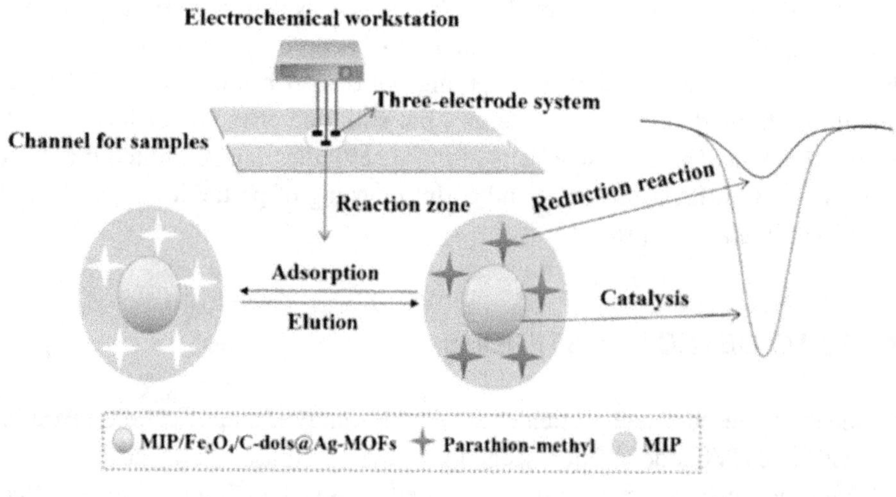

UV-Induced Fluorescence Spectroscopy

Fluorescence is a luminescence mechanism in which light is emitted by a compound that has already absorbed electromagnetic radiation or light. Fluorescence spectroscopy is primarily used to study structural variation in planar, rigid, conjugated, and aromatic compounds owing to changes in pH, ionic strength, ligands, solvents, and temperature (Gabbarini et al., 2019). The most common electromagnetic light used for fluorescence mechanism is UV light. When the sample is irradiated with UV light, then few molecules are absorbed while less-energetic radiations, i.e., visible region are re-emitted leading to a fluorescence effect (Figure 2). A detectable fluorescence takes place when emitted light lies in the visible region and absorbed light lies in the ultraviolet region, which consequently gives the fluorescent material a distinctive color that can specifically be perceived under UV exposure (Brling et al., 2011). Regarding the agricultural sector, fluorescence spectroscopy can be reliably applied as an analysis platform to detect pesticides through μ-PADs. UV-induced fingerprinting can selectively differentiate the presence of one pesticide from another as well as a pesticide molecule from the other organic compounds (Vodova et al., 2022). Emission via UV-fluorescence spectroscopy aids in specifically detecting the availability of the target body thereby increasing the chances of developing an efficacious sensing platform.

Ratiometric Fluorescence Imaging

Ratiometric fluorescent (RF) imaging is a commonly employed quantitative and visualization examination of the native microenvironment owing to its anti-interference and self-calibration ability (Bigdeli et al., 2019). In comparison to standard fluorescence intensity-dependent imaging, ratiometric fluorescence has a more positive influence such as microenvironment interference, and reducing background has no dependency on probe concentration. In ratiometric fluorescence intensities of two or more emissions/excitation, spectrum wavelengths are estimated to detect variation in the native environment (Figure 3) (L. Y. Liu et al., 2021). Alone fluorescent moieties can be utilized for rapid detection of variation in fluorescent intensity to track pesticides, however, the visual results cannot be perceived with the human eye. This integration of ratiometric fluorescence with μ-PADs presents a better analytical device with the ability to detect plant growth regulators without the use of complex external instruments. For real-time

Figure 2. The working principle of UV-based fluorescence spectroscopy. When UV-light is passed through the sample then a few particles are absorbed while a few particles are re-emitted these re-emitted particles lead to the fluorescence phenomenon.

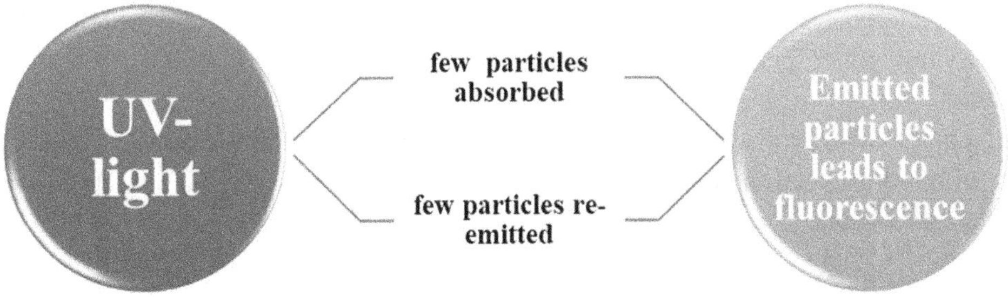

Figure 3. Basic principle of RF spectroscopy

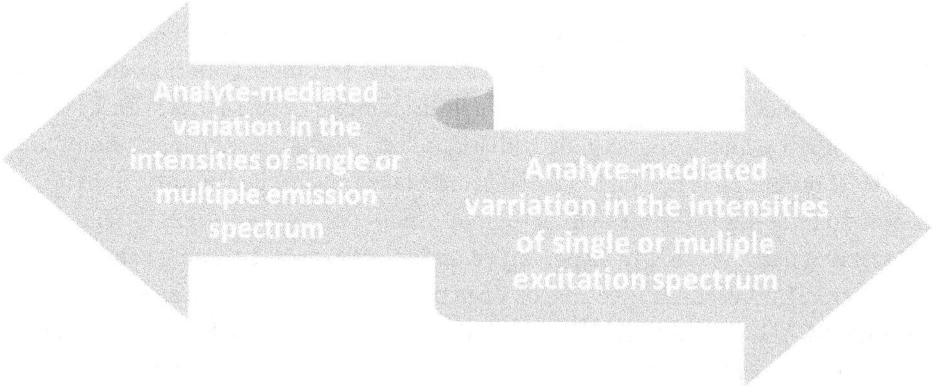

pesticide detection, Zhang et al. in the year 2019 prepared an analytical unit based on RF technology. The RF-dependent μ-PADs demonstrated a relative standard deviation (RSD) of <4.19%, LOF of 90 nM, and recovery rates ranging between 86.2% and 109.5%. RF imaging makes quantitative monitoring possible with the μ-PADs along with other analytical traits such as practical availability, good reproducibility, high precision, and fast response are also upheld by it (Z. Zhang et al., 2020). The shift in the spectra owing to a bit of variation in the local environment during RF imaging leads to sensitive detection of the target.

Electrochemical Sensing

Electrochemical strategy is the widely employed analytical approach to develop μ-PADs chips. It mainly involves an electrochemical reaction between the target and analyte (Fava et al., 2019). Scientists have been using the electrochemical approach to develop sensor units for a decade (Sunaina et al., 2022). A clinical and large-scale employable μ-PAD based on electrochemical sensing was reported by Fava et al. in the year 2019. The sensing unit consisted of a reference electrode, a working electrode, and sixteen independent channels, allowing the operation of either single or multiple electrochemical methods (Figure 4) (Fava et al., 2019). Also, the electrochemical strategy offers excellent paper-based electronics demonstrating various positive factors such as flexibility, biodegradability, portability, economical, easily available materials, and light weightiness. Jafry et al. in the year 2019 constructed a 2-D array of electrodes on a single paper sheet via electro hydrodynamic-jet printing. Additional use of silver nanoparticles permitted multilayered electrical connections to link individual electrodes (Jafry et al., 2019). The main advantage of using electrochemical sensing devices is that they require less operational time, can be used for on-site detection purposes and the reaction between target and analyte brings instantly noticeable changes which can be seen through human eyes.

Figure 4. Design of electrochemical sensing platform involving (a) working electrode and reference electrode and (b) counter electrode layers
Source: Fava et al. (2019)

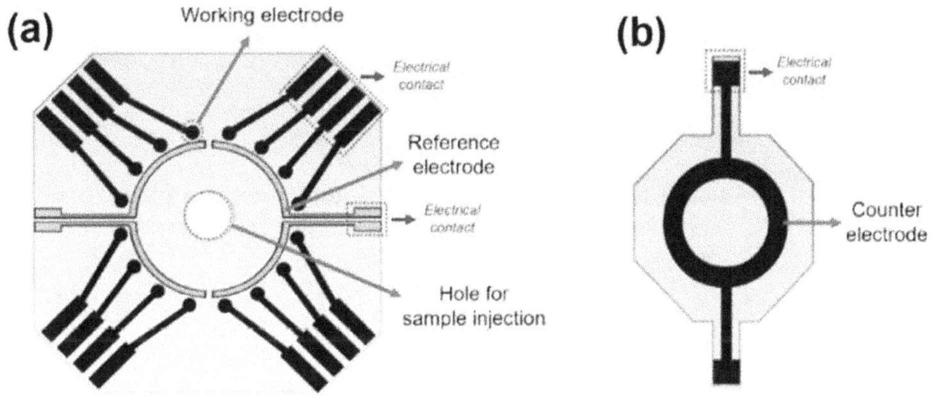

Electrospray Ionization

Electrospray ionization, i.e., ESI is based on a soft ionization approach that develops minimal fragments during the ionization process, applicable to analyze the mass of target molecules, which is practically not manageable for other standard characterization techniques. ESI-based technology utilizes electrical energy to support ion transfer from solution form to gaseous phase before mass spectroscopy (MS) analysis (Konermann et al., 2019). ESI-MS platform offers highly sensitive detection units which can be conveniently applied to detect organic molecules. Pesticide detection via ESI-MS has also gained required attention as researchers have been working on developing ESI-MS systems for efficient and selective pesticide detection (T. Chen et al., 2019). The platform comprising ESI-MS was presented by Chen et al. in the year 2017 as a possible alternative to conventional HPLC/MS unit to detect pesticides (Figure 5). In the case of ESI-MS, the aqueous sample was directly loaded on a cassette for separation with the paper chromatography technique and then supplemented with another paper chromatography for improving the sensing efficiency of ESI-MS system. ESI is a remarkable strategic approach as it can reduce the cost and time for food-safety assays and demonstrate the LOD at 1.0 ppb to detect vitamin b2 and acetaminophen and pesticides (C. H. Chen and Lin, 2017). The propensity of the ionization shown by the ESI-MS system preferably elevates the detection signals to sense the target body.

Competitive-Inhibiting Reaction

Competitive inhibition is a chemical route in which the inhibiting body attaches at the same region where the substrate is fated to attach. The molecular basis exhibited by competitive inhibitors that bind at the active region is that the inhibitor and substrate both are structural analogs, hence deceiving the enzyme in to perceiving and attaching the inhibitor (Engelking, 2015). This chemical pathway can also be applied to detect pesticides through the use of enzymes and their inhibitors. A similar approach was used by Kim et al. in the year 2018 to sense pesticides (Figure 6) (Kim et al., 2018). The working scheme involved an inhibition reaction between AChE and indoxyl acetate (IDA). When the analyte-

Figure 5. Functional process for the constructed μ-PADs for ESI-MS detection
Source: C. H. Chen and Lin (2017)

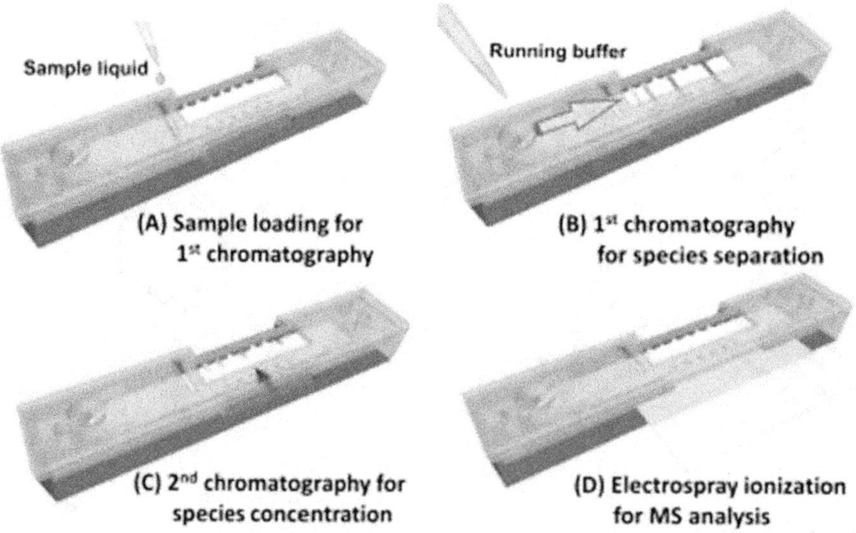

buffer solution was mixed with the competitive-inhibiting entities then the chromogenic precursor was converted in to a colored product. One indoxyl radical generated by acetylcholinesterase was oxidized and coupled with another indoxyl radical to form yellow colored 2,2′-biindoxyl which was further via competitive-inhibition reaction was converted to indigo dye (2,2′-biindoline)- 3,3′-dione. In the presence of analytes, an AChE-analyte complex is formed which hinders the IDA conversion hence depicting visible color changes and the presence of a target molecule (Md Meftaul et al. 2020). Colorimetric-based competitive-inhibiting μ-PADs devices offer a convenient, on-site, user-friendly though time-consuming analytical device for pesticide detection.

ENZYME INHIBITORY ACTIVITIES INVOLVED IN PESTICIDE MONITORING AND SENSING

Enzyme inhibition is a phenomenon in which enzyme-related processes such as enzyme activity and enzyme production are inhibited due to the attachment of the inhibitor with the enzyme. The most common enzyme inhibition employed for pesticide detection is AChE-dependent and BChE-dependent (Alsanosi, Skiffington, and Padmanabhan, 2014). Cholinesterase is the enzyme family, which can hydrolyze cholineesters. AChE is a member of cholinesterase which is present in central nerves and the peripheral nervous system, bringing about hydrolysis of neurotransmitter (acetylcholine) to terminate transmission of impulses at the cholinergic synapses (Scordo et al., 2018). BChE is another member of cholinesterase also referred to as pseudo cholinesterase is the nonspecific enzyme responsible for hydrolases of various choline esters demonstrating detoxification of noxious components involving OPs nerve agents (N. Yang et al., 2020). Enzyme inhibition produces reliable results which are easily readable however may be time-consuming as the production of color or signals is dependent on the enzyme reaction.

Figure 6. (a) Schematic representation for detecting pesticides. Blue arrows depict the expected flow of fluid on the channels of μ-PADs. (b) The phenomenon for AChEs inhibition by a sample contaminated with pesticides, with the deduced intensity of color in the exterior region. (c) The photographic image of constructed μ-PAD with pesticide (+) and (-) samples.
Source: Kim et al. (2018)

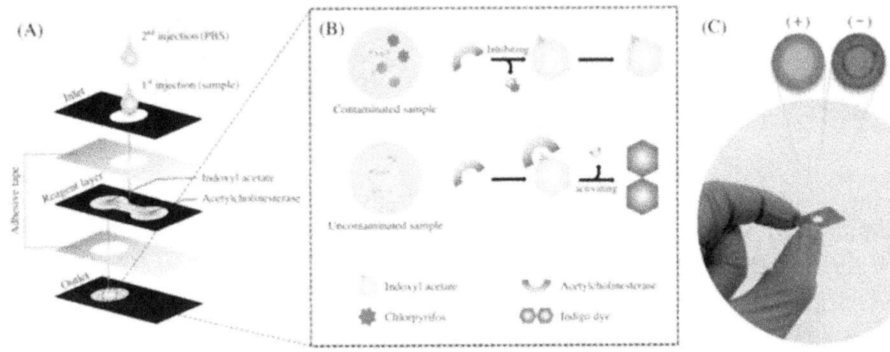

Acetylcholinesterase (AchE) Inhibition

AChE enzyme inhibition is an assay wisely used to detect pesticides via the colorimetric method. Carbamate (CM) and Organophosphate (OP) groups present in pesticides irreversibly inhibit the AChE enzyme (Singh et al., 2020). The enzyme activity is compared to the availability and unavailability of pesticides that provides pesticide concentration in the sample. Spectrophotometric assays specifically use acetylthiocholine (AcSCh) as a substrate for AChE and the resultant product, thiocholine is estimated via Ellman's method (Figure 7) (Fuyal and Giri, 2020). Time-sequence spectrum of the AChE can provide an ideal path for pesticide detection. The combination of indophenol acetate (substrate) and AChE requires only one drop of pesticide residue suspension sample to attain an enzymatic reaction. Taking into consideration the molecular aspect of AChE, it can improve the hydrolysis process of the substrate, indophenol acetate, and the subsequent addition of pesticide inhibits the enzyme activity which leads to electron transfer and variation in impedance value (Figure 7) (Luo et al., 2018). The rate of hydrolysis for indophenol acetate and AChE changes with time. The equilibrium state was reached after 15 minutes so 15 minutes can be selected as the perfect response interval (N. Yang et al., 2020). Mostly the μ-PADs detect the general presence and absence of the pesticide. However, a device that can provide the availability of specific pesticide genera can be more useful. One such concentration-detecting chip was reported by Yang et al. in the year 2019 (N. Yang et al., 2019). A seven-layered μ-PAD integrated with AChE and demonstrating chromogenic reaction identified pesticide through the spectrum of reflected light-intensity in time-sequence in accord with the various interaction efficiencies of organic pesticide and assured enzyme activity under optimum temperature (N. Yang et al., 2019). Foldable enzyme inhibition-dependent μ-PADs offer on-site rapid detection of pesticides. Lee et al. in the year 2018 reported a foldable μ-PAD having the capability to detect AChE inhibitors with a semi-quantitative method (Lee, Park, and Park, 2018). The use of an angle-based readout for a yellow region in a ring-kind paper channel unit was utilized for semi-quantitative examination via naked eyes. Paper channels indicated a color change to yellow due to the oxidation of cerium-oxide nanoparticles caused by hydrogen peroxide generated from the interactions of the enzyme (Lee, Park, and Park, 2018). Along

Figure 7. General mechanism for AChE-based enzyme inhibition involving hydrolysis of either acetylthiocholine or indophenol acetate by AchE enzyme leads to noticeable color change

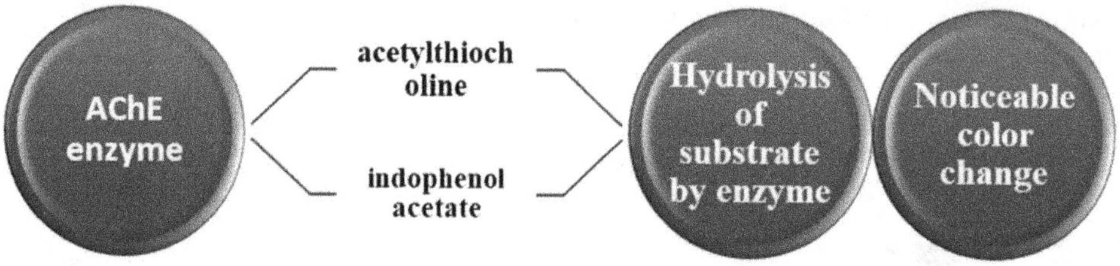

the same line, Aplux et al. combined a wet system with a paper-based sensor for quick and convenient screening of CM and OP pesticides based on AChE inhibition reaction (Apilux et al., 2015). The foldable device consisted of detection and cover sheets pre-loaded with AChE and indoxyl acetate, respectively. The decrease or absence of blue color generated through indoxyl acetate hydrolysis by AChE could be noticed in the availability of CM and OP in the given sample suspensions (Apilux et al., 2015). Further, a special quantitative paper-based detection unit dependent on AChE-assisted alginate hydrogel was developed by Xu et al. in the year 2019 to sense pesticides (Xu et al., 2019). The detection process was attained via the inhibition impact of the analyte on the AChE-catalyzed acetylcholine hydrolysis, which alters the pH value of the suspension to release calcium ions and induces alginate hydrogelation (Xu et al., 2019). Therefore, the enzyme inhibition pathway for the accurate detection of pesticides can be successfully employed on µ-PADs as inhibition reactions are capable of producing a visible color change which can easily be studied by unskilled workers and also can be effectively used for on-site determination of pesticides.

Butyrylcholinesterase Inhibition

The nerve agents irreversibly hinder the BChE enzyme which is capable to hydrolyze substrate butyrycholine, generating choline and butyric acid as products. Ferric hexacyanoferrate (Prussian blue) acts as a suitable coloring agent to detect thiocholine for it is capable of electro-catalyze the oxidation process of thiol-involving entities at low manifested potential, limiting fouling phenomenon taking place at bare carbon-electrodes (Figure 8). The enzymatic electroactive product, i.e., Thiocholine can be estimated in the availability and unavailability of nerve agent simulant utilizing screen-printed electrode enhanced with Prussian Blue (Cinti et al., 2017). Several BChE enzyme inhibition-based studies have been conducted to detect toxic nerve agents from the environmental samples (Hammoudi et al., 2021). For instance, Arduini et al. proposed a 3-D µ-PADs combined with an enzyme-inhibition sensor to detect paraoxon pesticides. The rate of inhibition related to the pesticide quantity can be estimated via chronoamperometrically analyzing the enzyme interactions in the presence or absence of a target through the use of a mobile potentiostat. The cathodic peak (Prussian blue) and anodic peak (Prussian white) obtained via cyclic voltammetry indicated the standard nature of oxidative interactions catalyzed through electrochemical assistance, depicting the suitability of 3-D-µ-PADs to detect pesticides based on thiocholine sensing (Vodova et al., 2022). In addition, grafting of BChE inhibition with potentiometric biosensing can improve the sensitivity to detect the analyte. Versatile potentiometric assays can be operated by unfolding and folding paper units. Also, the combination of paper and potentiometric assays increases

Figure 8. The hydrolysis of butyrylthiocholine substrate by BChE enzyme in presence of Prussian blue yields thiocholine by-product which aids in pesticide detection

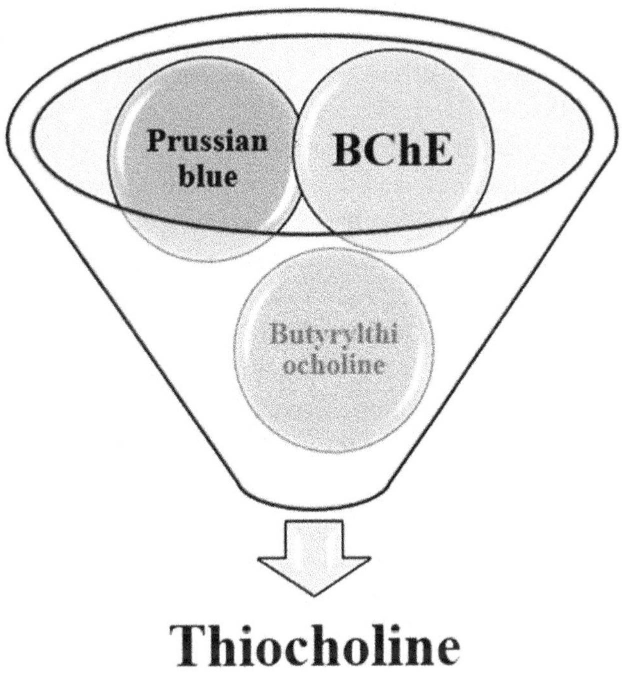

the sensitivity and precision to detect small protein molecules and enzyme-inhibition reactions (Ding et al., 2016). Though coupling the paper unit with the microfluidic platform can ease the flow rate of the sample on the matrix. Thus, BChE-based enzyme inhibition assays can perform well to sense the accurate presence of pesticide entities within the provided sample.

EXOTIC µ-PADS COUPLED PLATFORMS

Plastic

Coupling of plastic with µ-PADs increases stable flow operations as alone paper cannot promote stabilized flow owing to the presence of non-uniform cellulose fibers. In the year 2020, Jang et al. fabricated an analytical device by combining plastic and paper to detect pesticides (Jang et al., 2020). The sensor was developed by laminating-transparency film along with a double-sided adhesive and was made of an imbricating inlet-ending in the reaction region of a paper-based sensor. Integration of paper-based and lamination-based channels allowed stable flow functions as it reduces the wetting time for paper composite and promoted rapid mixing (Brling et al., 2011). Combination of plastic with the paper-enhanced pump-free capillary-driven µ-PAD platform.

Surface-Enhanced Raman Spectroscopy (SERS)

SERS is based on the enhancement effect of standard Raman scattering through the adsorption of pyridine on to copper, silver, and gold electrodes which consequently suggested an increase in the intensity of Raman scattering to about 10^4-10^6 (Q. Zhang et al., 2022). SERS can be considered as a potential candidate to couple μ-PADs to increase the efficiency of detection. Zhu et al. in the year 2017 coupled SERS with μ-PADs to detect pesticides. The device was prepared by pasting the hydrophilic region onto it. Additionally, nanoparticles were placed in one sample zone, and the target solution in another sample zone. Via capillary action, both solutions flowed through respective channels and were combined in the reaction region (J. Zhu et al., 2017). Such capillary action when combined with enhanced intensity producing SERS can lead to the detection of traces of the target molecule.

Dispersive Liquid-Liquid Microextraction (DLLM)

DLLM is the extraction method in which extraction solvent is dispersed in fine droplets within the aqueous sample. In DLLM partitioning process of analytes into the extraction form is instantaneous because of the more collective-surface area of the fine droplets. This increases the enrichment factor and reduces solvent consumption, in comparison to other solid/liquid phase extraction techniques (Tsagkaris et al., 2021). The coupling of DLLME with μ-PADs can lead to the development of a sensitive paper-based sensor system. Preconcentration via DLLME during the construction μ-PADs can preferably improve the sensitivity for trace examination of a particular pesticide. The efficacy of DLLME depends on four essential parameters which are, extraction time, NaCl amount, and the volume of acetone and n-hexane,

Table 1. Couples platforms and their properties

Coupling Platform	A Specialty of Coupling Platform	A Crucial Factor Contributed by Coupling Platform	Detection Method	Experimental Results	Ref.
Plastic	Laminating layers that increase morphology of paper-based sensor	Pump-free microfluidic sensor	Enzyme-based	linear calibration curve = 10 to 500 nmol L^{-1}	(Jang et al., 2020)
SERS	Enhancement of surface to increase the intensity of Raman scattering	Sensor system with increased intensity to detect target	Spectrophotometric analysis	LOD = 1.0×10^{-9} mol/L RSD = below 10% Recovery value = 95% to 110%	(J. Zhu et al., 2017)
DLLM	Extraction method with increased enrichment factor, and reduced solvent consumption	Preconcentration approach which increases the sensitivity of the sensor	Colorimetric method	LOD = 0.18–0.41 μg L^{-1} RSD = 2.2–6.01% Recovery value = 83–93% and 79-97% was estimated at 5 μg L^{-1} and 1 μg L^{-1} and	(Beshana et al., 2022)

and these parameters can be precisely optimized utilizing the response-surface technique (Beshana et al., 2022). A proper preconcentration method can lead to the fabrication of an efficient resultant product.

SMART SENSING μ-PADS PLATFORM

Smartphones are one of the best products of industrialization which have become part of human lifestyle over the years. The approach of the smartphone beyond communication has enhanced its utility as portable electronic equipment in this modernized era. Such portable devices have promoted smart management in fields like healthcare, food safety, environmental monitoring, and nature protection (Tong et al., 2022; Trofimchuk et al., 2020). Pesticide detection through smartphones allows an on-site, convenient, and user-friendly mode of screening. Smart platforms can be used as the primary system to detect pathogens in the provided sample before the definite long sensing lab processes (Table 2) (K. Yang et al., 2016). A novel three-layered μ-PAD chip with smartphone detection was proposed by Wu et al. in the year 2022. The biochemical variation caused via enzyme inhibition and the chromatic reaction was obtained by a smart device and analyzed via colorimetric quantitative examination (Wu et al., 2022). Presence of glyphosate a pesticide residue can have detrimental effects on living organisms and humans. An economical, selective, and fast detection platform can as a desirable sensing platform to monitor glyphosate in the environment. For the same Zhang et al. fabricated a new device that was enzyme-free and via ratiometric fluorescence, selectively sensed pesticide residues. Glyphosate when introduced undergoes aggregation-caused quenching with carbon dots fluorescence while gold nanoclusters aid in producing reference signals leading to quick ratiometric fluorescence (Q. Zhang et al., 2022). Smartphones can easily obtain the variation in color change and eventually depict the presence of the target in the practical sample (Figure 9). Using 3-D Origami Placer et al. speculated the construction of a μ-PAD device combined with a smartphone to detect iodine in consumable seaweed. Integration of methacrylate holder can improve connection in-between the layers of 3D-μ-PADs. Iodine detection is based on the formation of a triiodide-starch complex, a blue-colored composite (Placer et al., 2023).

Figure 9. Pictorial representation of mobile ratiometric fluorescent detection system for visually sensing glyphosate. Smartphones can easily obtain the variation in color change and eventually depict the presence of the target in the practical sample.
Source: Q. Zhang et al. (2022)

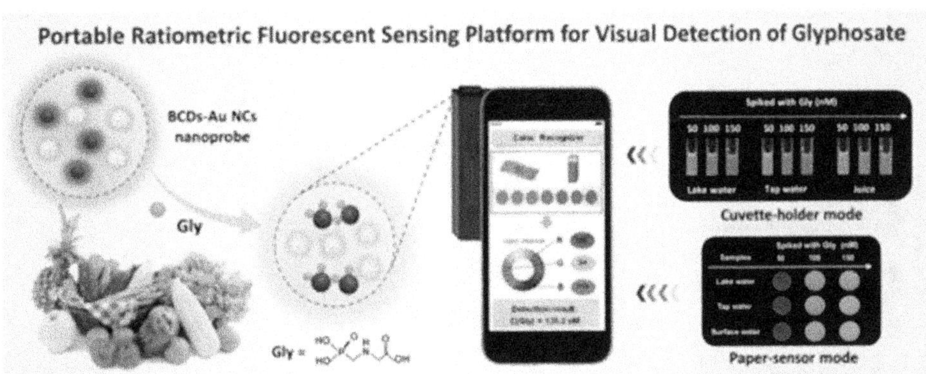

Chlorpyrifos is an organophosphate that is used as an insecticide and bears noxious effects on the environment. The first image-based sensing of chlorpyrifos was performed by Sankar et al. in the year 2020 (Sankar et al., 2020). The use of MATLAB software along with an Android smartphone can preferably enhance the selectivity for detection results generated by the sensing platform. The factors such as enzyme and substrate ratio and paper length can remarkably impact the analysis time and cost of the product (Sankar et al., 2020). With a similar approach to sense chlorpyrifos, Tsagkaris et al. constructed a device to detect residues of the pesticide in human fluids. The μ-PADs with an ability to detect exposure markers of the pesticide at an early stage can be of more value (Tsagkaris et al., 2021). A study based on the detection of cadmium ion, one of the component ions of pesticides was suggested by Attaallah et al in the year 2022 (Attaallah and Amine, 2022). The incorporation of μ-PADs with irreversible inhibition horseradish peroxidase enzyme inhibition and smartphone devices produces a cost-effective and selective platform for on-site quantitative sensing of cadmium ions (Attaallah and Amine, 2022). The enzyme inhibition reaction in the presence of target ions alters the color which can be conveniently examined through a smartphone indicating reliable results. Another study conducted by Zhou et al. to detect cadmium traced in the sample was based on coupling ZnSe quantum dots and the ion-imprinting method with a smartphone application (Zhou et al., 2020). Furthermore, the availability of antibiotics in the environment can also be an indication of pesticide contamination. Conventional detection methods for antibiotics require professionals and at times can be laborious. To this, the hybridization of μ-PADs and fluorescence with smartphones can offer a potential platform for rapid on-site detection of multi antibiotics simultaneously without requiring extra labor and time (Tong et al., 2022). The detection platform comprising of μ-PADs and coffee-ring effect with a smartphone detection system can be employed to sense norfloxacin in food samples (Trofimchuk et al., 2020). The coordination interaction resulted in color change which was obtained through a smart device and consequently examined by the colorimetric reaction.

However, most of the organic solvents are incompatible with AChE-based inhibition assays which generally reduces the efficiency of pesticide extraction in specimen pretreatment, which can lead to false negative results in real-time assessment of the sample. For this, Jin et al. proposed a novel approach

Figure 10. Schematic scheme presenting pesticide analysis by smartphone. (A) Commonly utilized platforms to detect pesticides in the provided sample, (B) data obtained on the smartphone, (C) mostly smartphone analyzed data is further detected via fluorescence or colorimetric methods.

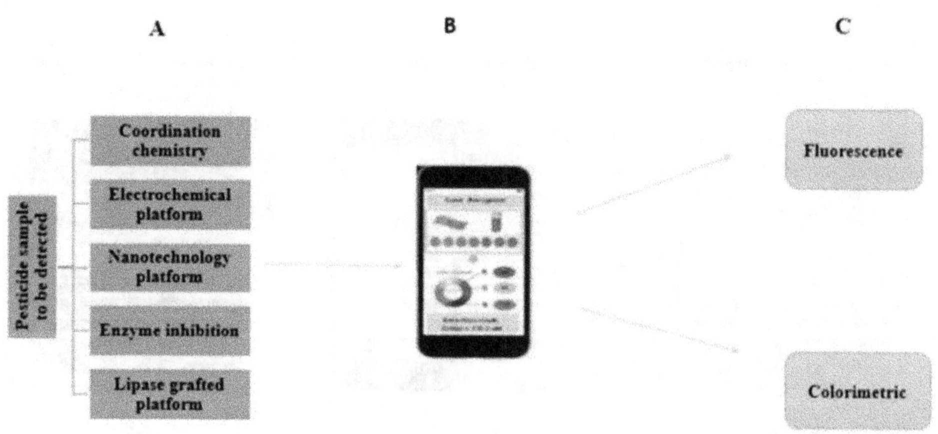

to improve AChE inhibition through the extraction of organic solvent combined with a spontaneous in-situ solvent evaporation process on μ-PAD and smartphone detection (Jin et al., 2020). As a result, smartphone-based μ-PADs promote an on-site, quick, convenient, and affordable sensing platform that can be used for the early detection of pesticides without the use of extra instrumentation.

Table 2. Smart sensing platforms and their properties

Detection of Pesticide/Ion Component	Unique Characteristic	LOD	Linear Range	Recovery Rate	Analyzing Method	References
Methomyl and Profenofo	three-layered μ-PAD chip with smartphone detection	34 nM and 55 nM	0.14–1.85 μmol L^{-1} and 0.27–2.1 μmol L^{-1}	-	Colorimetric	(Wu et al., 2022)
IO_3^- and I^-	3D-μ-PADs methacrylate holder + smart device	0.6 μM and 9.8 μM	-	90-109%	Colorimetric	(Placer et al., 2023)
Glyphosate	Integration of carbon dots, gold nanoclusters, and smartphone	4.19 nM	0 to 180 nM	-	ratiometric fluorescence	(Q. Zhang et al., 2022)
Chlorpyrifos	Augmentation of lipase	0.065 mgL^{-1}	-	-	MATLAB software	(Sankar et al., 2020)
Chlorpyrifos	Combination of BChE, wax-printing, and smartphone	10 μg L^{-1}	-	-	Colorimetric	(Tsagkaris et al., 2021)
Cadmium ions	Incorporation of horseradish peroxidase enzyme inhibition, and smartphone device	0.1 ppb	-	-	Colorimetric	(Attaallah and Amine, 2022)
Cadmium and lead ions	coupling ZnSe quantum dots and ion-imprinting method with smartphone	0.245 μg/L	1 to 70 μg/L	-	Fluorescence	(Zhou et al., 2020)
sulfamethazine, chloramphenicol, oxytetracycline,	Grafting carbon dots, nanosheets, aptamer, and smartphone	0.47, 0.34, and 0.48 ng/mL	-	95.2–101.2, 96.7–106.1, and 96.4–105%	Fluorescence	(Tong et al., 2022)
norfloxacin	Combining coordination chemistry with smartphone	5 ppm	-	-	Coffee-ring effect	(Trofimchuk et al., 2020)
OP and CAs	AchE-inhibition assay	0.02 mg L^{-1}	-	107.5±14.2%,	Colorimetric	(Jin et al., 2020)

QUANTUM DOTS HYBRIDIZATION ON μ-PADS

Quantum dots (QDs) also known as artificial atoms are dimensionally reduced semiconductors having optical and electrical importance. The nanometer-sized diameters of the semiconductor nanocrystalline demonstrate a quantum effect that consequently enhances electrical and optical properties (Cotta, 2020). QD material aids to attain core-shell structure as well as efficient and tunable photoluminescence with photochemical stability and narrow emission (Dewangan et al., 2022). Integration of QDs with μ-PADs can be a breakthrough for pesticide detection at a large-scale bringing forth precision, accuracy, and sensitivity. Fluorescence-based paper sensors have always indicated sensitive detection of the target molecule. With the use of fluorescence visualization, Wang et al. in the year 2019 detected three pesticides dimension, dichlorvos, and dimethoate through a double QDs-nanoporphyrins unit (Q. Wang et al., 2019). A novel approach involving a fluorescence sensor with molecular imprinting of silicon mesosphere on CdTe QDs can preferably sense 2, 4-dichlorophenoxyacetic acid in the provided sample (Figure 11) (Jia et al., 2017). Graphene QDs have also proved to be a potential option to detect pesticides in vegetables. Colorimetric assays when performed on graphene QDs can preferably detect chlorpyrifos pesticide. Thiocholine a product of acetylthiocholine hydrolysis via AChE causes aggregation of graphene-QDs capped with gold nanoparticles to generate purple-blue colored suspension (Chungchai et al., 2020). On the same line, Diduk et al. proposed the integration of graphene QDs over paper strips to detect phenols. The coupling of graphene and paper forms a disposable, cost-effective, rapid, and convenient sensor system. Reinforcement of nitrocellulose matrix with graphene QDs depicted the phenomenon of resonance energy transfer to sense the target (Álvarez-DIduk, Orozco, and Merkoçi, 2017). In addition, a photoelectrochemical (PEC)-molecularly imprinted mode of detection when combined with CdTe QDs can be utilized to detect S-fenvalerate through the μ-PAD sensor. (Y. Wang et al., 2013). Carbon dots are

Figure 11. Schematic description of the mechanism for the development of silicon mesosphere on CdTe QDs
Source: Jia et al. (2017)

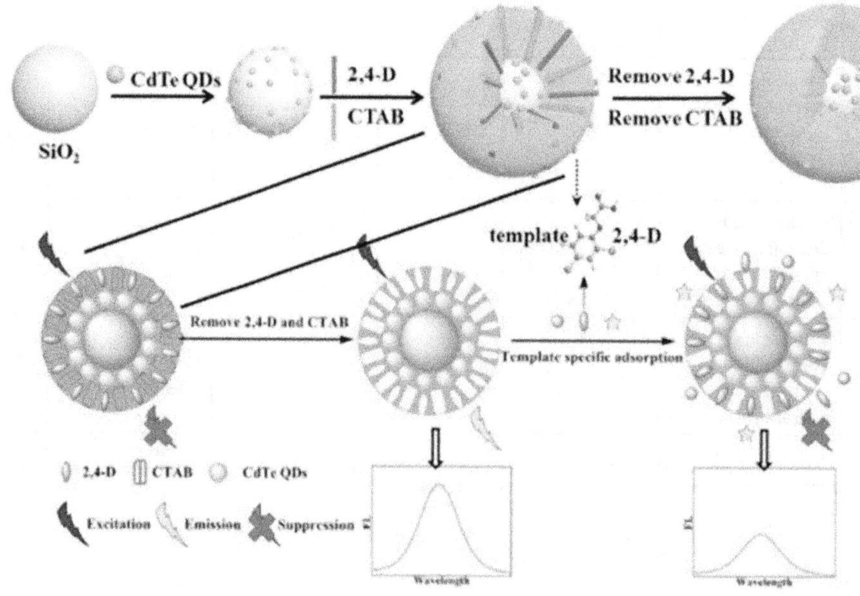

widely studied QDs due to their convenient availability and efficient electrical conductivity. With the use of carbon QDs, Yahyai et al. prepared a μ-PAD embedded with sulfur-nitrogen functionalized carbon dots to detect bendiocarb through the chemiluminescence (CL) method (Al Yahyai, Al-Lawati, and Hassanzadeh, 2021). Another preferred quantum dots unit for pesticide detection is Mn-ZnS augmented molecularly-imprinted polymers. The catalytic interaction of Mn-ZnS QDs@MIP during oxidation of 2, 2′-azino-bis (3-ethyl- benzothiazoline)-6-sulfonic acid (ABTS) by H_2O_2 aids in the detection process (Sawetwong et al. 2021).

Moreover, the functionalization of quantum dots with nitrogen can elevate the detection of p-nitroaniline. In such a sensing platform target is set as a template to construct polydopamine/N-dots imprinted substance via self-polymerization reaction of dopamine which entrapped analyte causing quenching of fluorescence intensity from the QDs present on paper (Nie et al., 2019). Additional quantum-based studies hybridized carbon dots with a molecular imprinting method (L. Zhu et al., 2022) and fluorimetric sensor (Chandra et al., 2022) to detect nitrophenol isomers and chlorpyrifos, respectively. The μ-PADs alone can have poor anti-interference capability and low accuracy under complex screening conditions limiting their clinical operations. To this carbon-dots coupling with fluorescence array and molecular imprinting can act as a possible solution (L. Zhu et al., 2022). For the first time, Khunkitchai et al. prepared ZnCd fluorescent QDs through an in-situ method within a paper matrix utilizing thiol-comprising precursor long with UV irradiation light (Khunkitchai et al., 2022). Furthermore, the combination of electrochemical detection and fluorescence sensing with CdTe/ZnS QDs can help to efficiently detect methyl paraoxon. The hydrolytic interaction of alkaline-phosphate was analyzed to generate a *p*-nitrophenol molecule in the availability of QDs (Dewangan et al., 2022).

Figure 12. Commonly utilized quantum dot platforms and their detection method

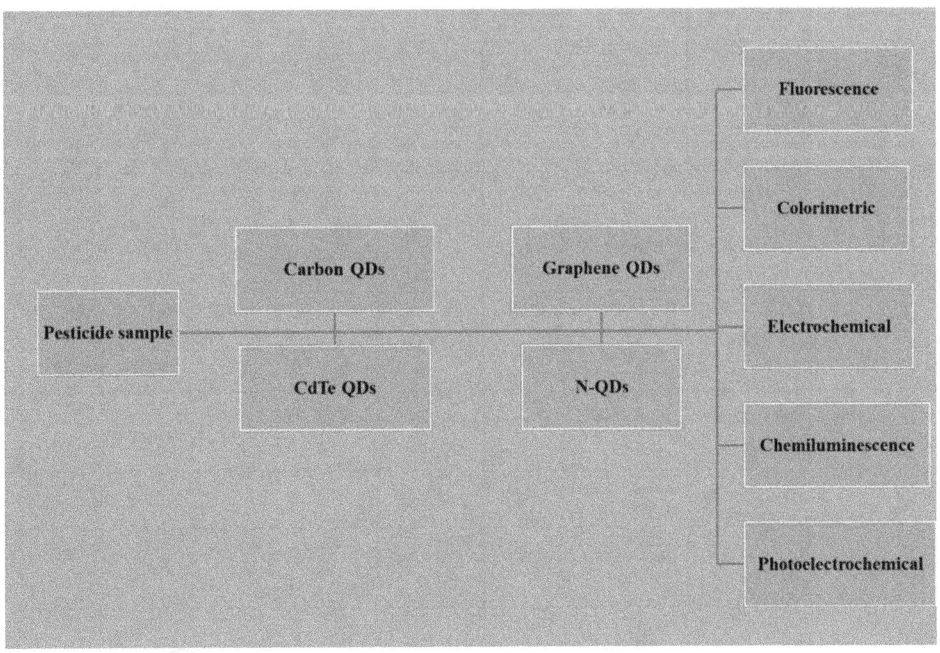

Table 3. Quantum-dots based analytical μ-PADs platform

Pesticide to Be Detected	Quantum Dot Platform	Working Principle	LOD	Other Analytical Parameters	Detection Method	Ref.
dementon, dichlorvos, and dimethoate	ZnCdSe and CdTe QDs-nanoporphyrins unit	PLSDA	-	Accuracy = 100%	Fluorescence	(Q. Wang et al., 2019)
2,4-dichlorophenoxyacetic acid	CdTe QDs	mechanism of electron-transfer-induced fluorescence quenching	2.1 nM	Linearity = 0.66–80 μM Recoveries = 95.0 to 110.1%	Fluorescence	(Jia et al., 2017)
chlorpyrifos	Graphene QDs	Thiocholine aggregation with QDs produces color.	0.0007 μg mL^{-1}	linear range = 0.001 to 1.0 μgmL^{-1} RSD = 0.3 to 1.6% Recoveries = 93.0% to 104.6%	Colorimetric	(Chungchai et al., 2020)
S-fenvalerate	CdTe QDs	Combination of standard ultraviolet lamp (~365 nm) with electrodes	3.2 x 10–9 mol L^{-1}	-	Photoelectrochemical	(Y. Wang et al., 2013)
bendiocarb	S, N-functionalized carbon dots	SO_3^{2-} and CL are oxidation by potassium permanganate	0.02 mgmL^{-1}	linear detection = 0.1–10 mgmL^{-1} recovery = 97.5–105.5%	chemiluminescence	(Al Yahyai, Al-Lawati, and Hassanzadeh, 2021)
glyphosate	Mn-ZnS QDs@MIP	oxidation of ABTS	0.002 μg mL^{-1}	operating range = 0.005–50 μg mL^{-1} RSD = 1.7%	Colorimetric	(Sawetwong et al., 2021)
OPs	CdTe/CdS	Capillary electrophoresis	50 to 180 mg/kg	RSD = 3% working ranges = 0.1–30 mg/kg recoveries = 88.7–96.1%	Laser-induced fluorescence	(Q. Chen and Fung, 2010)
phenols	graphene QDs	Resonance-energy transfer	-	Quenching = 80 ± 2% (at	Fluorescence	(Álvarez-DIduk, Orozco, and Merkoçi, 2017)
p-nitroaniline	N-QDs	Capturing of analyte leads to a quenching mechanism	1.65 nmol L^{-1}	Linear range = 5 nmol L^{-1} to 1 μmol L^{-1}	Fluorescence	(Nie et al., 2019)
Nitrophenol	Carbon-dots	cross-reactive linking affinity leading to fluorescence quenching	-	Accuracy = 100%	Fluorescence	(L. Zhu et al., 2022)
2,4-dichlorophenoxy-acetic acid	CdTe QDs	Fluorescence-resonance-energy transfer	90 nM	recovery rates 86.2% -109.5% RSD < 4.19% Linearity = 0.56–80 μM	ratiometric fluorescence	(Z. Zhang et al., 2020)
methyl paraoxon	CdTe/ZnS QDs	The hydrolytic interaction of alkaline-phosphate generated *p*-nitrophenol molecule	1.72 nM	-	electrochemical detection and fluorescence sensing	(Dewangan et al., 2022)
Thiram	Urchin-like nanocomposites	SERS technology	2.75 × 10^{-11} mol·L^{-1}	Correlation value = 1 × 10^{-10} to 1 × 10^{-7} mol·L^{-1}	Fluorescence	(H. Liu et al., 2022)
chlorpyrifos	Carbon-dots	irreversible catalytic enzyme inhibition	2.7 ng/mL	-	fluorescence	(Chandra et al., 2022)
parathion-methyl	Carbon-dots	Dual-catalytic amplification	1.16×10–11 mol L^{-1}	RSD < 5.0% recovery rates = 82.7 - 109%,	Electrochemical	(S. Li et al., 2021)

As a result, the fluorescence, colorimetric, and electrochemical signals generated via quantum dots are selective, sensitive, specific, and accurate as the electrochemical and optical properties of QDs structure enhance the detection phenomenon of the target molecule. Although a sophisticated amount of instrumentation is required to examine QDs-based results which in few cases can limit the use of QDs-based technology. But the integration of smartphones with QDs can be the best-suited solution to ease the instrumentation process.

ERROR ANALYSIS

Error analysis is an important parameter to achieve a perfect analytical platform with sensitive detection of the target analyte. All the possible errors that μ-P ADs can encounter are detection temperature, required time, chip structure, and detection wavelength. Several assays such as the Node test, Measurement test, and Global test can be utilized to decrease random error demonstrated during pesticide detection (Fava et al., 2019; Sunaina et al., 2022; N. Yang et al., 2017). Excellent chip structure contributes to developing color uniformity for optical detection so a double-channel designed chip demonstrated better detection results (Figure 13). The most suitable for pesticide detection was estimated to be 599.753 nm (Table 4). For a pesticide screening period of 1 min at 16°C served as the perfect detection criterion (N. Yang et al., 2017). Moreover, by using the mathematical formula,

$S = M - (0.0001244T2 - 0.0062T - 0.0009832) - (0.0161X - 0.251)$ (N. Yang et al., 2017)

Figure 13. A graph representing that double-channel structure demonstrated the highest optical visualization
Source: N. Yang et al. (2017)

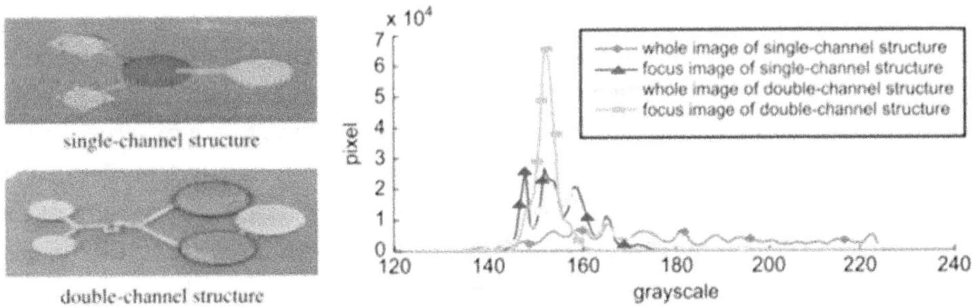

Table 4. Discussed error parameters with their best-suited value/morphology to remove the possibility of error in the fabricated μ-PADs unit

Error Parameter	Best-Suited Value/Morphology to Remove Error
Chip structure	Double-channel
Detection temperature	16°C
Required time	1 minute
Detection of error using the formula	<4.3%

Source: N. Yang et al. (2017)

Where S stands for optimized reflected-intensity, M stands for practiced reflected- intensity, T stands for preparation time, and lastly, X stands for detection temperature. Via this formula, detection error was calculated to be less than 4.3%. Therefore, by using the mathematical and arithmetic equation proposed by Yang et al. error analysis could be estimated for the constructed devices (N. Yang et al., 2017). A precise error analysis performed for a device can lead to the development of efficient μ-PADs for sensitive and selective detection of pesticides.

LIMITATION OF μ-PADS IN PESTICIDE SENSING

Though μ-PADs offer many advantages still there exist a few drawbacks such as onsite utilization, flow rate, sample retention, and capillary action are a few points which at some points hinder the commercialization applicability of μ-PADs. Onsite utilization refers to the use of an analytical platform on the ground zero of the sample to directly detect pathogens without wasting no time. Such platforms are yet under progress for large-scale utilization. The porous structure of paper hinders the flow rate of the solution which can decrease detection sensitivity. At the time it has been witnessed that sample solution evaporates from the paper surface causing a turn-off in the screening mechanism. A precise capillary action is also a point of concern during the paper-based analysis of pesticides. A possible solution to these referred problems can be the use of smartphone appliances, laminated sheets, specific organic solvents for particular samples or pesticides, and the fabrication of channels with the help of additional materials like polymers. With possible development in material science, more efficacious μ-PADs can be constructed for excellent detection of pesticides.

CONCLUSION AND PROSPECT

It can be stated that μ-PADs are efficient analytical platforms offering robust, economical, biodegradable, rapid, sensitive, precise, and selective detection of pesticides in the environment. The strategies that are generally used to sense pesticides are double-catalytic amplification which involved two current amplification entities, UV-induced fluorescence spectroscopy which is dependent on the absorbance of the particular pesticide, ratiometric fluorescence imaging which is based on wavelength intensities to detect variation in the native environment, electrochemical sensing which comprises of an electrochemical reaction between target and electrolyte, electrospray ionization electrically which includes assisted ion transfer for the analysis of target and competitive-inhibiting reaction consisting of an interrupted chemical pathway. The enzyme inhibition activities consisting of Acetylcholinesterase (a Cholinesterase enzyme responsible for neurotransmitter, acetylcholine break-down), and BChE a pseudocholinesterase nonspecific enzyme are vital for pesticide monitoring. Additionally, coupling of μ-PADs with plastic (laminating layers), SERS, and DLLM can enhance the sensitivity of paper-based devices and eventually promote the overall detection mechanism. The smartphone's incorporation with the μ-PADs increases the feasibility of easily assessable, and readable digital analysis thus offering a convenient on-site unit for sturdy and precise pesticide determination in nature. The use of quantum dots along with fluorescence spectroscopy elevates analyte detection as QDs have considerable electrical and optical characteristics with the capability to emit colored signals. The error analysis assays suggest that double-channel morphology exhibits the best detection efficiency of the μ-PADs. Although the commercial availability of μ-PADs is still under the path of development with overgrowing advancements, in the field of materials science this development will also be achieved in no time.

REFERENCES

Akita, T., Otsuka, Y., Irita, M., & Hayase, M. (2020). In-Situ Observation of Electrodeposition in a Microfluidic Device with through Silicon Via Model. *ECS Meeting Abstracts, MA2020-02*(25), 1790. 10.1149/MA2020-02251790mtgabs

Al Yahyai, I., Al-Lawati, H. A. J., & Hassanzadeh, J. (2021). A Paper-Based Chemiluminescence Detection Device Based on S,N-Doped Carbon Quantum Dots for the Selective and Highly Sensitive Recognition of Bendiocarb. *Analytical Methods*, *13*(31), 3461–3470. doi:10.1039/D1AY00728A PMID:34286731

Alizadeh, N., Salimi, A., & Hallaj, R. (2018). Mimicking Peroxidase Activity of Co2(OH)2CO3-CeO2 Nanocomposite for Smartphone Based Detection of Tumor Marker Using Paper-Based Microfluidic Immunodevice. *Talanta*, *189*, 100–110. doi:10.1016/j.talanta.2018.06.034 PMID:30086892

Alsanosi, Skiffington, & Padmanabhan. (2014). Pharmacokinetic Pharmacogenomics. In *Handbook of Pharmacogenomics and Stratified Medicine*. Academic Press. https://www.sciencedirect.com/science/article/pii/B9780123868824000177

Álvarez-DIduk, Orozco, & Merkoçi. (2017). Paper Strip-Embedded Graphene Quantum Dots: A Screening Device with a Smartphone Readout. *Scientific Reports*, *7*(1), 1–9. PMID:28127051

Apilux, A., Chartchalerm, I. N. A., Tantimongcolwat, T., & Prachayasittikul, V. (2015). Paper-Based Acetylcholinesterase Inhibition Assay Combining a Wet System for Organophosphate and Carbamate Pesticides Detection. *EXCLI Journal*, *14*, 307–319. PMID:26417364

Attaallah, R., & Amine, A. (2022). Highly Selective and Sensitive Detection of Cadmium Ions by Horseradish Peroxidase Enzyme Inhibition Using a Colorimetric Microplate Reader and Smartphone Paper-Based Analytical Device. *Microchemical Journal*, *172*, 106940. doi:10.1016/j.microc.2021.106940

Beshana, S., Hussen, A., Leta, S., & Kaneta, T. (2022). Dispersive Liquid–Liquid Microextraction Coupled with Microfluidic Paper-Based Analytical Device for the Determination of Organophosphate and Carbamate Pesticides in the Water Sample. *Analytical Sciences*, *38*(10), 1359–1367. doi:10.100744211-022-00167-7 PMID:35908131

Bigdeli, A., Ghasemi, F., Abbasi-Moayed, S., Shahrajabian, M., Fahimi-Kashani, N., Jafarinejad, S., Farahmand Nejad, M. A., & Hormozi-Nezhad, M. R. (2019). Ratiometric Fluorescent Nanoprobes for Visual Detection: Design Principles and Recent Advances - A Review. *Analytica Chimica Acta*, *1079*, 30–58. doi:10.1016/j.aca.2019.06.035 PMID:31387719

Brling, K. (2011). UV-Induced Fluorescence Spectra and Lifetime Determination for Detection of Leaf Rust (Puccinia Triticina) in Susceptible and Resistant Wheat (Triticum Aestivum) Cultivars. *Functional Plant Biology*, *38*(4), 337–345. doi:10.1071/FP10171 PMID:32480889

Chakraborty, S., Kumar, A., & Sen, P. (2018). Special Issue on Microfluidics: Theory and Applications. *Journal of the Indian Institute of Science*, *98*(2), 83–84. doi:10.100741745-018-0078-0

Chandra, S., Bano, D., Sahoo, K., Kumar, D., Kumar, V., Kumar Yadav, P., & Hadi Hasan, S. (2022). Synthesis of Fluorescent Carbon Quantum Dots from Jatropha Fruits and Their Application in Fluorometric Sensor for the Detection of Chlorpyrifos. *Microchemical Journal, 172*, 106953. doi:10.1016/j.microc.2021.106953

Chen, C. H., & Che, H. L. (2017). Mass Spectrometry Pesticide Screening with Paper-Based Microfluidic Cassette for 2D Paper Chromatography and Electrospray Ionization. *TRANSDUCERS 2017 - 19th International Conference on Solid-State Sensors, Actuators and Microsystems*, 1483–86. 10.1109/TRANSDUCERS.2017.7994339

Chen, Q., & Fung, Y. (2010). Capillary Electrophoresis with Immobilized Quantum Dot Fluorescence Detection for Rapid Determination of Organophosphorus Pesticides in Vegetables. *Electrophoresis, 31*(18), 3107–3114. doi:10.1002/elps.201000260 PMID:20872612

Chen, T., Yao, Q., Nasaruddin, R. R., & Xie, J. (2019). Electrospray Ionization Mass Spectrometry: A Powerful Platform for Noble-Metal Nanocluster Analysis. *Angewandte Chemie International Edition, 58*(35), 11967–11977. doi:10.1002/anie.201901970 PMID:30924248

Chungchai, W., Amatatongchai, M., Meelapsom, R., Seebunrueng, K., Suparsorn, S., & Jarujamrus, P. (2020). Development of a Novel Three-Dimensional Microfluidic Paper-Based Analytical Device (3D-MPAD) for Chlorpyrifos Detection Using Graphene Quantum-Dot Capped Gold Nanocomposite for Colorimetric Assay. *International Journal of Environmental Analytical Chemistry, 100*(10), 1160–1178. doi:10.1080/03067319.2019.1650921

Cinti, S., Minotti, C., Moscone, D., Palleschi, G., & Arduini, F. (2017). Fully Integrated Ready-to-Use Paper-Based Electrochemical Biosensor to Detect Nerve Agents. *Biosensors & Bioelectronics, 93*, 46–51. doi:10.1016/j.bios.2016.10.091 PMID:27839735

Cotta, M. A. (2020). Quantum Dots and Their Applications: What Lies Ahead? *ACS Applied Nano Materials, 3*(6), 4920–4924. doi:10.1021/acsanm.0c01386

Dabbagh, S. R., Becher, E., Ghaderinezhad, F., Havlucu, H., Ozcan, O., Ozkan, M., Yetisen, A. K., & Tasoglu, S. (2021). Increasing the Packing Density of Assays in Paper-Based Microfluidic Devices. *Biomicrofluidics, 15*(1), 011502. doi:10.1063/5.0042816 PMID:33569089

Dewangan, L., Korram, J., Karbhal, I., Nagwanshi, R., Ghosh, K. K., Pervez, S., & Satnami, M. L. (2022). Alkaline Phosphatase Immobilized CdTe/ZnS Quantum Dots for Dual-Purpose Fluorescent and Electrochemical Detection of Methyl Paraoxon. *Industrial & Engineering Chemistry Research, 61*(10), 3636–3646. doi:10.1021/acs.iecr.1c05065

Ding, J., Li, B., Chen, L., & Qin, W. (2016). A Three-Dimensional Origami Paper-Based Device for Potentiometric Biosensing. *Angewandte Chemie International Edition, 55*(42), 13033–13037. doi:10.1002/anie.201606268 PMID:27634584

Engelking, L. R. (2015). Enzyme Kinetics. In *Textbook of Veterinary Physiological Chemistry* (3rd ed.). Academic Press. https://www.sciencedirect.com/science/article/pii/B9780123919090500062

Fava, E. L., Silva, T. A., Prado, T. M., Moraes, F. C., Faria, R. C., & Fatibello-Filho, O. (2019). Electrochemical Paper-Based Microfluidic Device for High Throughput Multiplexed Analysis. *Talanta, 203*(May), 280–286. doi:10.1016/j.talanta.2019.05.081 PMID:31202339

Fuyal, M., & Giri, B. (2020). A Combined System of Paper Device and Portable Spectrometer for the Detection of Pesticide Residues. *Food Analytical Methods, 13*(7), 1492–1502. doi:10.100712161-020-01770-y

Gabbarini, V., Rossi, R., Ciparisse, J.-F., Malizia, A., Divizia, A., De Filippis, P., Anselmi, M., Carestia, M., Palombi, L., Divizia, M., & Gaudio, P. (2019). Laser-Induced Fluorescence (LIF) as a Smart Method for Fast Environmental Virological Analyses: Validation on Picornaviruses. *Scientific Reports, 9*(1), 3–9. doi:10.103841598-019-49005-3 PMID:31467322

Goswami, S., & Manna, K. (2021). Organochlorine Pesticide Residues and Other Toxic Substances in Salted Tenualosa Ilisha L.: Northeastern Part of India. *Foods and Raw Materials, 9*(2), 201–206. doi:10.21603/2308-4057-2021-2-201-206

Hammoudi, N.-E.-H., Sobhi, W., Attoui, A., Lemaoui, T., Erto, A., & Benguerba, Y. (2021). In Silico Drug Discovery of Acetylcholinesterase and Butyrylcholinesterase Enzymes Inhibitors Based on Quantitative Structure-Activity Relationship (QSAR) and Drug-Likeness Evaluation. *Journal of Molecular Structure, 1229*, 129845. doi:10.1016/j.molstruc.2020.129845

Hoehr, L., Bavendiek, J., Sackmann, J., & Schomburg, W. K. (2019). Ultrasonic Welding of Polymer Micro Fluidic Devices by Inserting Metal Parts. *Microsystem Technologies, 25*(2), 673–681. doi:10.100700542-018-3995-y

Hui, Y., Liu, Tang, Song, Madou, Xia, & Wu. (2019). Determination of Mercury(II) on a Centrifugal Microfluidic Device Using Ionic Liquid Dispersive Liquid-Liquid Microextraction. *Micromachines, 10*(8), 523. doi:10.3390/mi10080523 PMID:31398936

Hwang, J., Cho, Y. H., Park, M. S., & Kim, B. H. (2019). Microchannel Fabrication on Glass Materials for Microfluidic Devices. *International Journal of Precision Engineering and Manufacturing, 20*(3), 479–495. doi:10.100712541-019-00103-2

Jafry, A. T., Lee, H., Tenggara, A. P., Lim, H., Moon, Y., Kim, S.-H., Lee, Y., Kim, S.-M., Park, S., Byun, D., & Lee, J. (2019). Double-Sided Electrohydrodynamic Jet Printing of Two-Dimensional Electrode Array in Paper-Based Digital Microfluidics. *Sensors and Actuators. B, Chemical, 282*, 831–837. doi:10.1016/j.snb.2018.11.135

Jang, I., Carrão, D. B., Menger, R. F., Moraes de Oliveira, A. R., & Henry, C. S. (2020). Pump-Free Microfluidic Rapid Mixer Combined with a Paper-Based Channel. *ACS Sensors, 5*(7), 2230–2238. doi:10.1021/acssensors.0c00937 PMID:32583663

Jia, M., Zhang, Z., Li, J., Shao, H., Chen, L., & Yang, X. (2017). A Molecular Imprinting Fluorescence Sensor Based on Quantum Dots and a Mesoporous Structure for Selective and Sensitive Detection of 2,4-Dichlorophenoxyacetic Acid. *Sensors and Actuators. B, Chemical, 252*, 934–943. doi:10.1016/j.snb.2017.06.090

Jin, L., Hao, Z., Zheng, Q., Chen, H., Zhu, L., Wang, C., Liu, X., & Lu, C. (2020). A Facile Microfluidic Paper-Based Analytical Device for Acetylcholinesterase Inhibition Assay Utilizing Organic Solvent Extraction in Rapid Detection of Pesticide Residues in Food. *Analytica Chimica Acta, 1100*, 215–224. doi:10.1016/j.aca.2019.11.067 PMID:31987143

Kane. (2020). Passive Controlled Flow for Parkinson's Disease Neuronal Cell Culture in 3D Microfluidic Devices. *Organs-on-a-Chip, 2*. https://www.sciencedirect.com/science/article/pii/S2666102020300057

Kasoju, A., Shahdeo, D., Khan, A. A., Shrikrishna, N. S., Mahari, S., Alanazi, A. M., Bhat, M. A., Giri, J., & Gandhi, S. (2020). Author Correction: Fabrication of Microfluidic Device for Aflatoxin M1 Detection in Milk Samples with Specific Aptamers. *Scientific Reports, 10*(1), 9222. doi:10.103841598-020-64241-8 PMID:32494019

Kaur, H. (2021). Effect of Synthesis Methods and Conditions on Properties and Applications of Carbon Dots for the Detection of Potential Water Contaminants: A Review. *Critical Reviews in Analytical Chemistry*, 1–24. https://www.tandfonline.com/doi/abs/10.1080/10408347.2021.1977608 PMID:34605318

Khalil, A. T., Ayaz, M., Ovais, M., Wadood, A., Ali, M., Shinwari, Z. K., & Maaza, M. (2018). In Vitro Cholinesterase Enzymes Inhibitory Potential and in Silico Molecular Docking Studies of Biogenic Metal Oxides Nanoparticles. *Inorganic and Nano-Metal Chemistry, 48*(9), 441–448. doi:10.1080/24701556.2019.1569686

Khunkitchai, N., Nuchtavorn, N., Rypar, T., Vlcnovska, M., Nejdl, L., Vaculovicova, M., & Macka, M. (2022). UV-Light-Actuated in-Situ Preparation of Paper@ZnCd Quantum Dots for Paper-Based Enzymatic Nanoreactors. *Chemical Engineering Journal, 428*, 132508. doi:10.1016/j.cej.2021.132508

Kim, H. J., Kim, Y., Park, S. J., Kwon, C., & Noh, H. (2018). Development of Colorimetric Paper Sensor for Pesticide Detection Using Competitive-Inhibiting Reaction. *Biochip Journal, 12*(4), 326–331. doi:10.100713206-018-2404-z

Konermann, L., Metwally, H., Duez, Q., & Peters, I. (2019). Charging and Supercharging of Proteins for Mass Spectrometry: Recent Insights into the Mechanisms of Electrospray Ionization. *Analyst (London), 144*(21), 6157–6171. doi:10.1039/C9AN01201J PMID:31560020

Lee, S., Park, J., & Park, J. K. (2018). Foldable Paper-Based Analytical Device for the Detection of an Acetylcholinesterase Inhibitor Using an Angle-Based Readout. *Sensors and Actuators. B, Chemical, 273*, 322–327. doi:10.1016/j.snb.2018.06.059

Li, F., Guo, L., Hu, Y., Li, Z., Liu, J., He, J., & Cui, H. (2020). Multiplexed Chemiluminescence Determination of Three Acute Myocardial Infarction Biomarkers Based on Microfluidic Paper-Based Immunodevice Dual Amplified by Multifunctionalized Gold Nanoparticles. *Talanta, 207*, 120346. doi:10.1016/j.talanta.2019.120346 PMID:31594588

Li, S., Pang, C., Ma, X., Zhang, Y., Xu, Z., Li, J., Zhang, M., & Wang, M. (2021). Microfluidic Paper-Based Chip for Parathion-Methyl Detection Based on a Double Catalytic Amplification Strategy. *Mikrochimica Acta, 188*(12), 438. doi:10.100700604-021-05084-6 PMID:34839414

Liang, S., Wang, C., Li, Y., Xu, M., & Jia, H. (2021). Microfluidic Fabrication of ZrO2 Microspheres Using Improved External Gelation Aided by Polyacrylamide Networks. *Ceramics International, 47*(15), 21576–21581. doi:10.1016/j.ceramint.2021.04.169

Lim, W. Y., Thevarajah, T. M., Goh, B. T., & Khor, S. M. (2019). Corrigendum to: 'Paper Microfluidic Device for Early Diagnosis and Prognosis of Acute Myocardial Infarction via Quantitative Multiplex Cardiac Biomarker Detection' [Biosensors and Bioelectronics 128 (2019) 176-185]. *Biosensors & Bioelectronics, 129*, 299. doi:10.1016/j.bios.2019.01.036 PMID:30736959

Liu, C., Xu, T., Cheng, G., & Zhang, X. (2021). Target-Triggered Regioselective Assembly of Nanoprobes for Raman Imaging of Dual Cancer Biomarkers in Living Cells. *Sensors and Actuators. B, Chemical, 330*, 129319. doi:10.1016/j.snb.2020.129319

Liu, H., Zhao, P., Xiu, W., Zhang, L., Zhu, P., Ge, S., & Yu, J. (2022). SERS Paper Slip Based on 3D Dendritic Gold Nanomaterials Coupling with Urchin-like Nanoparticles for Rapid Detection of Thiram. *Sensors and Actuators. B, Chemical, 355*, 131264. doi:10.1016/j.snb.2021.131264

Liu, L. Y., Zhao, Y., Zhang, N., Wang, K.-N., Tian, M., Pan, Q., & Lin, W. (2021). Ratiometric Fluorescence Imaging for the Distribution of Nucleic Acid Content in Living Cells and Human Tissue Sections. *Analytical Chemistry, 93*(3), 1612–1619. doi:10.1021/acs.analchem.0c04064 PMID:33381958

Luo, R., Feng, Z., Shen, G., Xiu, Y., Zhou, Y., Niu, X., & Wang, H. (2018). Acetylcholinesterase Biosensor Based On Mesoporous Hollow Carbon Spheres/Core-Shell Magnetic Nanoparticles-Modified Electrode for the Detection of Organophosphorus Pesticides. *Sensors (Basel), 18*(12), 4429. doi:10.339018124429 PMID:30558201

Meftaul, M. (2020). Pesticides in the Urban Environment: A Potential Threat That Knocks at the Door. *Science of The Total Environment, 711*, 134612. https://www.sciencedirect.com/science/article/pii/S0048969719346030

Nak, H., Gurkan, I., Cebeci, H., & Ergene, A. F. (2020). Fabrication and Characterization of a Microfluidic Device with Vertically Aligned Multi Walled Carbon Nanotube Channels. *IFAC-PapersOnLine, 53*(2), 11761–11766. doi:10.1016/j.ifacol.2020.12.682

Niculescu, A. G., Chircov, C., Bîrcă, A. C., & Grumezescu, A. M. (2021). Fabrication and Applications of Microfluidic Devices: A Review. *International Journal of Molecular Sciences, 22*(4), 1–26. doi:10.3390/ijms22042011 PMID:33670545

Nie, Y., Liu, Y., Su, X., & Ma, Q. (2019). Nitrogen-Rich Quantum Dots-Based Fluorescence Molecularly Imprinted Paper Strip for p-Nitroaniline Detection. *Microchemical Journal, 148*, 162–168. doi:10.1016/j.microc.2019.04.080

Placer, L., Lavilla, I., Pena-Pereira, F., & Bendicho, C. (2023). A 3D Microfluidic Paper-Based Analytical Device with Smartphone-Assisted Colorimetric Detection for Iodine Speciation in Seaweed Samples. *Sensors and Actuators. B, Chemical, 377*, 133109. doi:10.1016/j.snb.2022.133109

Richardson, J. R., Fitsanakis, V., Westerink, R. H. S., & Kanthasamy, A. G. (2019). Neurotoxicity of Pesticides. *Acta Neuropathologica, 138*(3), 343–362. doi:10.100700401-019-02033-9 PMID:31197504

Sankar, K., Lenisha, D., Janaki, G., Juliana, J., Kumar, R. S., Selvi, M. C., & Srinivasan, G. (2020). Digital Image-Based Quantification of Chlorpyrifos in Water Samples Using a Lipase Embedded Paper Based Device. *Talanta, 208*, 120408. doi:10.1016/j.talanta.2019.120408 PMID:31816733

Sawetwong, P., Chairam, S., Jarujamrus, P., & Amatatongchai, M. (2020, December). 2021. "Enhanced Selectivity and Sensitivity for Colorimetric Determination of Glyphosate Using Mn–ZnS Quantum Dot Embedded Molecularly Imprinted Polymers Combined with a 3D-Microfluidic Paper-Based Analytical Device." *Talanta, 225*, 122077. doi:10.1016/j.talanta.2020.122077 PMID:33592801

Scordo, G., Moscone, D., Palleschi, G., & Arduini, F. (2018). A Reagent-Free Paper-Based Sensor Embedded in a 3D Printing Device for Cholinesterase Activity Measurement in Serum. *Sensors and Actuators. B, Chemical, 258*, 1015–1021. doi:10.1016/j.snb.2017.11.134

Singh, A. P., Balayan, S., Hooda, V., Sarin, R. K., & Chauhan, N. (2020). Nano-Interface Driven Electrochemical Sensor for Pesticides Detection Based on the Acetylcholinesterase Enzyme Inhibition. *International Journal of Biological Macromolecules, 164*, 3943–3952. doi:10.1016/j.ijbiomac.2020.08.215 PMID:32882280

Sunaina, Kaur, H., Kumari, N., Sharma, A., Sachdeva, M., & Mutreja, V. (2022). Optical and Electrochemical Microfluidic Sensors for Water Contaminants: A Short Review. *Materials Today: Proceedings, 48*, 1673–1679. doi:10.1016/j.matpr.2021.09.534

Tian, T., Bi, Y., Xu, X., Zhu, Z., & Yang, C. (2018). Integrated Paper-Based Microfluidic Devices for Point-of-Care Testing. *Analytical Methods, 10*(29), 3567–3581. doi:10.1039/C8AY00864G

Tong, X., Lin, X., Duan, N., Wang, Z., & Wu, S. (2022). Laser-Printed Paper-Based Microfluidic Chip Based on a Multicolor Fluorescence Carbon Dot Biosensor for Visual Determination of Multiantibiotics in Aquatic Products. *ACS Sensors, 7*(12), 3947–3955. Advance online publication. doi:10.1021/acssensors.2c02008 PMID:36454704

Trofimchuk, E., Nilghaz, A., Sun, S., & Lu, X. (2020). Determination of Norfloxacin Residues in Foods by Exploiting the Coffee-Ring Effect and Paper-Based Microfluidics Device Coupling with Smartphone-Based Detection. *Journal of Food Science, 85*(3), 736–743. doi:10.1111/1750-3841.15039 PMID:32017096

Tsagkaris, A. S., Migliorelli, D., Uttl, L., Filippini, D., Pulkrabova, J., & Hajslova, J. (2021). A Microfluidic Paper-Based Analytical Device (MPAD) with Smartphone Readout for Chlorpyrifos-Oxon Screening in Human Serum. *Talanta, 222*, 121535. doi:10.1016/j.talanta.2020.121535 PMID:33167243

Vodova, M., Nejdl, L., Pavelicova, K., Zemankova, K., Rrypar, T., Skopalova Sterbova, D., Bezdekova, J., Nuchtavorn, N., Macka, M., Adam, V., & Vaculovicova, M. (2022). Detection of Pesticides in Food Products Using Paper-Based Devices by UV-Induced Fluorescence Spectroscopy Combined with Molecularly Imprinted Polymers. *Food Chemistry, 380*, 132141. https://www.sciencedirect.com/science/article/pii/S0308814622001029. doi:10.1016/j.foodchem.2022.132141 PMID:35101791

Walgama, C., Pollok, N., Boatner, L., & Crooks, R. (2021). A Hybrid Paper and 3D Printed Microfluidic Platform for Electrochemical Detection of Ag-Labeled Magnetic Particles. *ECS Meeting Abstracts, MA2021-02*(57), 1858. 10.1149/MA2021-02571858mtgabs

Wang, C., Mu, X., Huo, J., Zhang, B., & Zhang, K. (2021). Highly-Efficient SERS Detection for E. Coli Using a Microfluidic Chip with Integrated NaYF4:Yb,Er@SiO2@Au under near-Infrared Laser Excitation. *Microsystem Technologies*, *27*(9), 3285–3291. doi:10.100700542-020-05087-y

Wang, Q., Yin, Q., Fan, Y., Zhang, L., Xu, Y., Hu, O., Guo, X., Shi, Q., Fu, H., & She, Y. (2019). Double Quantum Dots-Nanoporphyrin Fluorescence-Visualized Paper-Based Sensors for Detecting Organophosphorus Pesticides. *Talanta*, *199*(February), 46–53. doi:10.1016/j.talanta.2019.02.023 PMID:30952284

Wang, Y., Zang, D., Ge, S., Ge, L., Yu, J., & Yan, M. (2013). A Novel Microfluidic Origami Photoelectrochemical Sensor Based on CdTe Quantum Dots Modified Molecularly Imprinted Polymer and Its Highly Selective Detection of S-Fenvalerate. *Electrochimica Acta*, *107*, 147–154. doi:10.1016/j.electacta.2013.05.154

Wu, H., Chen, J., Yang, Y., Yu, W., Chen, Y., Lin, P., & Liang, K. (2022). Smartphone-Coupled Three-Layered Paper-Based Microfluidic Chips Demonstrating Stereoscopic Capillary-Driven Fluid Transport towards Colorimetric Detection of Pesticides. *Analytical and Bioanalytical Chemistry*, *414*(5), 1759–1772. doi:10.100700216-021-03839-x PMID:35059790

Xu, J., Hu, X., Khan, H., Tian, M., & Yang, L. (2019). Converting Solution Viscosity to Distance-Readout on Paper Substrates Based on Enzyme-Mediated Alginate Hydrogelation: Quantitative Determination of Organophosphorus Pesticides. *Analytica Chimica Acta*, *1071*, 1–7. doi:10.1016/j.aca.2019.04.017 PMID:31128750

Yang, K., Peretz-Soroka, H., Liu, Y., & Lin, F. (2016). Novel Developments in Mobile Sensing Based on the Integration of Microfluidic Devices and Smartphones. *Lab on a Chip*, *16*(6), 943–958. doi:10.1039/C5LC01524C PMID:26899264

Yang, N., Shaheen, N., Xie, L., Yu, J., Ahmad, H., & Mao, H. (2019). Pesticide residues identification by optical spectrum in the time-sequence of enzyme inhibitors performed on microfluidic paper-based analytical devices (μPADs). *Molecules (Basel, Switzerland)*, *24*(13), 2428. doi:10.3390/molecules24132428 PMID:31269660

Yang, N., Shen, K., Guo, J., Tao, X., Xu, P., & Mao, H. (2017). Error Analysis for Pesticide Detection Performed on Paper-Based Microfluidic Chip Devices. *Modern Physics Letters B*, *31*(19–21), 1–6. doi:10.1142/S0217984917400243

Yang, N., Zhou, X., Yu, D., Jiao, S., Han, X., Zhang, S., Yin, H., & Mao, H. (2020). Pesticide residues identification by impedance time-sequence spectrum of enzyme inhibition on multilayer paper-based microfluidic chip. *Journal of Food Process Engineering*, *43*(12), e13544. doi:10.1111/jfpe.13544

Yang, W., Zhou, X., Zhao, J., & Xu, W. (2018). A Cascade Amplification Strategy of Catalytic Hairpin Assembly and Hybridization Chain Reaction for the Sensitive Fluorescent Assay of the Model Protein Carcinoembryonic Antigen. *Mikrochimica Acta*, *185*(2), 100. doi:10.100700604-017-2620-6 PMID:29594400

Zhang, Q., Zhang, Z., Xu, S., Da, L., Lin, D., & Jiang, C. (2022). Enzyme-Free and Rapid Visual Quantitative Detection for Pesticide Residues Utilizing Portable Smartphone Integrated Paper Sensor. *Journal of Hazardous Materials*, *436*, 129320. doi:10.1016/j.jhazmat.2022.129320 PMID:35739808

Zhang, Z., Ma, X., Li, B., Zhao, J., Qi, J., Hao, G., Jianhui, R., & Yang, X. (2020). Fluorescence Detection of 2,4-Dichlorophenoxyacetic Acid by Ratiometric Fluorescence Imaging on Paper-Based Microfluidic Chips. *Analyst (London)*, *145*(3), 963–974. doi:10.1039/C9AN01798D PMID:31828260

Zhou, J., Li, B., Qi, A., Shi, Y., Qi, J., Xu, H., & Chen, L. (2020). ZnSe Quantum Dot Based Ion Imprinting Technology for Fluorescence Detecting Cadmium and Lead Ions on a Three-Dimensional Rotary Paper-Based Microfluidic Chip. *Sensors and Actuators. B, Chemical*, *305*, 305. doi:10.1016/j.snb.2019.127462

Zhu, J., Chen, Q., Kutsanedzie, F. Y. H., Yang, M., Ouyang, Q., & Jiang, H. (2017). Highly Sensitive and Label-Free Determination of Thiram Residue Using Surface-Enhanced Raman Spectroscopy (SERS) Coupled with Paper-Based Microfluidics. *Analytical Methods*, *9*(43), 6186–6193. doi:10.1039/C7AY01637A

Zhu, L., Mei, X., Peng, Z., Yang, J., & Li, Y. (2022). A Paper-Based Microfluidic Sensor Array Combining Molecular Imprinting Technology and Carbon Quantum Dots for the Discrimination of Nitrophenol Isomers. *Journal of Hazardous Materials*, *435*, 129012. doi:10.1016/j.jhazmat.2022.129012 PMID:35504132

Chapter 8
Application of MEMS Sensors in IoT-Integrated Frameworks

Lipsa Das
Amity University, Noida, India

Ayushi Thakur
Amity University, Noida, India

Pooja Anand
Amity University, Noida, India

ABSTRACT

Future IoT-based hardware may greatly benefit from the potential of MEMS sensors. MEMS sensors are perfect for use in smaller spaces due to their small size, low cost, and low power requirements, which makes them suitable for IoT applications. These sensors have a high degree of sensitivity, sufficient dependability, and perfect accuracy, which ensures that your data is always accurate. MEMS technology is transforming how many things are produced. The high tolerances present in semiconductor process technology enable them to perform with excellent repeatability and astounding accuracy. The most current development in the electronics industry, nanotechnology, is used to design and create advanced sensors. These sensors offer a better degree of precision and accuracy than ever before because of nanotechnology. It has uses in every sphere of social and economic life and ranges in size from a few millimetres to a few nanometers. Consequently, MEMS sensors and IoT systems are growing in popularity as a result of their capacity to do effective and balanced work across a range of industries.

INTRODUCTION

The Internet of Things is in charge of the internet's future. The environment is becoming more beneficial and interconnected in our daily lives. The possibilities are unlimited because there are billions of gadgets connected to the internet. Our interactions with the world around us are altering as a result of the IoT. Individuals who make use of internet of things can lead smarter lives, do better work, and have complete

DOI: 10.4018/978-1-6684-6952-1.ch008

control of their lives. The environment is becoming more beneficial and interconnected in our daily lives. The word "clever" is increasingly frequently used these days. Everything in the environment can now have a digital identity in the online world thanks to the Internet of Things (IOT). Smart objects include the following features: Data collection from the environment, Data transfer, and Self-identification. In an Internet of Things (IoT) world, we might use integrated devices to connect everyday objects to the internet, allowing convenient interaction between people, processes, and objects like thermostats, cars, kitchen appliances, and baby monitors. IoT, or the "Internet of Things," is a framework that unites web-enabled items or users with distinctive identities and has the capability to receive/transfer information across the internet with or without human participation (Lawton, 2015).

IoT devices combine the use of software, sensors, and other technologies on a single platform to carry out the desired outcome. IoT has many uses, including automation, healthcare, traffic management, autonomous vehicles, smart homes and appliances, smart irrigation systems, disaster avoidance, tourism, local businesses, and a lot more (Gubbi et al., 2013). The sensor is a highly important component of an IoT device because it helps develop solutions. Applications for the Internet of Things (IoT) using a range of sensors, including those for pressure, humidity, proximity, and acceleration, are urgently needed. Thermal sensors are used by IOT applications for a variety of purposes, including wearable technology, temperature sensors, and human intruder alarm systems. These IoT application devices are constructed from capacitive sensing components. These multi-sensor IoT application devices must feature an energy-efficient capacitive circuit due to the limited budget. Figure 1 illustrates several IoT device applications across many industries.

The way we connect with the environment around us is being revolutionised by IoT and MEMS sensor technology. MEMS sensors provide for increasingly precise and reliable measurements, enabling ever-sophisticated interactions between devices and their environment. MEMS sensors are essential to IoT because they link systems and equipment to the internet, enabling real-time data collection and analysis. The potential for these technologies' future development are virtually limitless.

The Internet of Things (IoT) connects many sensing gadgets to give them intelligence and decision-making autonomy. MEMS provide a number of benefits, including as low weight, low cost, high power, and small size. These traits help IOT get physical data from which to connect with the internet and the object. Since their invention, sensors have enabled data collection in a wide range of conditions. Various sensor applications using the IoT paradigm are shown in Figure 2.

IoT-based solutions are in high demand in today's technological world, and this demand is now fueling the growth of organisations. The Internet of Things (IoT) uses a variety of sensors, including those that sense proximity, humidity, pressure (Zhang et al., 2014), and acceleration. These Internet of Things application devices also have capacitive sensing components. The ability to perceive a variety of stimuli, including pressure, temperature, mechanical deformation, and pH, makes these smart materials useful as responsive materials. Such smart materials are suitable for usage in such IoT-based sensors and equipment due to their shape-changing and self-healing properties. One of the main objectives in this scenario is the accessibility of sensors with the characteristics of compact size, low energy consumption, and cost - effectiveness. These requirements are naturally met by MEMS (Micro-Electro-Mechanical System) technologies, which allow the incorporation of multiple sensors into a single chip (SoC) or a single package (SiP), along with challenging electronic linkages. However, with the development of MEMS technology, current MEMS sensors have significantly improved in terms of performance-based power and cost. Nowadays, the semiconductor industries are moving to create and manufacture high-quality sensors that are small and consume less power. The major metrics, including resolution,

Figure 1. IoT-based devices and technologies

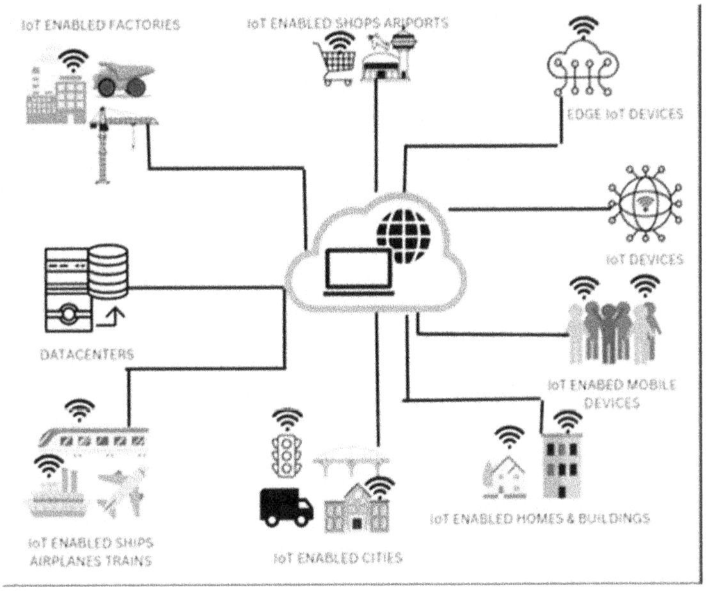

Figure 2. Application of sensors with IoT paradigm

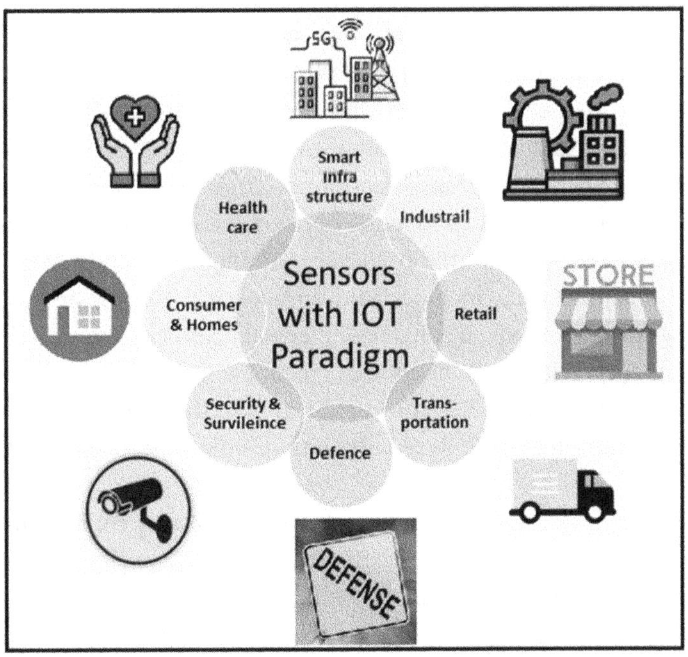

repeatability, dependability, selectivity, power consumption, and miniaturisation, are the main points of emphasis. The use of conventional sensors will be less prevalent as a result of the development of such technologies in the field of sensors.

Future IoT-based hardware may greatly benefit from the potential of MEMS sensors (Atzori et al., 2010). MEMS sensors are perfect for use in smaller spaces due to their small size, low cost, and low power requirements, which makes them suitable for IoT applications. These sensors have a high degree of sensitivity, sufficient dependability, and perfect accuracy, which ensures that your data is always accurate. MEMS technology is transforming how many things are produced. The high tolerances present in semiconductor process technology enable them to perform with excellent repeatability and astounding accuracy.

The most current development in the electronics industry, nanotechnology, is used to design and create advanced sensors. These sensors offer a better degree of precision and accuracy than ever before because to nanotechnology. It has uses in every sphere of social and economic life and ranges in size from a few millimetres to a few nanometers. Consequently, MEMS sensors and IoT systems are growing in popularity as a result of their capacity to do effective and balanced work across a range of industries. Their outcomes are shown to be proficient in numerous engineering sectors and other application areas.

INTRODUCTION OF MEMS

With MEMS, both electrical and mechanical elements can be combined to create small, interconnected gadgets or systems. Its dimensions range from a few micrometres to millimetres, and is produced utilising ICs that is integrated circuits unit production procedures. These things (or systems) are capable of microscale sensing, control, and actuation as well as macroscale effects production. MEMS are micromechanical devices with integrated mechanical and electrical components. One of the most promising technologies for the future has been named as MEMS. It has countless potential applications, and it will undoubtedly have a significant impact on how we live. MEMS is destined to have a major impact, whether it is applied to the healthcare, transportation, or communication industries. Because MEMS is a multidisciplinary topic, it draws on knowledge in design, engineering, and production from many other technological disciplines. A few of the fields that are incorporated in MEMS technology for design, engineering, and manufacturing include fluid engineering, optics, instrumentation, packaging, embedded circuits fabrication methods, mechanical, electrical and chemical engineering and materials science. Making small gadgets with a combination of mechanical and electrical components requires the use of a process approach called micro-electromechanical systems (MEMS). Its sizes can range from a few micrometres to millimetres, and is produced utilising ICs that is integrated circuit, batch production techniques.

MEMS, one of the most intriguing emerging technologies of this century and it integrates micromachining and silicon-based microelectronics both and has the capacity to transform consumer and industrial goods both. Its methods and technology, which is based on microsystems, have the ability to fundamentally alter how we live and how everyone lives. If semiconductor microfabrication was the first revolution in micromanufacturing, MEMS is the second.

In its simplest form, MEMS is composed of mechanical microstructures, microsensors, micro actuators, and microelectronics combined onto a single silicon chip. Figure 3 shows this conceptually.

Figure 3. Systematic illustration of components of MEMS

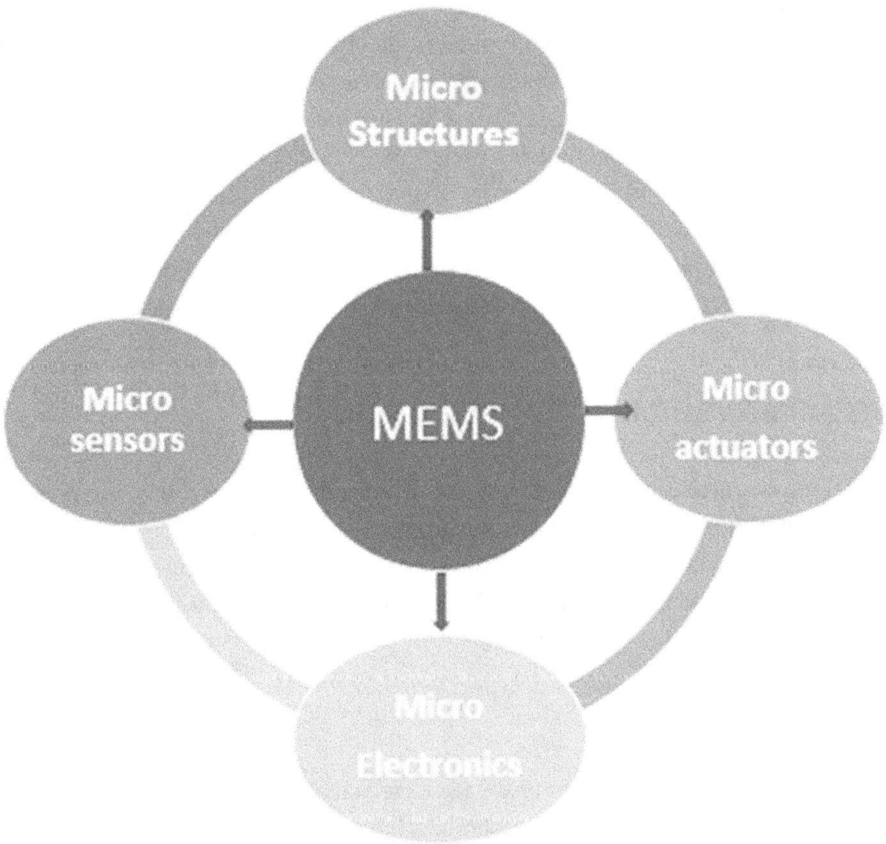

Both the formation of macro-scale effects and micro-scale sensing, control, and actuation are capabilities of these objects (or systems) (Ioan Raicu, 2004).

- **Mechanical Microstructures:** At the microscopic level, microstructures are material structures. They are the inside details of an object, creature, or material that can be seen with a magnification of over 25 times in a microscope. When occurrence of flaws, contaminants, grains, and grain borders are mentioned, the microstructure is meant to be understood.
- **Microsensors:** Microsensors are two- and three-dimensional micromachined structures that are smaller, more effective, more dependable, and have lower production costs than many other forms of sensors. They are a subset of the larger class of MEMS (micro-electro-mechanical systems) devices, which also includes micro actuators.
- **Micro Actuators:** A tiny servomechanism called a micro actuator transmits and transfers a precise amount of energy to another mechanism or system. The following conditions must be fulfilled for it to qualify as a generic actuator:
    - Important travel
    - Extreme precision

- Quick switching
- Power consumption is quite low.
- Sustainability of free force power

- **Microelectronics:** Microelectronics is a branch of electronics. Microelectronics, as the name implies, is the research and production (or microfabrication) of exceedingly tiny electrical designs and elements. Usually, but not always, this refers to scales smaller than a micrometre. These devices are often built with semiconductor materials. Many components used in traditional electronics design have microelectronic analogues. Capacitors, diodes, inductors, transistors, resistors and (naturally) insulators and conductors are frequently used in electronic gadgets. Because the components, leads, and pads are so incredibly tiny, new wiring methods like wire bonding are necessary. The equipment needed for this procedure is expensive (Maluf, 2000).

TYPES OF MEMS SENSOR IN IOT APPLICATION

Traditional MEMS sensors are entering a new stage of development as a result of the increased demand for sensor technology development to satisfy IoT requirements under 5G infrastructure. In order to actually produce all devices, MEMS fabrication technology must be considered first. The three main MEMS sensor types that will be utilised almost universally for IoT applications are then covered in the section after, in ascending order of working frequency.

- **Accelerometers:**

This particular type of MEMS physical sensor uses electrical signals to detect deflections or stress variations caused by an external acceleration. Researchers have explored a few The resonant accelerometer (Flader et al., 2019), the piezoelectric accelerometer (Yang et al., 2019), the piezoresistive accelerometer (Amarasinghe et al., 2007), and the capacitive accelerometer (Tsai et al., 2012) are a few examples of MEMS accelerometers built on different operating principles. Shin et al. differential resonant beams accelerometer (DRBA) (Shin et al., 2017). It is highly sensitive, stable, and resilient to stresses brought on by packaging as well as temperature variations. Due to their near proximity, the two rays will produce a differential frequency signal in reaction to an external acceleration from the proof mass while also cancelling out the effects of package stress and temperature. Also, it was observed that resonant accelerometers based on double-ended tuning forks (DETF) have mode localisation and micro-leverage (Flader et al., 2019). The paired MEMS beams would produce a voltage output when the centre proof mass is subjected to an external acceleration. As a result, the applied accelerations may be calculated using the deflection-output voltage mathematical relationship. The variation in resistance matched the material deflection directly (Partridge et al., 2000). The MEMS device's dependence on an external voltage was this mechanism's defect. Because of its long lifespan, the capacitive device was commonly utilised in an industrial setting.

- **Gyroscopes:**

It is the sensor that seeks to gauge angular rate or variance using the Coriolis force. Many diverse architectures, such as the vibratory gyroscope (Sung et al., 2008) and the resonant gyroscope (Nitzan et al., 2015), have been suggested for MEMS gyroscopes, which are frequently employed in a variety of electronics and navigation systems. When voltage is added, the disc will originally vibrate in either the drive mode or the 0 mode. The Coriolis effect will efficiently transfer power to the secondary mode and spin the vibration pattern when the device has an angular speed. Through this device's angular shift, the angular rate can be determined. It normally consists of three frames: a motors framework, a Coriolis structure, and a sensor structure (Giner et al., 2019). The drive frame will initially continue buzzing in one manner via electrostatic actuation by providing an ac source, much like a resonant gyroscope, while the detecting frame stays the same. The Coriolis frame will be powered by the motor frame as well. Due to the force created by the angular speed, the detecting frame will travel along another in-plane direction perpendicular to the driving direction, which will cause the capacitance to fluctuate. Additionally, we can foresee that the upcoming MEMS gyroscope and accelerometer will be able to sense in several directions and have higher sensitivity and lower noise levels (Zhang et al., 2019). Additionally, there has been nonstop research on new MEMS accelerometer and gyroscope mechanisms, system integration with IoTs, and ways to further reduce power usage. A number of designs that can measure both the amplitude and the energy of a movement have also been put forth in connection to the advancement of accelerometers. By determining whether a device's vibration is due to unbalance or an impact, these devices can be used in IoTs for mechanical tracking.

- **Acoustic Sensor:**

The MEMS market is dominated by MEMS acoustic sensors, which have substantially higher operating frequencies than normal MEMS physical sensors and are used primarily in consumer devices like mobile phones and smart speakers. The traditional MEMS acoustic sensor, commonly known as a microphone, has a flexible membrane and a back-plate with a bias voltage. This sensor uses a capacitive sensing device. Because of the limits of high signal level and sensitivity to environmental circumstances, many sensing strategies have been researched to enhance acoustic sensor performance (Leinders et al., 2015). These mechanisms, among others, use piezoelectric sensing to enable low-wattage directional detection, back-plate-less design to eliminate air dampening, and optical sensing to perform in hostile conditions. According to Lo et alback-plate, there are 42 sets of planar interdigitated electrodes (IDTs) organized in the same plane. Because there was no back-plate in a linear output and there was low damping noise, the out-of-plane capacitance change provided high-performance sensing.. Due to its low power requirement, the piezoelectric mechanism in sound detection was another fascinating subject. A piezoelectric microphone for sound detection has been developed which has inspired by biology (Chang et al., 2015). In order to use the d33 mode to identify the acoustic signal, the IDT structures were positioned above the aluminium nitride sheet. A reduced noise floor, no bias voltage, and directional sound detection were advantages of this design. Ishfaque et al. have described an optimised design with a circular diaphragm (Rahaman et al., 2019). The researcher (Ishfaque et al., 2019) presented an integration of the capacitive sensing and piezoelectric techniques for detection of acoustics to enhance its acoustic electrical response. This combination is depicted in Figure 4d. According to the statement, the asymmetrical design widened its frequency response range, and a subsequent refined cross design enabled spatial sound detection (Zhang et al., 2018). Because of its exceptionally broad frequency response, high sensitivity, tolerance to electromagnetic interference and excellent signal-to-noise ratio, optical MEMS microphones have drawn increasing attention. Fast demodulated white-light interferometry is used to measure the length change of the Fabry-Parot cavity that is created between the fibre tip and cantilever, which is subsequently transformed into electrical signals. In order to achieve high sensitivity and low noise detection, direct connection among acoustic waves and light is used in highly complicated sensing systems (Wu et al., 2017). Because of to its exceptional less power consumption that enables continuous standby, MEMS piezoelectric microphones have recently been marketed and are progressively gaining market share, especially in consumer electronics. To address the use in the harsh environment and maintain a high level of measurement precision, optical microphones were also created for non-destructive testing operation in the ultrasound spectrum. We may anticipate that as 5G technology advances, commercial MEMS microphones will become smaller and use nearly no power when integrated with IoT applications.

- **Magnetic Field Sensors:**

Small-scale microelectromechanical systems called MEMS magnetic field sensors aid in the detection and measurement of magnetic fields. Force variations are detected by sensors, allowing for simple electronic voltage frequency measurement. It can be positioned near the measurement site to increase spatial resolution. To lessen temperature-related issues related to silicon Hall cell features, it integrates instrumentation circuits and incorporated bulk Hall cell technology. MEMS magnetic field sensors are used to track linear angles, velocity, speed of rotation, positioning, and linear position in industry, consumer, and automobile industries.

- **Gas Sensor:**

In many different environments, MEMS gas sensors are commonly employed to evaluate the gas quality. The future MEMS gas sensor's objectives are tiny size, high sense, small concentrations sensing, selectivity to gas mixtures, sensor array, and multi-gas sensing (Cho et al., 2019). In MEMS gas monitors, various metal oxide materials were frequently used. This band-gap energy-based method for this gas or vapour sensor caused a shift in resistance. The researcher (Kang et al., 2017) described a SnO2(tin oxide) MEMS gas sensor with a local production of tin oxide nanotubes on the hung microheater as well as a SnO2 MEMS gas sensor with a local production of tin oxide nanotubes on the hanging microheater. They reportedly have a rapid response time (22 s) and consume very little energy (700 at 0.5percentage H2 concentrations) when operating at room temp (RT). Waveguide gas sensing's main benefits are its responsiveness and sensitivity (Hasan et al., 2018). A carbon dioxide gas sensor for 40 ppm measurement at air temperature was presented by Dihan et al. for the process using mid-Infrared radiation. A gas-selective trapping polymer was used in the sensing method to increase absorbance in the mid-IR regions based on the IR responses. Chang et al surface-enhanced IR absorption with a detection limit of 20 ppm to carbon dioxide was seen during a brief period of time (Chang et al., 2018). ZIF-8 and ZIF-67, zeolitic imidazolate framework nanostructures, were created (Leidinger et al., 2016) to sense gas at significantly lower concentrations. They asserted that the sensors exhibit a considerable resistance change and can identify ethanol, carbon monoxide gas, toluene, hydrogen (H2), and nitrogen oxides (NO2) at concentrations as low as 10 ppm. 2D materials are a new trend for MEMS gas sensors (Bouchaala et al., 2016). The biomimetic gas sensor, according to them, displayed linear hydrogen detection. Furthermore, they claimed that for the identical dimension, the bioinspired gas sensor performed better than gas sensors built using CVD graphene.

- **Pressure Sensor:**

Piezoresistive pressure sensing die are used in pressure sensors and are processed using MEMS technology. There are numerous pressure sensor applications available; the best sensor must be chosen from the many options in accordance with the intended purpose.

There are various pressure sensor kinds, and they are categorised based on the measurements they can take and the types of pressures they can monitor, such as high or low pressure.

MEMS pressure detectors are used in a several applications, including those in the military, health-care, industry, beverage and washing machines, aerospace, industrial machinery, and automotive. To detect and track the tyre pressure, sensors are built into the tyres of a vehicle, either internal or external.

Types of pressure sensors:

- Vacuum pressure sensor
- Sealed pressure sensor
- Gauge pressure sensor
- Absolute pressure sensor
- Differential pressure sensor

MEMS COMPONENT IN FUTURE SENSOR

The creation of sensor networks with a single sensor node is one of the most obvious developments pointing in the direction of the future sensor. Under a topology like an integrated sensor system, future sensors will undoubtedly battle with low power consumption, wireless transmission, and self-power generation Each sensor nodes will be equipped with MEMS components such MEMS switches, SAW and PMUT devices, and energy harvesters to provide full functionality and network connectivity.

MEMS Switch

MEMS switches, also known as micro-electrical contacts, have a variety of current as well as future uses with MEMS devices. MEMS switches are more power-efficient, have lower insertion loss than conventional field-effect transistors (Iannacci et al., 2017) and greater insolation. MEMS switches have various advantages because of the mechanical actuation, which genuinely opens or closes the circuit. Electrostatic actuation (Soon et al., 2013), electromagnetic actuation (Cho et al., 2010), piezoelectric actuation (Sinha et al., 2009), and electrothermal actuation (Rajaram et al., 2018) are the four fundamental actuation concepts. The piezoelectric principle, thermoelectric principle, electrostatic principle, and electromagnetic principle are four common principles used in MEMS switches.

MEMS Surface Acoustic Wave (SAW) and Piezoelectric Micromachined Ultrasonic Transducer (PMUT)

MEMS surface acoustic wave (SAW) devices are widely used because they are simple to fabricate and compatible with CMOS. SAW found use in faster data transmission and wireless contact with the advent of the 5G network.

MEMS Energy Harvesting

One of the MEMS devices that can gather diverse environmental energy sources and convert them into electricity is the MEMS energy harvester. In general, thermal energy, mechanical energy, biochemical energy, and radiant energy are used to harvest energy from the world (Gupta et al., 2017). Due to its widespread dispersal, mechanical energy was the best option among all of these energy sources for MEMS energy harvesting for sensor nodes. Numerous scholars have looked into MEMS energy harvesting using various conversion techniques, including piezoelectric transduction (Gu et al., 2010), electromagnetic induction (Yang et al., 2010), and electrostatic transduction (Zhang et al., 2018). The area of the comb structure or the distance varied with the variation of capacitance when an external vibration was applied to the electrostatic energy harvester (Lu et al., 2016). Another popular subject today is electret material-based energy harvesting, where electricity can be produced using a pre-charged electret material's capacitance variation. It stated that a low resonance of 95 Hz was used to reach the optimisation output power of 0.95 W (Tao et al., 2015). The study project focused on piezoelectric MEMS energy harvesters, which can convert mechanical strain force from structural vibration to electricity (Toprak et al., 2015). For a cantilever with a 1200 mm 300 mm cross section, a maximum power output of 35.1 pW was determined for a 500 mm tip displacement. Higher energy production is a benefit of hybrid energy harvesting.

The MEMS energy harvester's greatest flaw was that it operated at fixed or limited frequencies, whereas ambient vibrations fluctuated over a wide range. Liu et al. suggested the cantilevers array, which tuned to the frequency and extended the frequency bandwidth (Liu et al., 2008), to harvest energy at a wider range. The effective electrical power measurement output of the prototype generator was 3.98 mW. The constraint of space, which led to a system that was dramatically nonlinear, was the alternative technique for harvesting energy at broadband frequencies. The carrier, which has a restricted downward area and can serve as a stopper to increase the operational bandwidth for this energy harvester, was coupled with the piezoelectric MEMS energy harvesters (Liu et al., 2008). Due to its innovative structure, this energy harvester can now vibrate at 1g and produce power across a wide frequency range, ranging from 30-47 Hz and 19.4- 51.3 nW. The clamped-clamped beam was also given duffing stiffening by three dispersed resonances (Zhu et al., 2017), which claimed that the microfabricated device could enlarge the working bandwidth from 62.9 Hz to 383.7 Hz in order to harvest energy at an ultra-wide bandwidth frequency. A 3D dynamic behaviour for the three tremor modes at 1285, 1470, and 1550 Hz is also proposed by researcher (Liu et al., 2012). With an excitation acceleration of 1 g, the measurement of this MEMS device produced maximum power densities of 0.444, 0.242, and 0.125 W/cm3 at their vibration modesA frequency-up energy generator was developed by authors (Zhu et al., 2018) using nonlinear components and MEMS capacitor stoppers. Despite having a very low frequency and a modest amplitude, it was nevertheless able to gather energy (much lower than its MEMS resonance frequency. A flow sensing system and its energy harvesting were examined using the external broad range excitation (Liu et al., 2012). The output voltage and output power were measured to be 18.1 mV and 3.3 nW, respectively, at a flow velocity of 15.6 m/s.

SMART MATERIALS IN MEMS TECHNOLOGY

A new path for the advancement of MEMS technology was recently opened by a discovery in the study of smart materials (Hao et al., 2022). Micro electromechanical systems are a new class of precision devices that have emerged as a result of scientific advancements in the area of electronic miniaturization. These miniature precision devices are made using a micromachining process that combines mechanical and electrical components to complete the job on a microscopic scale (Chorsi et al., 2019). Smart materials are the materials which are manipulated to respond in a controllable and reversible manner, one or more properties that can be modified by some external stimuli such as stress, electric or magnetic field, light, certain temperature pH, chemical compounds (Zhang Z et al., 2019) (Fei Long et al., 2021). Smart materials that exhibit shape-changing abilities through a variety of mechanisms, including active origami and self-folding techniques, active self-folding structures, stimuli-based simulation, thermally induced actuation, moisture or solvent based actuation, and magnetically induced actuation, can exhibit some structural deformation that can be used to support the operation of MEM sensors (Ayushi et al., 2021) Such materials have large deformation, complex movement, and a wide range of versatility, making them flexible and suitable for MEMS uses (Fei Long et al., 2021). They inherit the property of responsiveness, so these are also known by responsive materials. One another synonym of smart materials is ACTIVE materials. These explicit properties of smart materials can be used as the basis of functioning of such applications. Various domain that uses smart materials in the field of IoT paradigm are shown in Figure 4.

Figure 4. Types of smart materials

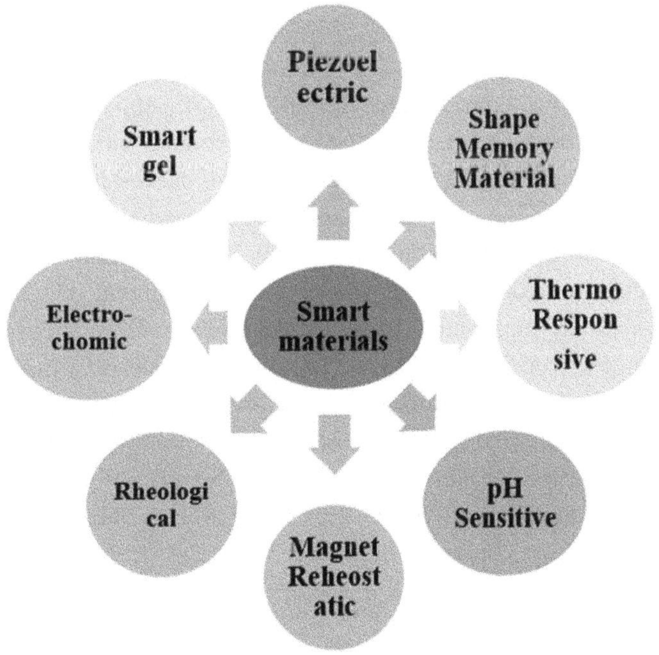

Class of Materials Used in MEMS: Many materials have been explored to be used as sensing and actuating elements in MEMS applications. The majority of integrated circuits in contemporary consumer devices are made of silicon. This super material is ideal for MEMS devices because of its low cost, high quality, and capacity to integrate electronic functionality. The fabrication of MEMS devices such as integrated circuit fields, commonly used material is silica (Daoudi et al., 2017), specially dopped single crystalline silicon, and polycrystalline silicon (Polysilicon) are the preferably used representative materials (Horst et al., 2020). Also Silicon can be used in the form of silicon nitride and silicon dioxide for insulation purpose and for passivation. Silica is present abundantly in quantity, but it is becoming insufficient for latest functionalities and complexities newly developed MEMs. Therefore, new materials have started involving in the fabrication of MEMs. Polymers have been growing and replacing silicon, possessing additional advantages like, biological compatibility, biological functionalities, economic cost, easy availability, easy processing, easy to fabricate, and flexibility. Polymers can not only apply in pure form they can be used in the form of composites when they can be mixed up with different filler materials to alter their mechanical, chemical and electrical properties (Kenawy et al., 2020) (Funano et al., 2021). In recent research It has been observed that application of glass has also increased for MEMs fabrication. The properties such as optical transparency, insulating properties, high solvent compatibility have made them suitable for MEMs recent innovations (Wang et al., 2018) (Wlodarczyk et al., 2019) (Ayushi et al., 2021).

Desirable Mechanical Properties of MEMS Materials: Micro electronic and mechanical systems (MEMS) is the integrated approach of making such micro systems that can have advantages of the technology related to electronics and mechanical all together at a common platform. MEMS designers increasingly leveraging the existing processes and technologies to make new innovative products. The performance of MEMS highly depends upon the responsive nature of materials used. The analysis and parametric evaluation of MEMS material is highly desirable for designing the MEMS Devices (Ayushi et al., 2021). There are mechanical properties which works as the basic mechanism for functioning of MEMS, such as strength, elastic properties, fatigue, internal strength, endurance power. In case of smart materials, they exhibit the smart features, such as stimulated response with respect to pH, temperature, stress, moisture, electrical or magnetic fields (Fei Long et al., 2021). The one desirable mechanical property like elasticity gives the prediction of the amount of deflection from an applied load and the mechanical strength of used materials restrains the operational limits of the device. Small size of the MEMS devices often leads to the mechanical failure due to the harsh working environment. The proper selection of the material associated with desirable mechanical properties can surely eliminate the mechanical failure of such devices even in extreme working conditions. The characterization of mechanical properties of used material for MEMS application has become the latest branch of research to fulfil the requirement of optimized results. However, some materials like silicon, one of the highly used materials in MEMS applications, has only been analyzed for electrical properties, there is still a wide scope of research in the characterization of silicon and many other thin films based on internal stress.

MEMS VS. AI AND ITS VR AND AR

MEMS (Micro-Electro-Mechanical Systems) and AI (Artificial Intelligence) are two different technologies that have different applications and purposes. MEMS refers to the technology of creating tiny devices and systems that have both electronic and mechanical components, usually on a micro scale.

MEMS devices, which can be used in a variety of uses including consumer electronics, medical devices, aerospace and defence, and larger systems, include sensors, actuators, and other parts. The capacity of machines to carry out tasks that ordinarily require human intelligence, such as learning, reasoning, and problem-solving, is referred to as artificial intelligence, or AI. With the help of algorithms and data, AI enables machines to learn and develop over time, becoming more precise and effective at carrying out particular duties. While MEMS and AI are different technologies, they can also be complementary. For example, MEMS sensors can be used to gather data that can be processed by AI algorithms to make predictions or provide insights. MEMS devices can also be controlled by AI systems to perform specific tasks, such as autonomous vehicles or robotics.

Overall, MEMS and AI are two distinct technologies with different applications, but they can work together to enable new capabilities and improve performance in various fields. MEMS (Micro Electronics and Microwave Integrated Circuit) is quickly becoming the future of AI. It's a low-power, lightweight technology that can be integrated into devices to bring about massive improvements in processing power and data storage. And it doesn't stop there: MEMS is also being used to create new devices such as prosthetic limbs, eyeglasses, and other electronic equipment.

MEMS Is the Future of AI

1. MEMS is the future of artificial intelligence.

MEMS technology is the future of artificial intelligence, because it has the potential to create more complex and intelligent machines than ever before. MEMS chips can process and learn quickly, leading to faster and more accurate responses in a wide range of applications. This could lead to significant improvements in machine learning, natural language processing, predictive maintenance, and other AI-related fields.

2. MEMS will make machines smarter faster.

MEMS technology can accelerate the learning process by allowing digital devices to communicate with each other directly. As these devices get smarter, they will be able to better identify patterns and understand complex instructions more quickly than ever before. This will allow machines to become even more intelligent at an accelerated pace, leading to continued advancements in AI development over time.

3. MEMS will help reduce energy consumption by 50%.

MEMS technology has the potential to reduce energy consumption by up to 50%. By using MEMS chips instead of traditional CPUs or GPUs, developers are hopeful that we can create systems that are less reliant on electricity and can operate for longer periods without recharging or recalibrating. This would mean significant reductions in CO2 emissions as well – potentially making MEMS technology one of the most sustainable technologies out there!

MEMS (microelectromechanical systems) are miniature machines that can sense and actuate in the physical world, and they could play an important role in the future of artificial intelligence (AI). MEMS are increasingly being used in AI-related applications such as autonomous vehicles, robotics, and healthcare. MEMS have the advantage of being able to sense physical attributes such as pressure, temperature, and acceleration, as well as provide feedback to the environment. This could enable AI systems to better understand and react to their environment. Additionally, MEMS are more power efficient than traditional sensors and can be miniaturized, which makes them more suitable for use in many applications. MEMS could also make AI systems more reliable and robust, as they are less prone to noise and interference than traditional sensors.

MEMS (Microelectromechanical Systems) and AI (Artificial Intelligence) are two different technologies that are used in a variety of applications. MEMS are micro-scale systems that use mechanical and electrical components to sense and respond to changes in the environment. AI is a form of computer science that works to mimic human cognitive functions, such as decision making and problem solving. MEMS are primarily used to measure physical parameters such as temperature, pressure, and acceleration, and can be used in a variety of applications, including automotive, medical, and industrial. AI, on the other hand, is used to process large amounts of data and optimize decision-making and problem-solving tasks. AI is used in many applications, from self-driving cars to medical diagnosis and facial recognition. MEMS and AI can be used together to create powerful and efficient systems. For example, AI can be used to analyse the data collected by MEMS, allowing for more accurate predictions and decisions. Additionally, MEMS can be used to provide physical feedback to AI-driven systems, allowing for more responsive and adaptive behaviour.

Toward VR and AR

VR and augmented reality (AR) technologies have become popular research topics due to the recent accelerated development of the information sector and human-machine interface technology when combined with MEMS/NEMS wearable devices to build an entire interactive system in 3D space. This type of interactive technology provides users with a more lifelike experience and has the capability to be used in a range of settings, including entertainment, medical rehabilitation, athletics training simulation, and more (Liao et al., 2019). In addition to the software component that generates a real interactive 3D world, the integrated sensors that collect data on user motion, force, and tactile sensation play a vital part in the total VR/AR interactive system (Dhakar et al., 2016). An object's attitude in 3D virtual space, including its position and rotation, can be controlled using a novel self-powered VR 3D-control sensor based on a triboelectric mechanism, according to Chen et al. (2018)'s work. It was effective to use two contacting spheres as the sensing point with human fingers for a "Virtual Assembly" application and to detect 3D force information. He et al. (2019) developed a self-powered glove-based intuitive interface for a range of control applications in real/cyber space by fusing triboelectric mechanisms and wearable technology. In contrast to rigid and heavy wearable human-machine interfaces, all of the sensory components in their design were made of elastic and flexible triboelectric materials, giving it extraordinary flexibility Similar to this, Liao developed a flexible and translucent tactile sensor based on the hetero-contact microstructure (Liao et al., 2019). These fingertip touch sensors worked well at turning fingertip events into interactive feedback and visual control in virtual reality because of their high sensitivity and short reaction times. The finger bending strain sensor proposed by Dejace et al. is based on stretchy gallium-based conductors, and they demonstrated how to precisely detect human hand kinematics using a virtual reality scenario (Dejace et al., 2019). These wearable human motion sensors have exceptional elasticity and stretchability, but due to their straightforward structural design, they can only collect a certain amount of sensory information. In order to develop more detailed and immersive interactive systems for VR and AR, many groups have chosen to combine different MEMS sensors and actuators to perform more complex tasks, such as environment temperature simulation, human motion tracking, haptic force feedback, and so forth. To track hand movements and identify collisions in virtual reality, Martnez et al. created an optical tracking system using LED markers strewn around the glove (Martinez et al., 2016). By adding 12 vibrotactile actuators that produce vibration feedback when in touch with virtual objects, it is also possible to recognise virtual 3D geometric shapes. In order to provide users with accurate continuous lateral stroking and surface roughness tactile sensations in virtual environments, Lin et al. published a method to identify the optimum driving voltages of piezoelectric actuators using machine learning technologies (Lin et al., 2018). A strategy for training baseball pitches that is more thorough and creative was developed by Tsai et al. (2019). They implanted a pressure-sensitive resistor (PSR) on each fingertip to measure the force of contact, and they employed changing resistance in their design to simulate finger bending. For 3D positioning, both a vive lighthouse and a wristband with IR receivers were employed. Electric shocks and servo motors can be used to imitate the actual contact and constraint of each finger. The player was able to rehearse handling a baseball in the virtual environment thanks to a full VR interactive system made up of all these sensors and actuators. It is simple to build an immersive VR/AR system with a variety of applications using these commonly accessible MEMS sensors and actuators. The flexibility and stretchability of the VR/AR system can be increased while the system's power consumption is reduced by replacing or combining these conventional stiff MEMS sensors/actuators with flexible NEMS wearable devices. More study is still needed in this field, which is still active.

CONCLUSION

MEMS's small size, ability to be made in vast quantities, and integration with electronics are revolutionising the way sensor, actuator, and control systems are implemented. MEMS technology has made it feasible to build innovative systems on a circuit with excellent functionalities, efficiency, and endurance in minuscule nanoscale chips at very cheap-priced levels. The primary comercial drivers for the MEMS industry at the moment are inertial detectors for consumer electronics, clinical and industrial control applications silicon-based pressure sensors for automobiles, microphones for phones and computers crash airbag inertial sensors,DLP for displays and projectors, and ink-jet printer cartridges. In the commercial, healthcare,industrial and defence sectors, MEMS gadgets are becoming recognised as high-quality product differentiators in a range of applications. Due to these commercial accomplishments and the many uses for the technology, MEMS has a very promising future in the market. MEMS is a young and very synergistic technology, thus it is anticipated that many new applications will arise, extending the market beyond those that are now recognised or characterised, especially as the Internet of Things develops and evolves. As a result, MEMS is a cutting-edge technology with a very promising future. MEMS is the future of technology because of the variety, importance to the economy, and range of potential uses for the IoT.

REFERENCES

Amarasinghe, R., Dao, D., Toriyama, T., & Sugiyama, S. (2007). Development of miniaturized 6-axis accelerometer utilizing piezoresistive sensing elements. *Sensors and Actuators. A, Physical, 134*(2), 310–320. doi:10.1016/j.sna.2006.05.044

Atzori, L., Lera, A., & Morabito, G. (2010). The Internet of Things: A survey. *Computer Networks, 54*(15), 2787–2805. doi:10.1016/j.comnet.2010.05.010

Ayushi, U. K. V., Mishra, S., & Kanu, N. J. (2021). Biomimetic 4D printed materials: A state-of-the-art review on concepts, opportunities, and challenges. *Materials Today: Proceedings, 47*(11), 3313-3319. doi:10.1016/j.matpr.2021.07.148

Bouchaala, A., Jaber, N., Yassine, O., Shekhah, O., Chernikova, V., Eddaoudi, M., & Younis, M. (2016). Nonlinear-based MEMS sensors and active switches for gas detection. *Sensors (Basel), 16*(6), 758. doi:10.339016060758 PMID:27231914

Chang, Y., Hasan, D., Dong, B., Wei, J., Ma, Y., Zhou, G., Ang, K., & Lee, C. (2018). All-dielectric surface-enhanced infrared absorption-based gas sensor using guided resonance. *ACS Applied Materials & Interfaces, 10*(44), 38272–38279. doi:10.1021/acsami.8b16623 PMID:30360088

Chen, T., Zhao, M., Shi, Q., Yang, Z., Liu, H., Sum, L., Ouyang, J., & Lee, C. (2018). Novel augmented reality interface using a self-powered triboelectric based virtual reality 3D-control sensor. *Nano Energy, 51*, 162–172. doi:10.1016/j.nanoen.2018.06.022

Cho, I., Kang, K., Yang, D., Yun, J., & Park, I. (2017). Localized liquid-phase synthesis of porous SnO_2 nanotubes on MEMS platform for low-power, high performance gas sensors. *ACS Applied Materials & Interfaces, 9*(32), 27111–27119. doi:10.1021/acsami.7b04850 PMID:28714311

Cho, I., & Yoon, E. (2010). Design and fabrication of a single membrane push-pull SPDT RF MEMS switch operated by electromagnetic actuation and electrostatic hold. *Journal of Micromechanics and Microengineering, 20*(3), 035028. doi:10.1088/0960-1317/20/3/035028

Cho, M., Zhu, J., Kang, K., & Park, I. (2019). Half-pipe palladium nanotube network hydrogen sensor based on electrospun nanofiber scaffolds. *Proceedings of the 2019 20th International Conference on Solid-State Sensors, Actuators and Microsystems & Eurosensors*, 1183–1186. 10.1109/TRANSDUCERS.2019.8808214

Chorsi, M. T., Curry, E. J., Chorsi, H. T., Das, R., Baroody, J., Purohit, P. K., Ilies, H., & Nguyen, T. D. (2019). Piezoelectric Biomaterials for Sensors and Actuators. *Advanced Materials, 31*(1), e1802084. doi:10.1002/adma.201802084 PMID:30294947

Daoudi, K., Gaidi, M., & Columbus, S. (2017). Biointerface Research in Applied Chemistry. *Ultrasound, 69*, 70. doi:10.3390/mi13020164

Dejace, L., Laubeuf, N., Furfaro, I., & Lacour, S. (2019). Gallium-based thin films for wearable human Motion sensors. *Advanced Intelligent Systems, 1*(5), 1900079. doi:10.1002/aisy.201900079

Dhakar, L., Gudla, S., Shan, X., Wang, Z., Tay, F. E. H., Heng, C.-H., & Lee, C. (2016). Large scale triboelectric nanogenerator and self-powered pressure sensor array using low cost roll-to-roll UV embossing. *Scientific Reports, 6*(1), 22253. doi:10.1038rep22253 PMID:26905285

Flader, I. B., Chen, Y., Yang, Y., Ng, E. J., Shin, D. D., Heinz, D. B., Ortiz, L. C., Alter, A. L., Park, W., Goodson, K. E., & .... (2019). Micro-tethering for fabrication of encapsulated inertial sensors with high sensitivity. *Journal of Microelectromechanical Systems, 28*(3), 372–381. doi:10.1109/JMEMS.2019.2900761

Flader, I. B., Chen, Y., Yang, Y., Ng, E. J., Shin, D. D., Heinz, D. B., Ortiz, L. C., Alter, A. L., Park, W., Goodson, K. E., & .... (2019). Micro-tethering for fabrication of encapsulated inertial sensors with high sensitivity. *Journal of Microelectromechanical Systems, 28*(3), 372–381. doi:10.1109/JMEMS.2019.2900761

Funano, S.-i., Ota, N., & Tanaka, Y. (2021). A simple and reversible glass–glass bonding method to construct a microfluidic device and its application for cell recovery. *Lab on a Chip, 21*(11), 2244–2254. doi:10.1039/D1LC00058F PMID:33908537

Giner, J., Maeda, D., Ono, K., Shkel, A., & Sekiguchi, T. (2019). MEMS gyroscope with concentrated springs suspensions demonstrating single digit frequency split and temperature robustness. *Journal of Microelectromechanical Systems, 28*(1), 25–35. doi:10.1109/JMEMS.2018.2881209

Gu, L., & Livermore, C. (2010). Passive self-tuning energy harvester for extracting energy from rotational motion. *Applied Physics Letters, 97*(8), 081904. doi:10.1063/1.3481689

Gubbi, J., Buyya, R., Marusic, S., & Palaniswami, M. (2013). Internet of Things (IoT): A vision, architectural elements, and future directions. *Future Generation Computer Systems, 29*(7), 1645–1660. doi:10.1016/j.future.2013.01.010

Gupta, R. K., Shi, Q., Dhakar, L., Wang, T., Heng, C. H., & Lee, C. (2017). Broadband energy harvester using non-linear polymer spring and electromagnetic/triboelectric hybrid mechanism. *Scientific Reports, 7*(1), 41396. doi:10.1038rep41396 PMID:28120924

Hao, Y., Zhang, S., Fang, B., Sun, F., Liu, H., & Li, H. (2022). A Review of Smart Materials for the Boost of Soft Actuators, Soft Sensors, and Robotics Applications. *Chinese Journal of Mechanical Engineering, 35*(1), 37. doi:10.118610033-022-00707-2

Hasan, D., & Lee, C. (2018). Hybrid metamaterial absorber platform for sensing of CO_2 gas at mid-IR. *Advancement of Science, 5*(5), 1700581. doi:10.1002/advs.201700581 PMID:29876204

He, T., Sun, Z., Shi, Q., Zhu, M., Anaya, D. V., Chen, T., Yuce, M. R., Thean, A. V.-Y., & Lee, C. (2019). Self-powered glove-based intuitive interface for diversified control applications in real/cyber space. *Nano Energy, 58*, 641–651. doi:10.1016/j.nanoen.2019.01.091

Horst, J., de Andreade, P., Duvoisin, C. A., & Vieira, R. (2020). Fabrication of conductive filaments for 3D-printing: Polymer nanocomposites. *Biointerface Research in Applied Chemistry, 10*(6), 6577–6586. doi:10.33263/BRIAC106.65776586

Iannacci, J. (2017). Introduction to MEMS and RF-MEMS: From the early days of microsystems to modern RF-MEMS passives. In *Rf-Mems Technology for High-Performance Passives* (pp. 2053–2563). IOP Publishing Ltd.

Ishfaque, A., Rahaman, A., & Kim, B. (2019). Bioinspired low noise circular-shaped MEMS directional microphone. *Journal of Micro/Nanolithography, MEMS, and MOEMS : JM3, 18*(1), 1. doi:10.1117/1.JMM.18.1.010501

Kenawy, S., & Khalil, A. (2020). Advanced ceramics and relevant polymers for environmental and biomedical applications. *Biointerface Research in Applied Chemistry, 10*(4), 5747–5754. doi:10.33263/BRIAC104.747754

Lawton, M. (2015). *The symbiosis of digital and physical security*. Available: https://futurelab.assaabloy.com/en/thesymbiosis-of-digital-and-physical-security/

Leidinger, M., Rieger, M., Sauerwald, T., Alépée, C., & Schütze, A. (2016). Integrated pre-concentrator gas sensor microsystem for ppb level benzene detection. *Sensors and Actuators. B, Chemical, 236*, 988–996. doi:10.1016/j.snb.2016.04.064

Leinders, S. M., Westerveld, W. J., Pozo, J., Van Neer, P. L. M. J., Snyder, B., O'Brien, P., Urbach, H. P., De Jong, N., & Verweij, M. D. (2015). A sensitive optical micro-machined ultrasound sensor (OMUS) based on a silicon photonic ring resonator on an acoustical membrane. *Scientific Reports, 5*(1), 1–10. doi:10.1038rep14328 PMID:26392386

Liao, X., Song, W., Zhang, X., Zhan, H., Liu, Y., Wang, Y., & Zheng, Y. (2019). Hetero-contact microstructure to program discerning tactile interactions for virtual reality. *Nano Energy, 60*, 127–136. doi:10.1016/j.nanoen.2019.03.048

Lin, P., & Smith, S. (2018). A tactile feedback glove for reproducing realistic surface roughness and continual lateral stroking perception. In D. Prattichizzo, H. Shinoda, H. Tan, E. Ruffaldi, & A. Frisoli (Eds.), *Haptic Sensing and Touch Enabled Computer Applications* (Vol. 10894, pp. 169–179). Springer. doi:10.1007/978-3-319-93399-3_16

Liu, H., Soon, B., Wang, N., Tay, C., Quan, C., & Lee, C. (2012). Feasibility study of a 3D vibration-driven electromagnetic MEMS energy harvester with multiple vibration modes. *Journal of Micromechanics and Microengineering, 22*(12), 125020. doi:10.1088/0960-1317/22/12/125020

Liu, H., Zhang, S., Kathiresan, R., Kobayashi, T., & Lee, C. (2012). Development of piezoelectric microcantilever flow sensor with wind-driven energy harvesting capability. *Applied Physics Letters, 100*(22), 223905. doi:10.1063/1.4723846

Liu, J., Fang, H., Xu, Z., Mao, X., Shen, X., Chen, D., Liao, H., & Cai, B. (2008). A MEMS-based piezoelectric power generator array for vibration energy harvesting. *Microelectronics, 39*(5), 802–806. doi:10.1016/j.mejo.2007.12.017

Lo, S. C., Lai, W. C., Chang, C. I., Lo, Y. Y., Wang, C., Bai, M. R., & Fang, W. (2015). Development of a no-back-plate SOI MEMS condenser microphone. *Proceedings of the Transducers—2015 18th International Conference on Solid-State Sensors, Actuators and Microsystems (TRANSDUCERS)*, 1085–1088. 10.1109/TRANSDUCERS.2015.7181115

Lu, Y., O'Riordan, E., Cottone, F., Boisseau, S., Galayko, D., Blokhina, E., Marty, F., & Basset, P. (2016). A batch-fabricated electret-biased wideband MEMS vibration energy harvester with frequency-up conversion behavior powering a UHF wireless sensor node. *Journal of Micromechanics and Microengineering, 26*(12), 124004. doi:10.1088/0960-1317/26/12/124004

Maluf, N. (2000). *An Introduction to Microelectromechanical Systems Engineering*. Artech House.

Martinez, J., Garcia, A., Oliver, M., Molina, J., & Gonzalez, P. (2016). Identifying virtual 3D geometric shapes with a vibrotactile glove. *IEEE Computer Graphics and Applications, 36*(1), 42–51. doi:10.1109/MCG.2014.81 PMID:25137722

Nitzan, S., Zega, V., Li, M., Ahn, C., Corigliano, A., Kenny, T., & Horsley, D. (2015). Self-induced parametric amplification arising from nonlinear elastic coupling in a micromechanical resonating disk gyroscope. *Scientific Reports*, *5*(1), 9036. doi:10.1038rep09036 PMID:25762243

Partridge, A., Reynolds, J., Chui, B., Chow, E., Fitzgerald, A., Zhang, L., Maluf, N., & Kenny, T. (2000). A high-performance planar piezoresistive accelerometer. *Journal of Microelectromechanical Systems*, *9*(1), 58–66. doi:10.1109/84.825778

Rahaman, A., Ishfaque, A., Jung, H., & Kim, B. (2019). Bio-inspired rectangular shaped piezoelectric MEMS directional microphone. *IEEE Sensors Journal*, *19*(1), 88–96. doi:10.1109/JSEN.2018.2873781

Rajaram, V., Qian, Z., Kang, S., Calisgan, S., McGruer, N., & Rinaldi, M. (2018). Zero-power electrically tunable micromechanical photoswitches. *IEEE Sensors Journal*, *18*(19), 7833–7841. doi:10.1109/JSEN.2018.2850898

Shin, D., Ahn, C., Chen, Y., Christensen, D., Flader, I., & Kenny, T. Environmentally robust differential resonant accelerometer in a wafer-scale encapsulation process. *Proceedings of the IEEE 30th International Conference on Micro Electro Mechanical Systems (MEMS)*, 17–20. 10.1109/MEMSYS.2017.7863328

Sinha, N., Wabiszewski, G., Mahameed, R., Felmetsger, V., Tanner, S., Carpick, R., & Piazza, G. (2009). Piezoelectric aluminum nitride nanoelectromechanical actuators. *Applied Physics Letters*, *95*(5), 4–7. doi:10.1063/1.3194148

Soon, B. W., Singh, N., Tsai, J. M., & Lee, C. (2013). A Bistable silicon nanofin: An ideal device for nonvolatile memory applications. *IEEE Nanotechnology Magazine*, *7*(2), 24–28. doi:10.1109/MNANO.2013.2260461

Sung, W., Sung, S., Lee, J., Kang, T., Lee, Y., & Lee, J. (2008). Development of a lateral velocity-controlled MEMS vibratory gyroscope and its performance test. *Journal of Micromechanics and Microengineering*, *18*(5), 055028. doi:10.1088/0960-1317/18/5/055028

Tao, K., Lye, S., Miao, J., Tang, L., & Hu, X. (2015). Out-of-plane electret-based MEMS energy harvester with the combined nonlinear effect from electrostatic force and a mechanical elastic stopper. *Journal of Micromechanics and Microengineering*, *25*(10), 104014. doi:10.1088/0960-1317/25/10/104014

Toprak, A., & Tigli, O. (2015). MEMS scale PVDF-TrFE-based piezoelectric energy harvesters. *Journal of Microelectromechanical Systems*, *24*(6), 1989–1997. doi:10.1109/JMEMS.2015.2457782

Tsai, M., Liu, Y., & Fang, W. (2012). A three-axis CMOS-MEMS accelerometer structure with vertically integrated fully differential sensing electrodes. *Journal of Microelectromechanical Systems*, *21*(6), 1329–1337. doi:10.1109/JMEMS.2012.2205904

Tsai, Y., Jhu, W., Chen, C., Kao, C., & Chen, C. (2019). Unity game engine: Interactive software design using digital glove for virtual reality baseball pitch training. *Microsystem Technologies*, *9*, 1–17.

Wang, T., Chen, J., Zhou, T., & Song, L. (2018). Fabricating Microstructures on Glass for Microfluidic Chips by Glass Molding Process. *Micromachines*, *9*(6), 269. doi:10.3390/mi9060269 PMID:30424202

Wlodarczyk, K. L., Hand, D. P., & Maroto-Valer, M. M. (2019). Maskless, rapid manufacturing of glass microfluidic devices using a picosecond pulsed laser. *Scientific Reports*, *9*(1), 20215. doi:10.103841598-019-56711-5 PMID:31882878

Wu, Y., Yu, C., Wu, F., Li, C., Zhou, J., Gong, Y., Rao, Y., & Chen, Y. (2017). A highly sensitive fiber-optic microphone based on graphene oxide membrane. *Journal of Lightwave Technology*, *35*(19), 4344–4349. doi:10.1109/JLT.2017.2737639

Yang, B., & Lee, C. (2010). Non-resonant electromagnetic wideband energy harvesting mechanism for low frequency vibrations. *Microsystem Technologies*, *16*(6), 961–966. doi:10.100700542-010-1059-z

Yang, J., Zhang, M., Si, C., Han, G., Ning, J., Yang, F., & Wang, X. (2019). A T-shape aluminum nitride thin-film piezoelectric MEMS resonant accelerometer. *Journal of Microelectromechanical Systems*, *28*(5), 776–781. doi:10.1109/JMEMS.2019.2924956

Zhang, H., Wei, X., Ding, Y., Jiang, Z., & Ren, J. (2019). A low noise capacitive MEMS accelerometer with anti-spring structure. *Sensors and Actuators. A, Physical*, *296*, 79–86. doi:10.1016/j.sna.2019.06.051

Zhang, S., Wang, T., Lou, L., Tsang, W., Sawada, R., Kwong, D., & Lee, C. (2014). Annularly grooved diaphragm pressure sensor with embedded silicon nanowires for low pressure application. *Journal of Microelectromechanical Systems*, *23*(6), 1396–1407. doi:10.1109/JMEMS.2014.2313635

Zhang, Y., Bauer, R., Whitmer, W. M., Jackson, J. C., Windmill, J. F. C., & Uttamchandani, D. (2018). A MEMS microphone inspired by ormia for spatial sound detection. *Proceedings of the 2018 IEEE Micro Electro Mechanical Systems (MEMS)*, 253–256. 10.1109/MEMSYS.2018.8346532

Zhang, Y., Wang, T., Luo, A., Hu, Y., Li, X., & Wang, F. (2018). Micro electrostatic energy harvester with both broad bandwidth and high normalized power density. *Applied Energy*, *212*, 362–371. doi:10.1016/j.apenergy.2017.12.053

Zhang, Z., Demir, K.G., & Gu, G.X. (2019). Developments in 4D-printing: a review on current smart materials, technologies, and applications. *IJSNM*, 205-224. doi:10.1080/19475411.2019.1591541

Zhu, J., Guo, X., & Huang, R. (2018). Asymmetric disappearance and periodic asymmetric phenomena of rocking dynamics in micro dual-capacitive energy harvester. *AIP Conference Proceedings*, *1971*. doi:10.1063/1.5041144

Zhu, J., Wang, A., Hu, H., & Zhu, H. (2017). Hybrid electromagnetic and triboelectric nanogenerators with multi-impact for wideband frequency energy harvesting. *Energies*, *10*(12), 2024. doi:10.3390/en10122024

Chapter 9
Microfluidic Synthesis of Nanoparticles

Ashish Kapoor
https://orcid.org/0000-0002-4686-394X
Department of Chemical Engineering, Harcourt Butler Technical University, India

Vijay Vaishampayan
https://orcid.org/0000-0002-9418-4247
Department of Chemical Engineering, Indian Institute of Technology, Ropar, India

Anjali Awasthi
Department of Chemical Engineering, Harcourt Butler Technical University, India

Soma Banerjee
Department of Plastic Technology, Harcourt Butler Technical University, India

ABSTRACT

Microfluidic systems are promising alternatives for continuous synthesis of nanoparticles in controlled environments, overcoming challenges of conventional methods. Stable nanoparticles within a narrow range of size and shape distribution can be reliably produced by controlling geometrical configuration of microfluidic devices and optimizing various operational parameters. This chapter provides an overview of fundamentals and advances in microfluidic synthesis of nanoparticles, categorizing devices into continuous and segmented flow microfluidic platforms. Strategies for functionalization of nanoparticles are explained, and their diverse applications in various fields are elaborated. Future prospects and underlying challenges are also examined for the commercialization of microfluidic devices for nanoparticle synthesis.

INTRODUCTION

Microfluidics is a field that deals with the study and manipulation of fluids at a small scale, typically within channels with dimensions ranging from tens to hundreds of micrometers (Luo et al., 2019). It is an interdisciplinary field that combines aspects of engineering, physics, chemistry, and biology. Micro-

DOI: 10.4018/978-1-6684-6952-1.ch009

fluidics has become a powerful tool for researchers to understand and control fluid behavior, enabling the development of new technologies for a wide range of applications (Carvalho et al., 2022; Gao et al., 2020). Microfluidic operations offer several advantages by virtue of functioning at miniaturized scale, such as improved control over reaction conditions, reduced reagent usage, dominant viscous effects, and faster reaction times (Y. Zhao et al., 2015). Microfluidic devices also allow for the integration of multiple processes on a single platform, enabling complex assays and analysis to be performed in a single step (Kovarik et al., 2012). Furthermore, microfluidic devices can be easily automated, making them useful for high-throughput applications (Postek & Garstecki, 2022; Rizkin et al., 2020). Microfluidics has found numerous applications in various fields, including chemical synthesis, drug delivery, biomolecular analysis, environmental monitoring and point-of-care diagnostics (Y. Liu et al., 2021; Mahdavi et al., 2020; Otzen et al., 2021; Sri Sruthi et al., 2021; Zhou et al., 2020). The unique features of microfluidic systems, such as effective control over fluid behavior and the ability to perform complex assays in a single step, have made them a powerful tool for developing new technologies in these fields (Gervais et al., 2011; Tang & Shum, 2016).

In this chapter, the focus is given on the application of microfluidics in nanoparticle (NP) synthesis, exploring the fundamentals of microfluidic nanoparticle synthesis, characterization techniques, and applications. The chapter highlights the advantages of microfluidic systems over traditional synthesis methods and discusses the challenges and emerging trends in the field. Overall, the chapter aims to provide a comprehensive understanding of microfluidic nanoparticle synthesis and its potential to revolutionize the field of nanotechnology.

CONVENTIONAL NANOPARTICLE SYNTHESIS METHODS

Nanotechnology has grabbed the attention of the researchers due to immense potential in numerous fields such as electronics, coatings, cosmetics, energy devices, and sensors (Banerjee & Kar, 2016a). The exceptional properties of the nanomaterials are due to high surface to volume ratio. The synthesis of the nanomaterials follows two basic approaches: top-down and bottom-up. In top-down approach, bulk material is converted into the small particles by several methods such as mechanical grinding, laser ablation, etching, and sputtering. In other approach, the nanomaterials are fabricated utilizing different routes such as sol-gel, chemical vapor deposition, solvothermal, hydrothermal, and reverse micelles. Overall, the synthesis of nanomaterials can be classified in three broad methodologies, namely, physical, chemical, and biological methods. In physical methods, ball milling, melt mixing, chemical vapor deposition, and arc discharge are common whereas, in chemical methods the nanomaterials are synthesized by sol-gel, co-precipitation, and solvothermal method of synthesis (Baig et al., 2021). In biological routes of synthesis, the nanomaterials are synthesized using microorganisms, algae and plant extracts.

Top-Down Approaches

In various top-down approaches, mechanical milling is one of the common methods for production of nanomaterials from bulk scale. This type of milling comes under the category of wet milling approach (Zhuang et al., 2016). During the process of ball milling, a cylinder with base fluid and bulk material is fed where balls or the other grinding medium is contained inside the cylinder. The gravitational or rotational impact of balls pursues grinding for conversion of bulk material to nano scale particles. This method is effectively utilized for the preparation of oxides, carbides, nanoalloys and nanocomposites.

With the help of beam of electrons, various nanoarchitectures are fabricated in lithographic technique. Lithographic techniques can again be sub categorized in two different classes namely, masked and maskless techniques often utilized in synthesis of nano patterning or micro-nano fabrication. The 3D nano patterning fabrication is obtained by ion implantation in combination with wet chemical synthesis approach (Khanna et al., 2019). Sputtering is another common method of fabrication of thin film of nanomaterial. In this process, nanofilms are fabricated by bombarding substrate surfaces with high energy particles. The potential advantage of this process is that the composition of thin film remains exactly the same with the target material. Sputtering is a cost-effective technique for nano pattern formations as compared to other lithographic techniques (Ayyub et al., 2001).

For the fabrication of carbon-based nanomaterials, arc discharge method is one of the favorable choices (D. Zhang et al., 2019). In this method, two graphite rods are placed in the chamber with a preset pressure of inert gas inside it. The inert gas helps in the development of proper carbon-based architectures and opens the window for fabrication of new architectures of nanomaterials. Laser ablation synthesis methodology leads to the generation of nanomaterials using powerful laser striking on the target (Ismail et al., 2020). During the process of hitting, the precursor vaporizes due to impact of high energy and forms nanoparticles. Metal particles are preferably synthesized by this technique without need of stabilizing agents. The nanomaterials frequently manufactured by this procedure are ceramics, carbon nanomaterials, and oxides. Pulsed wire discharge is another physical method utilized for the synthesis of the nanoparticles and mostly metal nanoparticles are prepared by this method. In this method, the metal wire is vaporized at first by using a beam of current to generate vapors which in the subsequent step is converted in to the nanoparticles when comes in contact with the ambient. Nitride and metal nanoparticles of uniform morphology have been synthesized by this technique.

Bottom-Up Approaches

Chemical vapor deposition (CVD) is a technique vastly explored for the synthesis of nanomaterials. In this method, a thin film of reactant in gaseous form is deposited on the surface of the substrate (Yadav et al., 2018). The substrate at first is placed in an oven and heated inside the tubular furnace. Subsequently, the hydrocarbon, as precursor is slowly introduced inside the chamber. At higher temperature, decomposition proceeds and the carbon atoms recombine to generate carbon nanotubes. The choice of catalyst decides the morphological features of the nanomaterials. The yield of the process is dependent on the substrate temperature and pressure of the gaseous phase inside the CVD in controlling the overall growth process. Solvothermal process is also utilized for the development of nanostructured materials. In a sealed vessel under high temperature and pressure, a heterogeneous reaction is carried out to generate nanoparticles (Banerjee et al., 2018). The solvothermal process is also similar to hydrothermal method, the difference remains only in the selection of the medium of reaction. In solvothermal process, non-aqueous medium is utilized as medium. Sol-gel is the most common wet-chemical synthesis technique widely explored for the development of the nanomaterials (Banerjee & Kar, 2016b). This method produces high quality metal oxide nanoparticles. Metal alkoxides are the precursor from which metal nanoparticles are formed by the process of sol-gel. In the first step, metal oxide precursor is hydrolyzed in presence of either water or alcohol to form a sol. Further, condensation takes place leading to the increase in overall viscosity of the reaction mixture. During the process of condensation, hydroxo and oxo- bridges are formed that further leads to the formation of metal-hydroxo or metal-oxo structures in the solution phase. Further with ageing, polycondensation continues with variation in pore structure and volume. After ageing, it is dried to remove organic solvents followed by calcination to develop nanoparticles. The advantages of this method are low process temperature, economic, homogeneity in formed particles, and easy generation of complex nanostructures.

Reverse micelle is a chemical method utilized for making nanomaterials of desired size and shape (Shiri et al., 2019). An emulsion of oil-in-water type forms micelles where hydrophobic tails are projected towards the core that has confined oil droplets inside it. In case of reverse micelle method, water-in-oil emulsion is used where hydrophilic heads face towards the core that has trapped water inside it. The core plays the role of the nanoreactor where synthesis of particles proceeds. The size of the nanoreactor is controlled by alteration of the water to surfactant ratio. Hard and soft template method is another modern process for the synthesis of nanoporous materials. This technique is considered to be advantageous due to ease of controlling the particle architecture and provisions of multiple ranges of morphologies. In soft template approach, block copolymers are utilized as templates for the synthesis of nanomaterials (W. Li & Zhao, 2013). A number of factors contribute to the formation of the mesoporous structures such as surfactant to precursor ratio, surfactant chemical structure, and experimental parameters. Various nanostructures such as nanospheres and nanorods can be synthesized by this approach. In case of hard template approach, a hard template is filled with the precursor which is subsequently converted to the nanostructures (Poolakkandy & Menamparambath, 2020). The hard template assists the formation of the mesoporous architecture. The selection of hard template remains a vital factor for success of formation of ordered mesopores. The materials commonly utilized as template include silica, carbon nanotubes, wood, colloid crystals, and carbon black. The hard template is to be selected in such a way that it can be easily removed, once the desired mesoporous nanostructure is developed.

Nanoparticle synthesis using batch processes presents several challenges that can affect the quality and consistency of the particles produced. These challenges include limited scalability due to the need for careful control of reaction conditions, variability in particle size and shape, increased risk of agglomeration and aggregation, and higher susceptibility to contamination. Controlling the pH and ensuring that it is uniform is a hurdle in preparation processes like sol-gel technology. Also, it is often encountered during the synthesis of nanoparticle grains that are not stable and grow during the process. These issues can make it difficult to achieve uniform particle properties and reproducibility from one batch to the next.

FUNDAMENTALS OF MICROFLUIDIC SYNTHESIS

Microfluidics is a promising technique that allows for the controlled fabrication of nanoparticles with well-defined properties (L.-L. Li et al., 2017). In this process, tiny amounts of reagents flow through microchannels. The reaction takes place within these channels, and the nanoparticles are formed as a result of the reaction.

The operating principles of microfluidic synthesis of nanoparticles involve several key factors such as fluid flow behavior, reaction kinetics, interfacial interactions, concentration, molar ratios of reactants, temperature and surface chemistry (Boken et al., 2016; Kašpar et al., 2019; H. Liu et al., 2012). The size, shape, and composition of the nanoparticles are largely determined by these factors, and controlling them can result in the production of nanoparticles with desirable properties (Chung et al., 2011). The fluid dynamic behavior in microfluidic devices play a crucial role in determining the properties of the synthesized nanoparticles. The flow rate, viscosity, and shear rate of the fluid are important parameters that influence the nucleation, growth, and agglomeration of nanoparticles. By finely controlling these parameters, it is possible to optimize the synthesis process to obtain nanoparticles with the desired size and shape.

The reaction kinetics involved in the microfluidic synthesis process can significantly impact the properties of the nanoparticles. In microfluidic devices, the reactants are quickly mixed due to the small dimensions of the channels, which leads to a rapid reaction rate. This high reaction rate can result in the formation of homogeneous nanoparticles with a narrow size distribution. Additionally, the reaction temperature can be easily controlled in microfluidic devices, which allows for the production of nanoparticles with controlled crystal structure and phase. Further, the surface chemistry of the microfluidic channels and the reactants can also play a critical role in determining the properties of the synthesized nanoparticles. The surface chemistry of the microfluidic channels can affect the flow behavior of the reactants, which in turn affects the nanoparticle formation. Microfluidics allows good control over flow behavior, which in turn translates into good control over physicochemical characteristics of nanoparticles. The microscale flow behavior in the channel can be captured in terms of Reynolds number (Re) and Peclet number (Pe). Typically, Re remains low in microchannels with streamlined flow that facilitates attainment of good control of monodisperse droplets. Pe designates the ratio between the advection and diffusion rates for flow in a microchannel. The molecular transport in microscale droplets is primarily governed by diffusion. Thus, their transfer from disperse to continuous phase is minimal, which makes drug encapsulation highly efficient (Fabozzi et al., 2023).

MICROFLUIDIC DEVICES AND THEIR PERFORMANCE

Microfluidic devices for nanoparticle synthesis can be broadly classified into two categories: continuous and segmented flow type microfluidic devices (Fabozzi et al., 2023; Gonidec & Puigmartí-Luis, 2018). Salient developments in these devices are presented here. Although both configurations appear quite similar from a broad perspective as they offer highly laminar flow conditions with fairly low values of Re, each of them is based on a different approach. Continuous flow type devices are characterized by spatio-temporal control of reactant concentrations via molecular diffusion, which establishes a controlled reaction-diffusion zone inside the microchannel as illustrated in Figure 1.

On the contrary, segmented flow systems rely on chaotic advection to attain swift mixing. The distinguishing features between continuous and segmented flow systems are highlighted in Table 1 (Gonidec & Puigmartí-Luis, 2018; Niculescu et al., 2022).

Continuous Flow Systems

Single phase continuous flow systems are most widely used in microfluidic nanoparticle synthesis due to their simplicity and effective control over fluid flow, temperature and duration of reaction. The nanoparticle synthesis is performed by utilization of one or more miscible solvents. In this technique, precursors flow into the microchannel where synthesis of NPs take place. The mixing of reagents during the process is achieved primarily by molecular diffusion. In comparison to macroscale operations, the microscale dimensions allow mixing time to be reduced to the order of a few milliseconds during nucleation. The mixing time is usually slow in continuous flow devices. However, in cases that need

Figure 1. Mixing of reactants in continuous and segmented flow microfluidic devices

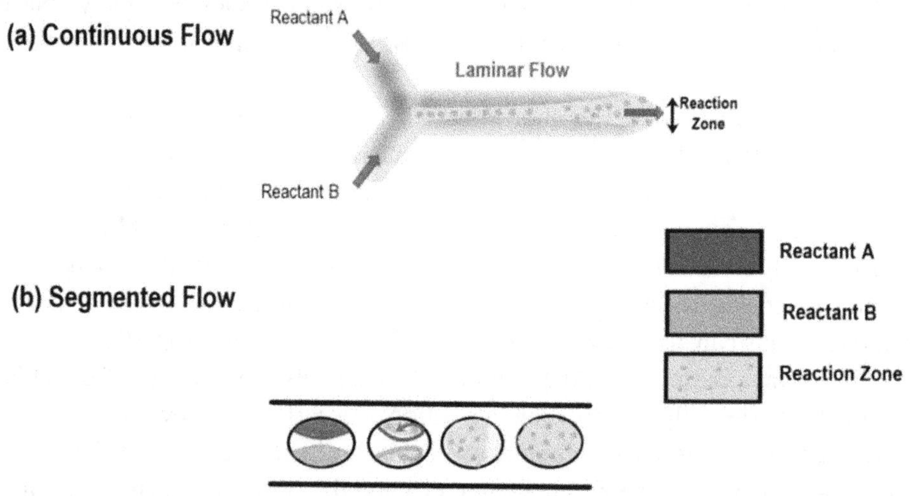

Table 1. Characteristic features of continuous and segmented flow microfluidic devices

Continuous Flow	Segmented Flow
Continuous fluid motion in microchannels	Droplet motion in microchannels formed using immiscible fluids
Slow mixing governed primarily by diffusion	Effective mixing possible by chaotic advection
Controlled reaction-diffusion environment	Programmable mixing
Parabolic velocity profile in microchannel	Continuous and dispersed phases in microchannel
Challenges in scalability	Challenges in implementation of stable gas-liquid systems

rapid interactions, active mixing approaches or geometrical effects such as spiral shaped channels or herringbone structures may be used to enhance mixing. The simplicity of design and operation allows addition of new reagents during the reaction. Further, these systems offer merits in terms of scalability by parallelization of multiple microfluidic reaction units for large scale production of nanoparticles (Jeong et al., 2016).

One of the most commonly used approaches used in case of continuous flow nanoparticle synthesis is co-precipitation. Iron oxide core chitosan NPs were prepared in a microfluidic device fabricated using 3D printing (Aşık et al., 2021). The primary factors that influenced the NP characteristics were volumetric flow rate, concentration of reactants, pH and geometrical aspects of microchannel. The microfluidic chip design with hurdles were found to reduce aggregate formation and obtain optimum mixing. The resultant NPs displayed DLS measurements of 190 ± 15 nm in diameter under optimized conditions. Although NPs can be prepared in a reproducible manner by this approach, the synthesis of different NPs with other materials necessitate optimization trials prior to large-scale production.

Magnetic nanoparticles were prepared by co-precipitation approach in a microfluidic device with serpentine micromixer (Thu et al., 2016). The microfluidic device was made of polydimethylsiloxane (PDMS) material using soft lithography. The mixture of Fe(II) and Fe(III) acidic solutions and NaOH precursor solutions were introduced into the device from two inlets. The serpentine channel provided

Figure 2. Microfluidic synthesis of $Cu/ZnO/Al_2O_3$ nanoparticles in a silicon-based continuous flow chip
Source: Tofighi et al. (2022)

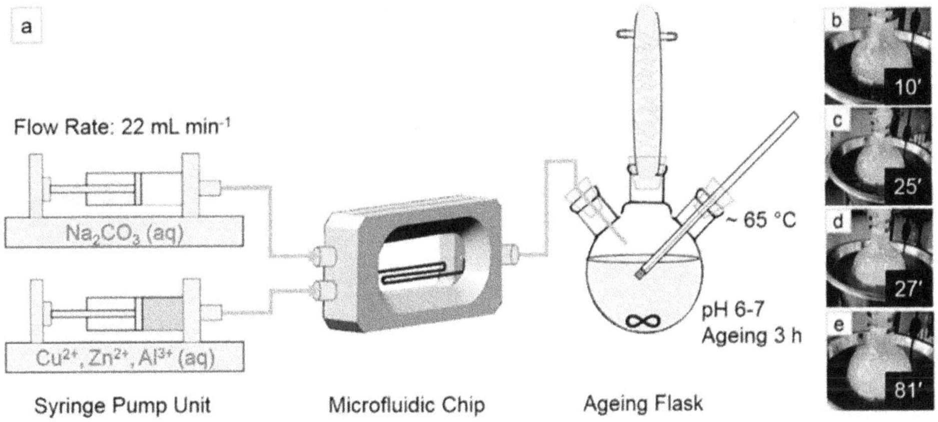

sufficient residence time for nanoparticle formation. In a recent study, $CuO/ZnO/Al_2O_3$ nanoparticles were synthesized via co-precipitation of a metal nitrate solution comprising of Cu^{2+}, Zn^{2+} and Al^{3+} with sodium carbonate solution using a microscale system as illustrate in Figure 2 (Tofighi et al., 2022). The chip was fabricated using Si-bonded glass. 40 mL of each reactant solution was introduced into the microfluidic device by means of syringe pump. The dilution of precursors prevented microchannels from getting clogged due to precipitate deposition. The output from the microfluidic device was subjected to ageing for 3 hours, washed, dried at ~343 K and calcined at ~603 K. The microfluidic synthesis resulted in the generation of 0.8 g of catalyst. The product exhibited small CuO crystallites resulting in better dispersion and more active Cu surface area.

Segmented Flow Systems

Even though microfluidic NP synthesis via co-precipitation approach in single phase continuous flow devices offer a simple, rapid and economical route, one of the major limitations of the method is clogging of microchannels. In some cases, such as magnetic NP synthesis, the issue gets aggravated due to extreme reactivity of magnetic precursors. In segmented flow systems, there are two or more immiscible fluid phases such that well-defined droplets are formed to carry out reactions in well-controlled microenvironments. These droplets could be visualized as compartmentalized units analogous to conventional batch processes, albeit with enhanced control over process parameters. Figure 3 illustrates several types of microfluidic device designs for segmented flow type operations.

Two-types of segmented flow devices have been developed, namely gas-liquid type and liquid-liquid type. Gas-liquid type systems consist of bubbly flow dispersed in a continuous liquid phase. Some of the most common types of liquid-liquid type devices use water-in-oil and oil-in-water dispersions. The design of microfluidic devices and flow rate play a crucial role in determining morphology of droplets formed.

A microreactor based on hydrodynamic flow focusing principle was demonstrated for synthesizing chitosan-coated magnetic NPs. These nanoparticles were designed for controlled release of cisdiammin-edichloroplatinum (II), a common therapeutic agent used in cancer treatment (Siavashy et al., 2021).These NPs were created by them utilizing a three-stage integrated microfluidic system. A superparamagnetic

Figure 3. Schematic illustration of segmented-flow type microfluidic device designs in various configurations, namely, (a-e) cross-flow, (f,g) co-flow, and (h-j) flow-focusing
Source: Abedini-Nassab et al. (2021)

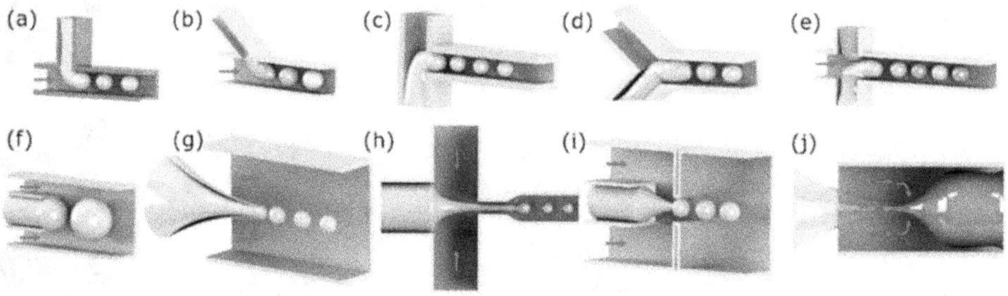

core was initially formed via precipitation. A polymer shell eventually developed on the magnetic core during the encapsulation phase. The therapeutic agent was then added after the core-shell particles became stabilized. Compared to traditional batch techniques, microfluidic platforms have a better ability to control the operational parameters. In this case, the central stream was focused and confined by two side streams under laminar flow conditions. The hydrodynamic flow focusing established homogenous mixing in the device via diffusive mechanism. The mixing rate was regulated by changing the flow rate ratios and stream widths. Subsequently, it controlled dimensions and surface characteristics of nanoparticles. The mixing duration was estimated by the following expression (equation 1).

$$t_{mix} \sim \frac{W_f^2}{4D} \approx \frac{W^2}{9D}\left(1+\frac{1}{FR}\right)^{-2} \tag{1}$$

W_f, W, D and FR designate central stream width, microchannel width, diffusion constant, and flow rate ratio of the focused to the side streams. The composite NPs formed using microfluidic approach were nearly monodispersed with dimension of 104 ± 14.59 nm (n = 3) with a polydispersity index of 0.029. Whereas, the conventional batch procedure resulted in formation of bigger sized particles of 423 ± 53.33 nm (n = 3) size having a polydispersity index of 0.384.

The improvement of design and potential extension of droplet type microfluidic devices depends on the extent of knowledge of hydrodynamics and mixing events within droplets. The impact of channel height on mixing effectiveness and iron oxide nanoparticle production using a droplet type microfluidic device was investigated using experimental and computational fluid dynamics simulation studies (Kašpar et al., 2020). As illustrated in Figure 4, the device had 4 inlets and 1 outlet. One of the inlets was for the continuous oil phase, and three others for dispersed phase including reactant 1, reactant 2 and blank. The device was fabricated in PDMS by soft lithography for three different heights of microchannels (20, 40 and 60 µm). Water served as the blank and facilitated separation of reactants prior to generation of droplets and also reduced the occurrences of fouling in the microscale device. The total residence time for a single droplet to travel through the entire microfluidic device ranged between 185 and 204 s for microchannels with various heights. The average dimensions of NPs obtained were 4.7 ± 0.9 nm, 6.2 ± 1.3 nm and 5.9 ± 1.1 nm for increasing microchannel heights. Their respective polydispersity indices were 0.037, 0.044 and 0.035. NPs synthesized using 20 µm high microchannel displayed a relatively

Figure 4. (a) Schematic drawing of microfluidic device for synthesis of iron oxide nanoparticles (A) and bright-field image of flow-focusing junction of microfluidic device fabricated using PDMS (B). (b) Size distribution and respective TEM images of nanoparticles produced by microfluidic devices with 20 µm (A and A'), 40 µm (B and B'), and 60 µm (C and C') microchannel heights.
Source: Kašpar et al. (2020)

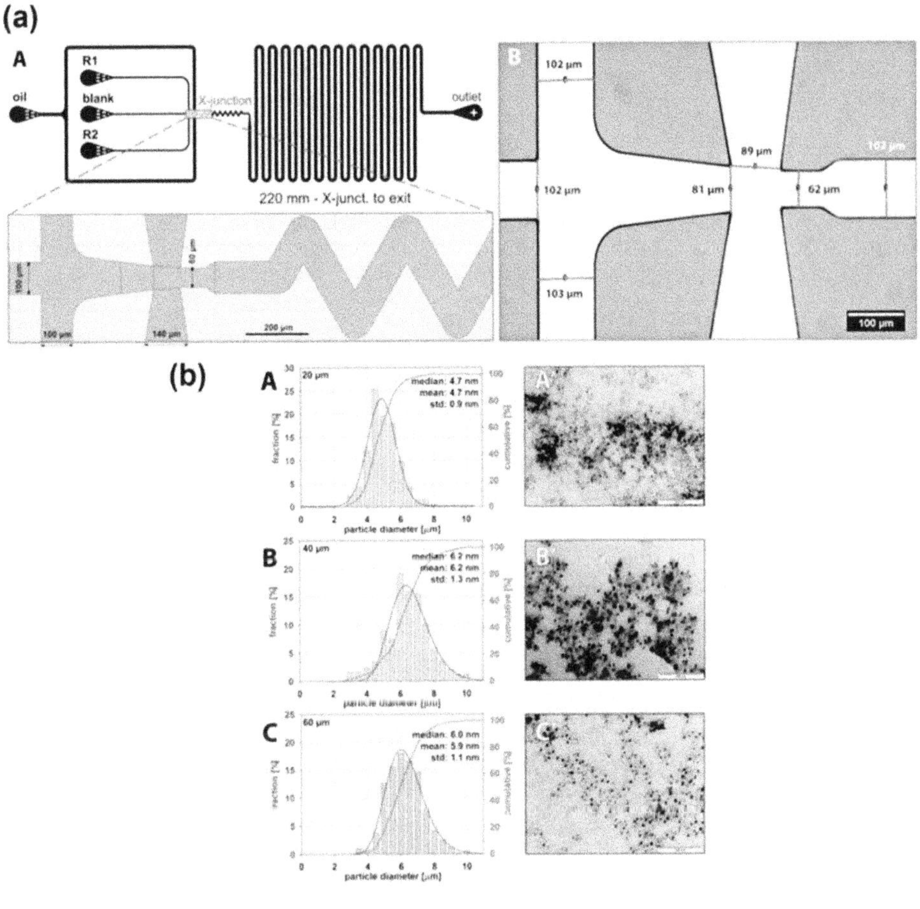

narrower particle size distribution that could be attributed to the faster homogenization compared to taller microchannels. The duration required to obtain mixing index ≥ 80% was determined as 19, 82 and 93 ms for 20, 40 and 60 µm high microchannels, respectively.

Enhancement of Performance in Microfluidic Nanoparticle Synthesis

The performance of microfluidic systems for nanoparticle synthesis can be improved by incorporating optimization and device design strategies. A practical demonstration of this was given in synthesis of lipid nanoparticles for in vivo delivery of small interfering ribonucleic acid (Belliveau et al., 2012). A syringe pump was used to pump lipid in ethanol and siRNA in aqueous solution into the two inlets of the microfluidic chip with staggered herringbone structures. The ethanol and aqueous phases quickly mixed as a result of the chaotic advection of the streams caused by herringbone structures. Hence, the

Figure 5. Schematic illustration of synthesis of lipid-polymer hybrid nanoparticles using a microfluidic device with staggered herringbone structures
Source: Gdowski et al. (2018)

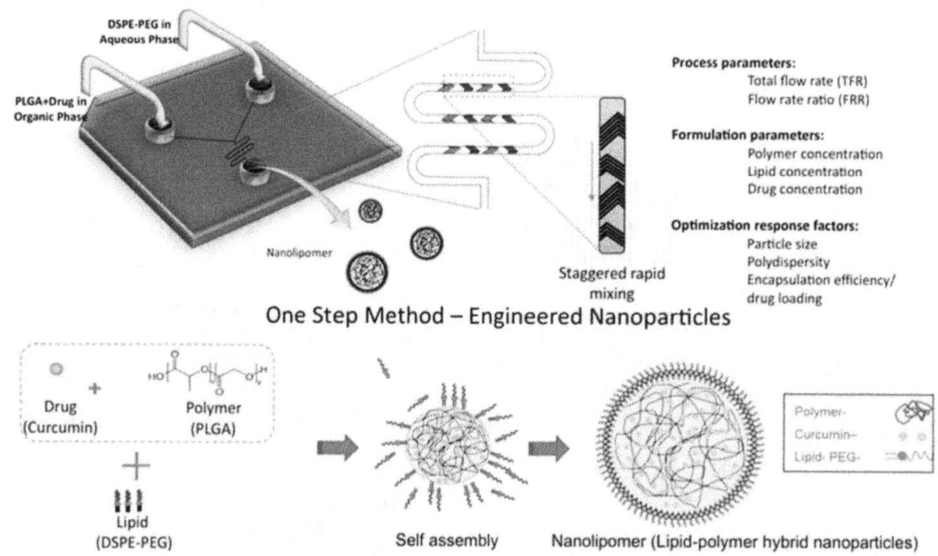

lipid solution polarity rapidly increased. Upon reaching a certain polarity, precipitates formed as lipid NPs. Figure 5 illustrates the formation of lipid-polymer hybrid nanoparticles designated as nanolipomers in a microfluidic device with staggered herringbone features for inducing rapid mixing (Gdowski et al., 2018). Poly(lactic-co-glycolic acid) (PLGA) and curcumin were injected from one of the inlets whereas 1,2-distearoyl-sn-glycero-3-phosphoethanolamine-N-amino(polyethylene glycol) (DSPE-PEG) solution was introduced from the other inlet. The process conditions were optimized and nanolipomers were obtained with average size of about 102 nm and polydispersity index of 0.126.

CHARACTERIZATION TECHNIQUES OF MICROFLUIDIC SYNTHESIZED NANOPARTICLES

Characterization of nanoparticles synthesized by microfluidics is crucial for understanding their properties and optimizing their performance. The common techniques used are mentioned as follows.

- **Dynamic Light Scattering (DLS):** This technique measures the size distribution of particles in a solution based on their Brownian motion. DLS is a widely used technique for measuring the hydrodynamic diameter and size distribution of nanoparticles. DLS approach was used to determine the lipid amount for complete covering on poly(lactide-co-glycolide) (PLGA) of monolayer and bilayer nanoparticles (L. Zhang et al., 2015).
- **Transmission Electron Microscopy (TEM):** TEM provides high-resolution images of individual nanoparticles, allowing for direct observation of their size, shape, and morphology. TEM can also be used to determine the crystal structure and lattice parameters of nanoparticles (Chakraborty et al., 2019; Mandal et al., 2022). TEM analysis was successfully demonstrated for studying -NH2-

and -COOH-functionalized magnetite nanoparticles (Chircov et al., 2022). NH2-functionalized NPs exhibited were uniform and quasi-spherical, whereas COOH-functionalized NPs exhibited polyhedral morphology. Further, crystallinity of NPs was also confirmed by high resolution-TEM images.

- **Scanning Electron Microscopy (SEM):** SEM provides high-resolution images of the surface of nanoparticles, allowing for observation of their morphology, size, and shape. SEM can also be used to determine the elemental composition of nanoparticles using energy-dispersive X-ray spectroscopy (EDS). The SEM images were used for characterization of Pd and Pd-Pt hybrid nanoparticles synthesized using a continuous flow microreactor. SEM image data revealed that the treatment process influenced the particle shape (Pekkari et al., 2019).
- **X-Ray Diffraction (XRD):** XRD is a technique used to determine the crystal structure of nanoparticles. X-rays are directed at the sample, and the diffraction pattern produced provides information about the crystal structure and other details of the nanoparticles (Roy et al., 2021). The crystallinity of metal oxide nanoparticles (TiO_2, CeO_2, ZnO) synthesized using a in a microfluidic flow-focusing device was confirmed by XRD analysis (Stolzenburg et al., 2018).
- **Fourier Transform Infrared Spectroscopy (FTIR):** FTIR measures the absorption of infrared radiation by the nanoparticles, providing information about their chemical composition and functional groups. FTIR spectra was used to gain information about functional groups during microscale synthesis of palladium nanoparticles via an eco-friendly protocol involving glucose and starch (Gioria et al., 2020).
- **Nuclear Magnetic Resonance (NMR) Spectroscopy:** NMR spectroscopy provides information about the chemical structure of nanoparticles, including their composition, bonding, and electronic properties. In a recent study, a flow-based microreactor and a miniature NMR relaxometer were designed and successfully demonstrated (Bemetz et al., 2018). The nanoparticles of approximately 25 nm primary particle dimension were prepared using 3D hydrodynamic flow focusing reactor and a 0.5 T permanent magnet was used for online measurement of NMR relaxation properties.
- **UV-Vis Spectroscopy:** UV-Vis spectroscopy measures the absorption of light by nanoparticles, providing insights in relation to their size and shape. An integrated system combining continuous gold nanoparticle synthesis with UV-Vis spectroscopic detection was exhibited for online process analysis (Yue et al., 2013). The spectra obtained at high time resolution (2 ms) allowed for the determination of liquid-phase concentration and multiphase flow features.

Here, it must be noted that there may be some differences in properties measured by different methods that should be accounted for. In a study involving synthesis of lignin-chitosan hybrid NPs using a microfluidic platform it was observed that particle size analysis using SEM, TEM and DLS indicated that particle size was smaller in case of SEM and TEM images compared to that estimated by DLS (Chai et al., 2021). The deviation could be due to the fact that SEM and TEM merely represent a part of the sample. On the other hand, DLS highlights large particles in the entire sample. Further, the differences could also be attributed to variation in operating conditions in different characterization techniques. While SEM and TEM estimate the size of NPs in dry state, DLS determines the hydrodynamic radius of NPs in the solution state, which include the dimensions of the particle and ion and solvent layer associated with the particle in the solution. Overall, a combination of several characterization techniques is typically used to obtain a comprehensive understanding of the properties of nanoparticles synthesized by microfluidics.

APPLICATIONS OF SYNTHESIZED NANOPARTICLES

Biomedical Applications

Drug Delivery

Microfluidics has emerged as a promising technique for the synthesis of nanoparticles with unique properties suitable for targeted drug delivery (Hasani-Sadrabadi et al., 2016). The development of drug delivery systems with enhanced efficacy and reduced toxicity is highly desirable. Here, it is discussed how microfluidic synthesized particles can be utilized for targeted drug delivery, which is a strategy that involves the selective delivery of drugs to specific sites in the body while minimizing their exposure to healthy tissues. Microfluidic synthesized particles can be functionalized with ligands or antibodies that selectively bind to specific receptors overexpressed on the surface of diseased cells. One example of targeted drug delivery using microfluidic synthesized particles is the delivery of chemotherapy drugs to cancer cells (Tahir et al., 2020). By functionalizing the nanoparticles with targeting moieties that specifically recognize cancer cells, these particles can selectively deliver the drug payload to cancer cells while minimizing toxicity to healthy tissues. Moreover, the small size of microfluidic synthesized nanoparticles enables them to penetrate deep into tumors, improving drug delivery to the tumor site. Another example of targeted drug delivery using microfluidic synthesized particles is the delivery of gene therapy to specific cells (Iliescu & Tresset, 2015). The functionalization of nanoparticles with targeting ligands specific for a particular cell type can enhance the uptake of gene therapy by those cells, thereby increasing the efficacy of the treatment. In addition to functionalizing nanoparticles with targeting ligands, microfluidic synthesized particles can also be engineered to respond to specific stimuli such as pH, temperature, or light. For instance, pH-responsive nanoparticles can release their drug payload in response to the acidic environment found in tumors, thereby improving drug delivery to cancer cells (Yan et al., 2020).

Imaging Agents

Microfluidic synthesis has also been utilized to develop imaging agents for various applications in medical diagnostics and research, with improved sensitivity and specificity. One example of imaging agents synthesized by microfluidics is magnetic nanoparticles for magnetic resonance imaging (MRI) (Wang et al., 2015). These nanoparticles can be synthesized with a specific size and shape to enhance their relaxivity, which is a measure of their ability to affect the magnetic resonance properties of water molecules surrounding them. Another example of imaging agents synthesized by microfluidics is quantum dots (QDs) for bioimaging (Hu et al., 2014). QDs are semiconductor nanoparticles that emit bright and stable fluorescence when excited by light. The size and composition of QDs can be precisely controlled during microfluidic synthesis to optimize their fluorescence properties. Furthermore, the surface of QDs can be functionalized with targeting ligands to enable the selective imaging. Microfluidic synthesis has been utilized to develop nanoparticles for photoacoustic imaging (PAI) (Cheheltani et al., 2016). These nanoparticles are designed to absorb light and convert it into heat, which is then detected by ultrasound imaging. The precise control over the size and shape of these nanoparticles provided by microfluidic synthesis can optimize their light absorption properties, enhancing their sensitivity and specificity for PAI.

Analytical Applications

The precise control over the physico-chemical characteristics of nanoparticles has enabled the development of sensitive and selective detection platforms (Vaishampayan et al., 2023). One example of analytical detection using microfluidic synthesized particles is the development of biosensors for the detection of biomolecules (Huang et al., 2015). These biosensors can be functionalized with ligands that selectively bind to the target biomolecules, enabling their detection with good precision and accuracy.

Another example of analytical detection using microfluidic synthesized particles is the detection of environmental pollutants or toxins (Mishra et al., 2022). Gold nanoparticles can be functionalized with aptamers that selectively bind to specific toxins or pollutants, enabling their detection with high sensitivity and specificity. The small size of these nanoparticles also allows for their detection using portable and low-cost devices such as lateral flow assays. Microfluidic synthesized nanoparticles have also been used for the development of immunoassays, which are analytical techniques that use the binding of antibodies to antigens to detect the presence of specific molecules in a sample (F. Li et al., 2020). By functionalizing nanoparticles with antibodies specific to a particular antigen, the specificity and detection limit of these immunoassays can be improved. Microfluidic synthesized nanoparticles are useful for the detection of pathogens. These nanoparticles can be functionalized with aptamers or antibodies that selectively bind to specific pathogens. Further, microfluidic synthesized nanoparticles can be used for the development of sensors for soil monitoring applications (Zhu et al., 2022). For example, nanoparticles synthesized by microfluidics can be used as the sensing element in electrochemical sensors for the detection of heavy metals and other contaminants in soil. The characteristics of nanoparticles can be leveraged to enhance the sensitivity and specificity of detection platforms for various applications in medical diagnostics, environmental monitoring, and food safety (Sridhar et al., 2022). Surface chemistry is one of the promising aspects in controlling the applications of nanoparticles in microfluidics. The surface chemistry of nanoparticles was tailored through fabrication of core-shell nanoparticles in a recent study (Mandal et al., 2022). Overall, microfluidic synthesis offers a promising platform for engineering nanoparticles for analytical detection applications.

Energy Applications

Solar Cells

The precise tuning of nanoparticle characteristics provided by microfluidic technology is beneficial for enhanced light absorption and energy conversion properties. QDs are examples of nanoparticles synthesized by microfluidics for solar cell applications (Chen et al., 2023). QDs can be tuned to absorb specific wavelengths of light, enabling their use in multi-junction solar cells that can harvest a larger portion of the solar spectrum. Perovskite materials have attracted significant interest for their potential to achieve high efficiencies in solar cell applications (Kulkarni et al., 2019). The precise control over the size and surface chemistry of perovskite nanomaterials provided by microfluidic synthesis can allow the optimization of their energy conversion properties and stability (Zou et al., 2022). Microfluidic synthesized nanoparticles can also be utilized as coatings in solar cells. For example, nanoparticles can be used in preparing anti-reflection coatings to enhance the light absorption of solar cells (Romasanta et al., 2019). The small size of these nanoparticles enables the formation of uniform and dense coatings that can improve the efficiency of solar cells. Finally, microfluidics can be used to construct nanomaterials with advanced capabilities for energy storage applications such as batteries or supercapacitors (Wu et al., 2022). Microfluidics can play a crucial role in the development of nanoparticles for solar cell and energy storage applications. The nanoparticle synthesis can be modulated to enhance the energy conversion and storage properties of these materials.

Fuel Cells

Microfluidic synthesis can be utilized to develop nanoparticles with catalytic properties for fuel cell reactions. Platinum nanoparticles find promising applications in fuel cells (Tahmasebi et al., 2018). Platinum is a highly effective catalyst for the oxygen reduction reaction (ORR) in fuel cells. Palladium nanoparticles also have utility in fuel cell applications (Chowdhury et al., 2016). Palladium-based materials are effective catalysts for the hydrogen oxidation reaction (HOR) in fuel cells (Miller et al., 2017). Microfluidic synthesized nanoparticles can also be utilized as supports for catalysts in fuel cells (Ran et al., 2015). For example, nanoparticles can be used as support materials for platinum catalysts in proton exchange membrane fuel cells (PEMFCs) (Higgins et al., 2010). The small size of these nanoparticles allows for their uniform dispersion, enhancing the stability and activity of the supported catalysts. Further, microfluidic synthesis can be employed to develop nanoparticles for hydrogen storage applications (Wang & Song, 2017).

Batteries

Microfluidic synthesis offers a plausible approach for the development of nanoparticles for battery applications. Transition metal oxide nanoparticles such as nickel oxide, cobalt oxide, and manganese oxide have potential in battery applications (S. Li et al., 2017). These nanoparticles have shown promise as anodes in lithium-ion batteries due to their high capacity and fast charge/discharge rates (Bresser et al., 2013). Sulfur nanoparticles on conductive matrices find viable usage in lithium-sulfur battery applications (Y. Li et al., 2021). Sulfur has a high theoretical capacity and can be utilized as a cathode material in lithium-sulfur batteries. Microfluidic synthesized nanoparticles can also be utilized as coatings or layers in batteries (T. Zhao et al., 2015). For example, nanoparticles synthesized by microfluidics can be used as protective coatings on electrode surfaces, improving their stability and performance. Additionally, nanoparticles can be used as conductive additives to improve the conductivity of electrode materials. Furthermore, microfluidic synthesis can also be utilized to develop nanoparticles for supercapacitor applications (Salve et al., 2020).

Environmental Applications

Water Treatment

Microfluidic synthesis has also been utilized to develop nanoparticles for water treatment applications (Ramezani et al., 2019). The controlled properties of nanoparticles provided by microfluidic technology has enabled the development of nanoparticles with enhanced adsorption and catalytic properties. Iron oxide nanoparticles have shown promise as adsorbents for the removal of pollutants from water (Getahun et al., 2022). Another example of nanoparticles synthesized by microfluidics for water treatment applications is silver nanoparticles (Khan et al., 2022). Silver nanoparticles exhibit good antimicrobial properties (Deshmukh et al., 2019). Microfluidic synthesized nanoparticles can also be utilized as catalysts for reactions involved in wastewater remediation. For example, nanoparticles synthesized can be used as catalysts for the degradation of organic pollutants in water (Rani et al., 2021). Finally, microfluidic synthesis can also be utilized to develop nanoparticles for membrane-based separation applications (Balta et al., 2012). Membrane-based separations have emerged as potential alternatives to conventional mass transfer operations in water pollution related applications (Dayanandan et al., 2021). Nanoparticles synthesized by microfluidics can be used as coatings or additives to improve the selectivity and permeability of membranes, enhancing their performance in water treatment applications (S. Zhang et al., 2013).

Soil Remediation

Microfluidics has the potential to produce nanoparticles with enhanced properties for soil remediation, such as zero-valent iron (ZVI) nanoparticles. The small size and high reactivity of ZVI nanoparticles can enable their effective adsorption and degradation of contaminants in soil, making them suitable for the remediation of soil contaminated with heavy metals, organic pollutants, and other contaminants (Galdames et al., 2020). Another example of nanoparticle application for soil remediation is titanium dioxide (TiO_2) nanoparticles (Irshad et al., 2021). These nanoparticles have photocatalytic properties that can be activated by light to produce reactive oxygen species that can degrade contaminants in soil, making them effective for the remediation of soil contaminated with organic pollutants and other contaminants. Nanoparticles can also be used for soil stabilization applications (Krishnan & Shukla, 2019).

Air Purification

Microfluidic synthesis has the potential to produce nanoparticles with enhanced properties for air purification applications (Peng et al., 2021). Metal-organic framework (MOF) nanoparticles are one example of materials that can be synthesized by microfluidics for air purification. MOFs have a porous structure that allows for high surface area and tunable properties, making them effective for gas adsorption and separation. Graphene oxide nanoparticles can also be synthesized by microfluidics for air purification. These nanoparticles have a large surface area and can be utilized for the adsorption of volatile organic compounds (VOCs) and other pollutants from air. Microfluidic synthesized nanoparticles can also be used as catalysts for air purification reactions, such as the degradation of VOCs and other air pollutants. This is due to the strong control over the surface chemistry of these nanoparticles, which allows for their optimization for catalytic activity in air purification applications.

FUTURE OUTLOOK AND CONCLUSION

Microfluidic nanoparticle synthesis has shown great potential for a wide range of applications, including drug delivery, imaging, energy storage, environmental remediation, and more. As the field of microfluidics continues to advance, there are several emerging trends and challenges that will shape the future of microfluidic nanoparticle synthesis. One of the key challenges in microfluidic nanoparticle synthesis is scaling up production to meet the demands of commercial applications. While microfluidic techniques offer effective control over nanoparticle properties, they are typically limited to small-scale production.

Researchers are currently exploring ways to increase the throughput of microfluidic systems, such as using parallel channels or larger reaction chambers. Another challenge in microfluidic nanoparticle synthesis is reproducibility. The precise control over nanoparticle properties provided by microfluidic technology is highly dependent on several parameters such as flow rates, temperature, and mixing conditions. Therefore, it is crucial to develop standardized protocols and quality control measures to ensure the reproducibility of nanoparticle synthesis across different systems and laboratories. Despite these challenges, there are several emerging trends in microfluidic nanoparticle synthesis that hold promise for the future. One trend is the development of multifunctional nanoparticles that can perform multiple tasks, such as drug delivery and imaging, in a single system. Another trend is the use of microfluidic synthesis to create complex, hierarchical structures of nanoparticles for advanced applications such as tissue engineering. In conclusion, microfluidic nanoparticle synthesis is a rapidly developing field with enormous potential for a wide range of applications. While there are still challenges to be overcome, such as scaling up production and ensuring reproducibility, emerging trends in the field offer exciting prospects for the future of microfluidic nanoparticle synthesis.

REFERENCES

Abedini-Nassab, R., Pouryosef Miandoab, M., & Şaşmaz, M. (2021). Microfluidic Synthesis, Control, and Sensing of Magnetic Nanoparticles: A Review. *Micromachines*, *12*(7), 768. doi:10.3390/mi12070768 PMID:34210058

Aşık, M. D., Kaplan, M., Çetin, B., & Sağlam, N. (2021). Synthesis of iron oxide core chitosan nanoparticles in a 3D printed microfluidic device. *Journal of Nanoparticle Research*, *23*(3), 62. doi:10.100711051-021-05171-y

Ayyub, P., Chandra, R., Taneja, P., Sharma, A. K., & Pinto, R. (2001). Synthesis of nanocrystalline material by sputtering and laser ablation at low temperatures. *Applied Physics. A, Materials Science & Processing*, *73*(1), 67–73. Advance online publication. doi:10.1007003390100833

Baig, N., Kammakakam, I., & Falath, W. (2021). Nanomaterials: A review of synthesis methods, properties, recent progress, and challenges. *Materials Advances*, *2*(6), 1821–1871. doi:10.1039/D0MA00807A

Balta, S., Sotto, A., Luis, P., Benea, L., Van der Bruggen, B., & Kim, J. (2012). A new outlook on membrane enhancement with nanoparticles: The alternative of ZnO. *Journal of Membrane Science*, *389*, 155–161. Advance online publication. doi:10.1016/j.memsci.2011.10.025

Banerjee, S., Benjwal, P., Singh, M., & Kar, K. K. (2018). Graphene oxide (rGO)-metal oxide (TiO2/Fe3O4) based nanocomposites for the removal of methylene blue. *Applied Surface Science*, *439*, 560–568. doi:10.1016/j.apsusc.2018.01.085

Banerjee, S., & Kar, K. K. (2016a). Superior water retention, ionic conductivity and thermal stability of sulfonated poly ether ether ketone/polypyrrole/aluminum phosphate nanocomposite based polymer electrolyte membrane. *Journal of Environmental Chemical Engineering*, *4*(1), 299–310. Advance online publication. doi:10.1016/j.jece.2015.11.033

Banerjee, S., & Kar, K. K. (2016b). Synergistic effect of aluminium phosphate and tungstophosphoric acid on the physicochemical properties of sulfonated poly ether ether ketone nanocomposite membrane. *Journal of Applied Polymer Science*, *133*(5), n/a. Advance online publication. doi:10.1002/app.42952

Belliveau, N. M., Huft, J., Lin, P. J., Chen, S., Leung, A. K., Leaver, T. J., Wild, A. W., Lee, J. B., Taylor, R. J., Tam, Y. K., Hansen, C. L., & Cullis, P. R. (2012). Microfluidic synthesis of highly potent limit-size lipid nanoparticles for in vivo delivery of siRNA. *Molecular Therapy. Nucleic Acids*, *1*(8), e37. Advance online publication. doi:10.1038/mtna.2012.28 PMID:23344179

Bemetz, J., Wegemann, A., Saatchi, K., Haase, A., Häfeli, U. O., Niessner, R., Gleich, B., & Seidel, M. (2018). Microfluidic-Based Synthesis of Magnetic Nanoparticles Coupled with Miniaturized NMR for Online Relaxation Studies. *Analytical Chemistry*, *90*(16), 9975–9982. Advance online publication. doi:10.1021/acs.analchem.8b02374 PMID:30044615

Boken, J., Soni, S. K., & Kumar, D. (2016). Microfluidic Synthesis of Nanoparticles and their Biosensing Applications. *Critical Reviews in Analytical Chemistry*, *46*(6), 538–561. doi:10.1080/10408347.2016.1169912 PMID:27070948

Bresser, D., Mueller, F., Fiedler, M., Krueger, S., Kloepsch, R., Baither, D., Winter, M., Paillard, E., & Passerini, S. (2013). Transition-Metal-Doped Zinc Oxide Nanoparticles as a New Lithium-Ion Anode Material. *Chemistry of Materials, 25*(24), 4977–4985. doi:10.1021/cm403443t

Carvalho, B. G., Ceccato, B. T., Michelon, M., Han, S. W., & de la Torre, L. G. (2022). Advanced Microfluidic Technologies for Lipid Nano-Microsystems from Synthesis to Biological Application. *Pharmaceutics, 14*(1), 141. doi:10.3390/pharmaceutics14010141 PMID:35057037

Chai, Y., Wang, Y., Li, B., Qi, W., Su, R., & He, Z. (2021). Microfluidic Synthesis of Lignin/Chitosan Nanoparticles for the pH-Responsive Delivery of Anticancer Drugs. *Langmuir, 37*(23), 7219–7226. doi:10.1021/acs.langmuir.1c00778 PMID:34078082

Chakraborty, S., Gupta, A. K., Roy, D., & Basumallick, A. (2019). Studies on nano-metal dispersed Cu-Cr matrix composite. *Materials Letters, 257*, 126739. Advance online publication. doi:10.1016/j.matlet.2019.126739

Cheheltani, R., Ezzibdeh, R. M., Chhour, P., Pulaparthi, K., Kim, J., Jurcova, M., Hsu, J. C., Blundell, C., Litt, H. I., Ferrari, V. A., Allcock, H. R., Sehgal, C. M., & Cormode, D. P. (2016). Tunable, biodegradable gold nanoparticles as contrast agents for computed tomography and photoacoustic imaging. *Biomaterials, 102*, 87–97. Advance online publication. doi:10.1016/j.biomaterials.2016.06.015 PMID:27322961

Chen, G., Zhu, X., Xing, C., Wang, Y., Xu, X., Bao, J., Huang, J., Zhao, Y., Wang, X., Zhou, X., Du, X., & Wang, X. (2023). Machine Learning-Assisted Microfluidic Synthesis of Perovskite Quantum Dots. *Advanced Photonics Research, 4*(1), 2200230. Advance online publication. doi:10.1002/adpr.202200230

Chircov, C., Bîrcă, A. C., Vasile, B. S., Oprea, O. C., Huang, K. S., & Grumezescu, A. M. (2022). Microfluidic Synthesis of -NH2- and -COOH-Functionalized Magnetite Nanoparticles. *Nanomaterials (Basel, Switzerland), 12*(18), 3160. Advance online publication. doi:10.3390/nano12183160 PMID:36144948

Chowdhury, S. R., Mukherjee, P., & Bhattachrya, S. (2016). Palladium and palladium–copper alloy nano particles as superior catalyst for electrochemical oxidation of methanol for fuel cell applications. *International Journal of Hydrogen Energy, 41*(38), 17072–17083. doi:10.1016/j.ijhydene.2016.05.239

Chung, C. K., Shih, T. R., Chang, C. K., Lai, C. W., & Wu, B. H. (2011). Design and experiments of a short-mixing-length baffled microreactor and its application to microfluidic synthesis of nanoparticles. *Chemical Engineering Journal, 168*(2), 790–798. doi:10.1016/j.cej.2010.12.035

Dayanandan, N., Kapoor, A., & Sivaraman, P. (2021). Studies on membrane distillation towards mitigating thermal pollution. *Chemical Papers, 75*(6), 2819–2833. doi:10.100711696-021-01525-x

Deshmukh, S. P., Patil, S. M., Mullani, S. B., & Delekar, S. D. (2019). Silver nanoparticles as an effective disinfectant: A review. *Materials Science and Engineering C, 97*, 954–965. doi:10.1016/j.msec.2018.12.102 PMID:30678983

Fabozzi, A., Della Sala, F., di Gennaro, M., Barretta, M., Longobardo, G., Solimando, N., Pagliuca, M., & Borzacchiello, A. (2023). Design of functional nanoparticles by microfluidic platforms as advanced drug delivery systems for cancer therapy. In Lab on a Chip (Vol. 23, Issue 5). doi:10.1039/D2LC00933A

Galdames, A., Ruiz-Rubio, L., Orueta, M., Sánchez-Arzalluz, M., & Vilas-Vilela, J. L. (2020). Zero-valent iron nanoparticles for soil and groundwater remediation. *International Journal of Environmental Research and Public Health, 17*(16), 5817. Advance online publication. doi:10.3390/ijerph17165817 PMID:32796749

Gao, Y., Wu, M., Lin, Y., & Xu, J. (2020). Trapping and control of bubbles in various microfluidic applications. *Lab on a Chip*, *20*(24), 4512–4527. doi:10.1039/D0LC00906G PMID:33232419

Gdowski, A., Johnson, K., Shah, S., Gryczynski, I., Vishwanatha, J., & Ranjan, A. (2018). Optimization and scale up of microfluidic nanolipomer production method for preclinical and potential clinical trials. *Journal of Nanobiotechnology*, *16*(1), 12. Advance online publication. doi:10.118612951-018-0339-0 PMID:29433518

Gervais, L., Hitzbleck, M., & Delamarche, E. (2011). Capillary-driven multiparametric microfluidic chips for one-step immunoassays. *Biosensors & Bioelectronics*, *27*(1), 64–70. doi:10.1016/j.bios.2011.06.016 PMID:21752632

Getahun, Y. W., Gardea-Torresdey, J., Manciu, F. S., Li, X., & El-Gendy, A. A. (2022). Green synthesized superparamagnetic iron oxide nanoparticles for water treatment with alternative recyclability. *Journal of Molecular Liquids*, *356*, 118983. Advance online publication. doi:10.1016/j.molliq.2022.118983

Gioria, E., Signorini, C., Wisniewski, F., & Gutierrez, L. (2020). Green synthesis of time-stable palladium nanoparticles using microfluidic devices. *Journal of Environmental Chemical Engineering*, *8*(5), 104096. Advance online publication. doi:10.1016/j.jece.2020.104096

Gonidec, M., & Puigmartí-Luis, J. (2018). Continuous- versus Segmented-Flow Microfluidic Synthesis in Materials Science. *Crystals*, *9*(1), 12. doi:10.3390/cryst9010012

Hasani-Sadrabadi, M. M., Taranejoo, S., Dashtimoghadam, E., Bahlakeh, G., Majedi, F. S., VanDersarl, J. J., Janmaleki, M., Sharifi, F., Bertsch, A., Hourigan, K., Tayebi, L., Renaud, P., & Jacob, K. I. (2016). Microfluidic Manipulation of Core/Shell Nanoparticles for Oral Delivery of Chemotherapeutics: A New Treatment Approach for Colorectal Cancer. *Advanced Materials*, *28*(21), 4134–4141. doi:10.1002/adma.201502697 PMID:27001745

Higgins, D. C., Meza, D., & Chen, Z. (2010). Nitrogen-Doped Carbon Nanotubes as Platinum Catalyst Supports for Oxygen Reduction Reaction in Proton Exchange Membrane Fuel Cells. *The Journal of Physical Chemistry C*, *114*(50), 21982–21988. doi:10.1021/jp106814j

Hu, S., Zeng, S., Zhang, B., Yang, C., Song, P., Hang Danny, T. J., Lin, G., Wang, Y., Anderson, T., Coquet, P., Liu, L., Zhang, X., & Yong, K. T. (2014). Preparation of biofunctionalized quantum dots using microfluidic chips for bioimaging. *Analyst (London)*, *139*(18), 4681–4690. Advance online publication. doi:10.1039/C4AN00773E PMID:25054471

Huang, Y., Shi, Y., Yang, H. Y., & Ai, Y. (2015). A novel single-layered MoS2 nanosheet based microfluidic biosensor for ultrasensitive detection of DNA. *Nanoscale*, *7*(6), 2245–2249. Advance online publication. doi:10.1039/C4NR07162J PMID:25567642

Iliescu, C., & Tresset, G. (2015). Microfluidics-Driven Strategy for Size-Controlled DNA Compaction by Slow Diffusion through Water Stream. *Chemistry of Materials, 27*(24), 8193–8197. doi:10.1021/acs.chemmater.5b04129

Irshad, M. A., Nawaz, R., Rehman, M. Z., Adrees, M., Rizwan, M., Ali, S., Ahmad, S., & Tasleem, S. (2021). Synthesis, characterization and advanced sustainable applications of titanium dioxide nanoparticles: A review. *Ecotoxicology and Environmental Safety, 212*, 111978. doi:10.1016/j.ecoenv.2021.111978 PMID:33561774

Ismail, R. A., Mohsin, M. H., Ali, A. K., Hassoon, K. I., & Erten-Ela, S. (2020). Preparation and characterization of carbon nanotubes by pulsed laser ablation in water for optoelectronic application. *Physica E, Low-Dimensional Systems and Nanostructures, 119*, 113997. Advance online publication. doi:10.1016/j.physe.2020.113997

Jeong, H.-H., Issadore, D., & Lee, D. (2016). Recent developments in scale-up of microfluidic emulsion generation via parallelization. *Korean Journal of Chemical Engineering, 33*(6), 1757–1766. doi:10.100711814-016-0041-6

Kašpar, O., Koyuncu, A. H., Hubatová-Vacková, A., Balouch, M., & Tokárová, V. (2020). Influence of channel height on mixing efficiency and synthesis of iron oxide nanoparticles using droplet-based microfluidics. *RSC Advances, 10*(26), 15179–15189. Advance online publication. doi:10.1039/D0RA02470H PMID:35495462

Kašpar, O., Koyuncu, A. H., Pittermannová, A., Ulbrich, P., & Tokárová, V. (2019). Governing factors for preparation of silver nanoparticles using droplet-based microfluidic device. *Biomedical Microdevices, 21*(4), 88. doi:10.100710544-019-0435-4 PMID:31571009

Khan, S. A., Jain, M., Pandey, A., Pant, K. K., Ziora, Z. M., Blaskovich, M. A. T., Shetti, N. P., & Aminabhavi, T. M. (2022). Leveraging the potential of silver nanoparticles-based materials towards sustainable water treatment. *Journal of Environmental Management, 319*, 115675. doi:10.1016/j.jenvman.2022.115675 PMID:35834856

Khanna, P., Kaur, A., & Goyal, D. (2019). Algae-based metallic nanoparticles: Synthesis, characterization and applications. *Journal of Microbiological Methods, 163*, 105656. doi:10.1016/j.mimet.2019.105656 PMID:31220512

Kovarik, M. L., Gach, P. C., Ornoff, D. M., Wang, Y., Balowski, J., Farrag, L., & Allbritton, N. L. (2012). Micro Total Analysis Systems for Cell Biology and Biochemical Assays. *Analytical Chemistry, 84*(2), 516–540. doi:10.1021/ac202611x PMID:21967743

Krishnan, J., & Shukla, S. (2019). The behaviour of soil stabilised with nanoparticles: an extensive review of the present status and its applications. In Arabian Journal of Geosciences (Vol. 12, Issue 14). doi:10.100712517-019-4595-6

Kulkarni, S. A., Mhaisalkar, S. G., Mathews, N., & Boix, P. P. (2019). Perovskite Nanoparticles: Synthesis, Properties, and Novel Applications in Photovoltaics and LEDs. *Small Methods, 3*(1), 1800231. doi:10.1002mtd.201800231

Li, F., Guo, L., Hu, Y., Li, Z., Liu, J., He, J., & Cui, H. (2020). Multiplexed chemiluminescence determination of three acute myocardial infarction biomarkers based on microfluidic paper-based immunodevice dual amplified by multifunctionalized gold nanoparticles. *Talanta*, *207*, 120346. Advance online publication. doi:10.1016/j.talanta.2019.120346 PMID:31594588

Li, L.-L., Li, X., & Wang, H. (2017). Microfluidic Synthesis of Nanomaterials for Biomedical Applications. *Small Methods*, *1*(8), 1700140. doi:10.1002mtd.201700140

Li, S., Liu, Y., Guo, P., & Wang, C. (2017). Self-Climbed Amorphous Carbon Nanotubes Filled with Transition Metal Oxide Nanoparticles for Large Rate and Long Lifespan Anode Materials in Lithium Ion Batteries. *ACS Applied Materials & Interfaces*, *9*(32), 26818–26825. doi:10.1021/acsami.7b06394 PMID:28745487

Li, W., & Zhao, D. (2013). An overview of the synthesis of ordered mesoporous materials. *Chemical Communications*, *49*(10), 943–946. Advance online publication. doi:10.1039/C2CC36964H PMID:23249963

Li, Y., Yao, C., Wang, C., Yang, P., Wu, R., Fei, L., Zhang, Y., & Jiang, Y. (2021). An approach through steam to form sulfur nanoparticles for lithium sulfur batteries. *Electrochemistry Communications*, *125*, 107010. doi:10.1016/j.elecom.2021.107010

Liu, H., Huang, J., Sun, D., Lin, L., Lin, W., Li, J., Jiang, X., Wu, W., & Li, Q. (2012). Microfluidic biosynthesis of silver nanoparticles: Effect of process parameters on size distribution. *Chemical Engineering Journal*, *209*, 568–576. doi:10.1016/j.cej.2012.08.049

Liu, Y., Sun, L., Zhang, H., Shang, L., & Zhao, Y. (2021). Microfluidics for Drug Development: From Synthesis to Evaluation. *Chemical Reviews*, *121*(13), 7468–7529. doi:10.1021/acs.chemrev.0c01289 PMID:34024093

Luo, T., Fan, L., Zhu, R., & Sun, D. (2019). Microfluidic Single-Cell Manipulation and Analysis: Methods and Applications. *Micromachines*, *10*(2), 104. doi:10.3390/mi10020104 PMID:30717128

Mahdavi, Z., Rezvani, H., & Keshavarz Moraveji, M. (2020). Core–shell nanoparticles used in drug delivery-microfluidics: A review. *RSC Advances*, *10*(31), 18280–18295. doi:10.1039/D0RA01032D PMID:35517190

Mandal, S., Kumar Gupta, A., Echeverria, E., McIlroy, D. N., Poplawsky, J. D., & Sachan, R. (2022). Laser-assisted nanofabrication of multielement complex oxide core–shell nanoparticles. *Materials & Design*, *220*, 110882. Advance online publication. doi:10.1016/j.matdes.2022.110882

Miller, H. A., Vizza, F., Marelli, M., Zadick, A., Dubau, L., Chatenet, M., Geiger, S., Cherevko, S., Doan, H., Pavlicek, R. K., Mukerjee, S., & Dekel, D. R. (2017). Highly active nanostructured palladium-ceria electrocatalysts for the hydrogen oxidation reaction in alkaline medium. *Nano Energy*, *33*, 293–305. doi:10.1016/j.nanoen.2017.01.051

Mishra, N., Dhwaj, A., Verma, D., & Prabhakar, A. (2022). Cost-effective microabsorbance detection based nanoparticle immobilized microfluidic system for potential investigation of diverse chemical contaminants present in drinking water. *Analytica Chimica Acta*, *1205*, 339734. Advance online publication. doi:10.1016/j.aca.2022.339734 PMID:35414395

Niculescu, A.-G., Mihaiescu, D. E., & Grumezescu, A. M. (2022). A Review of Microfluidic Experimental Designs for Nanoparticle Synthesis. *International Journal of Molecular Sciences*, *23*(15), 8293. doi:10.3390/ijms23158293 PMID:35955420

Otzen, D. E., Buell, A. K., & Jensen, H. (2021). Microfluidics and the quantification of biomolecular interactions. *Current Opinion in Structural Biology*, *70*, 8–15. doi:10.1016/j.sbi.2021.02.006 PMID:33831785

Pekkari, A., Say, Z., Susarrey-Arce, A., Langhammer, C., Härelind, H., Sebastian, V., & Moth-Poulsen, K. (2019). Continuous Microfluidic Synthesis of Pd Nanocubes and PdPt Core–Shell Nanoparticles and Their Catalysis of NO_2 Reduction. *ACS Applied Materials & Interfaces*, *11*(39), 36196–36204. doi:10.1021/acsami.9b09701 PMID:31418548

Peng, X., Jiang, M., Wang, X., Li, D., Sun, S., Shao, Y., & Zheng, Y. (2021). Photocatalytic purification of contaminated air in intensive care units by ZnSn(OH)6 nanoparticles. *Environmental Science and Pollution Research International*, *28*(24), 31770–31777. Advance online publication. doi:10.100711356-021-12970-2 PMID:33611731

Poolakkandy, R. R., & Menamparambath, M. M. (2020). Soft-template-assisted synthesis: A promising approach for the fabrication of transition metal oxides. *Nanoscale Advances*, *2*(11), 5015–5045. doi:10.1039/D0NA00599A PMID:36132034

Postek, W., & Garstecki, P. (2022). Droplet Microfluidics for High-Throughput Analysis of Antibiotic Susceptibility in Bacterial Cells and Populations. *Accounts of Chemical Research*, *55*(5), 605–615. doi:10.1021/acs.accounts.1c00729 PMID:35119826

Ramezani, S., Zahedi, P., Bahrami, S. H., & Nemati, Y. (2019). Microfluidic Fabrication of Nanoparticles Based on Ethyl Acrylate-Functionalized Chitosan for Adsorption of Methylene Blue from Aqueous Solutions. *Journal of Polymers and the Environment*, *27*(8), 1653–1665. Advance online publication. doi:10.100710924-019-01463-6

Ran, G., Fu, Q., & Xu, W. (2015). Microfluidic-based controllable synthesis of Pt nanocatalysts supported on carbon for fuel cells. *RSC Advances*, *5*(19), 14740–14746. Advance online publication. doi:10.1039/C4RA12145G

Rani, M., Keshu, & Shanker, U. (2021). Sunlight-induced photocatalytic degradation of organic pollutants by biosynthesized hetrometallic oxides nanoparticles. *Environmental Science and Pollution Research International*, *28*(43), 61760–61780. doi:10.100711356-021-15003-0 PMID:34189687

Rizkin, B. A., Shkolnik, A. S., Ferraro, N. J., & Hartman, R. L. (2020). Combining automated microfluidic experimentation with machine learning for efficient polymerization design. *Nature Machine Intelligence*, *2*(4), 200–209. doi:10.103842256-020-0166-5

Romasanta, L. J., D'Alençon, L., Kirchner, S., Pradère, C., & Leng, J. (2019). Thin Coatings of Cerium Oxide Nanoparticles with Anti-Reflective Properties. *Applied Sciences (Basel, Switzerland)*, *9*(18), 3886. doi:10.3390/app9183886

Roy, D., Chakraborty, S., Gupta, A. K., BasuMallick, A., Scattergood, R. O., & Koch, C. C. (2021). Synergistic effect of Nb and Zr additions on the structure-property relationships of nanocrystalline Cu processed by mechanical alloying and hot pressing. *Journal of Alloys and Compounds*, *854*, 157174. Advance online publication. doi:10.1016/j.jallcom.2020.157174

Salve, M., Mandal, A., Amreen, K., Pattnaik, P. K., & Goel, S. (2020). Greenly synthesized silver nanoparticles for supercapacitor and electrochemical sensing applications in a 3D printed microfluidic platform. *Microchemical Journal*, *157*, 104973. Advance online publication. doi:10.1016/j.microc.2020.104973

Shiri, M. S. Z., Henderson, W., & Mucalo, M. R. (2019). A review of the lesser-studied microemulsion-based synthesis methodologies used for preparing nanoparticle systems of the noble metals, Os, Re, Ir and Rh. In Materials (Vol. 12, Issue 12). doi:10.3390/ma12121896

Siavashy, S., Soltani, M., Ghorbani-Bidkorbeh, F., Fallah, N., Farnam, G., Mortazavi, S. A., Shirazi, F. H., Tehrani, M. H. H., & Hamedi, M. H. (2021). Microfluidic platform for synthesis and optimization of chitosan-coated magnetic nanoparticles in cisplatin delivery. *Carbohydrate Polymers*, *265*, 118027. doi:10.1016/j.carbpol.2021.118027 PMID:33966822

Sri Sruthi, P., Balasubramanian, S., Senthil Kumar, P., Kapoor, A., Ponnuchamy, M., Mariam Jacob, M., & Prabhakar, S. (2021). Eco-friendly pH detecting paper-based analytical device: Towards process intensification. *Analytica Chimica Acta*, *1182*, 338953. doi:10.1016/j.aca.2021.338953 PMID:34602199

Sridhar, A., Kapoor, A., Kumar, P. S., Ponnuchamy, M., Sivasamy, B., & Vo, D.-V. N. (2022). Lab-on-a-chip technologies for food safety, processing, and packaging applications: A review. *Environmental Chemistry Letters*, *20*(1), 901–927. doi:10.100710311-021-01342-4 PMID:34803553

Stolzenburg, P., Lorenz, T., Dietzel, A., & Garnweitner, G. (2018). Microfluidic synthesis of metal oxide nanoparticles via the nonaqueous method. *Chemical Engineering Science*, *191*, 500–510. Advance online publication. doi:10.1016/j.ces.2018.07.007

Tahir, N., Madni, A., Li, W., Correia, A., Khan, M. M., Rahim, M. A., & Santos, H. A. (2020). Microfluidic fabrication and characterization of Sorafenib-loaded lipid-polymer hybrid nanoparticles for controlled drug delivery. *International Journal of Pharmaceutics*, *581*, 119275. Advance online publication. doi:10.1016/j.ijpharm.2020.119275 PMID:32229283

Tahmasebi, S., Jerkiewicz, G., Baranton, S., Coutanceau, C., Furuya, Y., & Ohma, A. (2018). How Stable Are Spherical Platinum Nanoparticles Applied to Fuel Cells? *The Journal of Physical Chemistry C*, *122*(22), 11765–11776. Advance online publication. doi:10.1021/acs.jpcc.7b10617

Tang, M. Y. H., & Shum, H. C. (2016). One-step immunoassay of C-reactive protein using droplet microfluidics. *Lab on a Chip*, *16*(22), 4359–4365. doi:10.1039/C6LC01121G PMID:27738692

Thu, V. T., Mai, A. N., Van Trung, H., Thu, P. T., Tien, B. Q., Thuat, N. T., & Lam, T. D. (2016). Fabrication of PDMS-Based Microfluidic Devices: Application for Synthesis of Magnetic Nanoparticles. *Journal of Electronic Materials*, *45*(5), 2576–2581. Advance online publication. doi:10.100711664-016-4424-6

Tofighi, G., Lichtenberg, H., Gaur, A., Wang, W., Wild, S., Herrera Delgado, K., Pitter, S., Dittmeyer, R., Grunwaldt, J. D., & Doronkin, D. E. (2022). Continuous synthesis of Cu/ZnO/Al2O3nanoparticles in a co-precipitation reaction using a silicon based microfluidic reactor. *Reaction Chemistry & Engineering*, *7*(3), 730–740. Advance online publication. doi:10.1039/D1RE00499A

Vaishampayan, V., Kapoor, A., & Gumfekar, S. P. (2023). Enhancement in the limit of detection of lab-on-chip microfluidic devices using functional nanomaterials. *Canadian Journal of Chemical Engineering*, cjce.24915. Advance online publication. doi:10.1002/cjce.24915

Wang, J., & Song, Y. (2017). Microfluidic Synthesis of Nanohybrids. *Small*, *13*(18), 1604084. doi:10.1002mll.201604084 PMID:28256806

Wang, J., Zhao, K., Shen, X., Zhang, W., Ji, S., Song, Y., Zhang, X., Rong, R., & Wang, X. (2015). Microfluidic synthesis of ultra-small magnetic nanohybrids for enhanced magnetic resonance imaging. *Journal of Materials Chemistry. C, Materials for Optical and Electronic Devices*, *3*(48), 12418–12429. doi:10.1039/C5TC02279G

Wu, X., Sun, S., Yu, X., Zhu, X., Xu, M., Wu, G., & Xu, J. (2022). Review on Microfluidic Construction of Advanced Nanomaterials for High-Performance Energy Storage Applications. *Energy & Fuels*, *36*(9), 4708–4727. doi:10.1021/acs.energyfuels.2c00576

Yadav, A. K., Banerjee, S., Kumar, R., Kar, K. K., Ramkumar, J., & Dasgupta, K. (2018). Mechanical Analysis of Nickel Particle-Coated Carbon Fiber-Reinforced Epoxy Composites for Advanced Structural Applications. *ACS Applied Nano Materials*, *1*(8), 4332–4339. Advance online publication. doi:10.1021/acsanm.8b01193

Yan, J., Xu, X., Zhou, J., Liu, C., Zhang, L., Wang, D., Yang, F., & Zhang, H. (2020). Fabrication of a pH/Redox-Triggered Mesoporous Silica-Based Nanoparticle with Microfluidics for Anticancer Drugs Doxorubicin and Paclitaxel Codelivery. *ACS Applied Bio Materials*, *3*(2), 1216–1225. Advance online publication. doi:10.1021/acsabm.9b01111 PMID:35019322

Yue, J., Falke, F. H., Schouten, J. C., & Nijhuis, T. A. (2013). Microreactors with integrated UV/Vis spectroscopic detection for online process analysis under segmented flow. *Lab on a Chip*, *13*(24), 4855. Advance online publication. doi:10.1039/c3lc50876e PMID:24178763

Zhang, D., Ye, K., Yao, Y., Liang, F., Qu, T., Ma, W., Yang, B., Dai, Y., & Watanabe, T. (2019). Controllable synthesis of carbon nanomaterials by direct current arc discharge from the inner wall of the chamber. *Carbon*, *142*, 278–284. Advance online publication. doi:10.1016/j.carbon.2018.10.062

Zhang, L., Feng, Q., Wang, J., Zhang, S., Ding, B., Wei, Y., Dong, M., Ryu, J.-Y., Yoon, T.-Y., Shi, X., Sun, J., & Jiang, X. (2015). Microfluidic Synthesis of Hybrid Nanoparticles with Controlled Lipid Layers: Understanding Flexibility-Regulated Cell–Nanoparticle Interaction. *ACS Nano*, *9*(10), 9912–9921. doi:10.1021/acsnano.5b05792 PMID:26448362

Zhang, S., Qiu, G., Ting, Y. P., & Chung, T. S. (2013). Silver-PEGylated dendrimer nanocomposite coating for anti-fouling thin film composite membranes for water treatment. *Colloids and Surfaces. A, Physicochemical and Engineering Aspects*, *436*, 207–214. Advance online publication. doi:10.1016/j.colsurfa.2013.06.027

Zhao, T., Li, L., Chen, R., Wu, H., Zhang, X., Chen, S., Xie, M., Wu, F., Lu, J., & Amine, K. (2015). Design of surface protective layer of LiF/FeF3 nanoparticles in Li-rich cathode for high-capacity Li-ion batteries. *Nano Energy*, *15*, 164–176. Advance online publication. doi:10.1016/j.nanoen.2015.04.013

Zhao, Y., Cheng, Y., Shang, L., Wang, J., Xie, Z., & Gu, Z. (2015). Microfluidic Synthesis of Barcode Particles for Multiplex Assays. *Small*, *11*(2), 151–174. doi:10.1002mll.201401600 PMID:25331055

Zhou, L., Chen, Y., Fang, X., Liu, Y., Du, M., Lu, X., Li, Q., Sun, Y., Ma, J., & Lan, T. (2020). Microfluidic-RT-LAMP chip for the point-of-care detection of emerging and re-emerging enteric coronaviruses in swine. *Analytica Chimica Acta*, *1125*, 57–65. doi:10.1016/j.aca.2020.05.034 PMID:32674781

Zhu, X., Wang, K., Yan, H., Liu, C., Zhu, X., & Chen, B. (2022). Microfluidics as an Emerging Platform for Exploring Soil Environmental Processes: A Critical Review. In Environmental Science and Technology (Vol. 56, Issue 2). doi:10.1021/acs.est.1c03899

Zhuang, S., Lee, E. S., Lei, L., Nunna, B. B., Kuang, L., & Zhang, W. (2016). Synthesis of nitrogen-doped graphene catalyst by high-energy wet ball milling for electrochemical systems. *International Journal of Energy Research*, *40*(15), 2136–2149. Advance online publication. doi:10.1002/er.3595

Zou, S., Zhao, X., Ouyang, W., & Xu, S. (2022). Microfluidic Synthesis, Doping Strategy, and Optoelectronic Applications of Nanostructured Halide Perovskite Materials. *Micromachines*, *13*(10), 1647. doi:10.3390/mi13101647 PMID:36296000

KEY TERMS AND DEFINITIONS

Advection: Movement of molecules through a medium by bulk flow, typically driven by a pressure gradient or a concentration gradient.

Diffusion: Motion of molecules from high concentration to low concentration region, driven by the natural tendency of the molecules to distribute evenly in space.

Laminar Flow: Fluid flow in which the fluid particles move parallel to each other, with no mixing or crossflow between adjacent layers.

Microfluidics: Study and manipulation of fluids in channels with dimensions ranging from tens to hundreds of micrometers.

Nanoparticles: Nanoparticles are referred to particles having at least one dimension in the size scale of 10^{-9} m i.e., nano scale.

Nanostructures: This is a term indicating materials with structural elements, crystals or clusters having dimensions within the size range 1–100 nm.

Nanotechnology: Nanotechnology refers to a technology of materials having size in the nanometric range or it is the technology discussed on the properties of any material in the nanometric size scale.

Chapter 10
Design, Development, and Applications of Microfluidics for Rapid Detection and Quantization of Microparticles

Anil Baliram Ghubade
Lovely Professional University, India

Nitin Chauhan
Lovely Professional University, India

Parveen Kumar Sharma
Lovely Professional University, India

ABSTRACT

Rapid detection and quantization of microparticles whose size is comparable to biological entities provide a new avenue for rapid diagnosis and clinical applications. There are various methodologies that have been adopted for the detection and quantification of these micro entities like changes in permittivity, medium impedance, magnetic permeability, and other means. Recently, microfluidics plays an important role in the process of detection and diagnosis of these microparticles by using chip-based techniques. In this work, the authors reviewed the necessity of microfluidic devices, various designs for rapid detection, processes of device fabrication, and range of applicability. Recent research papers have been selected and studied. The outcomes of these papers are presented systematically in the chapter.

INTRODUCTION

The rapid and efficient detection of micro-organisms has put tremendous pressure to design and develop silicon based miniaturized devices for clinical applications. For the diagnosis of such micron entities, microfluidics plays an important role in transporting these biological entities through novel design of

DOI: 10.4018/978-1-6684-6952-1.ch010

microchannels. "microfluidics" is a term refers to transportation phenomenon at the microscopic length in the range of 1-500 um and design methodology to control and actuate the fluid (Addae-Mensah et al., 2010). The recent advancement and use of microfluidics is in detection of pathogens (Nasseri et al., 2018), analysis of viruses (Basiri et al., 2021), detection of foodborne pathogens (McDonnell et al., 2005), detection of nucleic acids (Zanoli & Spoto, 2012), detection of glucose in the blood sample (Noiphung et al., 2013), separation and quantification of cells (Cheng et al., 2007), COVID-19 antibody detection (Tayyab et al., 2020).

To mimic microfluidics applications, micro electromechanical devices (MEMS) assist microfluidics for rapid detection and quantification for biological entities in real time (Ghubade et al., 2009). Microfluidics manipulate a very small amount of fluids in the range of 10^{-9} to 10^{-18} liters through developing micro-channels with dimensions 10 to 100 microns results in rapid detection and separation of biological entities from the sample with high resolution, sensitivity, cost effective and short time of analysis (Whitesides, 2006). Microfluidics finds applications in four different domains namely molecular analysis, biodefence, molecular biology and microelectronics. Rapid development of microelectronic industry gives rise to the development of miniaturized silicon-based devices. The advent of silicon based microdevices provides the foundation for the development of micro electromechanical devices (MEMS) and henceforth the applications in medical field (Ho & Tai, 1998). The development of lab on chip (LOC) brings revolution in the field of microfluidics to integrate entire chemical and biological laboratory on a single platform for rapid and efficient diagnosis of micro entities as well as biological entities (Breslauer et al., 2006). Lab on chip is the concept to build the entire lab on a thin glass substrate, creating microchannels, interdigitated electrodes, sensors and electronic circuits for signal processing. The dimensions width or height of microchannels varies from 20 to 200 microns. Applying suitable low electrical potential between the electrodes can manipulate the micro entities within the microfluidic sample. Now a days, labs on chip helps researchers fulfill various specialized functions such as clinical diagnosis of bacteria and viruses, and DNA electrophoretic separation (Li, 2004). The Lab on chips must be able to perform microfluidic functions such as pumping, metering, mixing, flow switching, incubating, sample injection, separation of biological entities etc. The functioning of microfluidics depends on the fluid properties such as inertia and viscosity. The relationship between the flow and pressure is defined by viscosity and dimensions of the channel. The effect of viscosity, friction and electrostatic forces increases with decrease in size of microfluidic devices. Another important aspect of microfluidics is the surface tension which is primarily predominant. There are various terminologies involved in microfluidics namely laminar flow in which fluid flows side by side in parallel layers, the only mixing amongst the layer is through the process of diffusion. Electroosmosis and electrophoresis, these phenomena are realized when an electric potential is applied across the terminals of electrodes within microfluidic system. In electroosmosis, liquid within the microfluidic channels moves from one pole to another pole when electric field is applied across the two terminals of microchannels. if the electric field is applied through micro electrodes, net charge is induced in double layer to move it by resulting coulomb force. In Dielectrophoresis, when a neutral particle is immersed in fluid, the net charge is developed on it and under the influence of electrical potential, a dipole is created which moves the particles across the electrode. Understanding microfluidics and its fundamentals, the researchers are now interested in design and develop the miniaturized devices which can handle the fluid in micron scale and can manipulate the liquid. Therefore, For the microfluidic separation and quantification of micro particles requires the fabrication of microdevices, selection of appropriate methodology for separation and suitable technique for quantification. The current study emphases on fabrication of micro devices and methodology adopted for quantification of micro particles.

LITERATURE SEARCH METHODOLOGY

The literature search methodology used here is divided into two sections. First the Selection of keywords related to microfluidics and browsing of related research papers in the field of microfluidics and micro fabrications from 1998 to the present. The most relevant papers are selected and reviewed further to get the insight into the possible applications areas and recent applications in the various field of study. The ongoing section discusses the MEMS device fabrication methodologies for various microfluidic applications and design consideration for effective functioning of microfluidic devices.

DEVICE DESIGN AND FABRICATION OF MICROCHANNELS FOR MICROFLUIDIC APPLICATION.

Photolithography

Photolithography is most widely accepted technique used to create the patterns over the material. Photolithography plays a key role in semiconductor industries to produce Si based devices on large scale. The process consists of several stages to create the pattern over the substate which includes substrate cleaning, oxide formation, spin coat of photoresist material, preparation of mask, exposure to light, development and post baking. In first step, the silicon wafer substrate is thoroughly clean by acetone solution for 10 min and then rinse with DI water. The silicon substrate is then spin coated with negative photoresist.

Figure 1. Schematic representation of stages of photolithography
Source: Toropov and Vartanyan (2019)

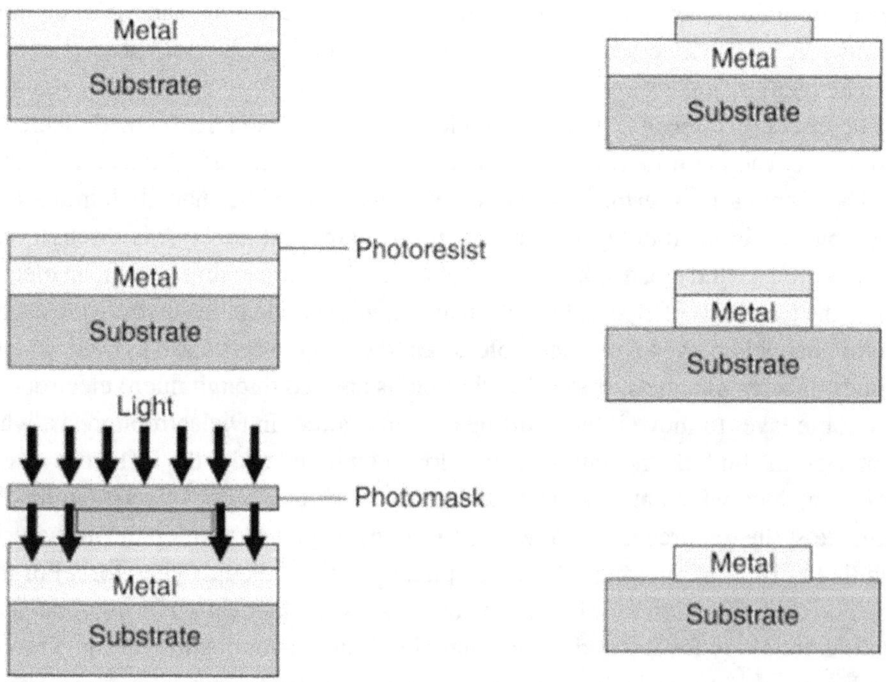

Simultaneously the required photomask needs to be design using CAD-CAM technique and place over the substrate for exposure to UV light. The photomask prevents the entire area of the photoresist and expose only the area which needs to be develop. Upon the exposure to UV light, the negative photoresist becomes hard and final develop in the developer solution to create the pattern over the substrate (Ghubade et al., 2009). Figure 1 shows the schematic representation of photolithography process.

Soft Lithography

Soft lithography is a technique used to create the microchannels in soft polymeric material such as polydimethylsiloxane (PDMS) using the mold which is created by photolithography technique. The soft lithography consists of preparation of soft polymeric material along with curing agent in the ratio of 10:1(PDMS: curing agent) (Singh et al., 2014). The master mold is then place in vacuum desiccator along with HMDS (hexa methyl disilazane) compound. The vapors of HMDS (hexa methyl disilazane) are formed over the master mold to create a very thin film. Then the prepared mixture of PDMS is poured over the master mold and degasify to remove the air bubbles from PDMS solution. The mold is then cured at 80 degrees Celsius in oven for 1 Hrs. Thereafter, the PDMS is peeled off to create the micro-channels for further microfluid application. The process flow of soft lithography is shown in Figure 2.

Figure 2. Schematic representation of soft lithography process
Source: Original, self

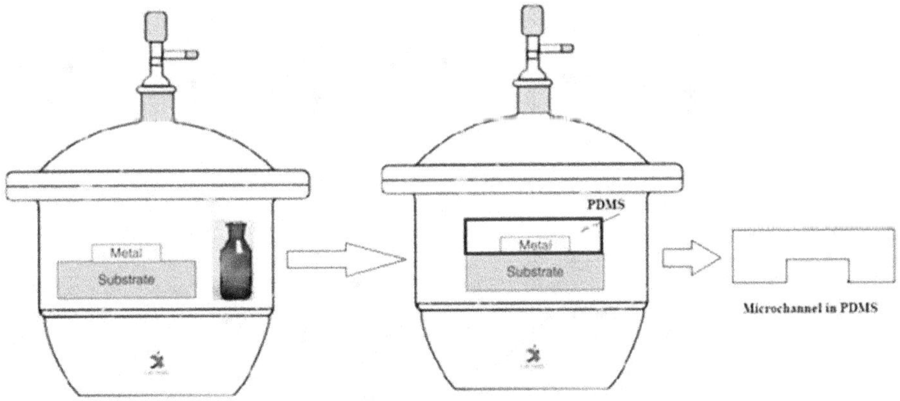

Chemical Vapor Deposition and Physical Vapor Deposition (PVD)

Chemical vapor deposition (CVD) is an additive process widely used technique to form a thin film on the substrate through chemical reaction of two gaseous phase precursor. CVD involves flowing of precursor gases in the chamber which contains the heated substrates over which the thin film to be deposited. The deposition accompanied by the production of chemical byproduct which taken out along with unreacted precursor gases as shown in Figure 3(a). Whereas physical vapor deposition uses principle of evaporation and subsequent condensation over the substrate to create a thin film over it (Park & Sudarshan, 2001).

Figure 3. a) Chemical vapor deposition, b) physical vapor deposition
Source: García-López and Marcì (2021) and Mittal, Sardar, and Jana (2021)

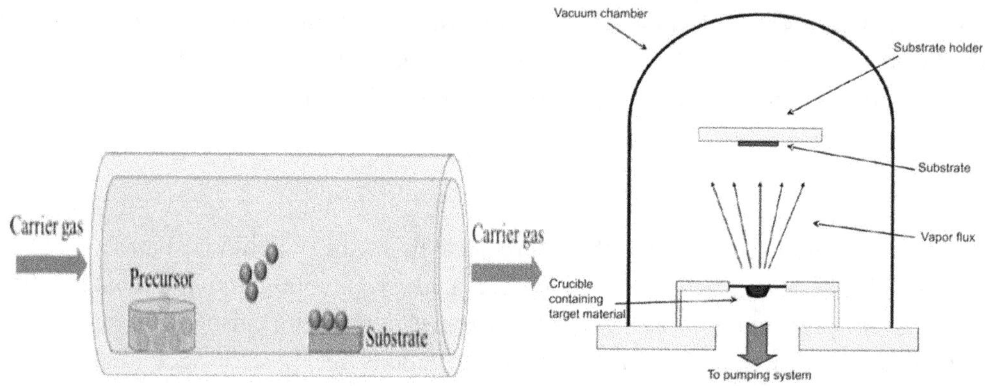

Silicon Wet Etching Process

Wet etching is a subtractive process to create microchannels for microfluidic applications. In this technique, the choice of substrate is exposed with alkaline or acidic solution(etchant) to remove the material from the surface of the substrate (Gosálvez et al., 2010). The wet etching consists of isotropic and anisotropic etching. In isotropic etching, the substrate is dissolved in hydrofluoric acid which removes the material equally in all direction cresting the undercuts. Whereas, in anisotropic etching, etchant such as potassium hydroxide (KOH) etches with different speeds in different directions. When properly designed, the anisotropic etch can produce cavities with straight sides and less undercutting of the mask as shown in Figure 4.

Figure 4. Isotropic and anisotropic etching
Source: Duty and Sandborn (2011)

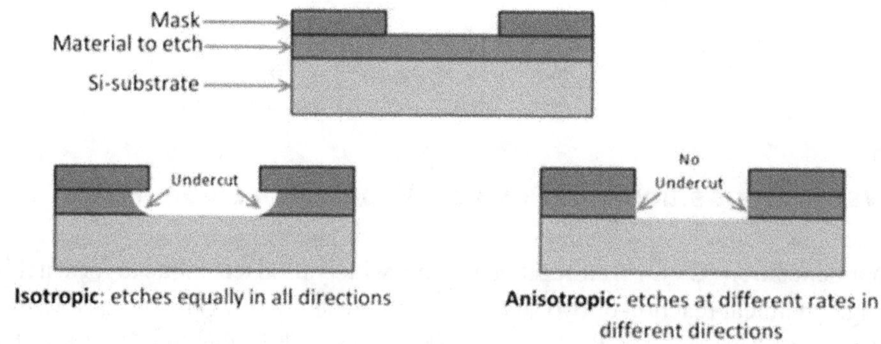

DESIGN OF MICROCHANNELS FOR MICROFLUIDIC APPLICATION

The author Weng et al. (Weng et al., 2016) design and develop the microfluidic chip for the rapid detection of hidden allergens in food as shown in Figure 5. The author implemented microfluidic enzyme immunosorbent assay (ELISA)platform combined with a custom-designed optical sensor for rapid detection of wheat gluten and Ara h 1 in food samples. Jiayu Liu et al. (Liu et al., 2021) used impedance based microfluidic Biosensor for selection capture and detection of salmonella serotypes B and D in ready-to-eat (RTE) Turkey matrix. The MEMS device consist of two detection regions, one is of SU8 microchannels and micro-gaped interdigitated electrode (IDE) as shown in Figure 6. Waseem Asghar et al. (Asghar et al., 2019) design and developed the microfluid device for rapid detection of Candida albicans from phosphate-buffered saline (PBS) and human whole blood. The microfluidic device shows the efficiency of 61–78% at various concentrations ranging from 10 to 10 5 colony-forming units per milliliter (cfu/mL). The microfluid device is fabricated using plastics layers and polymer adhesive. the devices consist of two halves, the microchannels were created on lower half which is double-sided adhesive whereas top face of double-sided adhesive is covered by PMMA as shown in Figure 7. The fluorescent bases quantification is implemented for detection of biological entity. Anil et al. (Ghubade et al., 2009) design and developed MEMS based device for effective detection and quantification of micro particles using principle of electrophoresis. The architecture consists of lower dye made up of silicon imprinted with a set of interdigitated electrodes (IDE) plasma bonded to a replicated piece of Polymethyl siloxane (PDMS). The fluorescent micro beads are used for rapid detection and sample was prepared by diluting a 109 beads/ml solution of fluorescent dye doped polystyrene microbeads. Dielectrophoretic capture of the micro-beads Fluorescence based characterization and quantification of the capture process have been performed using the Nikon microscope and the CCD camera. Liu et al. (Liu et al., 2021) developed the microfluidic chromatography detection (MCD) for the separation of Artificial sweeteners by UV microanalysis device as shown in Figure 9. Andreou et al. (Andreou et al., 2013) used the microfluids

Figure 5. Schematic representation of the microfluidic ELISA chip layout
Source: Weng et al. (2016)

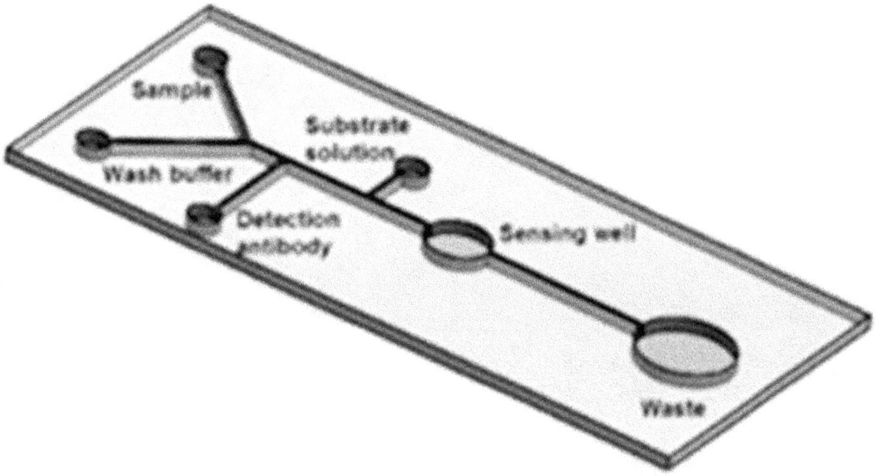

Figure 6. Schematic representation of microchannels
Source: Liu et al. (2021)

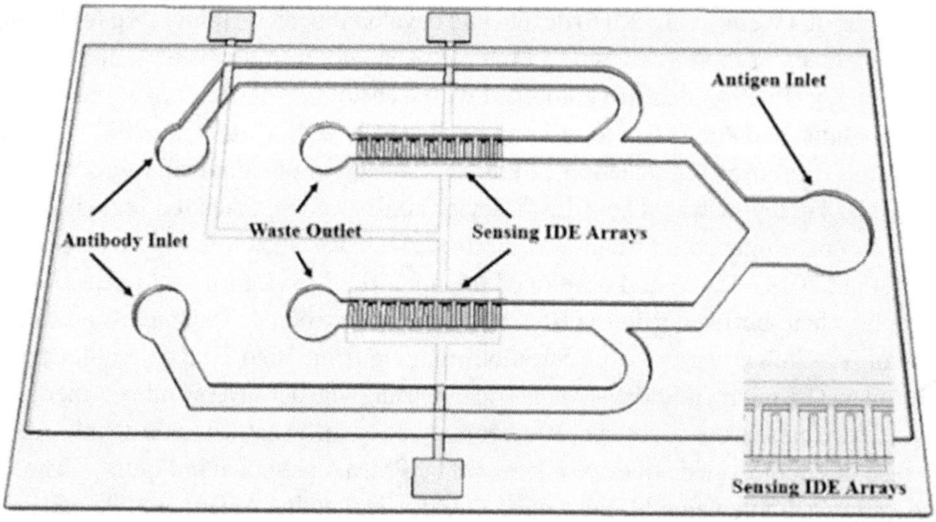

Figure 7. Microchannels in double-sided adhesive (DSA)
Source: Asghar et al. (2019)

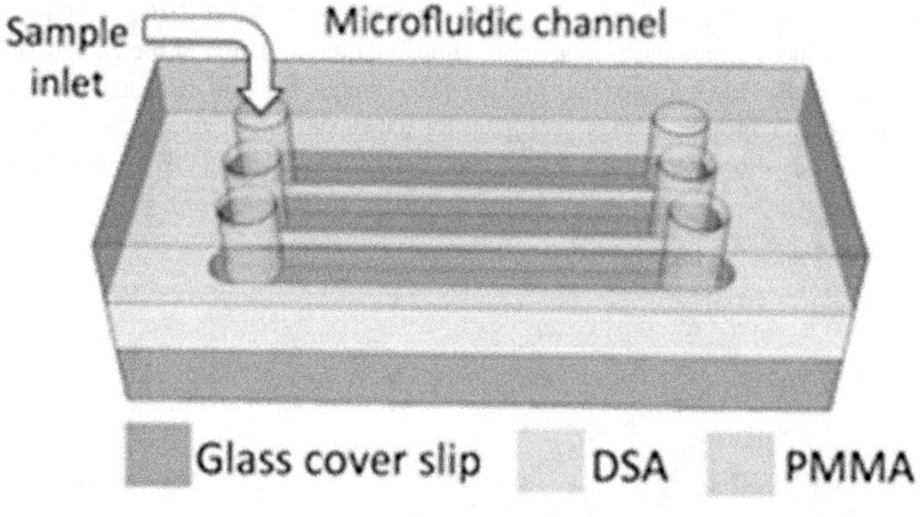

Figure 8. Schematic isometric representation of final assembly
Source: Ghubade et al. (2009)

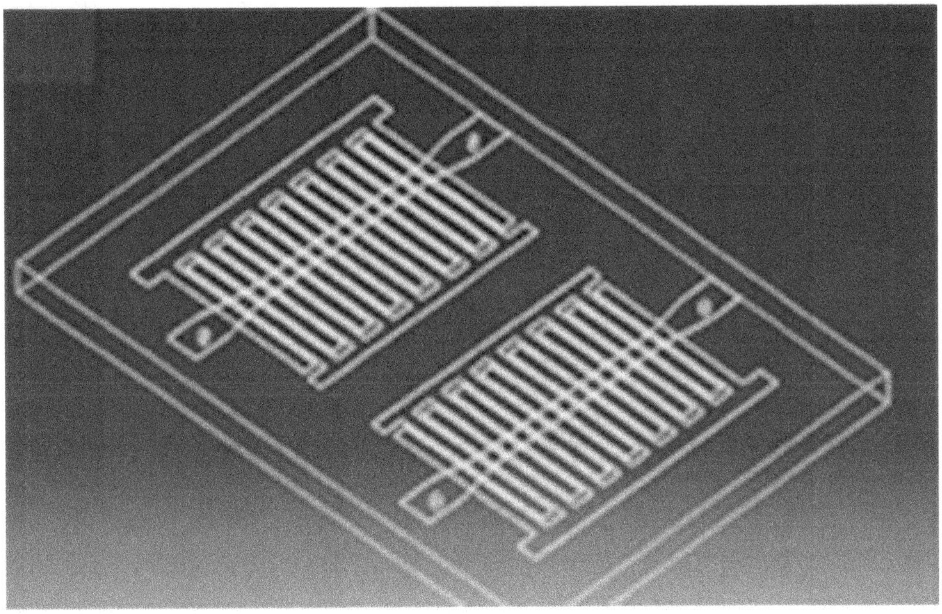

Figure 9. microfluidic MCD chip and UV microanalysis device
Source: Liu et al. (2021)

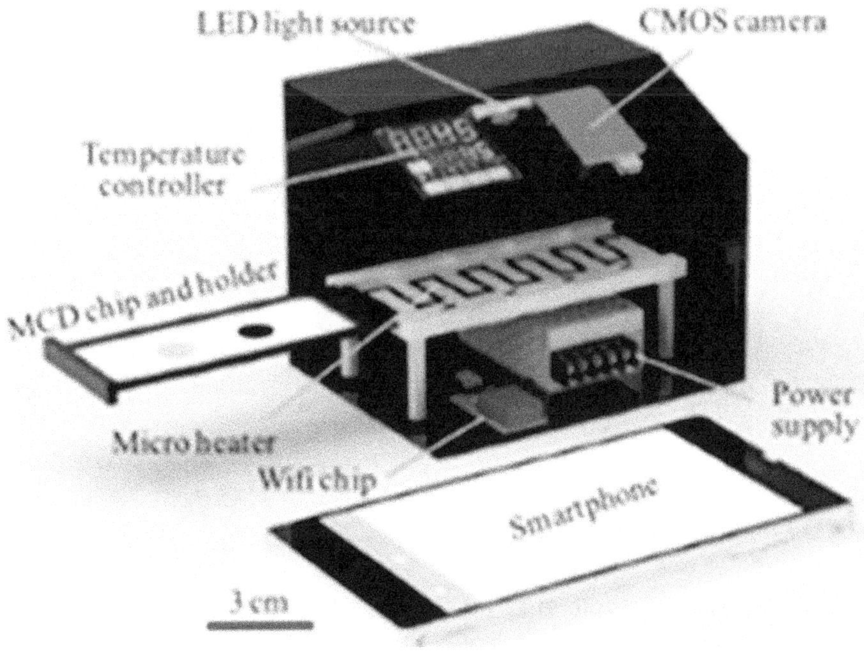

Figure 10. microfluidic device used for controlling Ag-NP aggregation
Source: Andreou et al. (2013)

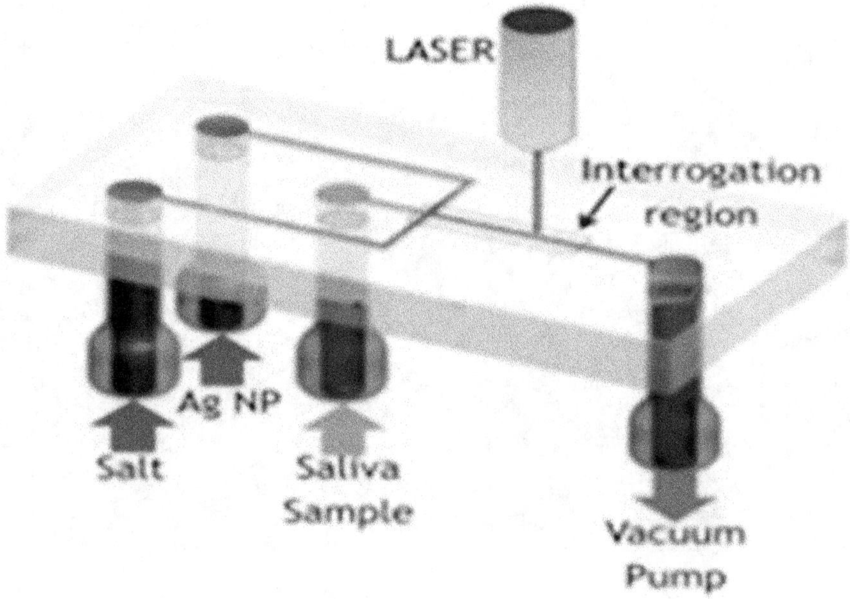

Table 1. Some microfluidic applications.

Author	Micro Particles/Biological Entities	Method of Rapid Detection	References
Weng et al.	wheat gluten and Ara h 1	Optical Sensor	(Weng et al., 2016)
Jiayu Liu et al.	salmonella serotypes	Impedance based microfluidic Biosensor	(Liu et al., 2021)
Waseem Asghar et al.	Candida albicans	Fluorescent image analysis	(Andreou et al., 2013)
Anil et al.	Micro beads	Dielectrophoresis	(Ghubade et al., 2009)
Liu et al.	Artificial sweeteners	microfluidic chromatography detection (MCD	(Liu et al., 2021)
Andreou et al.	Drugs of abuse in saliva	Raman spectroscopy	(Andreou et al., 2013)

for rapid detection of Drugs of Abuse in Saliva Using Surface Enhanced Raman Spectroscopy. The channels are created n PDMS (Sylgard 184) using standard soft lithography techniques and an SU-8 mold as shown in Figure 10.

RESULT AND DISCUSSION

Recently, microfluidics played an important role in rapid detection and quantization of micro particles or biological entities by manipulating fluids at the submillimetre scale. This is a very useful approach in improving diagnostics and biology research. Development of miniaturized devices brought the revolution in the microfluidic applications. Many researchers are developing the lab on chip platform to create the microchannels for effective manipulation of microfluidics through different rapid detection and sorting techniques such as dielectrophoretic assisted concentration of micro particles, optical sensors, impedance-based sorting and separation, Raman spectroscopy and many more. Extensive care needs to be taken while designing the devices and applying the required methodology for rapid detection. It is envisaged that microfluidics will play a significant role in the future medical and pharmaceutical technologies, including the preparation and administration of medicines coupled with intelligent analytics.

REFERENCES

Addae-Mensah, K. A., Wang, Z., Parsa, H., Chin, S. Y., Laksanasopin, T., & Sia, S. K. (2010). Fundamentals of microfluidics devices. *Microfluidic Devices in Nanotechnology: Fundamental Concepts*, 1-38.

Andreou, C., Hoonejani, M. R., Barmi, M. R., Moskovits, M., & Meinhart, C. D. (2013). Rapid detection of drugs of abuse in saliva using surface enhanced Raman spectroscopy and microfluidics. *ACS Nano*, *7*(8), 7157–7164. doi:10.1021/nn402563f PMID:23859441

Asghar, W., Sher, M., Khan, N. S., Vyas, J. M., & Demirci, U. (2019). Microfluidic chip for detection of fungal infections. *ACS Omega*, *4*(4), 7474–7481. doi:10.1021/acsomega.9b00499 PMID:31080939

Basiri, A., Heidari, A., Nadi, M. F., Fallahy, M. T. P., Nezamabadi, S. S., Sedighi, M., Saghazadeh, A., & Rezaei, N. (2021). Microfluidic devices for detection of RNA viruses. *Reviews in Medical Virology*, *31*(1), 1–11. doi:10.1002/rmv.2154 PMID:32844526

Breslauer, D. N., Lee, P. J., & Lee, L. P. (2006). Microfluidics-based systems biology. *Molecular BioSystems*, *2*(2), 97–112. doi:10.1039/b515632g PMID:16880927

Cheng, X., Liu, Y. S., Irimia, D., Demirci, U., Yang, L., Zamir, L., Rodríguez, W. R., Toner, M., & Bashir, R. (2007). Cell detection and counting through cell lysate impedance spectroscopy in microfluidic devices. *Lab on a Chip*, *7*(6), 746–755. doi:10.1039/B705082H PMID:17538717

Duty & Sandborn. (2011). *Wet Isotropic and Anisotropic Etching of Silicon Silicon wafer etching, a process used in the fabrication of microelectronic devices, is the removal*. Academic Press.

García-López, E. I., & Marcì, G. (2021). Preparation of photocatalysts by physical methodologies. In *Materials Science in Photocatalysis* (pp. 37–62). Elsevier. doi:10.1016/B978-0-12-821859-4.00007-6

Ghubade, A., Mandal, S., Chaudhury, R., Singh, R. K., & Bhattacharya, S. (2009). Dielectrophoresis assisted concentration of micro-particles and their rapid quantitation based on optical means. *Biomedical Microdevices*, *11*(5), 987–995. doi:10.100710544-009-9316-6 PMID:19415499

Gosálvez, M. A., Zubel, I., & Viinikka, E. (2010). Wet etching of silicon. In *Handbook of silicon based MEMS materials and technologies* (pp. 447–480). Elsevier. doi:10.1016/B978-0-12-817786-0.00017-7

Ho, C. M., & Tai, Y. C. (1998). Micro-electro-mechanical-systems (MEMS) and fluid flows. *Annual Review of Fluid Mechanics*, *30*(1), 579–612. doi:10.1146/annurev.fluid.30.1.579

Li, D. (2004). *Electrokinetics in microfluidics*. Elsevier.

Liu, C. C., Ko, C. H., Wang, Y. N., Fu, L. M., & Lee, S. Z. (2021). Rapid detection of artificial sweeteners in food using microfluidic chromatography detection system. *Chemical Engineering Journal*, *425*, 131528. doi:10.1016/j.cej.2021.131528

McDonnell, E. E., Han, S., Hilty, C., Pierce, K. L., & Pines, A. (2005). NMR analysis on microfluidic devices by remote detection. *Analytical Chemistry*, *77*(24), 8109–8114. doi:10.1021/ac051320+ PMID:16351162

Mittal, M., Sardar, S., & Jana, A. (2021). Nanofabrication techniques for semiconductor chemical sensors. In *Handbook of Nanomaterials for Sensing Applications* (pp. 119–137). Elsevier. doi:10.1016/B978-0-12-820783-3.00023-3

Nasseri, B., Soleimani, N., Rabiee, N., Kalbasi, A., Karimi, M., & Hamblin, M. R. (2018). Point-of-care microfluidic devices for pathogen detection. *Biosensors & Bioelectronics*, *117*, 112–128. doi:10.1016/j.bios.2018.05.050 PMID:29890393

Noiphung, J., Songjaroen, T., Dungchai, W., Henry, C. S., Chailapakul, O., & Laiwattanapaisal, W. (2013). Electrochemical detection of glucose from whole blood using paper-based microfluidic devices. *Analytica Chimica Acta*, *788*, 39–45. doi:10.1016/j.aca.2013.06.021 PMID:23845479

Park, J. H., & Sudarshan, T. S. (Eds.). (2001). *Chemical vapor deposition* (Vol. 2). ASM international.

Singh, R. K., Kumar, A., Kant, R., Gupta, A., Suresh, E., & Bhattacharya, S. (2014). Design and fabrication of 3-dimensional helical structures in polydimethylsiloxane for flow control applications. *Microsystem Technologies*, *20*(1), 101–111. doi:10.100700542-013-1738-7

Tayyab, M., Sami, M. A., Raji, H., Mushnoori, S., & Javanmard, M. (2020). Potential microfluidic devices for COVID-19 antibody detection at point-of-care (POC): A review. *IEEE Sensors Journal*, *21*(4), 4007–4017. doi:10.1109/JSEN.2020.3034892

Toropov, N., & Vartanyan, T. (2019). 1.03 Noble Metal Nanoparticles: Synthesis and Optical Properties. *Comprehensive Nanoscience and Nanotechnology*, 61.

Weng, X., Gaur, G., & Neethirajan, S. (2016). Rapid detection of food allergens by microfluidics ELISA-based optical sensor. *Biosensors (Basel)*, *6*(2), 24. doi:10.3390/bios6020024 PMID:27338488

Whitesides, G. M. (2006). The origins and the future of microfluidics. *Nature*, *442*(7101), 368-373.

Zanoli, L. M., & Spoto, G. (2012). Isothermal amplification methods for the detection of nucleic acids in microfluidic devices. *Biosensors (Basel)*, *3*(1), 18–43. doi:10.3390/bios3010018 PMID:25587397

Chapter 11
Fabrication Methods for Bio-MEMS

Sameer Saharan
Amity School of Applied Sciences, Amity University, India

Bhuvnesh Yadav
Amity School of Applied Sciences, Amity University, India

Aseem Grover
Amity School of Applied Sciences, Amity University, India

Shivam Saini
Amity School of Applied Sciences, Amity University, India

ABSTRACT

MEMS is the exotic relatives of semiconductors and integrated circuits (ICs), originally based on silicon wafer production processes but with the dimensions of space, flexion, and constantly varying output akin to analogue devices. Whereas BioMEMS refers to all interfaces and cross-sections of biological sciences and therapeutic disciplines with microsystems and nanotechnology, commercial microfluidic devices rely heavily on material selection and production techniques. Molding, replication, casting, and bonding procedures are being developed that is required for mass production with repeatability and functional dependability at a low cost, better design freedom, and manufacturing ease, which is critical in the medical disposable industry. Nowadays, BioMEMSs are the largest and most diversified applications of MEMS devices. Fabrication processes employed many steps process like wafer selection, lithography, etching, and substrate binding. Furthermore, packaging for safety and biocompatibility is a major problem for the BioMEMS engineer.

INTRODUCTION

A miniature machine with combination of mechanical and electrical components is called MEMS (micro-electromechanical system) (Badilescu & Packirisamy, 2016; Folch, 2016). The size of a MEMS device

DOI: 10.4018/978-1-6684-6952-1.ch011

can range from as little as one micrometre, which is considerably less than the width of a single human hair, to as much as several millimetres. MEMS is a term used to refer a class of micro mechatronic devices and their production procedures. Though some MEMS lack any mechanical components, they are nevertheless considered MEMS because they feature membranes, channels, cavities, and springs that are smaller than those found in traditional machinery. Some MEMS devices are also known as transducers because they transform a measured mechanical signal into an electrical or optical signal. In Europe, MEMS is more frequently referred as microsystems technology, whereas micromachines are the more prevalent name for MEMS in Japan is MST (Badilescu & Packirisamy, 2016; Folch, 2016).

Micro-electromechanical systems in the field of biomedicine are commonly referred as "Bio-MEMS". Lab-on-a-chip (LOC) and micro total analysis systems (μTAS) share many similarities with bio-MEMS and are frequently used interchangeably (Kharisov et al., 2016). The mechanical and microfabrication technologies that are well-suited to use in the life sciences are the primary emphasis of bio-MEMS. However, lab-on-a-chip research focuses on condensing many laboratory procedures and investigations onto a single (often microfluidic) chip. This definition states that lab-on-a-chip devices need not directly serve biological purposes in order to qualify as such. Similar to how micro total analysis systems are typically designed for chemical analysis rather than biological applications, biological uses may not be a primary consideration for these systems. Bio-MEMS may be thought of as a catch-all term for the study and development of microscale devices and systems for use in biological and biomedical research and practise; these devices and systems may or may not involve electrical or mechanical components.

Advancement in Bio-MEMS

The era of bio-MEMS begun with use of vaporized palladium islands as cell adhesion by Carter (1967) (Carter, 1967). Pregnancy test ClearBlue, introduced by Unipath Ltd., became the first microfluidic product to use paper (Olszynko-Gryn, 2017). Widmer and Andreas Manz introduced the term "micro total analysis system" in 1990. Micro total analysis systems were recommended by the study for application in chemical sensing. Miniaturised total chemical-analysis systems (μTAS) was then introduced due to three major reasons (Dempsey et al., 1997). Firstly, before the 1990s, lot of time, money, and expensive equipment were required to perform chromatographic tests, which was a major setback for drug discovery. The Human Genome Project, which began in October 1990, was a further aspect that contributed to advancements in DNA sequencing technology (Watson, 1990). This is very costly that is why capillary electrophoresis (cost effective) is now frequently used to separate DNA and other substances. Thirdly, the Defense Advanced Research Projects Agency of the Department of Defense (US) initiated support for the microfluidic research projects to develop field-deployable microsystems for the identification of biological and chemical agents that could be used by military to encounter terrorist threats. Photolithography tools were used in the microelectronics researchers for microfabrication of micro-electromechanical systems (Fedder, 1994; Maluf & Williams, 2004). MEMS technology was initially created for silicon or glass wafers and utilised solvent-based photoresists, both of which are incompatible with biological material, and therefore, the potential biological uses of this technology were initially limited. In 1993, Kumar and Whitesides revolutionised the bio-MEMS field by developing an relatively cheap method of Polydimethylsiloxane (PDMS) -based microfabrication (Kumar & Whitesides, 1993). The growth of bio-MEMS since then has been phenomenal. The first oligonucleotide chip was invented in 1991. The first solid microneedles for medication delivery were discovered in 1998 and in the same year, polymerase chain reaction chip with continuous flow was created. In 1999, microchannels with heterogeneous laminar flows was used for the selective treatment of cells. Hydrogels like agarose, biocompatible photoresists, and self-assembly are now at the forefront of research for enhancing bio-MEMS as potential alternatives to or supplements to PDMS.

Materials Used for Bio-MEMS

Silicon

A variety of high-precision microfabrication and micromachining processes can be used to create silicon, which performs well in the mechanical, electrical, optical, and with living organisms. The methods for making chips on a silicon substrate are not only accurate, but they can also make chips in batches that cost less. Silicon opens up a huge range of ways for microelectronics and micro photonics to work together. Because silicon is transparent in the near-infrared wavelength range, biochips can be made to work with silicon photonics elements like waveguides, couplers, gratings, interferometers, etc., all on the same chip. Due to its widespread use in microelectronics, silicon is an excellent material for biochips with electrical controllers for electrowetting, electroosmosis, etc. Silicon is appealing because it can be used in lithography, etching, and other ways to make microelectronics. Silicon has a high Young's modulus and thermal conductivity for deflection and heat transmission as a mechanical material. Despite being brittle, it possesses higher yield strength and is more elastic than metals. However, silicon is expensive, hard to make (cleaning, resist coating, photolithography, etc.), and takes a long time to make. Many research organisations are unable to produce many chips due to the requirement for a high-quality clean environment for the fabrication of silicon-based microfluidic devices (Desai et al., 2007; Szmigiel et al., 2007).

Glass

Glass is chosen over silicon in micro total analysis systems because of the optical transparency that it possesses, which makes it possible for visual examination and optical detection. it is a strong material with good chemical and thermal resistance. Glass microfluidic devices have a greater cost per function when manufactured in volume than polymer chips. Glass is typically used when sensitive chemical processes are involved. Glass is a substrate material that is commonly used in microsystem technology, particularly when creating microfluidic devices for biological research. It possesses advantageous material features, both in terms of its structure and its function. Glass is able to withstand the high voltages that are typically utilised in electrokinetically driven flows due to its strong fluorescence and dielectric properties (Khan Malek et al., 2007; Mund & Leib, 2004; Zhou et al., 2007). The most common substrate for the fabrication of DNA arrays is glass because of its exceptional chemical resistance, high temperature stability, chemical inertness, and established techniques for surface modification and functionalization (Badilescu & Packirisamy, 2016; Folch, 2016; Kentsch et al., 2006; Zhou et al., 2007). In comparison to silicon-based systems, the long-term chemical stability of devices may potentially be increased when glass substrates are used instead. The high mechanical strength and superior mechanical stability of glass are both utilised in a variety of applications. Glass is a brittle material that has poor electrical conductivity because of its structure. Although it is resistant to many strong chemicals, it is easily receptive to metallization, and it also has good optical characteristics. When it comes to electroosmosis applications that require high electric fields, it has proven to be a superior substrate over silicon. It is the ideal material for optical-based detection and flow visualisation since it is transparent in the visible and near-infrared (NIR) wavelength ranges. This material is ideal for creating biochips because of its good biocompatibility and high resistance to high-temperature sterilisation. Excellent chemical stability and solvent compatibility can be found in both silicon and glass. However, because to its amorphous nature, glass is challenging to microfabricate. Moreover, a clean room atmosphere is necessary for the glass chip bonding procedure.

Micro structuring glass (MSG) is a revolutionary new method that has the ability to fulfil requirements in microfluidic and microelectronic applications, including hermetic wafer level packaging. MSG has a number of advantages, some of which are chemical-resistant borosilicate glass layers, layered structuring with photoresist, and quick glass deposition at low temperatures (Mund & Leib, 2004).

SU-8

The thick negative photoresist known as SU-8 was developed by IBM for the purpose of creating moulds for electroplating. The epoxy-based negative photoresist exhibits various characteristics that are unexpectedly unique. It is available in a wide range (from hundreds of nanometres to hundreds of microns) of coating thickness. which can be deposited with only one step of standard spin coating and the appropriate dilution of SU-8 resin (Lorenz et al., 1997). Even more substantial layers can be effectively put and photopatterned. Secondly, by only exposing the SU-8 layer to UV light while the layer is covered with a contact mask, it is possible to create structures in the SU-8 layer that have almost straight sidewalls. The normal aspect ratio is approximately fifteen for measuring objects, but aspect ratio greater than 100 may be attainable through the application of specialised technologies in layers. As will be discussed in greater detail in the following section on material properties, the high degree of chemical functionality possessed by the monomers gives SU-8 its exceptional mechanical qualities and resistance to chemical attack. These attributes contribute to the fact that SU-8 is useful as a structural material in a diverse selection of applications (Mata et al., 2006; Nemani et al., 2013). Insulating in nature, SU-8 can be converted into a conductive material by the process of metal deposition. It can also be utilised as mould for electroplating for the production of substantial metal structures. Due to its translucency, investigation of underlying structures and fluorescence-based optical sensor data is quite easy. SU-8 processing requires minimum equipment and consumables. SU-8 has wide applications in micromechanics, labs-on-chip, integrated optics and packaging etc.(Dentinger et al., 2002; Keller et al., 2008; Lorenz et al., 1997; Mata et al., 2006; Mund & Leib, 2004; Nemani et al., 2013; Nordström et al., 2008; Sang, 2010; Szmigiel et al., 2007; Zhou et al., 2007).

Metal

MEMS components can also be synthesised from metals like gold, copper, nickel, chromium, aluminium, tungsten, titanium, silver, and platinum. These metals exhibit extremely high levels of reliability when employed within their limits, while not having some of the mechanical characteristics benefits that silicon does. Electroplating, evaporation, and sputtering methods can all be used to deposit metals. (Mourzina et al., 2005; Rotake et al., 2019).

Polymer

Polymers are capable of being manufactured in extremely large amounts and can take on a variety of forms in terms of their physical characteristics. MEMS devices are ideally suited for use in microfluidic applications such as disposable blood test cartridges. These devices can be fabricated from polymers via techniques such as injection moulding, embossing, or stereolithography (Al-Alawachi et al., 2022; Bhattacharyya et al., 2021; Rezai et al., 2012; Tokachichu & Bhushan, 2006).

Ceramics

Due to beneficial combinations of material properties, aluminium, silicon, silicon carbide, titanium nitrides, and other ceramics are being used frequently in the manufacturing of MEMS (Cros, 2021). Aluminium nitride (AlN) has pyroelectric and piezoelectric capabilities due to the wurtzite structure in which it crystallises, making sensors, for example, which are sensitive to normal and shear forces possible. On the other hand, Titanium nitride (TiN) has strong electrical conductivity and significant elastic elasticity, which enables the implementation of electrostatic MEMS actuation systems using extremely thin membranes. Due to its excellent resistance to biocorrosion, TiN is also well suited for use in biosensors and biogenic settings (Sadler et al., 2002, 2003; Saliterman, 2006).

Methods for Fabrication of Bio-MEMS

Silicon (both single-crystal and polycrystalline silicon materials) has been chosen as the basis for Bio-MEMS for long because same manufacturing techniques were used to create integrated circuits. However, these techniques are very costly and have restricted applications since they were developed for and optimised with a small subset of materials (mostly silicon and other semiconductors) and certain material features. Microfabrication techniques like photolithography, wet chemical etching, and fusion bonding are used to shape these materials. These techniques are also used to make microelectronic devices. The sequence of processes used in fabrication of Bio-MEMS is indicated in Figure 1.

The following steps are used for the fabrication of Bio-MEMS:

1. Thin-Film Deposition

The deposition of thin films of various materials onto a substrate can be accomplished by the application of a wide range of diverse methods. In the processes of microfabrication and micromachining, thin films may also serve as functional layers, masking layers, or sacrificial layers. A chemical reaction-driven process, an oxidation process, or a physical evaporation process, such as thermal evaporation, can be used to deposit thin films (Kotnala & Shah, 2015). Thermal oxidation, physical vapour deposition (PVD),

Figure 1. Procedure for fabrication of Bio-MEMS

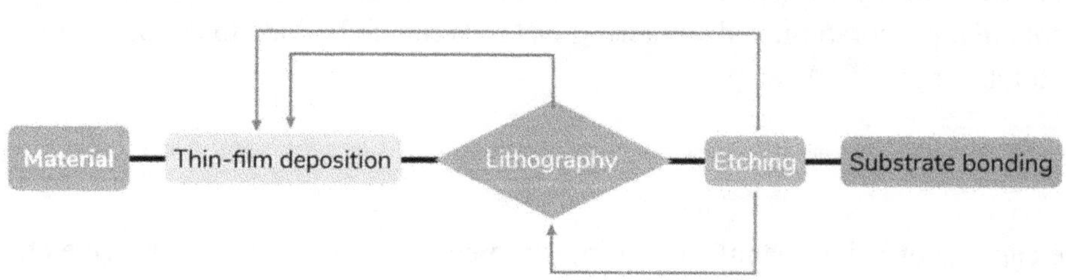

chemical vapour deposition (CVD), spin casting and dip casting are the commonly used deposition techniques (Bruck, 2017).

1.1. Thermal Oxidation

Silicon oxidises when subjected to high temperatures in the presence of oxygen, result in the formation of a very thin layer of silicon dioxide (SiO2) on the surface of the silicon. The oxidation can be dry or wet. In wet oxidation, the presence of humidity can fasten the formation of oxide. Silicon dioxide can be utilised either as an etching mask or for the purpose of electrical isolation in microstructures (Windhorn et al., 2003; Ziaie et al., 2004).

1.2. Chemical Vapour Deposition

Low-pressure chemical vapour deposition (LPCVD) at higher temperatures or plasma-enhanced CVD (PECVD) at lower temperatures are the most common methods for depositing inorganic materials such as silicon compounds and polysilicon. In the CVD process, reaction between gases and the substrate take place in a chamber at high temperature in vacuum to produce a solid film on the substrate. Thin films of amorphous and polycrystalline silicon are significant structural materials that are used in many microsystems. The deposition process is typically carried out at low pressures and at moderate temperatures. When the gas is pumped into the chamber, the molecules will react directly with the surface of the substrate, which will result in the formation of a thin layer of the desired substance. LPCVD is the name given to the reaction that occurs when it does so at high temperatures but under low pressure. In the case of plasma-enhanced chemical vapour deposition (PECVD), the reaction can take place at lower temperatures if the plasma is bolstered with additional radio frequency power. The thin films that are deposited using the LPCVD process have a higher level of residual stress than the thin films that are deposited using the PECVD process. (Chen et al., 2006; Lee & Yang, 2012; Mourzina et al., 2005; Sun et al., 2013; Vasudev et al., 2013).

1.3. Physical Vapour Deposition

The physical vapour deposition can be used to deposit metals and certain oxides of metals; and is based on thermal evaporation or sputtering at a high-vacuum environment. Metal films are typically used for making interconnections, ohmic contacts, and producing optical surfaces. Conductors, resistors, and dielectrics can all be deposited using physical vapour deposition while the process is carried out in a vacuum. The substrate is heated in a vacuum while the target material is vaporised using this technique, and the vapour is then allowed to deposit on the substrate. The use of resistive heating and electron beam heating are the two primary types of heat sources that are utilised in physical vapour deposition.

The deposition of metals that have low melting temperatures can be accomplished by the use of thermal evaporation. It is also feasible to deposit metal, ceramic, and diamond-like materials using a hybrid method that combines magnetron sputtering with pulsed laser ablation. When using the sputtering process, metals, complex materials, and refractive metals can be deposited by sputtering a cathode target with positive ions generated by an inert gas discharge. Plasma of inert gases such as argon is formed in a chamber by subjecting it to intense electric fields while maintaining a high level of vacuum. The positive argon ions are drawn to the target that is attached to the cathode when the gas is ionised in the plasma state. These gas ions assault the target, which results in the atoms of the target being freed. The anode exerts an attractive force on the target atoms that have been liberated, which causes the substrate to acquire a coating of an extremely thin layer. During the process of sputtering, the production of reactive sputtering will occur if non-inert gases are introduced into the surrounding atmosphere (Badilescu & Packirisamy, 2016; Folch, 2016).

1.4. Spin-Casting Process

The fabrication of films with a smooth surface can be done quickly and easily by a process called spin casting. The material is first dissolved in a volatile liquid solvent before it is used in the spin-casting process. The solution is put onto a wafer, and then the wafer is rotated very quickly. The liquid is spread out across the surface of the wafer, the solvent evaporates resulting in a thin layer of the solid substance. Altering the rotational speed of the liquid as well as the viscosity of the liquid allows for the thickness of the film to be adjusted. The deposition of photoresists and certain inorganic glasses can be done with spin casting process. The most significant drawback associated with spin-coat materials is that they have a high level of residual stress and are prone to shrinking after the solvent has been removed. Films that are produced using the spin-coating method have a lower density and are more vulnerable to chemical assault than films deposited using other methods (Rabarot et al., 2003; Saliterman, 2006; Sharma & Madou, 2012).

1.5. Dip Coating Technique

The dip coating is a process in which the substrate that is going to be coated is dipped into a liquid and then removed from the liquid at a speed that has been precisely determined at restricted temperature. The withdrawal speed, amount of solid material, and the viscosity of liquid are the primary factors that determine the coating thickness. The sol-gel manufacturing process is utilised in the production of ceramic and glass products using the technique of dip-coating. The sol-gel process, refers to the transformation of a system from a liquid sol (colloidal solution) into a solid gel phase. In a typical sol-gel process, the precursor (inorganic metal salt or a metal organic compound) is subjected to a sequence of hydrolysis and polymerization processes, which ultimately results in the formation of a sol (colloidal suspension). This technique can be used to deposit thin films of a variety of materials, including silicon oxide, titanium dioxide, indium tin oxide (ITO), and others (Jiazhen Zhang, 2022).

2. Lithography

Lithography is often defined as the selective exposure to a radiation source that transfers a pattern into a photosensitive material. When exposed to radiation, a substance is said to be photosensitive if its physical characteristics change. The pattern of the radiation on the material is transferred to the exposed material if a photosensitive material is selectively exposed to radiation (for example, by masking some of the radiation), as the characteristics of the exposed and unexposed portions differ. The exposed area can then be treated or removed, creating a mask for the substrate beneath (Badilescu & Packirisamy, 2016; Folch, 2016; Saliterman, 2006). Wet and dry etching, metal or other thin film deposition, and photolithography are common applications.

2.1. Electron Beam Lithography

It is also known as e-beam lithography. The goal of electron beam lithography is to create very small structures in the resist that may later be transferred to the substrate material, typically through etching, just like in photolithography. By dipping the resist in a solvent and selecting to remove only the exposed or unexposed portions of the resist, the electron beam makes the resist more or less soluble. This lithography technique has a modest throughput and great resolution. It is employed in the creation of photomasks, the production of semiconductor devices in small quantities, and in research and development.

E-beam lithography and optical lithography are currently considered to be state-of-the-art processes. The newly developed technology has the potential to significantly lower the cost of these processes. Electro lithography is the name given to the novel method that this technology brings to the table for etching fine lines onto a substrate.

1. The lithography process is an easy one to understand. First, a plate of transparent glass is taken and then a layer of the polymer is coated on it.
2. On top of this, a layer of chromium is placed, and then a trench is excavated to create the desired pattern on the chromium layer, exposing portions of the polymer layer.
3. The exposed polymer is dissolved with acetone, and those components of it are eliminated as a result.
4. As a result, this creates a space in the middle of the polymer-chromium sandwich.

This whole assembly functions in the same way as the negative of a processed photograph. When the selected metal is spluttered onto this sandwich, it will fill the space that has been created and then fall onto the glass plate in a straightforward manner. In this manner, the appropriate pattern can be made using a metal of choice. This pattern will have a width that is equivalent to both the width of the trench and the thickness of the polymer layer (Hosseini et al., 2019; Scholten & Meng, 2016).

Advantage of e-beam lithography:

The advantages of the e-beam lithography are following:

- This method has the potential to cut the cost of the equipment that is now being utilised.
- Since it does not require high currents or a vacuum atmosphere, it is a relatively more environmentally friendly method than those that are currently in use.

Disadvantage of e-beam lithography:

In spite of the high resolution that can be achieved using electron-beam lithography, users frequently do not take it into consideration due to data-related flaws and physical issues.

Data-related flaws of electron-beam lithography can be further categorized in two subcategories, depending on the nature of the problem. Mistakes of blanking or deflection occur when the electron beam is not deflected correctly when it is meant to, whereas errors of shaping occur in variable-shaped beam systems when the incorrect shape is projected onto the sample.

There are different types of physical flaws, such as sample charging (either negative or positive), errors in the calculations of backscattering, outgassing, contamination etc. As the "write time" for electron beam lithography can potentially surpass a day, randomly occurring flaws have a higher probability of occurring. Once again, larger data volumes can present an increased number of opportunities for errors to occur.

2.2. Ion-Beam Lithography

Ion-beam lithography, also known as ion-projection lithography, is the process of creating very small structures, such as integrated circuits or other nanostructures, by scanning a focussed beam of ions in a structured form across a surface. This is done in order to create very precise patterns on the surface.

It has been observed that ion-beam lithography is an effective method for the transfer of high-fidelity patterns onto three-dimensional surfaces. Ion-beam lithography provides patterning with a better resolution than ultraviolet lithography, x-ray lithography, and electron beam lithography because heavier particles have more momentum than the lighter particles. Because of this, the wavelength of the ion beam is even shorter than that of the electron beam, and as a result, there is essentially no diffraction. Additionally, the momentum reduces the amount of scattering that occurs both in the target and in any remaining gas. In comparison to x-ray and electron beam lithography, soft lithography has a lower potential for causing radiation damage to the underlying structures of the object being scanned (Brown et al., 1981).

Ion-beam lithography is a technique that is very comparable to electron beam lithography; however, it makes use of ions, which are much heavier charged particles. Ions travel in more direct paths than electrons through vacuum and matter resulting in very high resolution. This is in addition to the fact that diffraction is practically non-existent. As a result of the slower speed of the ions, secondary particles, which include electrons and atoms, have an extremely limited range. On the other side, producing powerful sources is more difficult, and for a given range, higher acceleration voltages are needed. Due of the higher energy loss rate, higher particle energy for a given range, and lack of significant space charge effects, shot noise will typically be larger. Shot noise is more apparent as a result of these elements together (Watt et al., 2005).

Ions moving at high speeds have distinct visual properties than electrons because of the greater momentum that they possess. Ions also interact with matter in a different way than electrons do. They are able to go much more directly through matter and have a considerably shorter range in it. At low energies, at the lower end of the range, they transfer more of their energy to the nuclei of the atoms than they do to the atoms themselves; as a result, the atoms become dislocated rather than ionised. They resort to doping the resist if the ions are unable to diffuse out of it. The Bragg curve that describes the way energy is lost in matter is one that has a lower statistical spread. They have a higher optical "stiffness," which means that in order to focus or bend them, larger fields or greater distances are required. The increased momentum is resistant to the effects of space charge (Joshi-Imre & Bauerdick, 2014).

2.3. X-Ray Lithography

X-ray lithography (XRL) exposes resists to an X-ray source through a mask, allowing them to be patterned. Resists are materials that are sensitive to X-rays and change their dissolution rate in a certain solvent after being irradiated. The X-rays may now penetrate the resist and generate the desired pattern. While the mask is composed of high-attenuation Z-elements (Au, W, etc.), the carrier substrate is composed of low-attenuation components (Si, diamond, SiNx, SiC, Be, etc.). As a result, the carrier substrate is transparent. Smith and Spears developed the concept of X-Ray Lithography (XRL) in 1972. Because XRL has a shorter wavelength and deeper penetration than standard UV lithography, it has stimulated the interest of micro and nanofabrication professionals. This is due to the fact that XRL opens up new possibilities for the design of microdevices with a high aspect ratio, thick resists, and practically vertical sidewalls. X-ray lithography, also known as LIGA (Lithography Galvanoformung Abformung), is the foundational stage of the LIGA process. This stage involves the electrodeposition of metal in the developed resist structure in order to obtain a mould or an electrode for a subsequent replication process, such as moulding or electro discharge machining. The energy range of soft X-rays is between 150 eV and 2 keV, while the energy range of hard X-rays is greater than 5 keV. X-rays are also known as "deep rays". The use of soft XRL is appropriate for producing high-resolution structures with a thickness of less than 50 nm of resist. Irradiating thick resists and the LIGA process are two popular applications of the Deep XRL (DXRL) technique. At this time, the XRL method can be performed at facilities that make use of synchrotron radiation. The semiconductor business is interested in XRL because it helps define technological nodes, which is why the industry is paying attention to it. Initially, the node number defined the gate length or half-pitch (HP), whereas at the present time (below 22 nm), it relates to a specific generation of chips that were fabricated with a particular technology. The term "node" refers to a specific manufacturing process for semiconductors as well as the rules that govern its design. X-ray light has a shorter wavelength than ultraviolet light, so it has the potential to guarantee a "resolution reserve" for all of the technological nodes. In addition, unlike UV lithography, it does not need a new facility at each technological node in order to function properly. Despite this, the technique's full potential was not realised since efforts were concentrated first on ultraviolet (UV) lithography and then on extreme ultraviolet (EUV) lithography. More recently, XRL has attracted attention for the next 2 nodes (beyond the 10-nm technological node) onward, primarily due to the potential of soft X-ray interference lithography. This is because soft X-ray interference lithography could be used to create ever smaller and smaller features on semiconductors (Bharti et al., 2022; Spille & Feder, 1977).

After first version of the technique many upgrades are done in this technique which are following:

- Dynamic XRL:

It is a method in which both the sample holder and the mask can be driven independently, which makes it possible to carry out a grayscale dosage distribution exposure. The sample stage is moved in alternating vertical and horizontal positions when using the beam scan method. This allows for enormous areas to be scanned, despite the fact that the beam size used is rather modest (Pindyurin et al., 2008).

- Multiple Mask Lithography:

The method works by using a number of masks that are superimposed on one another, each of which has a position that can be adjusted in real time during the exposure phase. This results in a grayscale dose distribution throughout a particular area (Bharti et al., 2022).

- Pencil Beam Mode Exposure:

This method makes use of a pre-collimator, which narrows the field of irradiation; meanwhile, the movement of the sample holder during the exposure makes it possible to "draw" on the resist layer. In addition to this, an array structure's exposure time can be cut down significantly by dividing the beam into numerous microbeams and utilising a mask (Bharti et al., 2022).

3. Etching

After the hard bake, a procedure known as selective etching is used to move the desired pattern from the photoresist to the underlying film or wafer so that it may be replicated. Etching is the process of eliminating undesired portions of a film or substrate in a selective manner. Etching is used not only to produce patterns on mask layers but also to remove the surface damage caused by the previous process steps and to fabricate both two-dimensional and three-dimensional structures. Etching is used to remove the surface damage induced by the previous process steps. Etching is a process that can be performed on a variety of materials, including insulators, metals, and semiconductors. There are two broad categories that can be used to classify the numerous etching processes (Stevanus et al., 2022):

3.1. Wet Etching

The process of material removal known as wet etching makes use of liquid chemicals or etchants to carry out the process of removing materials from a wafer. The masks that are placed on the wafer are what define the precise patterns. The masks serve to shield certain materials from the liquid chemicals that are used to etch away other materials. In an earlier step of the fabrication process, lithography is used to deposit and imprint these masks directly onto the wafers. Several chemical reactions take place throughout the wet etching process. These reactions use up the initial reactants and produce new reactants. The method known as wet etching can be broken down into three primary stages. (1) The spread of the enchanting solution over the surface of the structure that is to be dissolved. (2) The reaction that takes place between the liquid etchant and the substance that is being removed by the etching process. In most cases, a reduction-oxidation reaction, or redox reaction, take place. The first step in this reaction is the oxidation of the substance, followed by the dissolution of the oxidised material. (3) The movement of the products of the reaction away from the surface that has been reacted.

- Anisotropic Wet Etching:

Crystalline materials can be etched by liquid etchants at various speeds, and these rates vary depending on which face of the crystal is exposed to the etchant. There is a significant disparity in the rate of etching that occurs depending on the crystallographic plane of the silicon. This phenomenon can make it possible for certain materials, like silicon, to have an extremely high anisotropy. Potassium hydroxide (KOH), ethylenediamine pyrocatechol (EDP), and tetramethylammonium hydroxide are a few examples of the anisotropic wet etching agents that can be used on silicon (TMAH). A silicon wafer with a thickness of one hundred nanometre would produce an etch pit in the shape of a pyramid after being etched. Flat and at an angle will be the appearance of the etched wall.

- Isotropic Wet Etching:

The most popular type of etchant solvent used for isotropic wet etching on silicon is a mixture of hydrofluoric acid, nitric acid, and acetic acid, which is abbreviated as HNA. The etch rate can be altered by changing the concentration of each etchant. In most cases, silicon dioxide or silicon nitride are used as masking materials in opposition to HNA. While the reaction is taking place, the material is being removed laterally at a rate that is comparable to the speed at which it is being etched downward. Even with isotropic dry etching, which is explained in the section on dry etching, this lateral and downward etching process still takes place. The use of a mask during wet chemical etching does not affect the process's ability to produce isotropic results because the liquid etchant can pass through the mask. Wet chemical etching is not typically used in situations in which the importance of directionality for high-resolution pattern transfer is significant.

- Hydrofluoric (HF) Etching:

It is normal practise to utilise hydrofluoric acid as an aqueous etchant for silicon dioxide. The concentration of the acid used is typically 49%, and the dilution ratios employed are 5:1, 10:1, or 20:1 BOE (buffered oxide etchant) or BHF (Buffered HF). Etchings on glass were one of their early uses, which dates back to the Middle Ages. Prior to the introduction of RIE (Reactive Ion Etching), it was routinely applied in integrated circuit production for the purpose of patterning the gate oxide. In the cleanroom, hydrofluoric acid is regarded as one of the more hazardous acids to work with. The moment it makes contact with the skin, it immediately begins to penetrate, and it continues to penetrate all the way to the bone. As a result, the effects of the damage are not felt until it is too late.

- Electrochemical Etching:

Etching can often be automated and selectively controlled through the use of a technique called electrochemical etching (ECE), which removes silicon in a dopant-selective manner. It is necessary to have an active p-n diode junction, and any kind of dopant can serve as the material that is etch-resistant (also known as an "etch-stop"). Boron is the dopant that is most frequently used for etch-stopping. ECE has been successfully used for regulating the silicon diaphragm thickness in commercial piezoresistive silicon pressure sensors. This was accomplished by combining its use with the wet anisotropic etching technique that was discussed above. Implantation, diffusion, or epitaxial deposition of silicon are the three methods that can be utilised to produce selectively doped areas.

3.2. Dry Etching

During the dry etching process, the substrate material is removed using plasma or etchant gases. The reaction that takes place can be brought about by either the high kinetic energy of particle beams or a chemical reaction, or it can be brought about by a combination of the two.

- Physical Dry Etching:

In order to remove substrate atoms using physical dry etching, high-energy kinetic energy beams (such as ions, electrons, or photons) are required to be applied. When the high-energy particles knock away the atoms from the surface of the substrate, the substance will evaporate after it has been released from the substrate. Since there is no chemical reaction taking place, the only thing that will be removed is the material that has been unmasked.

- Chemical Dry Etching:

In the process known as chemical dry etching, also known as vapour phase etching, liquid chemicals or etchants are not used. In order to etch into the surface of the silicon, this procedure calls for a chemical reaction between the etchant gases. In most cases, the chemical dry etching technique is isotropic and has a high degree of selectivity. When compared to isotropic etching, anisotropic dry etching is capable of producing a greater aspect ratio and a finer resolution in the material that is being removed. Because dry etching is a directed process, it is possible to prevent undercutting from occurring. Tetrafluoromethane, sulphur hexafluoride, nitrogen trifluoride, chlorine gas, and fluorine are examples of some of the ions that may be utilised in the process of chemical dry etching.

- Reactive Ion Etching:

In order to obtain high degrees of resolution, the process known as reactive ion etching (RIE) employs both chemical and physical principles. The method is one of the most varied and extensively used methods in both industry and research. The fact that the process involves both chemical and physical interactions contribute to the fact that it is carried out significantly more quickly. Ionization causes high-energy collisions, which contribute to the dissociation of etchant molecules into more reactive species. During the RIE process, cations are formed from reactive gases, which are then driven with a high amount of energy towards the substrate, where they undergo a chemical reaction with the silicon.

4. Substrate Bonding

The fabrication of complicated systems, mostly from silicon and glass substrates, is accomplished through the process of bonding. When it comes to the fabrication of micromechanical devices, there are a few different bonding methods that can be applied. Fusion bonding is one of the most common methods for joining two different types of materials together. This method is utilised for the purpose of adhering surfaces together after they have been cleansed and made hydrophilic via the hydroxylation process. The surfaces that are going to be bonded need to be completely smooth and level. When the substrates are pressed together, the surface attraction of the two hydrophilic surfaces will bind them together to form a boundary. This method makes it possible to create bonds between glass and glass, as well as between glass and silicon as well as silicon and glass. In order to strengthen the link between the components, the device is often annealed at high temperatures.

Another technique for attaching substrates that is employed in the production of devices is known as anodic (electrostatic) bonding. Because of this method, bonding can take place at temperatures between 300ºC and 400ºC lower. After heating the glass-silicon sandwich to an approximate temperature of 200ºC, a high voltage ranging from 200 to 1,000V is then applied. Glass is connected to the cathode, and silicon is connected to the anode in this battery. At high temperatures, there is a migration of Na+ ions from the glass toward the cathode, and a strong electrostatic field at the interface stimulates the production of covalent silicon-oxygen bonds that drive the fusion. Both of these events take place simultaneously. When employing this method, it is essential to make use of glass that has a thermal coefficient of expansion that is comparably near to that of silicon. In addition, wafer-to-wafer bonding can be accomplished by utilising intermediary layers such as gold thin films (eutectic bonding), glass-frit bonding, solder bonding, and so on. For the purpose of putting polymer parts together, additional methods of bonding, including as ultrasonic welding, thermal pressure bonding, solvent bonding, and so on, are also utilised. There are a few different ways that the bonded state of materials can be defined. Infrared transmission, ultrasonic, and x-ray topography are all examples of non-destructive techniques that can be utilised in the monitoring of the bonding process. It has been claimed that microfluidic applications can benefit from an adhesive bonding method that employs the use of SU-8 negative photoresist as an intermediary layer. It is possible to bond using this method at temperatures lower than 200ºC, and the planarity of the wafers is not as important as it is when using other bonding methods. Because of the low temperature, the bonding process does not cause a significant amount of stress. The technology of adhesive bonding was utilised throughout the process of fabricating microfluidic devices (Ziaie et al., 2004).

BioMEMS devices have become an enabling technology for a variety of life science applications due to their size relative to biospecies, increased sensitivity, small volume requirement, high throughput, reduced test time, performance tenability, batch fabrication feasibility, low cost, portability, disposability, possibility of microelectronics and microphotonics integration, low power consumption, automation capability for remote and hazardous environments, and reduced manual labour. Because of their enormous potential, the applications have broadened beyond biosensing and now include medication delivery and single-cell manipulation as well. The capacity of BioMEMS devices to be both portable and disposable has been an enormous help in the development of diagnostic biochips for use in point-of-care testing (POCT) and point-of-need testing (PON) settings. The lab-on-a-chip (LOC), micro total analysis systems (μTAS), and biochips that are utilised for POCT and PON applications have a significant and immediate potential for poor countries that lack suitable laboratory and hospital facilities and have a high demand for medical care. BioMEMS has a wide range of potential applications, including but not limited to those in the fields of bioterrorism, defence, diagnosis, drug discovery and delivery, pharmaceuticals, nutrition, public health, agriculture, and other areas of life science.

REFERENCES

Al-Alawachi, S. F. A., Mohammed, K. A., Zabibah, R. S., Sadoon, J., Abed, A., & Al-Salam, M. A. (2022). Effect of copper oxide nanoparticles on some physical properties of PS/PEO polymer blend. *International Journal of Nanoscience, 21*(6), 2250044. doi:10.1142/S0219581X22500442

Badilescu, S., & Packirisamy, M. (2016). *BioMEMS: Science and engineering perspectives*. CRC Press. doi:10.1201/b12255

Bharti, A., Turchet, A., & Marmiroli, B. (2022). X-ray lithography for nanofabrication: Is there a future? *Frontiers in Nanotechnology, 4*, 835701. doi:10.3389/fnano.2022.835701

Bhattacharyya, R., Bhatt, G., Basu, A., & Basu, A. K. (2021). *Application of Polymers in Biomems Biomedical Devices and Related Challenges*. AIP Publishing. doi:10.1063/9780735423954_004

Brown, W. L., Venkatesan, T., & Wagner, A. (1981). Ion beam lithography. *Nuclear Instruments and Methods in Physics Research, 191*(1-3), 157–168. doi:10.1016/0029-554X(81)90999-X

Bruck, E. (2017). *Handbook of Magnetic Materials*. Elsevier.

Carter, S. B. (1967). Haptotactic islands: A method of confining single cells to study individual cell reactions and clone formation. *Experimental Cell Research, 48*(1), 189–193. doi:10.1016/0014-4827(67)90298-4 PMID:4862713

Chen, H.-Y., Elkasabi, Y., & Lahann, J. (2006). Surface modification of confined microgeometries via vapor-deposited polymer coatings. *Journal of the American Chemical Society, 128*(1), 374–380. doi:10.1021/ja057082h PMID:16390168

Cros, F. (2021). BioMEMS. *Biomedical Materials (Bristol, England)*, 581–620.

Dempsey, E., Diamond, D., Smyth, M. R., Urban, G., Jobst, G., Moser, I., Verpoorte, E. M., Manz, A., Widmer, H. M., & Rabenstein, K. (1997). Design and development of a miniaturised total chemical analysis system for on-line lactate and glucose monitoring in biological samples. *Analytica Chimica Acta, 346*(3), 341–349. doi:10.1016/S0003-2670(97)90075-1

Dentinger, P. M., Clift, W. M., & Goods, S. H. (2002). Removal of SU-8 photoresist for thick film applications. *Microelectronic Engineering, 61*, 993–1000. doi:10.1016/S0167-9317(02)00490-2

Desai, S. P., Taff, B. M., & Voldman, J. (2007). A photopatternable silicone for BioMEMS applications. Proc. Micro Total Anal. Syst.(-TAS).

Fedder, G. K. (1994). *Simulation of microelectromechanical systems*. University of California.

Folch, A. (2016). *Introduction to bioMEMS*. CRC Press. doi:10.1201/b12263

Hosseini, S., Espinoza-Hernandez, M., Garcia-Ramirez, R., Cerda-Kipper, A. S., Reveles-Huizar, S., & Acosta-Soto, L. (2019). *BioMEMS*. Springer.

Joshi-Imre, A., & Bauerdick, S. (2014). Direct-write ion beam lithography. *Journal of Nanotechnology, 2014*, 1–26. doi:10.1155/2014/170415

Keller, S., Blagoi, G., Lillemose, M., Haefliger, D., & Boisen, A. (2008). Processing of thin SU-8 films. *Journal of Micromechanics and Microengineering, 18*(12), 125020. doi:10.1088/0960-1317/18/12/125020

Kentsch, J., Breisch, S., & Stelzle, M. (2006). Low temperature adhesion bonding for BioMEMS. *Journal of Micromechanics and Microengineering, 16*(4), 802–807. doi:10.1088/0960-1317/16/4/017

Khan Malek, C., Robert, L., Boy, J.-J., & Blind, P. (2007). Deep microstructuring in glass for microfluidic applications. *Microsystem Technologies, 13*(5), 447–453. doi:10.100700542-006-0185-0

Kharisov, B. I., Kharissova, O. V., & Ortiz-Mendez, U. (2016). *CRC concise encyclopedia of nanotechnology*. CRC Press. doi:10.1201/b19457

Kotnala, R., & Shah, J. (2015). Ferrite materials: nano to spintronics regime. In *Handbook of magnetic materials* (Vol. 23, pp. 291–379). Elsevier.

Kumar, A., & Whitesides, G. M. (1993). Features of gold having micrometer to centimeter dimensions can be formed through a combination of stamping with an elastomeric stamp and an alkanethiol "ink"followed by chemical etching. *Applied Physics Letters, 63*(14), 2002–2004. doi:10.1063/1.110628

Lee, D., & Yang, S. (2012). Surface modification of PDMS by atmospheric-pressure plasma-enhanced chemical vapor deposition and analysis of long-lasting surface hydrophilicity. *Sensors and Actuators. B, Chemical, 162*(1), 425–434. doi:10.1016/j.snb.2011.12.017

Lorenz, H., Despont, M., Fahrni, N., LaBianca, N., Renaud, P., & Vettiger, P. (1997). SU-8: A low-cost negative resist for MEMS. *Journal of Micromechanics and Microengineering, 7*(3), 121–124. doi:10.1088/0960-1317/7/3/010

Maluf, N., & Williams, K. (2004). *An introduction to microelectromechanical systems engineering*. Artech House.

Mata, A., Fleischman, A. J., & Roy, S. (2006). Fabrication of multi-layer SU-8 microstructures. *Journal of Micromechanics and Microengineering, 16*(2), 276–284. doi:10.1088/0960-1317/16/2/012

Mourzina, Y., Steffen, A., & Offenhäusser, A. (2005). The evaporated metal masks for chemical glass etching for BioMEMS. *Microsystem Technologies, 11*(2), 135–140. doi:10.100700542-004-0430-3

Mund, D., & Leib, J. (2004). Novel microstructuring technology for glass on silicon and glass-substrates. *2004 Proceedings 54th Electronic Components and Technology Conference.*

Nemani, K. V., Moodie, K. L., Brennick, J. B., Su, A., & Gimi, B. (2013). In vitro and in vivo evaluation of SU-8 biocompatibility. *Materials Science and Engineering C, 33*(7), 4453–4459. doi:10.1016/j.msec.2013.07.001 PMID:23910365

Nordström, M., Keller, S., Lillemose, M., Johansson, A., Dohn, S., Haefliger, D., Blagoi, G., Havsteen-Jakobsen, M., & Boisen, A. (2008). SU-8 cantilevers for bio/chemical sensing; fabrication, characterisation and development of novel read-out methods. *Sensors (Basel), 8*(3), 1595–1612. doi:10.33908031595 PMID:27879783

Olszynko-Gryn, J. (2017). Thin blue lines: Product placement and the drama of pregnancy testing in British cinema and television. *British Journal for the History of Science*, *50*(3), 495–520. doi:10.1017/S0007087417000619 PMID:28923127

Pindyurin, V., Goldenberg, B., Petrova, E., Ancharova, U., Eliseev, V., Korolkov, V., & Nasyrov, R. (2008). Dynamic X-ray lithography for blazed diffractive optics fabrication. *Proceedings of SPIE - The International Society for Optical Engineering, 7102*. 10.1117/12.797728

Rabarot, M., Bablet, J., Ruty, M., Kipp, M., Chartier, I., & Dubarry, C. (2003). Thick SU-8 photolithography for BioMEMS. *Micromachining and Microfabrication Process Technology Viii*, *4979*, 382–393. doi:10.1117/12.478244

Rezai, P., Wu, W., & Selvaganapathy, P. (2012). Microfabrication of polymers for bioMEMS. In *MEMS for Biomedical Applications* (pp. 3–45). Elsevier. doi:10.1533/9780857096272.1.3

Rotake, D. R., Darji, A. D., & Kale, N. S. (2019). Design of microfluidic experimental setup for the detection of heavy metal ions using piezoresistive biomems sensor. *Microelectronics International*, *33*(1), 10–28. doi:10.1108/MI-05-2019-0025

Sadler, D. J., Changrani, R., Roberts, P., Chou, C.-F., & Zenhausern, F. (2002). Thermal management of BioMEMS. *ITherm 2002 Eighth Intersociety Conference on Thermal and Thermomechanical Phenomena in Electronic Systems.*

Sadler, D. J., Changrani, R., Roberts, P., Chou, C.-F., & Zenhausern, F. (2003). Thermal management of BioMEMS: Temperature control for ceramic-based PCR and DNA detection devices. *IEEE Transactions on Components and Packaging Technologies*, *26*(2), 309–316. doi:10.1109/TCAPT.2003.815093

Saliterman, S. (2006). *Fundamentals of BioMEMS and medical microdevices* (Vol. 153). SPIE press.

Sang, S. (2010). *An approach to the design of surface stress-based PDMS micro-membrane biosensors-concept, numerical simulations and prototypes*. Academic Press.

Scholten, K., & Meng, E. (2016). Electron-beam lithography for polymer bioMEMS with submicron features. *Microsystems & Nanoengineering*, *2*(1), 1–7. doi:10.1038/micronano.2016.53 PMID:31057839

Sharma, S., & Madou, M. (2012). Micro and nano patterning of carbon electrodes for bioMEMS. *Bio-inspired. Biomimetic and Nanobiomaterials*, *1*(4), 252–265. doi:10.1680/bbn.12.00010

Spille, E., & Feder, R. (1977). X-ray lithography. *X-Ray Optics*, 35-92.

Stevanus, J., Villanueva, A. H. R., Paliath, N., & Rifat, R. A. (2022). *JMN455–SSI Design Lab II: Manufacturing Processes for Smart Systems Lab Report*. Academic Press.

Sun, T., Park, W.-T., Tsang, J. W. M., Yee, T. B., & Je, M. (2013). Cytocompatibility assessment of Si, plasma enhanced chemical vapor deposition-formed SiO2 and Si3N4 used for neural prosthesis: A comparative study. *Nanoscience and Nanotechnology Letters*, *5*(8), 916–920. doi:10.1166/nnl.2013.1629

Szmigiel, D., Pamula, E., Sobczak, J. W., & Grabiec, P. (2007). Surface properties of silicone rubber for sensor applications. *SMCBS'2007 International Workshop.*

Tokachichu, D. R., & Bhushan, B. (2006). Bioadhesion of polymers for BioMEMS. *IEEE Transactions on Nanotechnology*, *5*(3), 228–231. doi:10.1109/TNANO.2006.874047

Vasudev, M. C., Anderson, K. D., Bunning, T. J., Tsukruk, V. V., & Naik, R. R. (2013). Exploration of plasma-enhanced chemical vapor deposition as a method for thin-film fabrication with biological applications. *ACS Applied Materials & Interfaces*, *5*(10), 3983–3994. doi:10.1021/am302989x PMID:23668863

Watson, J. D. (1990). The human genome project: Past, present, and future. *Science*, *248*(4951), 44–49. doi:10.1126cience.2181665 PMID:2181665

Watt, F., Bettiol, A., Van Kan, J., Teo, E. J., & Breese, M. (2005). Ion beam lithography and nanofabrication: A review. *International Journal of Nanoscience*, *4*(03), 269–286. doi:10.1142/S0219581X05003139

Windhorn, K., Meixner, L., Drost, S., Schroder, H., Kilgus, E., Scheel, W., Reichl, H., Ebling, F., Kuchler, M., & Nester, J. (2003). Polymer BioMEMs with integrated optical and fluidic functionality. *Proc. Micro Syst. Technol*, *2003*, 416–423.

Zhang, J., Yang, L., Xu, H., Zhou, J., Sang, Y., Cui, Z., Liu, C., Liu, J., Guo, T., Wang, X., Wang, L., Chen, G., & Chen, X. (2022). Dip-Coating Self-Assembly Fabrication and Polarization Sensitive Photoresponse of Aligned Single-Walled Carbon Nanotube Film. *Sensors (Basel)*, *22*(2), 490. doi:10.339022020490 PMID:35062451

Zhou, X., Poenar, D. P., Liu, K. Y., Li, W., Tse, M. S., Chen, H., Heng, C.-K., & Tan, S. N. (2007). Glass-based BioMEMS devices for optically excited cell impedance measurement. *Sensors and Actuators. A, Physical*, *133*(2), 301–310. doi:10.1016/j.sna.2006.06.037

Ziaie, B., Baldi, A., Lei, M., Gu, Y., & Siegel, R. A. (2004). Hard and soft micromachining for BioMEMS: Review of techniques and examples of applications in microfluidics and drug delivery. *Advanced Drug Delivery Reviews*, *56*(2), 145–172. doi:10.1016/j.addr.2003.09.001 PMID:14741113

Chapter 12
Microfluidic Technology for Advanced Drug Delivery Systems

Ankit Patil
https://orcid.org/0000-0002-3002-1107
Indian Institute of Technology, Jammu, India

Yogesh M. Nimdeo
Indian Institute of Technology, Jammu, India

ABSTRACT

Drug delivery is a procedure to transport drugs to an appropriate site for effective action, achieving the benefits of drugs. In the past few decades, nano-sized drug particles have attracted many researchers to develop an efficient drug delivery system. Microfluidic technology, for its ability to synthesize precise nanoparticles with uniform size, has emerged as a reliable technology for drug delivery. The microreactor has been established as a promising device due to its ability to provide intensive mixing in low-dimension microchannels having a high surface-to-volume ratio. It synthesizes drug particles with higher encapsulation efficiency and uniform particle size. The enhancement in the encapsulation efficiency results in the controlled release of drug particles at the appropriate site for effective medication. In addition, other applications of microreactors to synthesize various organic and inorganic nanoparticles are also discussed in this chapter.

INTRODUCTION

To achieve the benefits of drugs, targeting the drug particle at a suitable site is necessary, which can be achieved by the proper drug delivery system (Y. Zhang et al., 2013). The four key considerations for a successful drug delivery system are: retain, evade, target, and discharge (Mills & Needham, 1999). In recent years, researchers are very much interested in achieving an advanced targeted drug delivery system to enhance the effectiveness of drugs (Riahi et al., 2015). In targeted drug delivery, the medicine

DOI: 10.4018/978-1-6684-6952-1.ch012

is transported to a specific location to boost the medication concentration at the particular site. This increases the drug's effectiveness and reduces its adverse effects. Since the drug's influence on the rest of the body is reduced which lowers total toxicity. (Manzari et al., 2021). However, the fundamental problems during drug delivery are the safe transfer of medicines to pathogenic areas, solubility for water-insoluble drugs, and regulated release (Azimullah et al., 2019). In most of the research, nano-carriers are utilized for targeted drug delivery to overcome these issues. The nanocarriers are advantageous because of their lower size, enhanced solubility, bio-availability, high drug loading, drug-releasing capacity, and high surface-to-volume ratio (Gagliardi et al., 2021; Shah et al., 2021; Singh & Lillard, 2009). Several therapeutic nanoparticle systems are currently being investigated, but the polymer-based nanoparticle as a drug carrier is emerging as an adequate solution (Anju et al., 2020). Such drug-loaded nanocarriers have demonstrated good consistent drug release kinetics, a longer systemic circulation half-life, and low adverse effects (Farooq et al., 2019).

Several techniques for producing drug-loaded nanocarriers have been developed, including spray-drying, high-pressure homogenization, solvent evaporation, ultrasonication, and others (H. Zhao et al., 2007). Although these techniques were suitable for synthesizing nanoparticles, the commercial application remains a significant problem. This was because of the low repeatability of the processes, uncontrollable particle size, uneven morphology, limited batch-to-batch operations, and lower production capacity (Shrimal et al., 2019). Whereas microfluidic technology has emerged as an efficient technology to overcome these challenges through development of microreactor devices. Microreactor is a device in which the chemical reaction occurs on a continuous basis in small confinement called microchannels. Dimensions of these microchannels are as low as 1 mm. This low dimension gives a high surface area to volume ratio, resulting in better heat and mass transfer, and achieves effective and faster mixing, which leads to low particle size (PS) of nanoparticles with high repeatability (Sebastian et al., 2014, Mastroianni et al., 2022, J. Zhang et al., 2022). Microreactor handles toxic chemicals without requiring extensive inventory or operational volume (C. X. Zhao et al., 2011). It also provides easy handling for highly exothermic reactions due to the small dimension of microchannels. These microchannels are fabricated using different materials like PDMS (Poly-Di-Methyl-Siloxane), silicon, glass, and quartz (Bojang & Wu, 2020). Most notably, the residence time of reactions has been significantly reduced using microreactors, which allows hour-long operations to be performed in seconds (Pennemann et al., 2004). The microfluidic process can be scaled up by arranging microreactors in parallel arrays. But commercialization of this technique will require more scientific efforts. Microreactors are classified into two major types based on their flow pattern in channels throughout the nanoparticle synthesis process: continuous flow microreactors and segmented flow microreactors (Tian et al., 2018). A continuous flow microreactor is the most fundamental microreactor and is generally more productive and gives a uniform product (C. X. Zhao et al., 2011). Laminar flow is observed in a continuous flow microreactor, resulting in a low Reynolds number and parabolic velocity profile of fluid, leading to high residence time distribution and uneven mixing (Nguyen, 2012a). The segmented flow microreactor is promising when the system contains two immiscible liquids (Harries et al., 2003). So strategically, employing micro fluid segments in the continuous-flow synthesis of nanomaterials appears to be quite promising (Köhler et al., 2013). The essential components of a microreactor involve micromixers, pumps, and microchannels. The micromixers are the most crucial component of a microreactor system, which directly affects the efficiency of the microfluidic system. The author has briefly covered various types of micromixers in this chapter. Pictorial presentation of microreactor along with its component is shown in Figure 1.

Nanoparticles (NPs) and nanocomposites generated through microreactors exhibit distinctive physical and chemical characteristics, such as luminescence, photo electricity, catalysis, magnetic response, biocompatibility, etc., compared to macro-sized particles. As a result, there are several uses for nanoparticles and nanocomposites in various industries, including optical materials, semiconductors, catalysts, bio-labels, drug delivery systems, energy sector, and many more (Tian et al., 2018). It is also used in analytical chemistry to provide surfaces with enhanced Raman scattering effects (Bressan et al., 2019). Quantum dots, on the other hand, are simply semiconductor nanocrystal which has the potential to be used in solar cells, lasers, and light-emitting devices. Their applications are mainly dependent on size-related properties. Here the microfluidic reactors having good control over chemical processibility, and nanocrystal growth is favoured. As in segmented (or droplet-based) microfluidic reactors, ethanol quenches the reaction and cools the nanocrystals. Here, the chilling step, which is essential in batch processes for regulating the size of the nanocrystals, can be skipped (Kwak et al., 2017).

Overall, the development of microfluidic technology has opened up fascinating possibilities for regulated drug administration and prepared the path for the clinical application of nanomedicines. In this microreactors are of great importance because of their ability to obtain nanoparticles of uniform size, with good reproducibility and continuous nanoparticle synthesis. The microreactors loaded with drug particles and polymer have the unique ability to produce nanoparticles with high entrapment and loading efficiency. Specifically in drug delivery systems, proper encapsulation of drug particles in polymeric carriers plays a vital role in drug particle synthesis and its targeted delivery. This chapter describes the microfluidic techniques to develop drug nanoparticles using microreactors, which have

Figure 1. Schematic of microreactor and its component

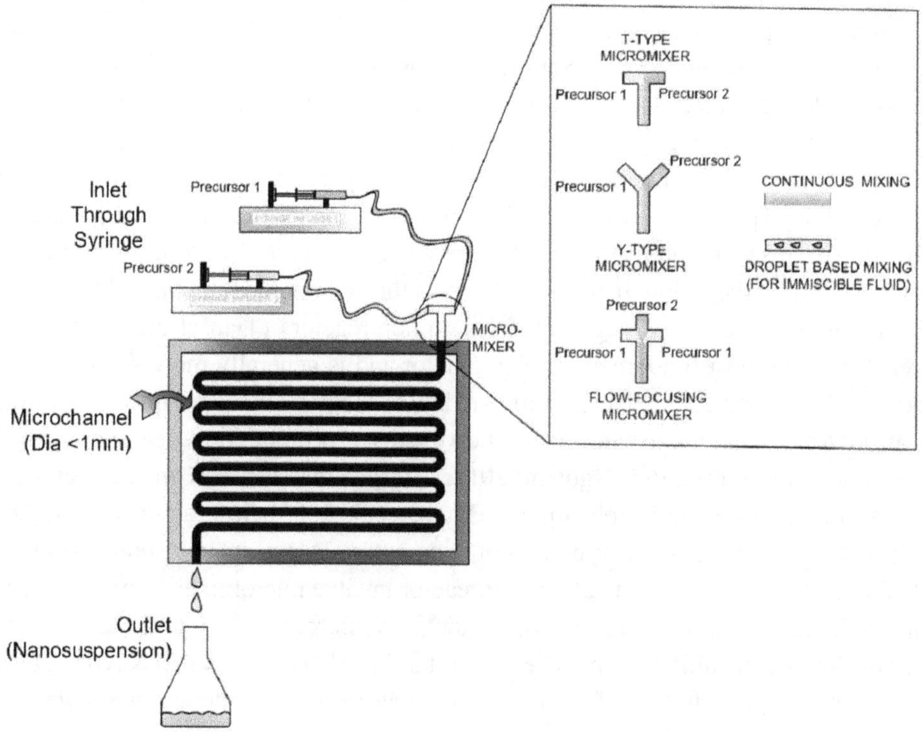

shown improvement over other available techniques in the literature. The author has also demonstrated the applications of the microfluidic approach to manufacture inorganic and organic nanoparticles to illustrate the usefulness of microfluidics. Also, the opportunities for the application of microfluidics platforms in drug delivery systems, including the therapeutic uses of the drug, and the current issues with the aforementioned approaches, are discussed.

MICROMIXERS

Micromixers are a crucial component when a reaction occurs in a microchannel; a key benefit of a micromixer is quick mixing. The principle of micromixers is based on molecular diffusion (Nguyen, 2012c). Since Reynold's number in the microfluidic system is minimal, and obtaining the turbulent flow required for efficient mixing is challenging (Nguyen, 2012b). According to Fick's law, higher the concentration gradient better will be the diffusive flux (Ulkir et al., 2021). So, for better mixing, a researcher needs to achieve a smaller mixing path and a higher interfacial area of contact between solute and solvent, which can be possible with the geometries of micromixers (Capretto et al., 2011). The short mixing path of the micromixer also results in a higher concentration gradient which is preferable for higher rates of mass diffusion to get better mixing (Qian et al., 2016). Based on the mixing methods, the micromixers are classified as passive and active micromixers. In passive micromixers, mixing achieved by pumping energy, whereas in active micromixers, mixing is carried out by the influence of external energy. (Bayareh et al., 2020). In addition, there are several subtypes of the passive and active micromixers which are described as follows.

Figure 2. The schematic of T type micromixers

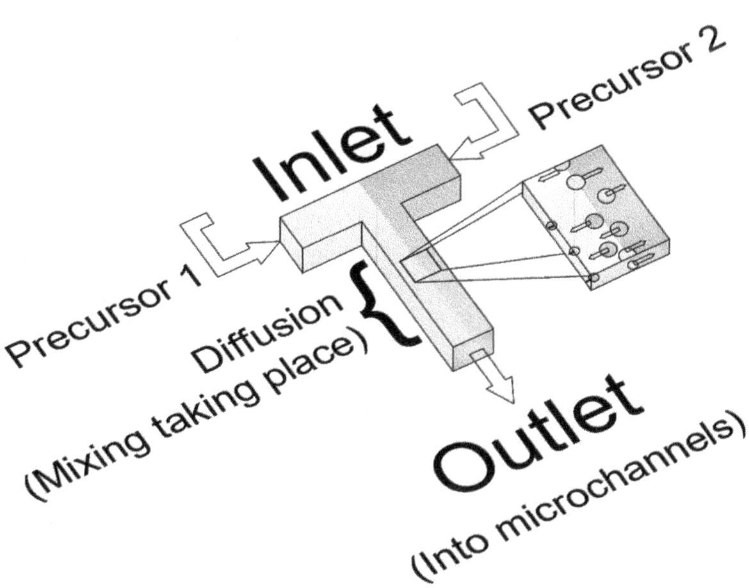

Figure 3. The schematic of Y type micromixers

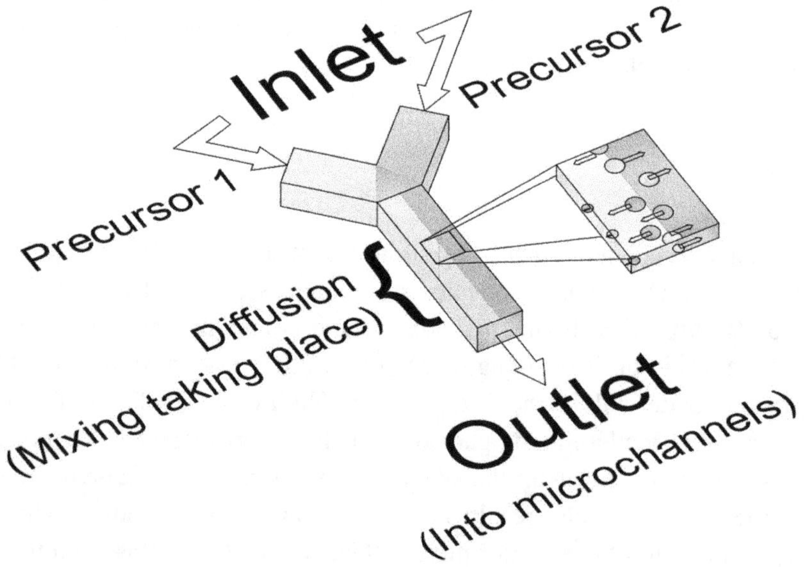

1. **Passive Micromixers:** The mixing is performed by pumping fluid into the micromixer. Here mixing occurs because of chaotic advection or molecular diffusion (Raza et al., 2020). Chaotic advection can be improved by altering the design of micromixers to adjust the laminar flow inside the channels (Baheri Islami et al., 2017). These are the most commonly used micromixers because of their simple structure and easy handling. It can be further classified based on the flow pattern and channel geometry (Capretto et al., 2011).
    a. **T- and Y-Shaped Micromixers:** These are the most basic and most straightforward designs for micromixers and are one of the first designed micromixers for mixing in microchannels. Two inlet streams of fluids are guided into two different T- or Y-channel inlet arms, which ultimately flow in a shared long microchannel (Udaya Kumar et al., 2022). In these micromixers, mixing depends only on diffusion, so the time required for mixing is more and requires a long channel. To overcome such issues, researchers have modified channels by adding obstacles in the pathway and roughening the channel walls. Also, shorten the diffusion length by narrowing the channel. Mixing can also be improved by replacing a straight channel with a square wavy channel (Clark et al., 2018). Structure of T- and Y- type micromixers is shown in Figure 2 and Figure 3 respectively.
    b. **Parallel Lamination Micromixers:** A parallel lamination mixer divides the fluid intake streams into numerous substreams and then re-joins them in a single stream. This may be accomplished with T-mixers and Y-mixers (Gambhire & Massai, 2011). Schematic shown in Figure 4.
    c. **Split-and-Recombine (SAR) Micromixers:** Also known as sequential lamination. As the diffusion time is directly proportional to the square of the diffusion distance. In this type of mixer, the mixed stream is divided into multiple smaller streams and re-joins in the next stage;

Figure 4. The schematic of parallel lamination micromixer

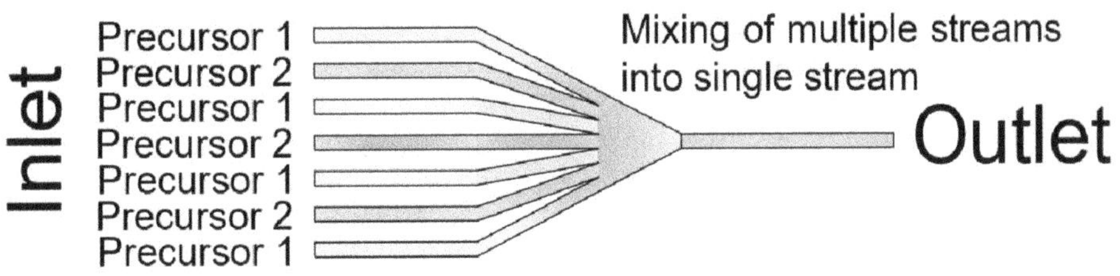

Figure 5. The schematic of split-and-recombine (SAR) micromixers

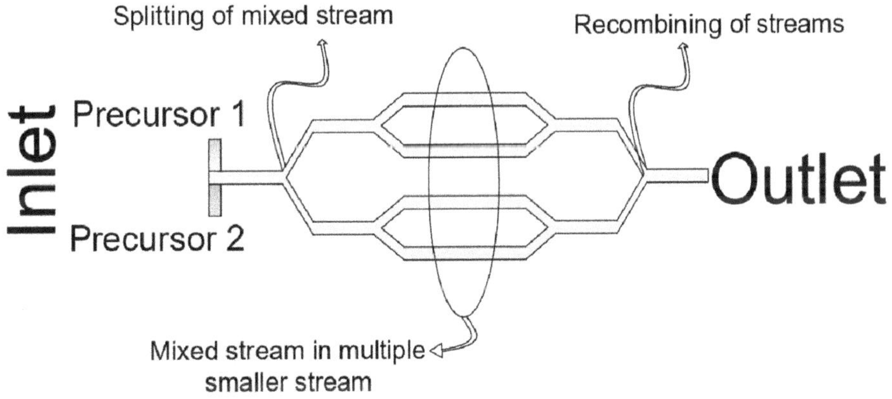

this significantly decreases the mixing time. Based on their channel layout, SAR micromixers are further classified as 3D SAR or planar SAR (P-SAR) mixers (Capretto et al., 2011; Qian et al., 2016). Schematic shown in Figure 5.

d. **Flow Focusing Mixing-Based Micromixers:** Another method for increasing the diffusion and decreasing the mixing path is reducing the solute inlet width. The most straightforward design for hydrodynamic mixing has three inlet channels in a long microchannel, having two solvent streams (R. Hu et al., 2019).

e. **Chaotic Advection Micromixers:** Chaotic advection is another mechanism used to increase the mass transfer in the flow, which has a low Reynolds number. The mass transfer can be increased by creating disturbances in the flow pattern through sudden bends or structural changes in the microchannels. Here instead of using straight microchannels, a spiral (Sudarsan & Ugaz, 2006a), splitting and recombining based structures (Simonnet & Groisman, 2005),

Figure 6. The schematic of flow focusing mixing based micromixers

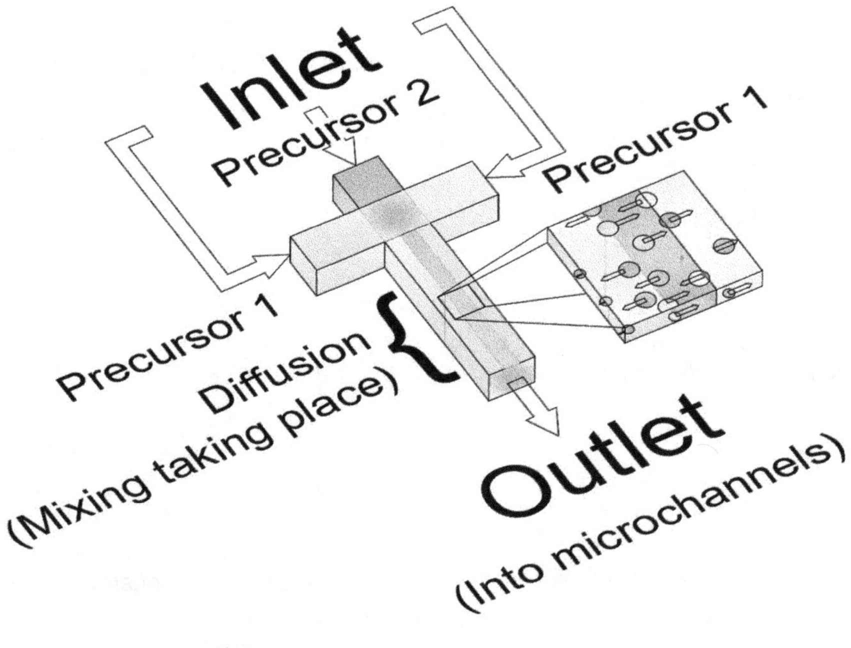

Figure 7. The schematic of spiral microchannels

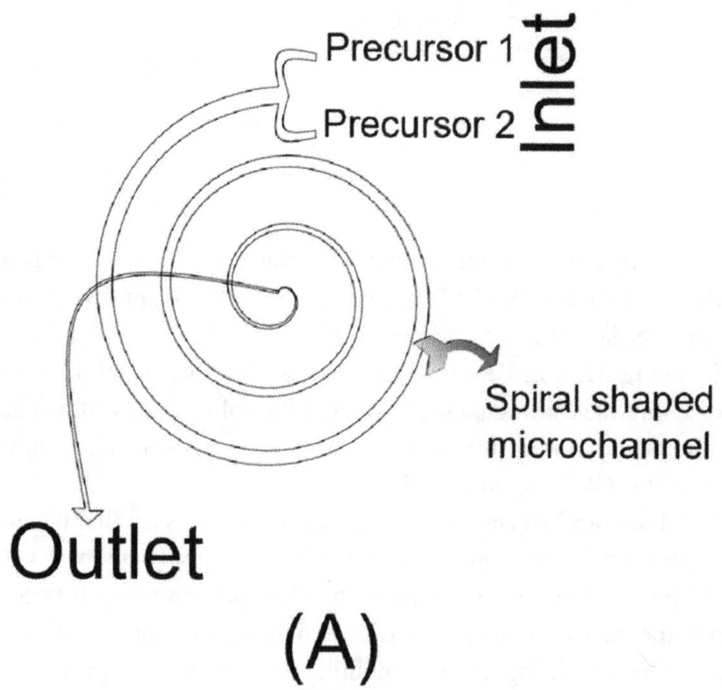

Figure 8. The schematic of square shaped microchannels

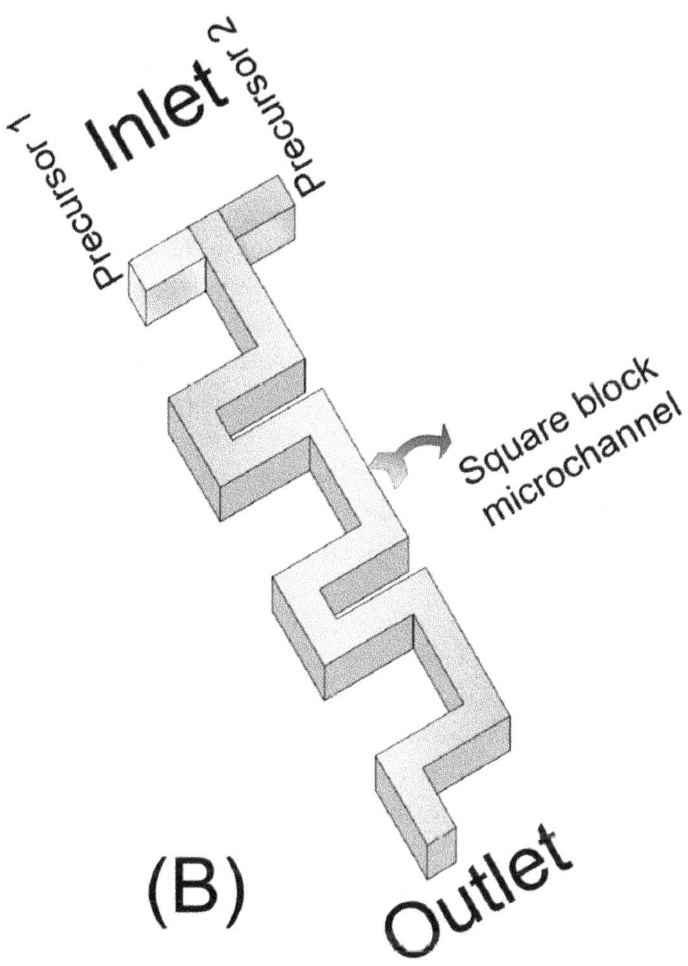

curved channels, or multi-vortex channels (Sudarsan & Ugaz, 2006b) can be used. By introducing such microchannels, researchers noticed high mass diffusion and better mixing (Nguyen, 2012a). Figure 7 and 8 shows the schematics of spiral and square shaped microchannels.

f. **Microdroplet-Based Micromixers:** Microdroplets are generated when two or more immiscible fluids interacts with each other and forms multiphase flow. The primary channel design used in the microdroplet-based mixers are T- shaped or flow-focusing mixing. In T- junction configuration dispersed phase interacts with the continuous phase channel perpendicularly, resulting in droplet formation at the interaction point. In the flow focusing mixing pattern, the dispersed phase flow through the middle stream, and the continuous phase flows through the other two lateral channels. Pressure and shear force on the inner fluid promotes the production of a narrow neck, which finally collapses, resulting in the formation of a droplet. In general, mixing inside microdroplets is facilitated by the reduction in diffusion path length and the interaction between the fluids to be mixed as a result of droplet's geometrical confinement (Capretto et al., 2011; Nguyen, 2012b).

2. **Active Micromixers:** In these micromixers external energy causes the flow disruption to accomplish efficient mixing (Capretto et al., 2011; Qian et al., 2016). On the basis of types of energy employed to disrupt the flow these micromixers are classified as:
    a. **Pressure Field Disturbance:** Inducing a pressure field disturbance is a straightforward approach for active mixing (Cai et al., 2017). The pressure field disturbance is created using two vortex micropumps at inlet of a microfluidic mixer working based on the same fluid variation or fluid stretching and folding principles. The different fluids of a finite volume that are to be mixed is pumped into an expansion chamber to enhance the interfacial surface area between the liquids. Here the Reynold's number of flows in the micromixer is generally maintained near about 30 (Bayareh et al., 2020; Gambhire & Massai, 2011).
    b. **Electrokinetic Disturbance:** Electrokinetic disturbance micromixers uses a fluctuating electric field to generate mixing in microfluidic channels. The fluctuating electric field causes a fast stretching and folding of fluid surfaces, allowing the fluid stream to be stirred in the laminar flow. Reynolds number in such micromixer is of order unity. Such type of micromixers is preferrable for microfluidic bioanalytical applications. The efficiency of electrokinetic mixing can be increased by combining it with a passive micromixing technique that employs channel shapes to promote secondary flow. (Capretto et al., 2011; Oddy et al., 2001; Sundaram & Tafti, 2004).
    c. **Dielectrophoretic Disturbance:** The mobility of neutral particles brought on by an alternating current electric field is referred to as dielectrophoresis. It causes the particles to become asymmetrically polarised. A dipole moment forms on the particles which produce a force that causes them to move either away from or toward the electrodes. The stretching and folding phenomena occur as a result of particle build-up in a quasi-novelocity area, resulting in chaotic motion and an increase in the mixing index. Kim et al, demonstrated that the mixing quality was improved with decreasing channel depth. But it is necessary for the channel depth to be more than 20 micrometres. (Bayareh et al., 2020; D. Kim et al., 2008; Kumar et al., 2009).
    d. **Electrowetting Shaking:** Electrowetting is referred as the change of a surface's wetting characteristics by an applied electric field. It is an active method of causing droplet mixing. This type of mixing is based on the use of an electric field to regulate the interfacial tension of a droplet (Capretto et al., 2011). The fluid flow patterns may be created by the movement of liquid droplets, which also facilitates the mixing of different species inside the droplets. The droplets serve as virtual mixing chambers, with mixing taking place by vibrating the droplet over a number of electrodes at different frequencies. However, this mixing process takes time. So, several researchers have suggested that, using electrowetting to shake, divide, and merge the droplets in order to produce recirculating patterns which increases the interfacial area between the two liquids to be mixed can results in speeding up the mixing process. (Gambhire & Massai, 2011).
    e. **Magneto-Hydrodynamic Disturbance (MHD):** A micromixer that creates complicated flow fields produced by Lorentz forces generated through combination of a magnetic and electric field. Here the electric field is generated by a series of electrodes. It is possible to create flow fields that are rather complicated since the electrodes may be patterned in a variety of ways. As a consequence, the flow fields regulated the mixing of a dye species in an electrolyte solution. These micromixers, however, are not quick enough when rapid mixing is required

in some biochemical studies, as their mixing durations are in the seconds range (Bau et al., 2001; WONG et al., 2004).

f. **Ultrasound Disturbance:** It is a novel sophisticated technology, used to increase different physical and chemical processes in liquid medium by creating cavitations and sonic streaming (Parvizian et al., 2014). Acoustic stirring caused by ultrasonic waves is used to achieve mixing. The ultrasonic action creates acoustic churning of the fluid perpendicular to the flow direction, which improves mixing inside the microfluidic channel (Zeng & Palmer, 2005). The benefits of ultrasonic energy include energy savings and its optimization, which improves product quality, and reduces manufacturing time (Lebon et al., 2018).

SYNTHESIS OF INORGANIC NANOPARTICLES USING MICROFLUIDIC DEVICES

Over the last few years, researchers are very much interested in the synthesis of inorganic nanoparticles like metals, semiconductors, and quantum dots because of their numerous applications in various fields of science and technology. The particle size of inorganic nanomaterials may affect their applications in catalysis, biosensors, imaging, optoelectronics, drug discovery, and many more (Bakunin et al., 2004).

Recently among available metal nanoparticles synthesizing copper (Cu) nanoparticles is of great interest for the researchers because of their excellent physical and chemical properties. The Cu nanoparticles are also preferred due to their high electrical and thermal conductivity, as well as low price compared to silver (Ag) and gold (Au) (Din & Rehan, 2017). Xu et al. synthesized Copper nanocolloids using a T-shaped continuous flow microreactor. Researchers used 0.02 mol/L Copper sulfate pentahydrate ($CuSO_4.5H_2O$) solution and (0.04M, 0.1M, 0.4M) sodium borohydride ($NaBH_4$) solution. They injected it into a T- shaped micromixer at the same flow rate where the reduction of metal salt occurred. Transmission electron microscopy (TEM) and X-ray diffraction were used to evaluate the diameter, shape, uniformity in particle size, and elemental compositions of Cu particles. UV-vis spectroscopy was utilised to confirm the chemical composition of nanocolloids and determine the average size of copper nanocolloids. Researchers were also interested to see the effect of the molar concentration of $NaBH_4$ solution on the particle size of copper nanocolloids. They noted that the mean diameter of copper nanocolloids decreases with an increased $NaBH_4$ molar concentration. At 0.04M $NaBH_4$ researchers got the particle's maximum mean diameter, i.e., 4.25 nm. Copper nanocolloid suspension was continually generated in this T-shaped microreactor and remained stable for up to 60 days (Xu et al., 2015).

Gao et al., analysed the effect of mixing on the particle size of metal nanoparticles synthesized in a microreactor. They synthesized silver (Ag) nanoparticles of different particle sizes by varying the mixing parameters of precursors in a 3D curved microreactor. They concluded that the particle size decreases with an increase in the mixing index. This is because of increased nucleation rate, which results in higher nuclei concentration. The curve in the microreactor channel led to the formation of the dean vortex where the mixing was varied by a change in the direction of the dean vortices. This was done by configuring the inlet streams relative to the curvature of the reactor. The experiment was helpful in designing microreactors and knowing that disturbance in flow could affect the reaction rate and hence can be used to control the nanoparticle size (Gao et al., 2022).

Consequently, it is crucial to understand the technique to synthesize Ag nanoparticles having uniform shape, size, and distribution. As they are mainly used in various sectors, including bactericidal medications, drug delivery, medical diagnostics, and cell labelling (Lee & Jun, 2019). Yang et al. developed a process for controllable synthesis of Ag nanoparticles. This process includes a multi-stage microreactor using two valveless piezoelectric pumps as its power source. To control the nanoparticles' size, the author conducted the experiment by adjusting particular operational parameters like driving voltage of pump and reductant concentration of precursors . Since the driving voltage has a significant impact on the size, concentration, and homogeneity of the produced Ag nanoparticles. Adjusting the driving voltage has changed the inlet flow rate and the effect of mixing on particle size is investigated. The reaction time is shortened as driving voltage increases due to an increase in flow rate. As a result, the concentration and average particle size of the Ag nanoparticles produced in the colloid has been increased and uniformity got improved. The experimental results were obtained at different voltage of 30, 50, 80, and 120 Vpp. At these voltages the inlet flow rate and average particle size along with deviation were 0.05, 0.10, 0.15, 0.20 ml/min, and 24.67±7.03, 24.98±5.42, 25.01±4.80, 25.93±4.15 nm respectively. The reductant concentration also significantly impacts the quality of synthesized Ag nanoparticles. It was observed that the reaction rate increases with reductant concentration of precursors, which increases the concentration of Ag nanoparticles as well as particle size with better uniformity in colloids up to a specific limit. After which the synthesized Ag nanoparticles get agglomerated. The experimental scheme and obtained results were at concentration ratios of $C_6H_{12}O_6/AgNO_3$ 0.5:1, 1:1, 2:1, and 4:1 average particle size along with deviation were 21.35±6.55, 22.16±5.88, 25.93±4.15, and 28.19±6.71 nm. This study demonstrated that the microfluidic system offers benefits like small size, simple integration, and low cost among others. It is possible to create Ag nanoparticles in a continuous, controlled, effective manner with high quality (Z. Yang et al., 2022).

Gold (Au) nanoparticles has remarkable catalytic, physicochemical, biological, and hypoallergenic characteristics and ease of functionalization (Amina & Guo, 2020). Gold (Au) nanoparticles have numerous applications in therapeutics, analytical research, catalysis, and medical diagnostics (X. Hu et al., 2020). Huang et al. synthesized ultrasmall gold nanoparticles using a capillary-based microreactor. Gold (III) chloride trihydrate and trisodium citrate are mixed at high temperatures in a capillary system. Their research examined the influence of surface-to-volume ratio, the material of the capillary channel, temperature (30-100^{0}C), and average residence time on particle size and uniformity. The surface-to-volume ratio was varied by varying the capillary diameter from 0.3-1 mm; capillary materials used were fluorinated ethylene propylene, fused silica, polytetrafluoroethylene (PTFE), and polyether-etherketone. The lowest particle size observed was 1.9±0.2 nm at residence time of 30 minutes at 100^{0}C, using a PTFE capillary tube of 0.3 mm inner diameter. In this study, researchers also compared the particle size of Au nanoparticles synthesized using a segmented flow microreactor. The particle size of Au nanoparticles obtained using a segmented flow reactor was 17.4±1.4 nm. This comparison emphasizes the importance of the capillary wall surface on uniformity and size of nanoparticles (Huang et al., 2018). Experiments with the same length (2.5 m) of microchannels but varying inner tube diameters were conducted to show that the PTFE tubing wall plays a vital role in generating ultrasmall particles. The relationship between the reactor surface-to-volume ratio and the trend of nanoparticle size is shown in Figure 9.

The use of metal oxide nanoparticles in biomedicine for drug administration, targeting, and imaging necessitates regulated size and uniform distribution (Nikolova & Chavali, 2020). For the controlled synthesis of metal oxide nanoparticles, microfluidic technology has proved to be more effective (James et al., 2020). Xu et al., synthesized cupric oxide (CuO) via controlled oxidation of copper (Cu) nanoparticles and the size of Cu-CuO composite crystallites was found to be 10 nm, which was proportionate to the size of copper nanoparticles. A stainless steel microreactor with counter-current flow and a channel

Figure 9. TEM images of gold nanoparticles synthesized at the same average residence time (30 min), with PTFE tubing of varying I.D. at 100°C. (a) Aqueous phase segmented by octane in capillary with an I.D. of 0.3 mm, and single-phase flow in capillary with I.D. of (b) 1 mm, (c) 0.8 mm, (d) 0.5 mm. The insert shows the larger size gold nanoparticles found on another area of the same TEM grid of (d). (e) Nanoparticle size obtained as a function of the reactor surface-to-volume ratio from (i) slug flow, (ii) 1 mm I.D. tube, (iii) 0.8 mm I.D. tube, (iv) and (v) 0.5 mm I.D. tube, and (vi) 0.3 mm I.D. tube.
Source: Huang et al. (2018)

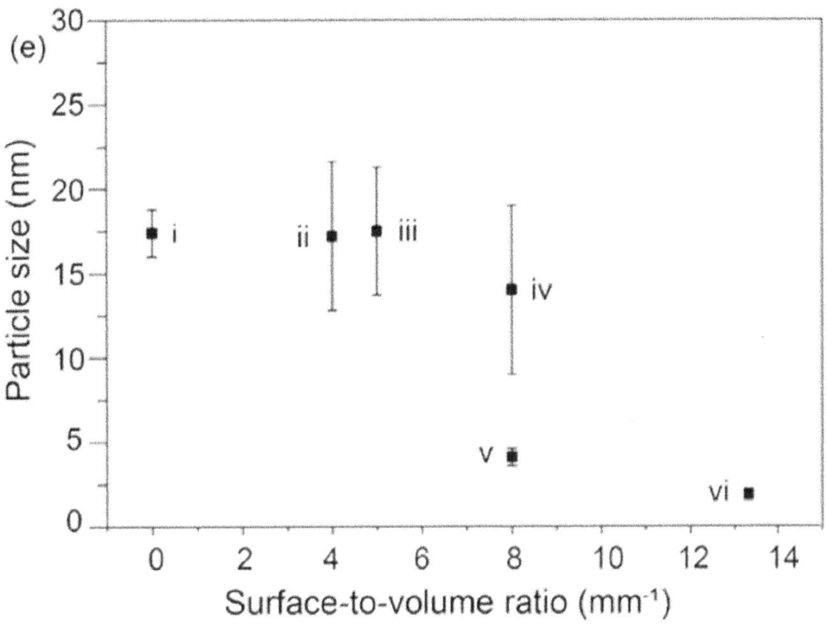

diameter of around 40 μm was used to continuously synthesize Cu nanoparticles by aqueous reduction with sodium borohydride ($NaBH_4$) when $CuSO_4$ and $NaBH_4$ were pumped at a flow rate of 20 ml/min. The Cu nanoparticles obtained were characterized by a high-resolution transmission electron microscope (HRTEM). It was found that the particles have a spherical morphology and uniform particle distribution with a mean diameter of 9.3 nm, where 90% of particles were in the range of 7-12 nm. The methylene blue (MB) degradation kinetics was observed to follow first order. When exposed to UV radiation, the Cu-CuO nanocomposite showed remarkable photocatalytic activity for the breakdown of MB (L. Xu et al., 2017).

In literature, some of the studies also investigated whether the shape of metal oxide nanoparticles affects their cytotoxicity, photocatalysis, and piezoelectric capabilities (Mclaren et al., 2009). Hao et al. performed an experiment in which they synthesized zinc oxide (ZnO) micro/nanoparticles of various shapes (sphere, short rod, ellipsoid, long rod, cube, platelet, and urchin) in a spiral microfluidic reactor. Zinc nitrate ($Zn(NO_3)_2$) and sodium hydroxide (NaOH) were mixed in a spiral microfluidic reactor having a microchannel length of 92 mm to synthesize the nanoparticles. The shape of ZnO micro/nanoparticles was altered by changing the reactant concentration and the flow rate of intake fluid. It was observed that metal oxide's properties also depend on the shape, with greater surface area ZnO structures usually displaying higher performance (Hao et al., 2019). Some more studies on nanoparticles synthesis using a microreactor are given below in Table 1.

Table 1. Inorganic nanoparticles synthesized using a microreactor

Nanoparticle	Method	Microreactor Type	Objective	Nano-Particle Size (nm)	Morphology	Observation	Ref.
Pt NP	Synthesis by gravity	Continuous flow microreactor	Nanosheet for enhancing catalytic activity.	Thickness 2.6±0.87 nm Width 557±112 nm	Nanosheet	Because of its huge surface area and dispersibility, Pt nanosheets perform better in catalysis and should function well in hydrogen generation, fuel cells, and sensors.	(Sasaki et al., 2022)
$BaSO_4$	Chaotic mixing mechanism	Four-stage asymmetric oscillating feedback microreactor	To synthesize uniformly sized aggregated primary nanoparticles of barium sulphate (BaSO4)	23-109 nm		It is possible to regulate the average primary particle size by changing the volumetric flow rate and reactant concentrations.	(H. Yang et al., 2022)
Bi_2O_3	T-shaped microreactor with uniform circular cross-sectional channel.	Polydimethylsiloxane (PDMS) and poly(methyl methacrylate) (PMMA)-based microreactor	Improved inline photocatalysis of nanoparticles of bismuth oxide (Bi2O3).	6.7 nm	Spherical	The high stability of Bi_2O_3 nanoparticles even after three cycles suggests that coated microreactors for photocatalysis can be reused. Escherichia coli bacterial pathogens were effectively inhibited by the tiny concentration of as-synthesised Bi_2O_3 nanoparticles.	(Katoch et al., 2021)
Si NP	T-junction served as a droplet producer.	Droplet-Based Microfluidics	The quality of the nanoparticles generated in the micro-fluidic reactor was compared to those made using the conventional approach.	5.76±1.27 nm	Spherical	When compared to the NPs generated from the bench scale system, which included highly monodispersed silica nanoparticles, the microreactor's NPs particle quality was greatly improved.	(Ling et al., 2021)
CeO_2	Hydrothermal methods	Continuous Microreactor	Fast Synthesis Using Deep Eutectic Reline as solvent.	12.5 nm	Nanorods	Compared to alternative processes, this produces CeO_3 nanoparticles much more quickly.	(Exposito et al., 2020)
Pd NP		Continuous flow microreactor	Green synthesis of time-stable palladium nanoparticles.	4.4±1.68 nm	Spherical	Turn over frequency remains unchanged till 7 months, which means particle obtained by microreactor are stable.	(Gioria et al., 2020)
α-Al_2O_3	Microreactor precipitation method	Stream microreactor	To compare homogeneity and dispersity of particle prepared by conventional parallel flow.	110 nm	Spherical nanoscaled	Ultrafine -Al_2O_3 powders that have good thermal expansion properties, high crystallinity, and uniform size distribution are synthesized.	(Yu et al., 2020)
$BiFeO_3$	Heat treatment of hydroxides co-precipitated	Impinging-jets microreactor	Formation of nanocrystalline BiFeO3	20 nm	Spherical	It was feasible to parametrize the influence of the forming particles size on the circumstances because of the hypothesised process of the $BiFeO_3$ nano powder creation.	(Proskurina et al., 2019)
Si NP	microreactor-assisted rapid mixing process	Microreactor with hydrodynamic focusing micromixer	Synthesise hollow silica nanoparticle (HSNP)	30 nm	Diverse shape, narrow size distribution	The customised HSNPs produced by this work's microreactor-assisted method might be employed for cell labelling and drug delivery applications where the particle size and configuration are critical.	(Apelgren et al., 2019)

Table 1 Continued

Nanoparticle	Method	Microreactor Type	Objective	Nano-Particle Size (nm)	Morphology	Observation	Ref.
CuS	Mixing an aqueous $CuBr_2$/ branched polyethyleneimine (BPEI) solution, an aqueous L-ascorbic acid solution, and an ethanol solution.	Continuous Couette-Taylor microreactor	Studies on morphology changes	Size: 105-161 nm Thickness: 83-63 nm.	Hexagonal nanoplates	Rotational speed, mean residence time, and feed solution provide simple modulation of the nanoparticles' morphologies and sizes.	(Tang et al., 2019)
Superparamagnetic Iron Oxide Nanoparticles (SPION)	Spiral microchannel, high production yields by coupling microreactors using stacked plates.	Single phase microreactor	The high temperature, continuous production of superparamagnetic iron oxide nanoparticles of size< 4nm	<4 nm		The microreactor used in this work allows for reproducible, continuous SPION production that is simple to scale up.	(Uson et al., 2018)
Zn doped Fe_3O_4		Continuous flow microreactor	To develop doped nanoparticles.	<5 nm	Spherical	Here, it has been demonstrated that using microreactors may successfully produce doped Fe_3O_4 nanoparticles with progressively higher Zn contents that are smaller than 5 nm.	(Simmons et al., 2015)
$CuCrO_4$	Mixing aqueous solutions of copper nitrate and chromium nitrate	Continuous Flow Microreactor	To investigate the impact of various Cu/Cr metal ion ratios on particle size.	192-300 nm	Cubical	The use of a microreactor reduced the residence time, hence crystal growth of copper chromite nanoparticles is restricted which causes the development of smaller copper chromite nanoparticles.	(Appalakutti et al., 2015)
Ni NP	Aqueous reduction	T-shaped continuous flow microreactor	To study the effect of flow rate on particle diameter	5.43±1 nm	Spherical	It clears that a higher flow rate causes the mean particle diameter to drop.	(L. Xu, Srinivasakannan, et al., 2015)
Si NP	Emulsification of the aqueous siliceous precursor	Droplet-Based microreactor	Synthesis of monodisperse mesoporous silica	5 μm	mesoporous structure	The pore shape and size will be easier to manage because of the slower kinetics.	(Carroll et al., 2008)

SYNTHESIS OF ORGANIC NANOPARTICLES USING MICROFLUIDIC DEVICES

Recently microreactor technique has been gaining demand because of its ability to synthesize uniform nanoparticles and good control over the process. However, most studies in this field focus on the technical features of microreactors rather than how they can be beneficial for organic chemistry. Consequently, not many studies have been done on synthesizing organic nanoparticles using microreactors. It is important to focus more on the studies to improve the efficiency of synthetic organic chemistry, as the raw material grow increasingly scarce and traditional method is causing many of them to be wasted.

Kamat et al. synthesized monodispersed chitosan (CS) nanoparticles using an active microreactor. The potential of CS nanoparticles as a drug delivery system is excellent due to their capacity to preserve drugs from deterioration, enhance loading, and sustain release. Due to the slow flow rate passive mixer faces few drawbacks because of the scaling effect, so the use of an active micromixer is preferred. Crosslinker tripolyphosphate (TPP) and polymer chitosan (CS) are fed to the well-designed microreactor to obtain

Figure 10. Chitosan nanoparticles synthesis by (a) conventional method and optimization of microreactor-based on-chip method by (b) altering the flow rate, (c) altering the chitosan concentrations at constant flow rate
Source: Kamat et al. (2015)

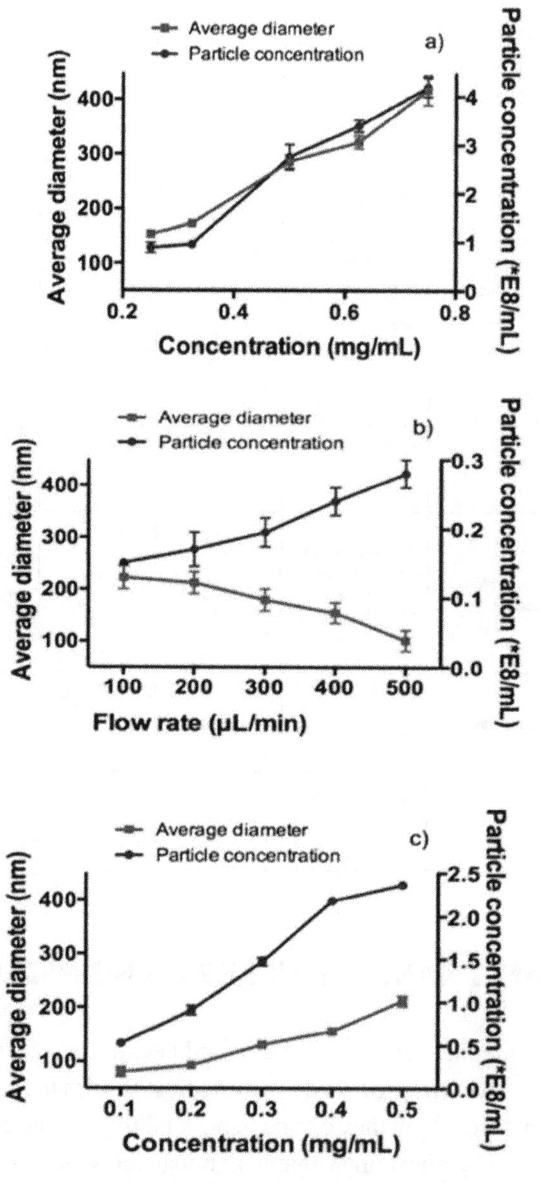

the nanoparticles. The size of CS nanoparticle obtained by the conventional synthesis method using 0.5 mg/ml CS was 250 ± 27 nm with a charge of +29.8±8 mV. When the same parameters were used for synthesizing nanoparticles using a microreactor, the obtained particle size was 154±20 nm. On further optimization at 0.4 mg/ml, the CS particle size was 130±9 nm with a higher charge of +39.8±5 mV. The smaller particle size of CS enhances the entrapment capacity and sustained release of drug particles

(Kamat et al., 2015). Figure 10 shows synthesis of chitosan nanoparticles using (a) standard methods and (b) flow rate and (c) chitosan concentration adjustments while maintaining a constant flow rate.

In another research, the researcher examined the effects of variables, including residence time, CS concentration, surfactant concentration, and pH of the CS solution, to analyse the nucleation rate and particle size growth. It has been found that a decrease in residence time causes a dramatic fall in particle growth rate and nucleation rate by 29 and 7 times, respectively. To study the impact of concentration, the CS concentration was increased from 0.05-0.1% W/V. Even though a high CS concentration produces a large number of nuclei, each nucleus expands at faster rates due to CS chain repulsion, which results in larger particles. For stabilizing the nanoparticle, a surfactant is preferred; here, sodium dodecyl sulfate (SDS) is used as a surfactant. To know the impact of SDS concentration on particle size, its concentration varied from 0.005, 0.008, and 0.035M, both below and above critical micellar concentration (CMC), by keeping residence time constant at 8.7s. Stable and smaller particles were produced when the surfactant SDS concentration was above its CMC. The optimum pH range in which the researcher got the minimum particle size was 4.5 to 5. Compared to those acquired from the batch reactor, the morphology of the particles produced from the microreactor shows that they are smaller, agglomerated and narrowly spread (Rathi & Gaikar, 2017). Some more application of organic nanoparticle synthesis using microreactor are given in Table 2.

Dobhal et al. synthesized mono-dispersed poly-methyl-methacrylate-methacrylic acid (PMMA) nanoparticles based on the principle of nanoprecipitation method using continuous microreactor. The main aim for conducting the experiment was to study the effect of parameters like flow rate of organic and aqueous phase polymer concentration and residence time on particle size and uniformity of particle size. As a result, it was noted that the particle size reduces with decrease in residence. To study the effect of flow rate the aqueous to organic flow rate ratio was changed from 9: 1, 4: 1, 7: 3, 3: 2, to 1: 1 (v/v) and found increase in particle size (Dobhal et al., 2017). A comparison of mean particle size obtained at various residence durations and aqueous to organic phase flow rate ratios is shown in Figure 11.

APPLICATION OF MICROFLUIDICS IN DRUG DELIVERY

Drug delivery techniques have significantly aided in transforming potential medicines into effective treatments (Vargason et al., 2021). The drug delivery technique's primary focus is enhancing the drugs' solubility, managing their release, enhancing their action, and modifying their pharmacokinetics (Tiwari et al., 2012). Microfluidics provides a novel way to synthesize NPs with specialised characteristics for various pharmaceutical applications. Its quick scaling indicates the viability of utilizing microfluidics on an industrial scale (Liu et al., 2022).

Chai et al. synthesized lignin (Lig)/chitosan (Chi) nanoparticles using a microreactor for sustained delivery of anticancer drugs. Results showed that Lig/Chi nanoparticles had an average particle size of around 180 nm, which may be employed as nanocarriers for drug delivery. Docetaxel (DTX) and curcumin (CCM) anticancer drug nanoparticles were created by assembling it with Lig/Chi NPs in a microreactor. It was discovered that the surface charge of lignin-based nanoparticles had changed from negative (-13 mV) to positive (+54.5 mV) when comparing the zeta potentials of Lig and Lig/Chi nanoparticles. These nanoparticles showed high drug loading efficiency, were biocompatible, and could release medications in response to changes in pH in the tumour's mildly acidic environment. Anticancer drug-loaded nanoparticles showed somewhat lower zeta potentials. Still, they remained strongly positively charged,

Figure 11. Comparative analysis of mean particle size obtained at different residence times and at different aqueous to organic phase flow rate ratios. Organic phase concentration is 3.75 mg mL^{-1} (a) and organic phase concentration is 1.875 mg mL^{-1} (b)
Source: Dobhal et al. (2017)

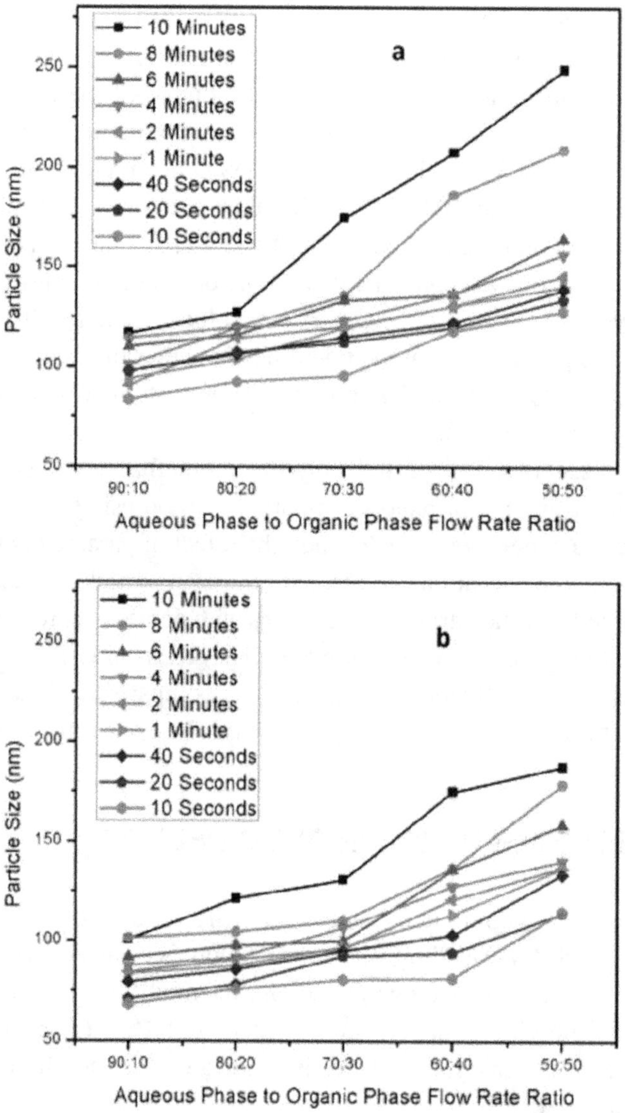

demonstrating enough of an electric double layer between the particles to prevent aggregation. This investigation has demonstrated the capability of Lig/Chi nanoparticles as pH-responsive drug delivery vehicles. (Chai et al., 2021).

Nanoscale drugs are an alternate method for speeding up the dissolving rate of therapeutic components having low solubility (Savjani et al., 2012). Low bioavailability is one of the main issues with poorly soluble drugs (Agrawal & Patel, 2011). One of the main barriers to the development of extremely effective pharmaceuticals is the limited water solubility of many medications. A pharmacological substance's

Table 2. Organic nanoparticles synthesized using a microreactor

Nanoparticles	Method	Microreactor Type	Objective	Yield/Particle Size	Observation	Ref.
Perbutyric acid	In the presence and absence of the heterogeneous catalyst amberlyst15 hydrogen, butyric acid and hydrogen peroxide are utilised as reactants.	Spiral capillary microreactor	Analyzing the influence of various factors such as temperature, hydrogen peroxide concentration, and catalyst on the synthesis of Perbutyric acid	56.2% conversion	The reaction rate increased as the concentration of hydrogen peroxide rose. Furthermore, it is possible to deduce that the solid catalyst has greater activity, resulting in the maximum synthesis of Perbutyric acid in short residence time.	(Katuri et al., 2022)
Polyacryl-amide (PAM)	Ex-situ and in-situ mixing	Microfluidic T-junction.	Effects of variables on the size, homogeneity, and shape of the beads, including the ratio of the volumetric flow rates of the organic carrier phase and the aqueous phase, microbore tube diameter, design of the microfluidic junction, and monomer concentration.		As some degree of polymerization in the aqueous phase in ex-situ mixing approach may result in pump choking, in-situ procedure is advised.	(Sen et al., 2020)
poly-methyl-methacrylate methacrylic acid (PMMA)	On the basis of the nanoprecipitation principle	Continuous flow microreactor	To determine the crucial factors that determine the size of nanoparticles produced in microreactors.	90 nm	The mean particle size increased with an increase in residence time and with an increase in the amount of the organic phase.	(Dobhal et al., 2017)
Ethyl diazoacetate	Polymeric micro separator-based dual channel membrane method.	Droplet based microreactor	The use of ethyl diazoacetate in a continuous flow process for the reactions and separations of toxic chemicals	99%	It has been demonstrated that the integrated microfluidic system is durable and robust in handling dangerous chemicals, even when operated continuously for extended periods of time.	(Maurya et al., 2014)
PEG-PLGA NP	Flow focusing geometry	3D flash flow microreactor	To build a unique and pressure-tolerant 3D-FFM device for scale-up polymer nanoparticle synthesis using a low-cost method.	50 nm and 85 nm	The 3D-FFM technology allows for the repeatable synthesis of polymer nanoparticles without aggregation at high polymer to solvent ratios and high flow rates.	(Min et al., 2014)
Rubrene	Non-solvent crystallization process	3D hydrodynamic focusing microreactor	Intended to prevent fouling and clogging inside the main channel. To control the supersaturation level and mixing conditions.	50 - 110 nm	The ability to produce rubrene nanocrystals of a certain size using this 3D hydrodynamic focusing apparatus was then shown. Reaching optimal mixing conditions takes less time with a greater focusing ratio.	(Génot et al., 2010)

Figure 12. Size of DOX-encapsulated PLGA NPs synthesized with precursor solvents of different flowrate ratios and different compositions VDCM/VDMSO. The solid markers represent the particle hydrodynamic sizes and the open ones the polydispersity index (PDI) as deduced by 2nd order cumulant fit of the correlogram curves.
Source: Xu et al. (2017)

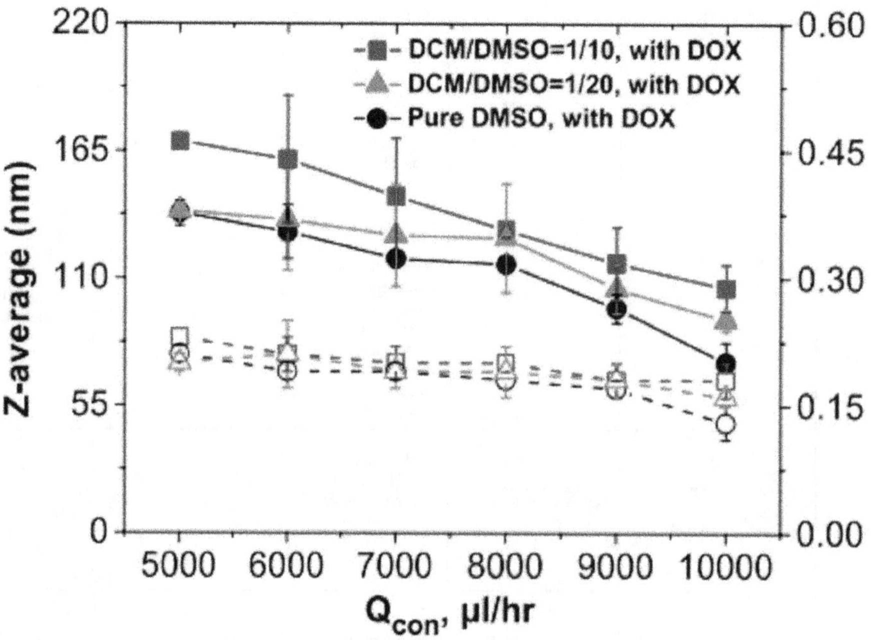

weak bioavailability may have a restricted therapeutic potential for clinical application, which might result in inadequate clinical results (Kakran et al., 2012). Microfluidics technology may create nanoparticles with distinctive sizes and precisely regulated shapes (Kulkarni & Goel, 2020). The application of microfluidic technology is restricted due to the clogging of microchannels. The performance of both natural and artificial systems is drastically changed by clog development. Microchannel clogging can occur as a result of the deposition and assembly of particles while they are in motion due to hydrodynamic, steric, and colloidal forces (Dressaire & Sauret, 2017).

Kim et al. investigated the use of a continuous droplet-based microreactor to synthesize medicinal nanoparticles without channel obstruction. This work suggested a continuous flow droplet-based microreactor for anti-solvent precipitation to produce itraconazole (ITZ) nanoparticles. In order to create ITZ nanoparticles, each droplet of the microliter scale functions as a separate microreactor and carries out a controlled mixing process. ITZ nanoparticles have been synthesized with an average particle size of 130-340 nm. The droplet-based microreactor offers a suitable mixing environment for obtaining smaller and more homogenous particles as compared to the T channel microfluidic device as the particle growth was restricted in droplets. Unlike the T channel, where the average particle size grows by 430 nm, the average size of particles formed in droplets only increases by 80 nm (S. Kim et al., 2020).

Table 3. Applications of microreactor in drug delivery

Nanoparticle	Method	Microreactor Type	Encapsulating Material	Particle Size	Encapsulation Efficiency	Drug Release	PDI	Morphology	Comments	Ref
Budesonide	Mixing in resonating membrane	3D microfluidic geometry		135.7 nm			0.044		To avoid the need for stabilisers and determine the smallest diameter particles that can be manufactured, the medication was examined at low doses. In order to maximise the production of the nanodrugs, numerous micromixers might be run simultaneously.	(Liu et al., 2022)
Rifampicin	Micromixing-assisted nanoprecipitation	Continuous flow microreactor	Eudragit® RS100	17.7 ± 5.0 nm	42.3 ± 2.4%	90.3 ± 1.8%	0.07	Spherical	A greater encapsulation efficiency, high productivity, and drug loading than those attained in a typical batch procedure were promoted by the micromixing-assisted technique.	(Yus et al., 2020)
doxorubicin (DOX) and paclitaxel (PTX)	Conjugation of Doxorubicin (DOX), a hydrophilic anticancer drug, and paclitaxel (PTX), a hydrophobic drug through a redox-sensitive disulfide bond	3D microfluidic co-flow focusing device		180 nm	89%		0.106	Spherical	Prodrug nanoparticles can selectively kill cancer cells that fulfil the redox requirements to rupture the disulfide link and release the medicines DOX and PTX, and they have no therapeutic impact on blood or healthy cells.	(Ma et al., 2019)
Curcumin	Nanoprecipitation method	Cross-channel microreactor	PLGA	40 - 160 nm	68%		0.7	Spherical	Controlling the mixing time has the most significant impact on the NP size	(Lababidi et al., 2019)
Ketorolac tromethamine (KT)	Continuous Mixing at constant flow rate.	Tubular microreactor	Chitosan	164.2 - 255 nm	76% - 96%	71.53% upto 12 h		Smooth and spherical	Chitosan concentration and encapsulation efficiency were shown to be directly correlated, with higher chitosan concentrations resulting in greater encapsulation efficiency.	(Patil et al., 2020)
Curcumin	Nanoprecipitation method	Flow focusing microreactor	PLGA	30 - 70 nm	55% - 67%		0.2	Smooth and spherical	Microfluidic produced Cur-PLGA NP a safe and efficient method of delivering curcumin. In instance, it increases curcumin's anticancer efficacy against leukaemia Jurkat cells when compared to natural curcumin.	(Leung & Shen, 2018)
Doxorubicin (DOX) or Tamoxifen (TAM)	Continuous flow-focusing method	coaxial flow microfluidic device	poly(D,L-lactic-co-glycolic acid) (PLGA)	DOX: 106-169 TAM: 196-211	DOX: 80 ± 2% TAM: 88 ± 6%	33.1% in 24 h		Spherical	In comparison to conventional approaches, the NPs created utilising the partly water-miscible precursor solvent in the flow-focusing microfluidic device showed much improved drug loading capability and a longer sustained release stage.	(J. Xu et al., 2017)

continued on following page

Table 3. Continued

Nanoparticle	Method	Microreactor Type	Encapsulating Material	Particle Size	Encapsulation Efficiency	Drug Release	PDI	Morphology	Comments	Ref
Mithramycin (MTH)	Hydrodynamic flow focusing	Continuous flow microreactor	Polymeric micelles, based on Pluronic	52-61 nm	83% - 85%				A helpful platform for the manufacturing of PM-MTH is provided by microfluidics, which has better controllability, repeatability, lower dimensions, and polydispersity.	(Capretto et al., 2012)

Wagh et al. synthesized ketorolac tromethamine (KTC) micro-/nanospheres utilizing four distinct methods: spray drying, high-pressure homogenization, single emulsion solvent evaporation, and a microreactor. Micro-/nanospheres were synthesized using ethyl cellulose (EC) as an encapsulating ingredient. The encapsulation efficiency of the micro-/nanospheres ranged from 42.34 to 89.33% by the solvent evaporation approach, from 76.36 to 91.13% by high-pressure homogenization, from 70.74 to 79.68% by spray drying, and from 79.00 to 89.49% by the microreactor technique. The particle size distribution of drug-loaded micro-/nanosphere formulations made in microreactors ranges from 600 to 1100 nm, whereas showed drug release from 39.56 – 49.34%. The obtained drug-loaded micro-/nanospheres have smooth surfaces and come in spherical and oval shapes (Wagh et al., 2019).

By using microdroplets of a partly water-miscible mixed solvent (dimethyl sulfoxide DMSO+ dichloromethane DCM) as a precursor, Xu et al. synthesized drug-loaded PLGA nanoparticles using continuous microfluidic flow-focusing method. The method was tested on hydrophilic (Doxorubicin) and hydrophobic (Tamoxifen) drug. According to the investigation, the microfluidic approach emerged as the promising technology for synthesizing nanoparticles and may be used for large-scale nanoparticle manufacturing. Researchers also studied the effect of different flowrate ratios and different compositions VDCM/VDMSO on the size of doxorubicin nanoparticles which is shown in Figure 12 (J. Xu et al., 2017).

Few applications of microreactor in drug delivery are given in Table 3.

CONCLUSION

Many innovative technologies for the synthesis of drug nanoparticles have been developed throughout the years. However, microfluidics is developing as the most dependable technology and is now the most extensively investigated technology. Microfluidic technology provides several benefits over conventional macroscale platforms for synthesizing nanomaterials, including far greater control over the produced nanomaterial's content, size distribution and morphology. The literature evaluation in this chapter reveals that recent years have seen substantial advancements in the production of micro/nanoparticles utilizing microfluidic systems. It is the most efficient method due to proper mixing, the minimal dimensional area accessible for reaction, and the development of thin-size nanoparticles. A variety of microparticles with unique compositions and structures may be synthesized using microfluidics, including spheres, nanorods, and cubical. In addition, microreactors offer exceptional controllability and repeatability for medication administration. Conventional methods such as spray drying and freeze-drying have been employed for decades to encapsulate pharmaceuticals in polymeric materials. Nonetheless, microreactor technology has overcome the limitations of these approaches.

In contrast, because of microfluidics distinctive characteristics, such as fast mixing, handling minute amounts of sample, rapid mass transfer, precise monitoring of reaction conditions and reagent addition, affordability, and minimal production times, this technology provides a novel method for synthesizing NPs with regulated properties for a variety of applications. Microreactors for the production of polymers, fine chemicals, and innovative materials are quickly evolving. These "miniature" reactors are intended to personalize goods and processes. The microfluidic method's ability to provide uniform reaction environments with rapid reagent mixing, continually modify reaction conditions, and regulate the reagent mixing order throughout the reaction process allows them to be most suitable for NP synthesis for drug delivery applications. The consensus is that microfluidic and microreactor technologies will revolutionize the drug delivery system. Microfluidic technology provides a potent tool for the regulated synthesis of drug delivery systems, accurate drug release, and continual tracking of drug transport to the targeted region at an adequate rate.

Although producing NPs for drug administration using microfluidics has several benefits, successfully synthesizing NPs with the desired characteristics needs skill and continual process optimization.

REFERENCES

Agrawal, Y., & Patel, V. (2011). Nanosuspension: An approach to enhance solubility of drugs. *Journal of Advanced Pharmaceutical Technology & Research*, *2*(2), 81. doi:10.4103/2231-4040.82950 PMID:22171298

Amina, S. J., & Guo, B. (2020). A Review on the Synthesis and Functionalization of Gold Nanoparticles as a Drug Delivery Vehicle. *International Journal of Nanomedicine*, *15*, 9823–9857. doi:10.2147/IJN.S279094 PMID:33324054

Anju, S., Prajitha, N., Sukanya, V. S., & Mohanan, P. V. (2020). Complicity of degradable polymers in health-care applications. *Materials Today. Chemistry*, *16*, 100236. doi:10.1016/j.mtchem.2019.100236

Apelgren, P., Amoroso, M., Säljö, K., Montelius, M., Lindahl, A., Stridh Orrhult, L., Gatenholm, P., Kölby, L., Arulkumar, S., Parthiban, S., Goswami, A., Varma, R. S., Naushad, M., Gawande, M. B., Lipskas, J., Deep, K., Yao, W., Grogan, S. P., Dorthé, E. W., … Salvatore, D. A. (2019). Continuous, size and shape-control synthesis of hollow silica nanoparticles enabled by a microreactor-assisted rapid mixing process. *Materials Today: Proceedings, 27*. https://doi.org// doi:10.1016/j.jare.2020.01.010%0Ahttps://

Appalakutti, S., Sonawane, S., Bhanvase, B. A., Mittal, V., & Ashokkumar, M. (2015). Process intensification of copper chromite (CuCr2O4) nanoparticle production using continuous flow microreactor. *Chemical Engineering and Processing*, *89*, 28–34. doi:10.1016/j.cep.2014.12.012

Azimullah, S., Vikrant, Sudhakar, C., Kumar, P., Patil, A., Md. Usman, M. R., & Shaikh Usman, M. Z. (2019). Nanosuspensions as a promising approach to enhance bioavailability of poorly soluble drugs: An update. *Journal of Drug Delivery and Therapeutics*, *9*(2), 574–582. doi:10.22270/jddt.v9i2.2436

Baheri Islami, S., Khezerloo, M., & Gharraei, R. (2017). The effect of chaotic advection on mixing degree and pressure drop of non-Newtonian fluids flow in curved micromixers. *Journal of the Brazilian Society of Mechanical Sciences and Engineering, 39*(3), 813–831. doi:10.100740430-016-0689-1

Bakunin, V. N., Suslov, A. Y., Kuzmina, G. N., Parenago, O. P., & Topchiev, A. V. (2004). Synthesis and application of inorganic nanoparticles as lubricant components-a review. *Journal of Nanoparticle Research, 6*(2), 273–284. doi:10.1023/B:NANO.0000034720.79452.e3

Bau, H. H., Zhong, J., & Yi, M. (2001). A minute magneto hydro dynamic (MHD) mixer. *Sensors and Actuators. B, Chemical, 79*(2–3), 207–215. doi:10.1016/S0925-4005(01)00851-6

Bayareh, M., Ashani, M. N., & Usefian, A. (2020). Active and passive micromixers: A comprehensive review. *Chemical Engineering and Processing, 147*, 107771. doi:10.1016/j.cep.2019.107771

Bojang, A. A., & Wu, H. S. (2020). Design, fundamental principles of fabrication and applications of microreactors. *Processes (Basel, Switzerland), 8*(8), 891. Advance online publication. doi:10.3390/pr8080891

Bressan, L. P., Robles-Najar, J., Adamo, C. B., Quero, R. F., Costa, B. M. C., de Jesus, D. P., & da Silva, J. A. F. (2019). 3D-printed microfluidic device for the synthesis of silver and gold nanoparticles. *Microchemical Journal, 146*, 1083–1089. doi:10.1016/j.microc.2019.02.043

Cai, G., Xue, L., Zhang, H., & Lin, J. (2017). A Review on Micromixers. *Micromachines, 8*(9), 274. doi:10.3390/mi8090274 PMID:30400464

Capretto, L., Cheng, W., Hill, M., & Zhang, X. (2011). *Micromixing Within Microfluidic Devices*. doi:10.1007/128_2011_150

Capretto, L., Mazzitelli, S., Brognara, E., Lampronti, I., Carugo, D., Hill, M., Zhang, X., Gambari, R., & Nastruzzi, C. (2012). Mithramycin encapsulated in polymeric micelles by microfuidic technology as novel therapeutic protocol for beta-thalassemia. *International Journal of Nanomedicine, 7*, 307–324. PMID:22287841

Carroll, N. J., Rathod, S. B., Derbins, E., Mendez, S., Weitz, D. A., & Petsev, D. N. (2008). Droplet-based microfluidics for emulsion and solvent evaporation synthesis of monodisperse mesoporous silica microspheres. *Langmuir, 24*(3), 658–661. doi:10.1021/la7032516 PMID:18171093

Chai, Y., Wang, Y., Li, B., Qi, W., Su, R., & He, Z. (2021). Microfluidic Synthesis of Lignin/Chitosan Nanoparticles for the pH-Responsive Delivery of Anticancer Drugs. *Langmuir, 37*(23), 7219–7226. doi:10.1021/acs.langmuir.1c00778 PMID:34078082

Clark, J., Kaufman, M., & Fodor, P. (2018). Mixing Enhancement in Serpentine Micromixers with a Non-Rectangular Cross-Section. *Micromachines, 9*(3), 107. doi:10.3390/mi9030107 PMID:30424041

Din, M. I., & Rehan, R. (2017). Synthesis, Characterization, and Applications of Copper Nanoparticles. *Analytical Letters, 50*(1), 50–62. doi:10.1080/00032719.2016.1172081

Dobhal, A., Kulkarni, A., Dandekar, P., & Jain, R. (2017). A microreactor-based continuous process for controlled synthesis of poly-methyl-methacrylate-methacrylic acid (PMMA) nanoparticles. *Journal of Materials Chemistry. B, Materials for Biology and Medicine*, *5*(18), 3404–3417. doi:10.1039/C7TB00560A PMID:32264406

Dressaire, E., & Sauret, A. (2017). Clogging of microfluidic systems. *Soft Matter*, *13*(1), 37–48. doi:10.1039/C6SM01879C PMID:27801463

Exposito, A. J., Barrie, P. J., & Torrente-Murciano, L. (2020). Fast Synthesis of CeO2Nanoparticles in a Continuous Microreactor Using Deep Eutectic Reline As Solvent. *ACS Sustainable Chemistry & Engineering*, *8*(49), 18297–18302. doi:10.1021/acssuschemeng.0c06949

Farooq, M. A., Aquib, M., Khan, D. H., Ghayas, S., Ahsan, A., Ijaz, M., Banerjee, P., Khan, M. A., Ahmad, M. M., & Wang, B. (2019). Nanocarrier-mediated co-delivery systems for lung cancer therapy: Recent developments and prospects. *Environmental Chemistry Letters*, *17*(4), 1565–1583. doi:10.100710311-019-00897-7

Gagliardi, A., Giuliano, E., Venkateswararao, E., Fresta, M., Bulotta, S., Awasthi, V., & Cosco, D. (2021). Biodegradable Polymeric Nanoparticles for Drug Delivery to Solid Tumors. *Frontiers in Pharmacology*, *12*(February), 1–24. doi:10.3389/fphar.2021.601626 PMID:33613290

Gambhire, G., & Massai, D. (2011). A Review on Different Micromixers and its Micromixing within Microchannel. *International Journal of Current Engineering and Technology*. doi:10.14741/Ijcet/22774106/spl.4.2016.83

Gao, Y., Pinho, B., & Torrente-Murciano, L. (2022). Tailoring the size of silver nanoparticles by controlling mixing in microreactors. *Chemical Engineering Journal*, *432*, 134112. doi:10.1016/j.cej.2021.134112

Génot, V., Desportes, S., Croushore, C., Lefèvre, J.-P., Pansu, R. B., Delaire, J. A., & von Rohr, P. R. (2010). Synthesis of organic nanoparticles in a 3D flow focusing microreactor. *Chemical Engineering Journal*, *161*(1–2), 234–239. doi:10.1016/j.cej.2010.04.029

Gioria, E., Signorini, C., Wisniewski, F., & Gutierrez, L. (2020). Green synthesis of time-stable palladium nanoparticles using microfluidic devices. *Journal of Environmental Chemical Engineering*, *8*(5), 104096. doi:10.1016/j.jece.2020.104096

Hao, N., Xu, Z., Nie, Y., Jin, C., Closson, A. B., Zhang, M., & Zhang, J. X. J. (2019). Microfluidics-enabled rational design of ZnO micro-/nanoparticles with enhanced photocatalysis, cytotoxicity, and piezoelectric properties. *Chemical Engineering Journal*, *378*(July), 122222. doi:10.1016/j.cej.2019.122222 PMID:32831625

Harries, N., Burns, J. R., Barrow, D. A., & Ramshaw, C. (2003). A numerical model for segmented flow in a microreactor. *International Journal of Heat and Mass Transfer*, *46*(17), 3313–3322. doi:10.1016/S0017-9310(03)00120-0

Hu, R., Liu, C., Xuan, J., Xu, Y., Li, T., Liu, B.-F., Li, Y., & Yang, Y. (2019). 3D hydrodynamic flow focusing-based micromixer enables high-resolution imaging for studying the early folding kinetics of G-quadruplex. *Sensors and Actuators. B, Chemical*, *293*, 312–320. doi:10.1016/j.snb.2019.05.026

Hu, X., Zhang, Y., Ding, T., Liu, J., & Zhao, H. (2020). Multifunctional Gold Nanoparticles: A Novel Nanomaterial for Various Medical Applications and Biological Activities. *Frontiers in Bioengineering and Biotechnology*, *8*, 990. Advance online publication. doi:10.3389/fbioe.2020.00990 PMID:32903562

Huang, H., du Toit, H., Besenhard, M. O., Ben-Jaber, S., Dobson, P., Parkin, I., & Gavriilidis, A. (2018). Continuous flow synthesis of ultrasmall gold nanoparticles in a microreactor using trisodium citrate and their SERS performance. *Chemical Engineering Science*, *189*, 422–430. doi:10.1016/j.ces.2018.06.050

James, M., Revia, R. A., Stephen, Z., & Zhang, M. (2020). Microfluidic Synthesis of Iron Oxide Nanoparticles. *Nanomaterials (Basel, Switzerland)*, *10*(11), 2113. doi:10.3390/nano10112113 PMID:33114204

Kakran, M., Li, L., & Müller, R. H. (2012). Overcoming the challenge of poor drug solubility. *Pharmaceutical Engineering*, *32*(4), 82–89.

Kamat, V., Marathe, I., Ghormade, V., Bodas, D., & Paknikar, K. (2015). Synthesis of Monodisperse Chitosan Nanoparticles and in Situ Drug Loading Using Active Microreactor. *ACS Applied Materials & Interfaces*, *7*(41), 22839–22847. doi:10.1021/acsami.5b05100 PMID:26448128

Katoch, V., Sharma, N., Sharma, M., Baghoria, M., Panda, J. J., Singh, M., & Prakash, B. (2021). Microflow synthesis and enhanced photocatalytic dye degradation performance of antibacterial Bi2O3 nanoparticles. *Environmental Science and Pollution Research International*, *28*(15), 19155–19165. doi:10.100711356-020-11711-1 PMID:33398764

Katuri, P., Maralla, Y., Sharada, S., & Narasaiah Tumma, B. (2022). Synthesis of perbutyric acid with amberlyst catalyst in a continuous microreactor. *Materials Today: Proceedings*. Advance online publication. doi:10.1016/j.matpr.2022.06.202

Kim, D., Raj, A., Zhu, L., Masel, R. I., & Shannon, M. A. (2008). Non-Equilibrium Electrokinetic Nanofluidic Mixers. *Electrochemistry (Tokyo, Japan)*, 1–3.

Kim, S., Wang, H., Yan, L., Zhang, X., & Cheng, Y. (2020). Continuous preparation of itraconazole nanoparticles using droplet-based microreactor. *Chemical Engineering Journal*, *393*(March), 124721. doi:10.1016/j.cej.2020.124721

Köhler, J. M., Li, S., & Knauer, A. (2013). Why is Micro Segmented Flow Particularly Promising for the Synthesis of Nanomaterials? *Chemical Engineering & Technology*, *36*(6), 887–899. doi:10.1002/ceat.201200695

Kulkarni, M. B., & Goel, S. (2020). Microfluidic devices for synthesizing nanomaterials—A review. *Nano Express*, *1*(3), 032004. doi:10.1088/2632-959X/abcca6

Kumar, G. N. S., Mitra, S. K., & Bhattacharjee, S. (2009). *Dielectrophoretic Mixing With Novel Electrode Geometry* (Vol. 2). Fora. doi:10.1115/FEDSM2009-78260

Kwak, C. H., Park, J. P., Lee, S. S., Muruganantham, R., Kwon, S., Roh, C., Kim, S.-W., & Huh, Y. S. (2017). Droplet-Based Microfluidic Reactor for Synthesis of Size-Controlled CdSe Quantum Dots. *Journal of Nanoscience and Nanotechnology*, *18*(2), 1339–1342. doi:10.1166/jnn.2018.14918 PMID:29448587

Lababidi, N., Sigal, V., Koenneke, A., Schwarzkopf, K., Manz, A., & Schneider, M. (2019). Microfluidics as tool to prepare size-tunable PLGA nanoparticles with high curcumin encapsulation for efficient mucus penetration. *Beilstein Journal of Nanotechnology*, *10*, 2280–2293. doi:10.3762/bjnano.10.220 PMID:31807413

Lebon, G. S. B., Tzanakis, I., Pericleous, K., & Eskin, D. (2018). Experimental and numerical investigation of acoustic pressures in different liquids. *Ultrasonics Sonochemistry*, *42*, 411–421. doi:10.1016/j.ultsonch.2017.12.002 PMID:29429686

Lee, S., & Jun, B.-H. (2019). Silver Nanoparticles: Synthesis and Application for Nanomedicine. *International Journal of Molecular Sciences*, *20*(4), 865. doi:10.3390/ijms20040865 PMID:30781560

Leung, M. H. M., & Shen, A. Q. (2018). Microfluidic Assisted Nanoprecipitation of PLGA Nanoparticles for Curcumin Delivery to Leukemia Jurkat Cells. *Langmuir*, *34*(13), 3961–3970. doi:10.1021/acs.langmuir.7b04335 PMID:29544247

Ling, F. W. M., Abdulbari, H. A., & Chin, S. Y. (2021). Microfluidic Chips for Formulation of Silica Nanoparticles and Enzyme Immobilization. *Chemical Engineering & Technology*, *44*(8), 1423–1431. doi:10.1002/ceat.202100098

Liu, Y., Yang, G., Hui, Y., Ranaweera, S., & Zhao, C. X. (2022). Microfluidic Nanoparticles for Drug Delivery. *Small*, *2106580*(36), 1–33. doi:10.1002mll.202106580 PMID:35396770

Ma, X., Özliseli, E., Zhang, Y., Pan, G., Wang, D., & Zhang, H. (2019). Fabrication of redox-responsive doxorubicin and paclitaxel prodrug nanoparticles with microfluidics for selective cancer therapy. *Biomaterials Science*, *7*(2), 634–644. doi:10.1039/C8BM01333K PMID:30534690

Manzari, M. T., Shamay, Y., Kiguchi, H., Rosen, N., Scaltriti, M., & Heller, D. A. (2021). Targeted drug delivery strategies for precision medicines. *Nature Reviews. Materials*, *6*(4), 351–370. doi:10.103841578-020-00269-6 PMID:34950512

Mastroianni, L., Vajglová, Z., Eränen, K., Peurla, M., Di Serio, M. Yu., Murzin, D., Russo, V., & Salmi, T. (2022). Microreactor technology in experimental and modelling study of alcohol oxidation on nanogold. *Chemical Engineering Science*, *260*, 117920. doi:10.1016/j.ces.2022.117920

Maurya, R. A., Min, K. I., & Kim, D. P. (2014). Continuous flow synthesis of toxic ethyl diazoacetate for utilization in an integrated microfluidic system. *Green Chemistry*, *16*(1), 116–120. doi:10.1039/C3GC41226A

Mclaren, A., Valdes-Solis, T., Li, G., & Tsang, S. C. (2009). Shape and Size Effects of ZnO Nanocrystals on Photocatalytic Activity. *Journal of the American Chemical Society*, *131*(35), 12540–12541. doi:10.1021/ja9052703 PMID:19685892

Mills, J. K., & Needham, D. (1999). Targeted drug delivery. *Expert Opinion on Therapeutic Patents*, *9*(11), 1499–1513. doi:10.1517/13543776.9.11.1499

Min, K. I., Im, D. J., Lee, H. J., & Kim, D. P. (2014). Three-dimensional flash flow microreactor for scale-up production of monodisperse PEG-PLGA nanoparticles. *Lab on a Chip*, *14*(20), 3987–3992. doi:10.1039/C4LC00700J PMID:25133684

Nguyen, N.-T. (2012a). Fundamentals of mass transport in the microscale. Micromixers. doi:10.1016/B978-1-4377-3520-8.00002-4

Nguyen, N.-T. (2012b). Micromixers based on chaotic advection. Micromixers. doi:10.1016/B978-1-4377-3520-8.00006-1

Nguyen, N.-T. (2012c). Micromixers based on molecular diffusion. Micromixers. doi:10.1016/B978-1-4377-3520-8.00005-X

Nikolova, M. P., & Chavali, M. S. (2020). Metal Oxide Nanoparticles as Biomedical Materials. *Biomimetics*, *5*(2), 27. doi:10.3390/biomimetics5020027 PMID:32521669

Oddy, M. H., Santiago, J. G., & Mikkelsen, J. C. (2001). Electrokinetic instability micromixing. *Analytical Chemistry*, *73*(24), 5822–5832. doi:10.1021/ac0155411 PMID:11791550

Parvizian, F., Rahimi, M., Azimi, N., & Alsairafi, A. A. (2014). CFD Modeling of Micromixing and Velocity Distribution in a 1.7-MHz Tubular Sonoreactor. *Chemical Engineering & Technology*, *37*(1), 113–122. doi:10.1002/ceat.201300231

Patil, J., Rajput, R., Patil, P., Mujumdar, A., & Naik, J. (2020). Generation of sustained release chitosan nanoparticles for delivery of ketorolac tromethamine: A tubular microreactor approach. *International Journal of Polymeric Materials*, *69*(8), 516–524. doi:10.1080/00914037.2019.1581201

Pennemann, H., Watts, P., Haswell, S. J., Hessel, V., & Löwe, H. (2004). Benchmarking of microreactor applications. *Organic Process Research & Development*, *8*(3), 422–439. doi:10.1021/op0341770

Proskurina, O. V., Abiev, R. S., Danilovich, D. P., Panchuk, V. V., Semenov, V. G., Nevedomsky, V. N., & Gusarov, V. V. (2019). Formation of nanocrystalline BiFeO3 during heat treatment of hydroxides co-precipitated in an impinging-jets microreactor. *Chemical Engineering and Processing*, *143*(July), 107598. doi:10.1016/j.cep.2019.107598

Qian, S., Duval, J. F. L., & Mahi, F. T. (2016). Mixers. In Reference Module in Materials Science and Materials Engineering (pp. 1–51). Elsevier. doi:10.1016/B978-0-12-803581-8.00547-6

Rathi, N., & Gaikar, V. G. (2017). Optimization of Continuous Synthesis of Cross-Linked Chitosan Nanoparticles Using Microreactors. *Chemical Engineering & Technology*, *40*(3), 506–513. doi:10.1002/ceat.201600339

Raza, W., Hossain, S., & Kim, K. Y. (2020). A review of passive micromixers with a comparative analysis. *Micromachines*, *11*(5), 455. Advance online publication. doi:10.3390/mi11050455 PMID:32349452

Riahi, R., Tamayol, A., Shaegh, S. A. M., Ghaemmaghami, A. M., Dokmeci, M. R., & Khademhosseini, A. (2015). Microfluidics for advanced drug delivery systems. *Current Opinion in Chemical Engineering*, *7*, 101–112. doi:10.1016/j.coche.2014.12.001 PMID:31692947

Sasaki, K., Miyake, K., Uchida, Y., & Nishiyama, N. (2022). Mechanochemical Synthesis of Dispersible Platinum Nanosheets for Enhanced Catalysis in a Microreactor. *ACS Applied Nano Materials*, *5*(4), 4998–5005. doi:10.1021/acsanm.1c04583

Savjani, K. T., Gajjar, A. K., & Savjani, J. K. (2012). Drug Solubility: Importance and Enhancement Techniques. *ISRN Pharmaceutics*, *2012*, 1–10. doi:10.5402/2012/195727 PMID:22830056

Sebastian, V., Arruebo, M., & Santamaria, J. (2014). Reaction Engineering Strategies for the Production of Inorganic Nanomaterials. *Small, 10*(5), 835–853. doi:10.1002mll.201301641 PMID:24123934

Sen, N., Shaikh, T., Singh, K. K., Sirsam, R., & Shenoy, K. T. (2020). Synthesis of polyacrylamide (PAM) beads in microreactors. *Chemical Engineering and Processing, 157*(April), 108105. doi:10.1016/j.cep.2020.108105

Shah, A., Aftab, S., Nisar, J., Ashiq, M. N., & Iftikhar, F. J. (2021). Nanocarriers for targeted drug delivery. *Journal of Drug Delivery Science and Technology, 62*(February), 102426. doi:10.1016/j.jddst.2021.102426

Shrimal, P., Jadeja, G., Naik, J., & Patel, S. (2019). Continuous microchannel precipitation to enhance the solubility of telmisartan with poloxamer 407 using Box-Behnken design approach. *Journal of Drug Delivery Science and Technology, 53*(August), 101225. doi:10.1016/j.jddst.2019.101225

Simmons, M. D., Jones, N., Evans, D. J., Wiles, C., Watts, P., Salamon, S., Escobar Castillo, M., Wende, H., Lupascu, D. C., & Francesconi, M. G. (2015). Doping of inorganic materials in microreactors-preparation of Zn doped Fe3O4 nanoparticles. *Lab on a Chip, 15*(15), 3154–3162. doi:10.1039/C5LC00287G PMID:26099495

Simonnet, C., & Groisman, A. (2005). Chaotic mixing in a steady flow in a microchannel. *Physical Review Letters, 94*(13), 8–11. doi:10.1103/PhysRevLett.94.134501 PMID:15903994

Singh, R., & Lillard, J. W. Jr. (2009). Nanoparticle-based targeted drug delivery. *Experimental and Molecular Pathology, 86*(3), 215–223. doi:10.1016/j.yexmp.2008.12.004 PMID:19186176

Sudarsan, A. P., & Ugaz, V. M. (2006a). Fluid mixing in planar spiral microchannels. *Lab on a Chip, 6*(1), 74–82. doi:10.1039/B511524H PMID:16372072

Sudarsan, A. P., & Ugaz, V. M. (2006b). Multivortex micromixing. *Proceedings of the National Academy of Sciences of the United States of America, 103*(19), 7228–7233. doi:10.1073/pnas.0507976103 PMID:16645036

Sundaram, N., & Tafti, D. K. (2004). Evaluation of microchamber geometries and surface conditions for electrokinetic driven mixing. *Analytical Chemistry, 76*(13), 3785–3793. doi:10.1021/ac0499591 PMID:15228355

Tang, Z., Kim, W. S., & Yu, T. (2019). Studies on morphology changes of copper sulfide nanoparticles in a continuous Couette-Taylor reactor. *Chemical Engineering Journal, 359*, 1436–1441. doi:10.1016/j.cej.2018.11.034

Tian, Z., Ge, X., Wang, Y., & Xu, J. (2018). Nanoparticles and nanocomposites with microfluidic technology. In *Polymer-Based Multifunctional Nanocomposites and Their Applications*. Elsevier Inc. doi:10.1016/B978-0-12-815067-2.00001-9

Tiwari, G., Tiwari, R., Bannerjee, S., Bhati, L., Pandey, S., Pandey, P., & Sriwastawa, B. (2012). Drug delivery systems: An updated review. *International Journal of Pharmaceutical Investigation, 2*(1), 2. doi:10.4103/2230-973X.96920 PMID:23071954

Udaya Kumar, A., Sai Ganesh, D., Vamsi Krishna, T., Sashank, B., & Satyanarayana, T. (2022). Modeling and investigation on mixing characteristics of T & Y-shaped micromixers for microfluidic devices. *Materials Today: Proceedings*, *59*, 501–505. doi:10.1016/j.matpr.2021.11.474

Ulkir, O., Girit, O., & Ertugrul, I. (2021). Design and Analysis of a Laminar Diffusion-Based Micromixer with Microfluidic Chip. *Journal of Nanomaterials*, *2021*, 1–10. Advance online publication. doi:10.1155/2021/6684068

Uson, L., Arruebo, M., Sebastian, V., & Santamaria, J. (2018). Single phase microreactor for the continuous, high-temperature synthesis of <4 nm superparamagnetic iron oxide nanoparticles. *Chemical Engineering Journal*, *340*, 66–72. doi:10.1016/j.cej.2017.12.024

Vargason, A. M., Anselmo, A. C., & Mitragotri, S. (2021). The evolution of commercial drug delivery technologies. *Nature Biomedical Engineering*, *5*(9), 951–967. doi:10.103841551-021-00698-w PMID:33795852

Wagh, P., Mujumdar, A., & Naik, J. B. (2019). Preparation and characterization of ketorolac tromethamine-loaded ethyl cellulose micro-/nanospheres using different techniques. *Particulate Science and Technology*, *37*(3), 347–357. doi:10.1080/02726351.2017.1383330

Wong, S., Ward, M., & Wharton, C. (2004). Micro T-mixer as a rapid mixing micromixer. *Sensors and Actuators. B, Chemical*, *100*(3), 359–379. doi:10.1016/j.snb.2004.02.008

Xu, J., Zhang, S., MacHado, A., Lecommandoux, S., Sandre, O., Gu, F., & Colin, A. (2017). Controllable Microfluidic Production of Drug-Loaded PLGA Nanoparticles Using Partially Water-Miscible Mixed Solvent Microdroplets as a Precursor. *Scientific Reports*, *7*(1), 1–12. doi:10.103841598-017-05184-5 PMID:28684775

Xu, L., Peng, J., Srinivasakannan, C., Chen, G., & Shen, A. Q. (2015). Synthesis of copper nanocolloids using a continuous flow based microreactor. *Applied Surface Science*, *355*, 1–6. doi:10.1016/j.apsusc.2015.07.070

Xu, L., Srinivasakannan, C., Peng, J., Zhang, D., & Chen, G. (2015). Synthesis of nickel nanoparticles by aqueous reduction in continuous flow microreactor. *Chemical Engineering and Processing*, *93*, 44–49. doi:10.1016/j.cep.2015.04.010

Xu, L., Srinivasakannan, C., Peng, J., Zhang, L., & Zhang, D. (2017). Synthesis of Cu-CuO nanocomposite in microreactor and its application to photocatalytic degradation. *Journal of Alloys and Compounds*, *695*, 263–269. doi:10.1016/j.jallcom.2016.10.195

Yang, H., Wei, S. X., Chen, H., Chen, L., Au, C. T., Xie, T. L., & Yin, S. F. (2022). A high-throughput chaotic advection microreactor for preparation of uniform and aggregated barium sulfate nanoparticles. *AIChE Journal. American Institute of Chemical Engineers*, *68*(April), 1–10. doi:10.1002/aic.17810

Yang, Z., Dong, L., Wang, M., Jia, Y., Wang, C., Li, P., & Liu, G. (2022). Sensors and Actuators : A. Physical Controllable synthesis of silver nanoparticles using a multi-stage microfluidic reactor driven by two valveless piezoelectric pumps. *Sensors and Actuators. A, Physical*, *346*(September), 113871. doi:10.1016/j.sna.2022.113871

Yu, H., Wang, M., Zhou, J., Yuan, B., Luo, J., Wu, W., Chen, Z., & Yu, R. (2020). Microreactor-assisted synthesis of α-alumina nanoparticles. *Ceramics International, 46*(9), 13272–13281. doi:10.1016/j.ceramint.2020.02.105

Yus, C., Arruebo, M., Irusta, S., & Sebastián, V. (2020). Microflow nanoprecipitation of positively charged gastroresistant polymer nanoparticles of Eudragit® RS100: A study of fluid dynamics and chemical parameters. *Materials (Basel), 13*(13), 2925. Advance online publication. doi:10.3390/ma13132925 PMID:32629799

Zeng, L., & Palmer, J. (2005). Enhancement of Micromixing Tees Using Ultrasound Energy. In Microreactor Technology and Process Intensification (pp. 322–333). doi:10.1021/bk-2005-0914.ch020

Zhang, J., Zhu, L., Jia, H., Wei, K., & Wen, L. (2022). Microreactor facilitated preparation and Ni-doping of MnO2 nanoparticlcs for supercapacitors. *Journal of Alloys and Compounds, 889*, 161772. doi:10.1016/j.jallcom.2021.161772

Zhang, Y., Chan, H. F., & Leong, K. W. (2013). Advanced materials and processing for drug delivery: The past and the future. *Advanced Drug Delivery Reviews, 65*(1), 104–120. doi:10.1016/j.addr.2012.10.003 PMID:23088863

Zhao, C. X., He, L., Qiao, S. Z., & Middelberg, A. P. J. (2011). Nanoparticle synthesis in microreactors. *Chemical Engineering Science, 66*(7), 1463–1479. doi:10.1016/j.ces.2010.08.039

Zhao, H., Wang, J. X., Wang, Q. A., Chen, J. F., & Yun, J. (2007). Controlled liquid antisolvent precipitation of hydrophobic pharmaceutical nanoparticles in a MicroChannel reactor. *Industrial & Engineering Chemistry Research, 46*(24), 8229–8235. doi:10.1021/ie070498e

KEY TERMS AND DEFINITIONS

Drug-Delivery: The term "drug delivery" describes the methods, formulations, manufacturing processes, storage arrangements, and technology used to transfer a pharmaceutical molecule to its intended target location.

Evaporation: Evaporation takes place on the surface of a liquid as it transitions into the gas phase.

Inorganic Compound: Inorganic matter is defined as material that is not formed from living things and does not include any carbon that was created biologically.

Microfluidics: Microfluidics is the study of how fluids behave, are precisely controlled, and are moved on a microscopic scale where surface forces outweigh volumetric forces.

Nanoparticles: A matter particle with a dimension of between one and one hundred nanometres is referred to as a nanoparticle or ultrafine particle.

Organic Compound: Any chemical substance that has carbon-hydrogen or carbon-carbon bonds is considered an organic compound.

Polymers: A polymer is a substance or material made up of macromolecules, which are very big molecules made up of several repeating subunits.

Chapter 13
Modern Lab-on-Chip Biosensors Application on Infectious COVID-19 Detection

Ranjit Barua
Indian Institute of Engineering Science and Technology, Shibpur, India

Anwita Sarkar
Indian Institute of Engineering Science and Technology, Shibpur, India

Sudipto Datta
Indian Institute of Science, Bangalore, India

ABSTRACT

The unexpected increase of the global SARS-CoV-2 or COVID-19 pandemic in early 2020 has required exceptional methods to accomplish instant detection and decrease the spread rate. This chapter highlights the advance of lab-on-chip biosensing devices for the identification of COVID-19 infection. In this chapter, the authors study the roles of using lab-on-chip equipment in response to the present global COVID-19 pandemic, comprising their potential for rapid sample-to-answer processing times and facility of incorporation into a variety of healthcare backgrounds.

INTRODUCTION

In the 21st century, the rise of new respiratory tract infections has been a growing alarm that has turned into a most important reason for alarm, and this increasing alert became a worldwide COVID-19 pandemic or also known as severe acute respiratory syndrome coronavirus-2 (SARS-CoV-2) (Asghari et al., 2021; Barua et al., 2021; Hsieh et al., 2020). The new, imperceptible battle against the COVID-19 pandemic can stimulate main developments in the progress of identification, treatment, and diagnosis methods. The extremely contagious COVID-19 virus is tough to identify, in patients with clinically apparent signs of cough, fever, and breathing problems (Barua et al., 2021). Figure 1 shows the systematic

DOI: 10.4018/978-1-6684-6952-1.ch013

Figure 1. The systematic and respiratory problems due to COVID-19 infection

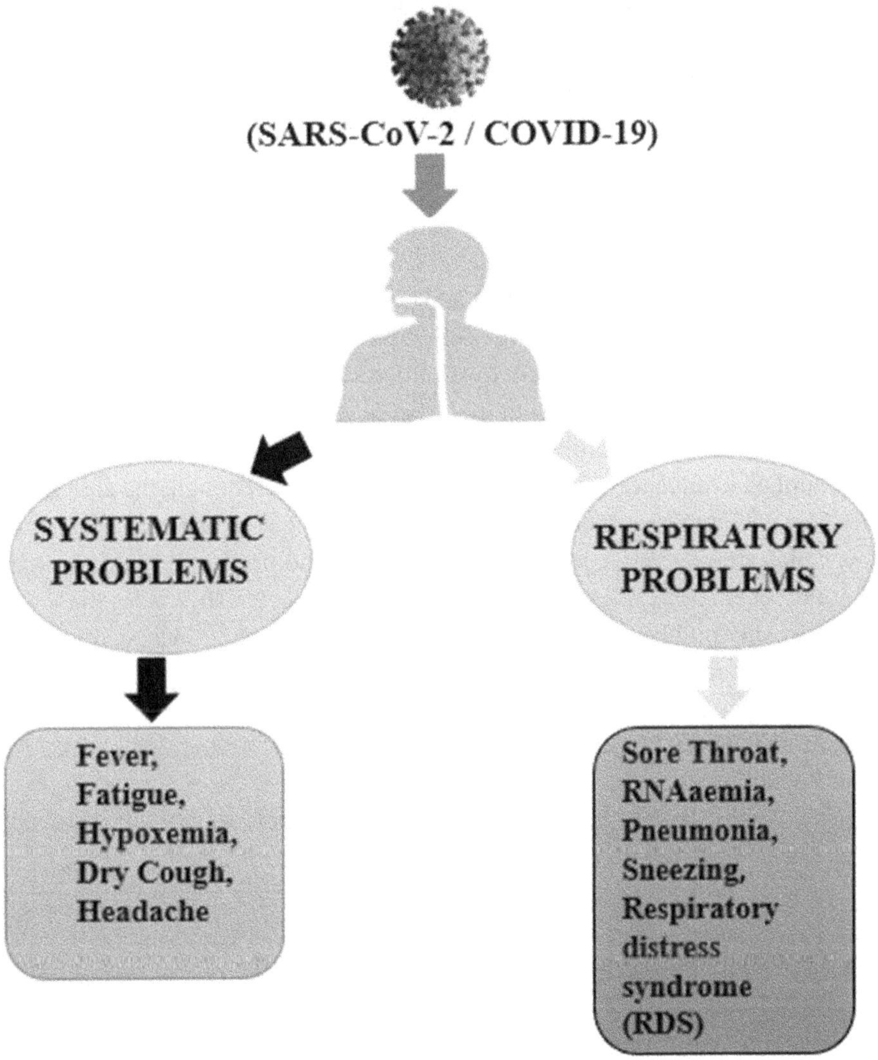

Figure 2. The COVID-19 virus
Source: Kitchen et al. (2022)

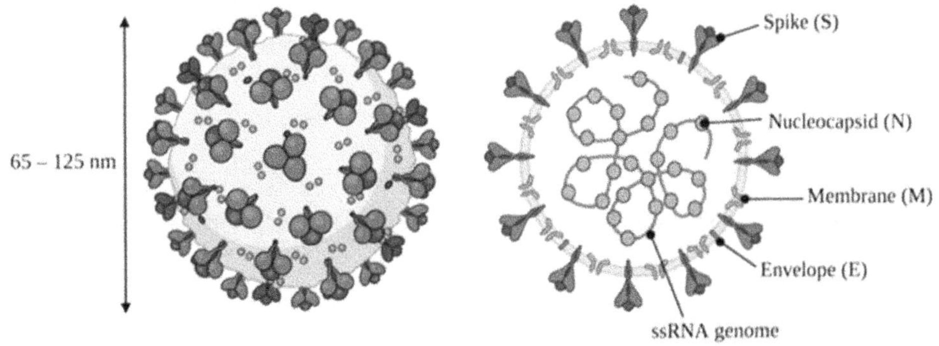

and respiratory problems due to COVID-19 infection. COVID-19 is a single-stranded RNA virus (Figure 2) which is existent in animals and humans and is generally spread through sneezing and close social contact (Singhal et al., 2020; Barua et al., 2021; Adhikari et al., 2020; Luan et al., 2020). The genome size of COVID-19 varies from 29.8 kb-29.9 kb and its structure genome followed the particular gene physiognomies to identified CoVs, and COVID-19 keeps the largest genome size of any recognized RNA virus (Crits-Christoph et al., 2021).

The most recognized and normally used technique for recognizing pathogens like COVID-19 or SARS-CoV-2 depends on real-time reverse transcription polymerase chain reaction (RT-PCR) (Wikramaratna et al., 2020). The high accurateness and exactness of the clinical RT-PCR test at the initial phases of infection in a patient (particularly symptomatic) make it the utmost consistent process to identify COVID-19 up to the present time. The test is precisely done for a qualitative investigation of the nucleic acid of the pathogen in persons who meet COVID-19 infection symptoms and signs (Dinnes et al., 2021). For infected persons who are recognized as an incorrect negative or asymptomatic or false result with a viral load underneath the recognition limit of RT-PCR tests, the second-best alternative is a serological test and it can be applied to advance control the COVID-19 spread. Though, the above-mentioned techniques and alike medical analysis all necessitate advanced clinical investigation, expensive apparatus, and proficiency, which are ineffective for the determination of point-of-care rapid identification (Barua et al., 2021; Pecoraro et al., 2022; Sharif et al., 2021; Padhy et al., 2020).

The rapidly increasing number of COVID-19 positive cases and the incapacity to mass scale tests because of the unaffordable price and time-consuming clinical diagnosis processes have enforced the essential to advance techniques that are more advantageous to rapid, economical, and efficient time-saving detection of COVID-19 (Barua et al., 2022). Point-of-care biosensors can offer the upcoming

Figure 3. Different test procedures of COVID-19 detection
Source: Barua et al. (2021)

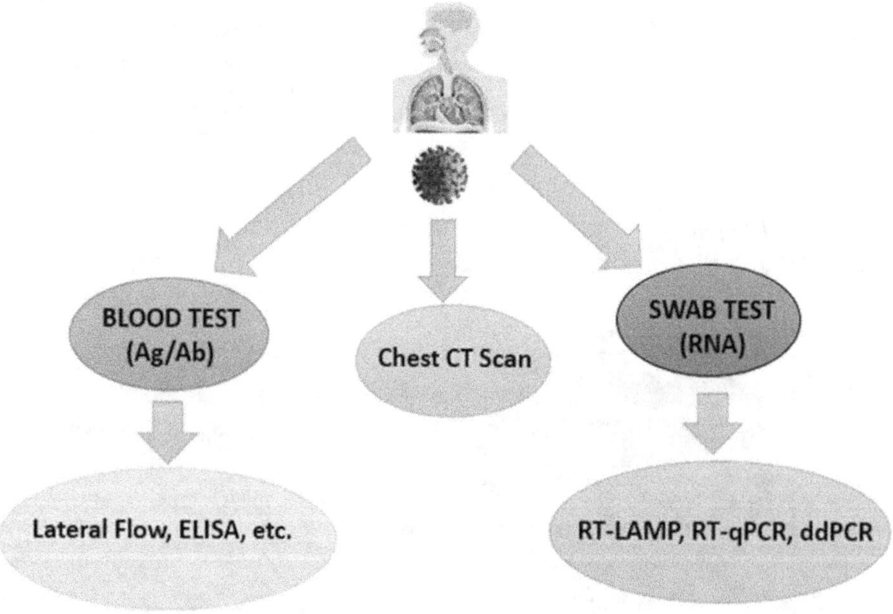

Figure 4. Promising lab-on-chip for COVID-19 detection

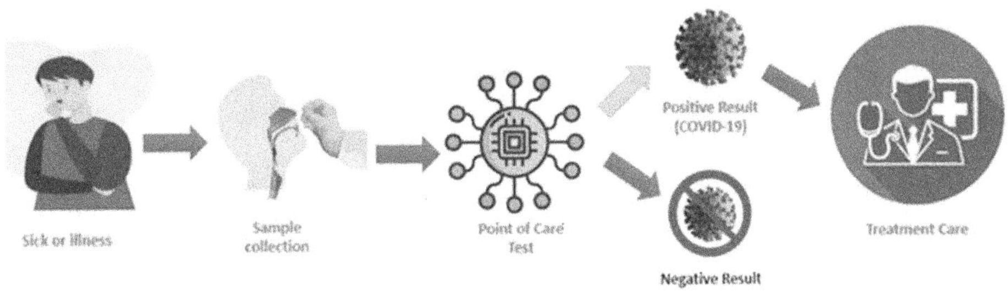

best alternative consistent outcome to clinical analysis with a considerably quicker real-time recognition deprived of conceding accuracy and sensitivity (Adil et al., 2021; Tsang et al., 2021). Figure 3 shows the different test for COVD-19 detection. The WHO (World Health Organization) has described and specified the criteria and necessities of a suitable biosensor to be applied as affordable, sensitive, specific, user-friendly, rapid and robust, equipment-free, and deliverable to end-users (ASSURED) (Kilic et al., 2018). Companionable to these necessities, there have been many point-of-care nucleic acid biosensor stands presented, advanced, or adapted to identify COVID-19 by ASSURED criteria (Qiu et al., 2020; Ji et al., 2020; Kilic et al., 2018). Though, the main test for all accessible biosensors is first verification of COVID-19 with a low level of viral RNA at the beginning of the contamination to circumvent advance spread (Vermisoglou et al., 2020; Nguyen et al., 2020).

EMERGENT LAB-ON-CHIP FOR COVID-19 DIAGNOSTICS

Here we will discuss novel lab-on-chip devices for the detection of COVID-19, in addition, to studying high-impact other research articles which could be significant to COVID-19 disease. These viral identification approaches make use of electrical, optical, magnetic, colorimetric, and lateral flow-based characteristics in technology to apprehend and transduce investigative signs. A summary of every modality will be specified, followed by specimens of detailed analysis that engage every method to COVID-19 or SARS-CoV-2 viruses (Tymm et al., 2020). The explanation of how these methods have been useful for other viruses may be beneficial for adjusting and increasing new Lab-on-Chip biosensors for the detection of COVID-19 disease. The detection limit is a very significant standard for these COVID-19 clinical tests. Figure 4 shows the schematic diagram of modern Lab-on-Chip application for COVID-19 detection.

COLORIMETRIC AND FLUORESCENT BIOSENSORS

Colorimetric and fluorescent biosensors are a usual identification technique by which change of spectral is frequently applied to transduce the existence of a specific biomarker (Scullion et al., 2011). Figure 5 shows the different types of colorimetric biosensors used in virus detection. LAMP (loop-mediated isothermal amplification) or isothermal amplification reactions can be detected in real-time by inserting the fluorescent dye CV (crystal violet) (Passaro et al., 2007), which shows a violet color solution

Figure 5. Different types of colorimetric biosensors

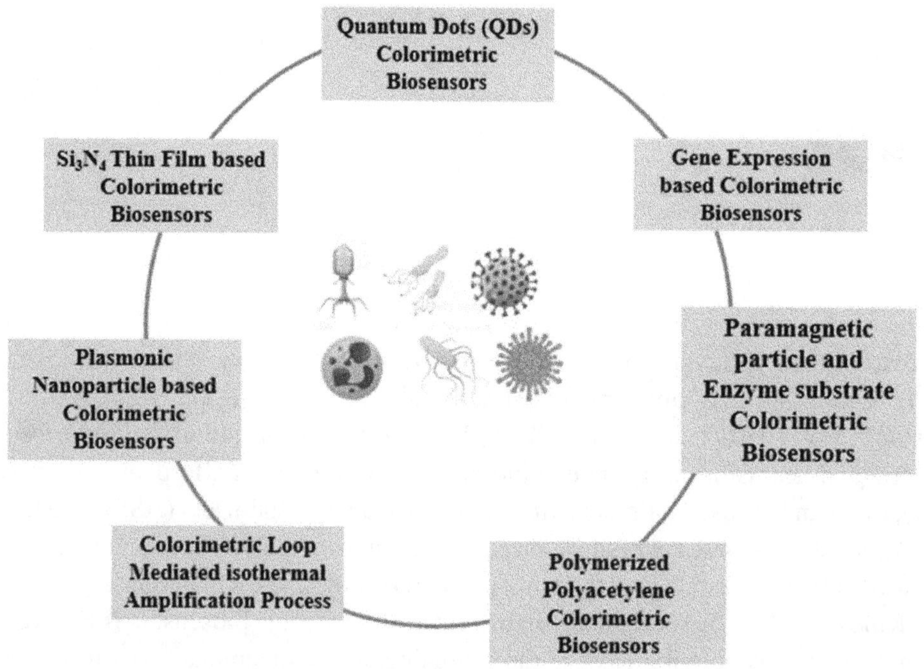

(aqueous). Crystal violet assigns to the main groove of dsDNA and transforms into the LCV (leuco type) which is colorless (Armani et al., 2003). The reaction solution for loop-mediated isothermal amplification develops colored only in the existence of dsDNA (Homola et al., 1999; Mauriz et al., 2020).

MAGNETIC NANOPARTICLES (MNPS) TECHNOLOGIES

In recent times, magnetic nanoparticles (iron oxide based) have established importance as encouraging materials for high-sensitive and quick diagnosis approaches because of their exclusive properties (Xu et al., 2019). Magnetic nanoparticles act as "nano-magnets" with the existence of an outer magnet and it moves quickly in the direction of the outside magnet whereas their magnetism is instantaneously lost as soon as the magnetic field is eliminated (Ha et al., 2018). Also, it is well admitted that magnetic nanoparticles are steadily conjugated with numerous biomaterials deprived of dropping their magnetic characteristics. Due to these proficiencies, magnetic nanoparticles permit magnetic pre-concentration of aim materials from an identical dilute attentiveness through two phases: (a) object materials cooperate with bio-conjugated magnetic nanoparticles, and (b) objective material-bio-conjugated magnetic nanoparticles are composed with an outer magnet and re-detached into a minor volume of matrix for the pre-concentration of models to recover the recognition sensitivity. Furthermore, magnetic nanoparticles compromise small circumstantial noise, meanwhile the organic molecules that do not act together with magnetic nanoparticles are non-magnetic (Gowri et al., 2021). Additionally, magnetic nanoparticles offer great surface area, which can permit rapid recognition time through quick interface between the bio-conjugated magnetic nanoparticles and objective materials. Certainly, magnetic nanoparticles have been efficiently useful to fast and extremely sensitive recognition for a wide range of target analyses from bacteria (Mun et al., 2015), and viruses (Sánchez-Cano et al, 2021) to food allergen (Yin et al., 2022). In recent times, Zhao et al. 2020 established magnetic nanoparticles coated with PC (poly (amino ester)-carboxyl groups) for effective extraction of COVID-19 RNA, combining the lysis, extraction, and necessary stages of viral genetic material into a single phase for RT-PCR responses.90 Because of a strong interaction between the nucleic acids and the carboxyl groups, RNA molecules are absorbed onto the nanoparticles, and then it can be directly placed into an RT-PCR without any supplementary elution phase. This extraction technique is an enhancement over conventional silica-based spin column RNA extractions, wherever testers involve pre-lysis to discharge the nucleic acids from viral particles and multiple centrifugation phases are essential. Viral RNA can be cleaned within twenty minutes by this technology (Chacón-Torres et al., 2020).

PLASMONIC SENSORS

A dual-function plasmonic biosensor was explained by Qiu et al., 2020, which combined the LSPR (localized surface plasmon resonance) and PPT (plasmonic photothermal effect) sensing transduction for COVID-19 RNA without the requirement for RT-PCR. Because of the exceptional characteristics of plasmonic biosensors nanoparticles, heating energy is confined near the nanoparticles which can be applied as a heat basis for thermal processing (Ahmadivand et al., 2021). Hybridization happens between the aim and the conjugated probe, but a single mismatch can reason the melting temperature to decrease pointedly. The plasmonic photothermal effect rises the hybridization rate and localized surface plasmon resonance sensing comeback, in the case of rapid and sensitive recognition of nucleic acids by refining hybridization kinetics (Labouta et al., 2021). The biosensor confirmed high specificity and a low recognition limit of COVID-19 classifications down to 0.22 pM (Qiu et al., 2020).

Lateral Flow Assays (LFA)

Lateral flow identifies the existence of target material as a liquid sample runs alongside the surface of a functionalized pad with sensitive molecules which can make a piece of noticeable information (Figure 6). A lateral flow investigation has been advanced for the recognition of COVID-19 antibodies (Zhou et al., 2021). An LNPs (lanthanide-doped polystyrene nanoparticle) was developed based on LFIA for screening anti-COVID-19 IgG in human serum (Figure 7), where, the target IgG was captured by recombinant nucleocapsid COVID-19 phosphoprotein coated on the experiment zone (Chen et al., 2020). A fluorescent signal from an LNP-labeled mouse anti-human IgG antibody was dignified, permitting measurable recognition. Though this investigation was advanced for antibody recognition, this technology is also capable of COVID-19 antigen detection (Kim et al., 2022).

ELECTRICAL AND ELECTROCHEMICAL SENSORS

Electrochemical sensors use resistive or capacitive variations to identify binding changes of applicable analytes (Das et al., 2016). A biosensing device named 'eCovSens' was made by Mahari et al., 2020,

Figure 6. Graphical demonstration of antibody detection by LFA system (lateral flow assays)
Source: Premraj et al. (2020)

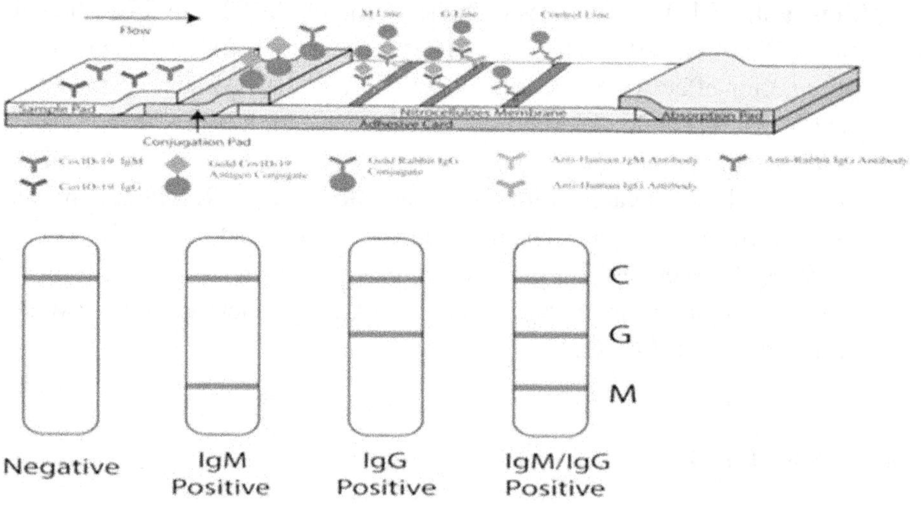

Figure 7. LNPs system
Source: Chen et al. (2020)

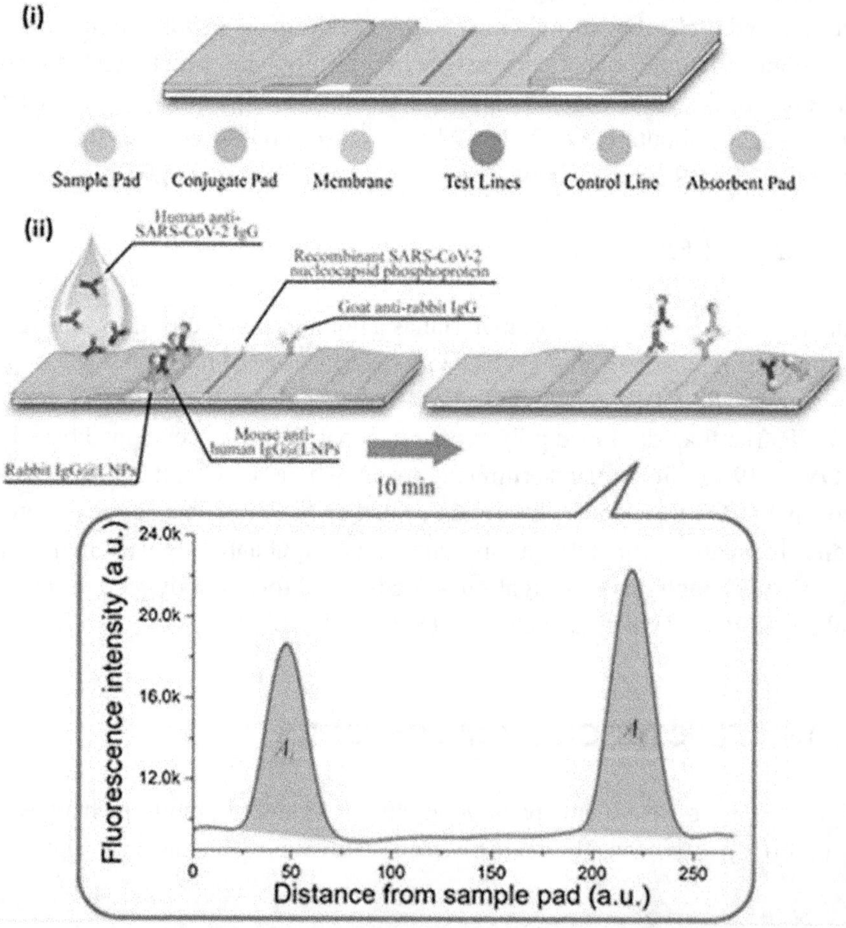

where an SPCE (screen-printed carbon electrode) and likened it with a commercial potentiostat containing FTO electrode (fluorine-doped tin oxide). They estimated their novel method in terms of compassion, identification, specificity, rate of detection, sample capacity, transferability, and permanency for the detection of SARS-CoV-2 (Mahari et al., 2020). This device contained of a bio-recognition component, a carbon electrode transducer, and an electronic arrangement to identify variations in the voltage. Conjugated gold nanoparticles both identify the viral particles and performance as a catalyst to amplify the electrochemical signal by augmenting electrical conductivity. A FET (field-effect transistor) device was developed by Seo et al., 2020, to distinguish the spike protein. Field-effect transistors have high selectivity and sensitivity and are characteristically arranged to apprehension analytes from side to side biorecognition on the guiding channel. Graphene sheets on a field-effect transistor were functionalized with an anti-COVID-19 spike protein antibody. This sensor alignment permitted for recognition of spike protein of SARS-COV-2 in purified antigen, cultured virus, and the specimens of nasopharyngeal swab with a limit of recognition of 1 FG/mL in phosphate-buffered saline and 100 FG/mL in clinical transport medium (Seo et al., 2020).

CONCLUSION

In this chapter, we have studied present advances in the area of Lab-on-Chip diagnostic biosensors for COVID-19. We also highlight the advance of lab-on-chip apparatuses for detecting COVID-19. We emphasize rapid, point-of-care equipment that can be prescribed at scale. Such devices could be capable replacements for the present advanced typical reverse transcription-polymerase chain reaction (RT-PCR) investigative testing. Noteworthy development has been prepared using these modest integrated arrangements for the recognition of both protein biomarkers and SARS-CoV-2 or COVID-19 RNA. Further work en route for a finding of other viruses may come to be applied as we carry on to revolutionize and advance novel platforms for SARS-CoV-2 or COVID-19. With private and public maintenance together with new allowances, advanced assessment, and rapid endorsements, we hope that innovative machinery carries on to be completely obtainable to increase this response to this COVID-19 pandemic.

ACKNOWLEDGMENT

The authors would like to thank Mechanical Engineering Dept. OmDayal Group of Institutions, also Biomedical Engineering Department, IIEST Shibpur, Howrah, and thanks to Mrs. Nibedita Bardhan for language proof reading.

REFERENCES

Adhikari, S. P., Meng, S., Wu, Y. J., Mao, Y. P., Ye, R. X., Wang, Q. Z., Sun, C., Sylvia, S., Rozelle, S., Raat, H., & Zhou, H. (2020). Epidemiology, causes, clinical manifestation and diagnosis, prevention and control of coronavirus disease (COVID-19) during the early outbreak period: A scoping review. *Infectious Diseases of Poverty*, *9*(1), 29. doi:10.118640249-020-00646-x PMID:32183901

Adil, M. T., Rahman, R., Whitelaw, D., Jain, V., Al-Taan, O., Rashid, F., Munasinghe, A., & Jambulingam, P. (2021). SARS-CoV-2 and the pandemic of COVID-19. *Postgraduate Medical Journal*, *97*(1144), 110–116. doi:10.1136/postgradmedj-2020-138386 PMID:32788312

Ahmadivand, A., Gerislioglu, B., Ramezani, Z., Kaushik, A., Manickam, P., & Ghoreishi, S. A. (2021). Functionalized terahertz plasmonic metasensors: Femtomolar-level detection of SARS-CoV-2 spike proteins. *Biosensors & Bioelectronics*, *177*, 112971. doi:10.1016/j.bios.2021.112971 PMID:33434777

Armani, D. K., Kippenberg, T. J., Spillane, S. M., & Vahala, K. J. (2003). Ultra-high-Q toroid microcavity on a chip. *Nature*, *421*(6926), 925–928. doi:10.1038/nature01371 PMID:12606995

Asghari, A., Wang, C., Yoo, K. M., Rostamian, A., Xu, X., Shin, J. D., Dalir, H., & Chen, R. T. (2021). Fast, accurate, point-of-care COVID-19 pandemic diagnosis enabled through advanced lab-on-chip optical biosensors: Opportunities and challenges. *Applied Physics Reviews*, *8*(3), 031313. doi:10.1063/5.0022211 PMID:34552683

Barua, R., Datta, P., & Chowdhury, A. R. (2022). The Impact of 3D Printing Technology on the COVID-19 Pandemic. In Quality of Healthcare in the Aftermath of the COVID-19 Pandemic (pp. 135-154). IGI Global.

Barua, R., Datta, S., Datta, P., & Chowdhury, A. R. (2022). The Study of Traditional Medicine for the Treatment of COVID-19. In *Handbook of Research on Pathophysiology and Strategies for the Management of COVID-19* (pp. 221–241). IGI Global. doi:10.4018/978-1-7998-8225-1.ch014

Barua, R., Datta, S., Dutta, P., & Chowdhury, A. R. (2021). Machine learning and AI application of COVID-19 pandemic outbreak. *Impacts and Implications of COVID-19*, 2.

Barua, R., Datta, S., Roychowdhury, A., & Datta, P. (2021). The study of the epidemiology and clinical features of the novel coronavirus (COVID-19). In Epidemiological Research Applications for Public Health Measurement and Intervention (pp. 25-39). IGI Global.

Barua, R., Datta, S., RoyChowdhury, A., & Datta, P. (2021). A Review on the epidemiology and characteristics of COVID-19. *Journal of Current Biomedical Reports*, *2*(3), 95–106. doi:10.52547/10.52547/JCBioR.2.3.95

Chacón-Torres, J. C., Reinoso, C., Navas-León, D. G., Briceño, S., & González, G. (2020). Optimized and scalable synthesis of magnetic nanoparticles for RNA extraction in response to developing countries' needs in the detection and control of SARS-CoV-2. *Scientific Reports*, *10*(1), 19004. doi:10.103841598-020-75798-9 PMID:33149153

Chen, Z., Zhang, Z., Zhai, X., Li, Y., Lin, L., Zhao, H., Bian, L., Li, P., Yu, L., Wu, Y., & Lin, G. (2020). Rapid and Sensitive Detection of anti-SARS-CoV-2 IgG, Using Lanthanide-Doped Nanoparticles-Based Lateral Flow Immunoassay. *Analytical Chemistry*, *92*(10), 7226–7231. doi:10.1021/acs.analchem.0c00784 PMID:32323974

Crits-Christoph, A., Kantor, R. S., Olm, M. R., Whitney, O. N., Al-Shayeb, B., Lou, Y. C., Flamholz, A., Kennedy, L. C., Greenwald, H., Hinkle, A., Hetzel, J., Spitzer, S., Koble, J., Tan, A., Hyde, F., Schroth, G., Kuersten, S., Banfield, J. F., & Nelson, K. L. (2021). Genome Sequencing of Sewage Detects Regionally Prevalent SARS-CoV-2 Variants. *mBio*, *12*(1), e02703–e02720. doi:10.1128/mBio.02703-20 PMID:33468686

Das, J., Ivanov, I., Sargent, E. H., & Kelley, S. O. (2016). DNA Clutch Probes for Circulating Tumor DNA Analysis. *Journal of the American Chemical Society*, *138*(34), 11009–11016. doi:10.1021/jacs.6b05679 PMID:27513828

Dinnes, J., Deeks, J. J., Berhane, S., Taylor, M., Adriano, A., Davenport, C., Dittrich, S., Emperador, D., Takwoingi, Y., Cunningham, J., Beese, S., Domen, J., Dretzke, J., Ferrante di Ruffano, L., Harris, I. M., Price, M. J., Taylor-Phillips, S., Hooft, L., Leeflang, M. M., ... Van den Bruel, A.Cochrane COVID-19 Diagnostic Test Accuracy Group. (2021). Rapid, point-of-care antigen and molecular-based tests for diagnosis of SARS-CoV-2 infection. *Cochrane Database of Systematic Reviews*, *3*(3), CD013705. Advance online publication. doi:10.1002/14651858.CD013705.pub2 PMID:33760236

Gowri, A., Kumar, N. A., & Anand, B. S. (2021). Recent advances in nanomaterials based biosensors for point of care (PoC) diagnosis of COVID-19–a minireview. *Trends in Analytical Chemistry*, *137*, 116205. doi:10.1016/j.trac.2021.116205 PMID:33531721

Ha, Y., Ko, S., Kim, I., Huang, Y., Mohanty, K., Huh, C., & Maynard, J. A. (2018). Recent Advances Incorporating Superparamagnetic Nanoparticles into Immunoassays. *ACS Applied Nano Materials*, *1*(2), 512–521. doi:10.1021/acsanm.7b00025 PMID:29911680

Homola, J., Yee, S. S., & Gauglitz, G. (1999). Surface plasmon resonance sensors. *Sensors and Actuators. B, Chemical*, *54*(1-2), 3–15. doi:10.1016/S0925-4005(98)00321-9

Hsieh, C. L., Goldsmith, J. A., Schaub, J. M., DiVenere, A. M., Kuo, H. C., Javanmardi, K., Le, K. C., Wrapp, D., Lee, A. G., Liu, Y., Chou, C. W., Byrne, P. O., Hjorth, C. K., Johnson, N. V., Ludes-Meyers, J., Nguyen, A. W., Park, J., Wang, N., Amengor, D., ... McLellan, J. S. (2020). Structure-based design of prefusion-stabilized SARS-CoV-2 spikes. *Science*, *369*(6510), 1501–1505. doi:10.1126cience.abd0826 PMID:32703906

Ji, T., Liu, Z., Wang, G., Guo, X., Akbar Khan, S., Lai, C., Chen, H., Huang, S., Xia, S., Chen, B., Jia, H., Chen, Y., & Zhou, Q. (2020). Detection of COVID-19: A review of the current literature and future perspectives. *Biosensors & Bioelectronics*, *166*, 112455. doi:10.1016/j.bios.2020.112455 PMID:32739797

Kilic, T., Erdem, A., Ozsoz, M., & Carrara, S. (2018). microRNA biosensors: Opportunities and challenges among conventional and commercially available techniques. *Biosensors & Bioelectronics*, *99*, 525–546. doi:10.1016/j.bios.2017.08.007 PMID:28823978

Kim, C., Yoo, Y. K., Lee, N. E., Lee, J., Kim, K. H., Lee, S., Kim, J., Park, S. J., Lee, D., Lee, S. W., Hwang, K. S., Han, S. I., Lee, D., Yoon, D. S., & Lee, J. H. (2022). Nanoelectrokinetic-assisted lateral flow assay for COVID-19 antibody test. *Biosensors & Bioelectronics*, *212*, 114385. doi:10.1016/j.bios.2022.114385 PMID:35623254

Kitchen, L. C., Berman, M., Halper, J., & Chazot, P. (2022). Rationale for 1068 nm Photobiomodulation Therapy (PBMT) as a Novel, Non-Invasive Treatment for COVID-19 and Other Coronaviruses: Roles of NO and Hsp70. *International Journal of Molecular Sciences*, *23*(9), 5221. doi:10.3390/ijms23095221 PMID:35563611

Labouta, H. I., Hooshmand, N., Upreti, T., & El-Sayed, M. A. (2021). Localized Plasmonic Photothermal Therapy as a Life-saving Treatment Paradigm for Hospitalized COVID-19 Patients. *Plasmonics*, *16*(4), 1029–1033. doi:10.100711468-020-01353-x PMID:33495692

Luan, R. S., Wang, X., Sun, X., Chen, X. S., Zhou, T., Liu, Q. H., Lü, X., Wu, X. P., Gu, D. Q., Tang, M. S., Cui, H. J., Shan, X. F., Ouyang, J., Zhang, B., & Zhang, W., & Sichuan University Covid-, E. (2020). Sichuan da xue xue bao. Yi xue ban =. *Sichuan Da Xue Xue Bao. Yi Xue Ban = Journal of Sichuan University. Medical Science Edition*, *51*(2), 131–138. doi:10.12182/20200360505 PMID:32220177

MahariS.RobertsA.ShahdeoD.GandhiS. eCovSens-ultrasensitive novel in-house built printed circuit board based electrochemical device for rapid detection of nCovid-19. bioRxiv 2020. doi:10.1101/2020.04.24.059204

Mauriz, E. (2020). Recent Progress in Plasmonic Biosensing Schemes for Virus Detection. *Sensors (Basel)*, *20*(17), 4745. doi:10.339020174745 PMID:32842601

Mun, S., & Choi, S. J. (2015). Detection of Salmonella typhimurium by antibody/enzyme-conjugated magnetic nanoparticles. *Biochip Journal*, *9*(1), 10–15. doi:10.100713206-014-9102-2

Nguyen, H. (2020). Editorial of Special Issue "Nanostructured Light-Emitters". *Micromachines*, *11*(6), 601. doi:10.3390/mi11060601 PMID:32575847

. Padhy, B. M., Mohanty, R. R., Das, S., & Meher, B. R. (2020). Therapeutic potential of ivermectin as add on treatment in COVID 19: A systematic review and meta-analysis. *Journal of Pharmacy & Pharmaceutical Sciences, 23*, 462–469. doi:10.18433/jpps31457

Passaro, V., Dell'Olio, F., Casamassima, B., & De Leonardis, F. (2007). Guided-Wave Optical Biosensors. *Sensors (Basel)*, *7*(4), 508–536. doi:10.33907040508

Pecoraro, V., Negro, A., Pirotti, T., & Trenti, T. (2022). Estimate false-negative RT-PCR rates for SARS-CoV-2. A systematic review and meta-analysis. *European Journal of Clinical Investigation*, *52*(2), e13706. doi:10.1111/eci.13706 PMID:34741305

Premraj, A., Aleyas, A. G., Nautiyal, B., & Rasool, T. J. (2020). Nucleic Acid and Immunological Diagnostics for SARS-CoV-2: Processes, Platforms and Pitfalls. *Diagnostics (Basel)*, *10*(11), 866. doi:10.3390/diagnostics10110866 PMID:33114057

Qiu, G., Gai, Z., Tao, Y., Schmitt, J., Kullak-Ublick, G. A., & Wang, J. (2020). Dual-Functional Plasmonic Photothermal Biosensors for Highly Accurate Severe Acute Respiratory Syndrome Coronavirus 2 Detection. *ACS Nano*, *14*(5), 5268–5277. doi:10.1021/acsnano.0c02439 PMID:32281785

Sánchez-Cano, A., Ruiz-Vega, G., Vicente-Gómez, S., de la Serna, E., Sulleiro, E., Molina, I., Sánchez-Montalvá, A., & Baldrich, E. (2021). Development of a Fast Chemiluminescent Magneto-Immunoassay for Sensitive *Plasmodium falciparum* Detection in Whole Blood. *Analytical Chemistry*, *93*(37), 12793–12800. doi:10.1021/acs.analchem.1c03242 PMID:34496566

Scullion, M. G., Di Falco, A., & Krauss, T. F. (2011). Slotted photonic crystal cavities with integrated microfluidics for biosensing applications. *Biosensors & Bioelectronics*, *27*(1), 101–105. doi:10.1016/j.bios.2011.06.023 PMID:21764290

Seo, G., Lee, G., Kim, M. J., Back, S. H., Choi, M., Ku, K. B., Lee, C. S., Jun, S., Park, D., Kim, H. G., Kim, S. J., Lee, J. O., Kim, B. T., Park, E. C., & Kim, S. I. (2020). Rapid Detection of COVID-19 Causative Virus (SARS-CoV-2) in Human Nasopharyngeal Swab Specimens Using Field-Effect Transistor-Based Biosensor. *ACS Nano*, *14*(4), 5135–5142. doi:10.1021/acsnano.0c02823 PMID:32293168

Sharif, N., Alzahrani, K. J., Ahmed, S. N., & Dey, S. K. (2021). Efficacy, Immunogenicity and Safety of COVID-19 Vaccines: A Systematic Review and Meta-Analysis. *Frontiers in Immunology*, *12*, 714170. doi:10.3389/fimmu.2021.714170 PMID:34707602

Singhal, T. (2020). A Review of Coronavirus Disease-2019 (COVID-19). *Indian Journal of Pediatrics*, *87*(4), 281–286. doi:10.100712098-020-03263-6 PMID:32166607

Tsang, H. F., Chan, L., Cho, W., Yu, A., Yim, A., Chan, A., Ng, L., Wong, Y., Pei, X. M., Li, M., & Wong, S. C. (2021). An update on COVID-19 pandemic: The epidemiology, pathogenesis, prevention and treatment strategies. *Expert Review of Anti-Infective Therapy*, *19*(7), 877–888. doi:10.1080/14787210.2021.1863146 PMID:33306423

Tymm, C., Zhou, J., Tadimety, A., Burklund, A., & Zhang, J. (2020). Scalable COVID-19 Detection Enabled by Lab-on-Chip Biosensors. *Cellular and Molecular Bioengineering*, *13*(4), 313–329. doi:10.100712195-020-00642-z PMID:32837587

Vermisoglou, E., Panáček, D., Jayaramulu, K., Pykal, M., Frébort, I., Kolář, M., Hajdúch, M., Zbořil, R., & Otyepka, M. (2020). Human virus detection with graphene-based materials. *Biosensors & Bioelectronics*, *166*, 112436. doi:10.1016/j.bios.2020.112436 PMID:32750677

Wikramaratna, P. S., Paton, R. S., Ghafari, M., & Lourenço, J. (2020). Estimating the false-negative test probability of SARS-CoV-2 by RT-PCR. Euro surveillance: bulletin Europeen sur les maladies transmissibles = European communicable disease bulletin, 25(50), 2000568. doi:10.2807/1560-7917.ES.2020.25.50.2000568

Xu, C., Akakuru, O. U., Zheng, J., & Wu, A. (2019). Applications of Iron Oxide-Based Magnetic Nanoparticles in the Diagnosis and Treatment of Bacterial Infections. *Frontiers in Bioengineering and Biotechnology*, *7*, 141. doi:10.3389/fbioe.2019.00141 PMID:31275930

Yin, H. Y., Li, Y. T., Tsai, W. C., Dai, H. Y., & Wen, H. W. (2022). An immunochromatographic assay utilizing magnetic nanoparticles to detect major peanut allergen Ara h 1 in processed foods. *Food Chemistry*, *375*, 131844. doi:10.1016/j.foodchem.2021.131844 PMID:34952385

ZhaoZ.. A simple magnetic nanoparticles-based viral RNA extraction method for efficient detection of SARS-CoV-2. bioRxiv 2020.02.22.961268 (2020). doi:10.1101/2020.02.22.961268

Zhou, Y., Wu, Y., Ding, L., Huang, X., & Xiong, Y. (2021). Point-of-care COVID-19 diagnostics powered by lateral flow assay. *Trends in analytical chemistry. Trends in Analytical Chemistry*, *145*, 116452. doi:10.1016/j.trac.2021.116452 PMID:34629572

KEY TERMS AND DEFINITIONS

COVID-19: The new coronavirus sars-cov-2 is the cause of the extremely contagious respiratory disease COVID-19. It was first identified in Wuhan, China, in December 2019, and it soon spread throughout the world, causing a global pandemic. The signs of COVID-19 include fever, coughing, and breathing difficulties, and they can be minor to severe. When an infected individual coughs, sneezes, talks, or breathes, respiratory droplets are essentially how the virus is transmitted. The prevention of COVID-19's spread by practices such mask use, excellent hand cleanliness, social seclusion, and immunization is essential.

Lab-on-Chip: The term "lab-on-a-chip" (LoC) describes the fusion of various laboratory operations onto a tiny microchip or a miniature device. The preparation, analysis, and detection of samples are only a few of the duties that LoC devices are made to handle in the lab. The devices can be used for a range of applications in disciplines like biomedical engineering, chemistry, and microbiology. They are typically very small, occasionally measuring only a few millimeters.

Lateral Flow Assays: An easy and quick diagnostic test called a lateral flow assay (lfa) is used to find out whether a target molecule is present or absent in a sample, usually a liquid like blood or urine. Lfas are based on the immunoassay theory, which use antibodies to find particular antigens.

Magnetic Nanoparticles: Nano-scale particles with magnetic properties are known as magnetic nanoparticles; their sizes typically range from 1 to 100 nanometers. Iron, nickel, cobalt, or their alloys, which are magnetic materials, are frequently used in their construction. These nanoparticles' magnetic characteristics make them suitable for a range of biomedical, materials science, and environmental remediation applications.

Plasmonic Sensors: Biosensors known as plasmonic sensors use the plasmonics principles to track changes in a sample's refractive index. Plasmon resonance happens when light energy is absorbed by metal nanoparticles, producing oscillations of the conduction electrons at the metal's surface. Plasmonics is the study of the interaction between light and metal surfaces.

Compilation of References

Abedini-Nassab, R., Pouryosef Miandoab, M., & Şaşmaz, M. (2021). Microfluidic Synthesis, Control, and Sensing of Magnetic Nanoparticles: A Review. *Micromachines*, *12*(7), 768. doi:10.3390/mi12070768 PMID:34210058

Abgrall, P., & Gue, A. M. (2007). Lab-on-chip technologies: Making a microfluidic network and coupling it into a complete microsystem—a review. *Journal of Micromechanics and Microengineering*, *17*(5), R15–R49. doi:10.1088/0960-1317/17/5/R01

Addae-Mensah, K. A., Wang, Z., Parsa, H., Chin, S. Y., Laksanasopin, T., & Sia, S. K. (2010). Fundamentals of microfluidics devices. *Microfluidic Devices in Nanotechnology: Fundamental Concepts*, 1-38.

Adhikari, S. P., Meng, S., Wu, Y. J., Mao, Y. P., Ye, R. X., Wang, Q. Z., Sun, C., Sylvia, S., Rozelle, S., Raat, H., & Zhou, H. (2020). Epidemiology, causes, clinical manifestation and diagnosis, prevention and control of coronavirus disease (COVID-19) during the early outbreak period: A scoping review. *Infectious Diseases of Poverty*, *9*(1), 29. doi:10.118640249-020-00646-x PMID:32183901

Adil, M. T., Rahman, R., Whitelaw, D., Jain, V., Al-Taan, O., Rashid, F., Munasinghe, A., & Jambulingam, P. (2021). SARS-CoV-2 and the pandemic of COVID-19. *Postgraduate Medical Journal*, *97*(1144), 110–116. doi:10.1136/postgradmedj-2020-138386 PMID:32788312

Agrawal, Y., & Patel, V. (2011). Nanosuspension: An approach to enhance solubility of drugs. *Journal of Advanced Pharmaceutical Technology & Research*, *2*(2), 81. doi:10.4103/2231-4040.82950 PMID:22171298

Ahmadivand, A., Gerislioglu, B., Ramezani, Z., Kaushik, A., Manickam, P., & Ghoreishi, S. A. (2021). Functionalized terahertz plasmonic metasensors: Femtomolar-level detection of SARS-CoV-2 spike proteins. *Biosensors & Bioelectronics*, *177*, 112971. doi:10.1016/j.bios.2021.112971 PMID:33434777

Ahmed, T., Shimizu, T.S., Stocker, R. (2010). Microfluidics for bacterial chemotaxis. *Integrative Biology, 2*(11–12), 604–629 . . hdl:1721.1/66851. PMID 20967322 doi:10.1039/c0ib00049c

Aishan, Y., Yalikun, Y., Shen, Y., Yuan, Y., Amaya, S., Okutaki, T., Osaki, A., Maeda, S., & Tanaka, Y. (2021). A chemical micropump actuated by self-oscillating polymer gel. *Sensors and Actuators B: Chemical, 337*. doi:10.1016/j.snb.2021.129769

Akita, T., Otsuka, Y., Irita, M., & Hayase, M. (2020). In-Situ Observation of Electrodeposition in a Microfluidic Device with through Silicon Via Model. *ECS Meeting Abstracts, MA2020-02*(25), 1790. 10.1149/MA2020-02251790mtgabs

Al Yahyai, I., Al-Lawati, H. A. J., & Hassanzadeh, J. (2021). A Paper-Based Chemiluminescence Detection Device Based on S,N-Doped Carbon Quantum Dots for the Selective and Highly Sensitive Recognition of Bendiocarb. *Analytical Methods*, *13*(31), 3461–3470. doi:10.1039/D1AY00728A PMID:34286731

Al-Alawachi, S. F. A., Mohammed, K. A., Zabibah, R. S., Sadoon, J., Abed, A., & Al-Salam, M. A. (2022). Effect of copper oxide nanoparticles on some physical properties of PS/PEO polymer blend. *International Journal of Nanoscience, 21*(6), 2250044. doi:10.1142/S0219581X22500442

Ali & Takahata. (2011). Wireless microfluidic control with integrated shape-memory-alloy actuators operated by field frequency modulation. *J Micromech Microeng, 21*(7).

Alizadeh, N., Salimi, A., & Hallaj, R. (2018). Mimicking Peroxidase Activity of Co2(OH)2CO3-CeO2 Nanocomposite for Smartphone Based Detection of Tumor Marker Using Paper-Based Microfluidic Immunodevice. *Talanta, 189*, 100–110. doi:10.1016/j.talanta.2018.06.034 PMID:30086892

Alsanosi, Skiffington, & Padmanabhan. (2014). Pharmacokinetic Pharmacogenomics. In *Handbook of Pharmacogenomics and Stratified Medicine*. Academic Press. https://www.sciencedirect.com/science/article/pii/B9780123868824000177

Alvarez-Braña, Y., Etxebarria-Elezgarai, J., Ruiz de Larrinaga-Vicente, L., Benito-Lopez, F., & Basabe-Desmonts, L. (2021). Modular micropumps fabricated by 3D printed technologies for polymeric microfluidic device applications. *Sensors and Actuators B: Chemical, 342*. doi:10.1016/j.snb.2021.129991

Álvarez-DIduk, Orozco, & Merkoçi. (2017). Paper Strip-Embedded Graphene Quantum Dots: A Screening Device with a Smartphone Readout. *Scientific Reports, 7*(1), 1–9. PMID:28127051

Amarasinghe, R., Dao, D., Toriyama, T., & Sugiyama, S. (2007). Development of miniaturized 6-axis accelerometer utilizing piezoresistive sensing elements. *Sensors and Actuators. A, Physical, 134*(2), 310–320. doi:10.1016/j.sna.2006.05.044

Amina, S. J., & Guo, B. (2020). A Review on the Synthesis and Functionalization of Gold Nanoparticles as a Drug Delivery Vehicle. *International Journal of Nanomedicine, 15*, 9823–9857. doi:10.2147/IJN.S279094 PMID:33324054

Amini, A., Guijt, R. M., Themelis, T., De Vos, J., & Eeltink, S. (2023). Recent developments in digital light processing 3D-printing techniques for microfluidic analytical devices. *Journal of Chromatography A, 1692*. doi:10.1016/j.chroma.2023.463842

Andreou, C., Hoonejani, M. R., Barmi, M. R., Moskovits, M., & Meinhart, C. D. (2013). Rapid detection of drugs of abuse in saliva using surface enhanced Raman spectroscopy and microfluidics. *ACS Nano, 7*(8), 7157–7164. doi:10.1021/nn402563f PMID:23859441

Andrews, J. R., & Gerner, B. (2004). Microfluidics, BioMEMS, and Medical Microsystems II. SPIE.

Angell, J. B., Terry, S. C., & Barth, P. W. (1983). Silicon Micromechanical Devices. *Scientific American, 248*(4), 44–55. doi:10.1038cientificamerican0483-44

Anju, S., Prajitha, N., Sukanya, V. S., & Mohanan, P. V. (2020). Complicity of degradable polymers in health-care applications. *Materials Today. Chemistry, 16*, 100236. doi:10.1016/j.mtchem.2019.100236

Anushka, B., Bandopadhyay, A., & Das, P. K. (2022). Paper based microfluidic devices: A review of fabrication techniques and applications. *The European Physical Journal. Special Topics*. Advance online publication. doi:10.1140/epjs11734-022-00727-y PMID:36532608

Apelgren, P., Amoroso, M., Säljö, K., Montelius, M., Lindahl, A., Stridh Orrhult, L., Gatenholm, P., Kölby, L., Arulkumar, S., Parthiban, S., Goswami, A., Varma, R. S., Naushad, M., Gawande, M. B., Lipskas, J., Deep, K., Yao, W., Grogan, S. P., Dorthé, E. W., … Salvatore, D. A. (2019). Continuous, size and shape-control synthesis of hollow silica nanoparticles enabled by a microreactor-assisted rapid mixing process. *Materials Today: Proceedings, 27*. https://doi.org// doi:10.1016/j.jare.2020.01.010%0Ahttps://

Apilux, A., Chartchalerm, I. N. A., Tantimongcolwat, T., & Prachayasittikul, V. (2015). Paper-Based Acetylcholinesterase Inhibition Assay Combining a Wet System for Organophosphate and Carbamate Pesticides Detection. *EXCLI Journal*, *14*, 307–319. PMID:26417364

Appalakutti, S., Sonawane, S., Bhanvase, B. A., Mittal, V., & Ashokkumar, M. (2015). Process intensification of copper chromite (CuCr2O4) nanoparticle production using continuous flow microreactor. *Chemical Engineering and Processing*, *89*, 28–34. doi:10.1016/j.cep.2014.12.012

Apple. (2022). *Apple unveils M2, taking the breakthrough performance and capabilities of M1 even further*. Apple. https://www.apple.com/in/newsroom/2022/06/apple-unveils-m2-with-breakthrough-performance-and-capabilities/

Armani, D. K., Kippenberg, T. J., Spillane, S. M., & Vahala, K. J. (2003). Ultra-high-Q toroid microcavity on a chip. *Nature*, *421*(6926), 925–928. doi:10.1038/nature01371 PMID:12606995

Asadi Dereshgi, H., & Yildiz, M. Z. (2018a). Investigation of electromechanical factors effecting piezoelectric actuator for valveless micropump characteristics. *J Eng Sci Technol*, *13*, 2843–2856.

Asadi Dereshgi, H., & Yildiz, M. Z. (2018b). A novel micropump design, Investigation of the voltage effect on the net flow rate. *Sakarya Univ J Sci*, *22*(4), 1152–1156. doi:10.16984aufenbilder.388658

Asghari, A., Wang, C., Yoo, K. M., Rostamian, A., Xu, X., Shin, J. D., Dalir, H., & Chen, R. T. (2021). Fast, accurate, point-of-care COVID-19 pandemic diagnosis enabled through advanced lab-on-chip optical biosensors: Opportunities and challenges. *Applied Physics Reviews*, *8*(3), 031313. doi:10.1063/5.0022211 PMID:34552683

Asghar, W., Sher, M., Khan, N. S., Vyas, J. M., & Demirci, U. (2019). Microfluidic chip for detection of fungal infections. *ACS Omega*, *4*(4), 7474–7481. doi:10.1021/acsomega.9b00499 PMID:31080939

Aşık, M. D., Kaplan, M., Çetin, B., & Sağlam, N. (2021). Synthesis of iron oxide core chitosan nanoparticles in a 3D printed microfluidic device. *Journal of Nanoparticle Research*, *23*(3), 62. doi:10.100711051-021-05171-y

Atkin, M., Poetter, K., Cattrall, R., & Harvey, E. (2004). In P. Woias & I. Papautsky (Eds.), *Microfluidics, BioMEMS, and Medical Microsystems II* (Vol. 5345, pp. 138–146). Spie-Int Soc Optical Engineering. doi:10.1117/12.524321

Attaallah, R., & Amine, A. (2022). Highly Selective and Sensitive Detection of Cadmium Ions by Horseradish Peroxidase Enzyme Inhibition Using a Colorimetric Microplate Reader and Smartphone Paper-Based Analytical Device. *Microchemical Journal*, *172*, 106940. doi:10.1016/j.microc.2021.106940

Attia, U. M., Marson, S., & Alcock, J. R. (2009). Micro-injection moulding of polymer microfluidic devices. *Microfluidics and Nanofluidics*, *7*(1), 1–28. doi:10.100710404-009-0421-x

Atzori, L., Lera, A., & Morabito, G. (2010). The Internet of Things: A survey. *Computer Networks*, *54*(15), 2787–2805. doi:10.1016/j.comnet.2010.05.010

Augustine, S., Paul, L. J., & Paulo, D. (2021). Relevance of micromachining in microfluidics and biomedical devices. In Handbooks in Advanced Manufacturing, Advanced Machining and Finishing (pp. 339-373). Elsevier. doi:10.1016/B978-0-12-817452-4.00009-9

Ayushi, U. K. V., Mishra, S., & Kanu, N. J. (2021). Biomimetic 4D printed materials: A state-of-the-art review on concepts, opportunities, and challenges. *Materials Today: Proceedings*, *47*(11), 3313-3319. doi:10.1016/j.matpr.2021.07.148

Ayyub, P., Chandra, R., Taneja, P., Sharma, A. K., & Pinto, R. (2001). Synthesis of nanocrystalline material by sputtering and laser ablation at low temperatures. *Applied Physics. A, Materials Science & Processing*, *73*(1), 67–73. Advance online publication. doi:10.1007003390100833

Azimullah, S., Vikrant, Sudhakar, C., Kumar, P., Patil, A., Md. Usman, M. R., & Shaikh Usman, M. Z. (2019). Nanosuspensions as a promising approach to enhance bioavailability of poorly soluble drugs: An update. *Journal of Drug Delivery and Therapeutics*, *9*(2), 574–582. doi:10.22270/jddt.v9i2.2436

Aziz, A. A., Azmi, M. A. M., Nazri, M. N. A. M., Zainal, M. H. F., Khalfan, M. A., & Bahrain, A. K. (2019). Fabrication of micropump for microfluidics application. *Applied Physics of Condensed Matter*. doi:10.1063/1.5118148

Badilescu, S., & Packirisamy, M. (2016). *BioMEMS: Science and engineering perspectives*. CRC Press. doi:10.1201/b12255

Baheri Islami, S., Khezerloo, M., & Gharraei, R. (2017). The effect of chaotic advection on mixing degree and pressure drop of non-Newtonian fluids flow in curved micromixers. *Journal of the Brazilian Society of Mechanical Sciences and Engineering*, *39*(3), 813–831. doi:10.100740430-016-0689-1

Baig, N., Kammakakam, I., & Falath, W. (2021). Nanomaterials: A review of synthesis methods, properties, recent progress, and challenges. *Materials Advances*, *2*(6), 1821–1871. doi:10.1039/D0MA00807A

Bakunin, V. N., Suslov, A. Y., Kuzmina, G. N., Parenago, O. P., & Topchiev, A. V. (2004). Synthesis and application of inorganic nanoparticles as lubricant components-a review. *Journal of Nanoparticle Research*, *6*(2), 273–284. doi:10.1023/B:NANO.0000034720.79452.e3

Balani, R. S., Patel, D., Barthwal, S., & Arun, J. (2021). Fabrication of Air Conditioning System Using the Engine Exhaust Gas. In P. K. Kapur, G. Singh, & S. Panwar (Eds.), *Advances in Interdisciplinary Research in Engineering and Business Management. Asset Analytics*. Springer., doi:10.1007/978-981-16-0037-1_29

Balta, S., Sotto, A., Luis, P., Benea, L., Van der Bruggen, B., & Kim, J. (2012). A new outlook on membrane enhancement with nanoparticles: The alternative of ZnO. *Journal of Membrane Science*, *389*, 155–161. Advance online publication. doi:10.1016/j.memsci.2011.10.025

Banerjee, S., Benjwal, P., Singh, M., & Kar, K. K. (2018). Graphene oxide (rGO)-metal oxide (TiO2/Fe3O4) based nanocomposites for the removal of methylene blue. *Applied Surface Science*, *439*, 560–568. doi:10.1016/j.apsusc.2018.01.085

Banerjee, S., & Kar, K. K. (2016a). Superior water retention, ionic conductivity and thermal stability of sulfonated poly ether ether ketone/polypyrrole/aluminum phosphate nanocomposite based polymer electrolyte membrane. *Journal of Environmental Chemical Engineering*, *4*(1), 299–310. Advance online publication. doi:10.1016/j.jece.2015.11.033

Banerjee, S., & Kar, K. K. (2016b). Synergistic effect of aluminium phosphate and tungstophosphoric acid on the physicochemical properties of sulfonated poly ether ether ketone nanocomposite membrane. *Journal of Applied Polymer Science*, *133*(5), n/a. Advance online publication. doi:10.1002/app.42952

Bao, L. (2020). Reinfection could not occur in SARS-CoV-2 infected rhesus macaques. bioRxiv 2020.03.13.990226

Barrett, M. P., Cooper, J. M., Regnault, C., Holm, S. H., Beech, J. P., Tegenfeldt, J. O., & Hochstetter, A. (2017, October). Microfluidics-Based Approaches to the Isolation of African Trypanosomes. *Pathogens (Basel, Switzerland)*, *6*(4), 47. doi:10.3390/pathogens6040047 PMID:28981471

Barua, R., Datta, P., & Chowdhury, A. R. (2022). The Impact of 3D Printing Technology on the COVID-19 Pandemic. In Quality of Healthcare in the Aftermath of the COVID-19 Pandemic (pp. 135-154). IGI Global.

Barua, R., Datta, S., Roychowdhury, A., & Datta, P. (2021). The study of the epidemiology and clinical features of the novel coronavirus (COVID-19). In Epidemiological Research Applications for Public Health Measurement and Intervention (pp. 25-39). IGI Global.

Barua, R., Datta, S., Sengupta, P., Chowdhury, A. R., & Datta, P. (2021). Advances in MEMS micropumps and their emerging drug delivery and biomedical applications. In Advances and Challenges in Pharmaceutical Technology. Academic Press. doi:10.1016/B978-0-12-820043-8.00002-5

Barua, R., Datta, S., Datta, P., & Chowdhury, A. R. (2022). The Study of Traditional Medicine for the Treatment of COVID-19. In *Handbook of Research on Pathophysiology and Strategies for the Management of COVID-19* (pp. 221–241). IGI Global. doi:10.4018/978-1-7998-8225-1.ch014

Barua, R., Datta, S., Dutta, P., & Chowdhury, A. R. (2021). Machine learning and AI application of COVID-19 pandemic outbreak. *Impacts and Implications of COVID-19*, 2.

Barua, R., Datta, S., RoyChowdhury, A., & Datta, P. (2021). A Review on the epidemiology and characteristics of COVID-19. *Journal of Current Biomedical Reports*, *2*(3), 95–106. doi:10.52547/10.52547/JCBioR.2.3.95

Basiri, A., Heidari, A., Nadi, M. F., Fallahy, M. T. P., Nezamabadi, S. S., Sedighi, M., Saghazadeh, A., & Rezaei, N. (2021). Microfluidic devices for detection of RNA viruses. *Reviews in Medical Virology*, *31*(1), 1–11. doi:10.1002/rmv.2154 PMID:32844526

Bau, H. H., Zhong, J., & Yi, M. (2001). A minute magneto hydro dynamic (MHD) mixer. *Sensors and Actuators. B, Chemical*, *79*(2–3), 207–215. doi:10.1016/S0925-4005(01)00851-6

Bayareh, M., Ashani, M. N., & Usefian, A. (2020). Active and passive micromixers: A comprehensive review. *Chemical Engineering and Processing*, *147*, 107771. doi:10.1016/j.cep.2019.107771

Belliveau, N. M., Huft, J., Lin, P. J., Chen, S., Leung, A. K., Leaver, T. J., Wild, A. W., Lee, J. B., Taylor, R. J., Tam, Y. K., Hansen, C. L., & Cullis, P. R. (2012). Microfluidic synthesis of highly potent limit-size lipid nanoparticles for in vivo delivery of siRNA. *Molecular Therapy. Nucleic Acids*, *1*(8), e37. Advance online publication. doi:10.1038/mtna.2012.28 PMID:23344179

Bemetz, J., Wegemann, A., Saatchi, K., Haase, A., Häfeli, U. O., Niessner, R., Gleich, B., & Seidel, M. (2018). Microfluidic-Based Synthesis of Magnetic Nanoparticles Coupled with Miniaturized NMR for Online Relaxation Studies. *Analytical Chemistry*, *90*(16), 9975–9982. Advance online publication. doi:10.1021/acs.analchem.8b02374 PMID:30044615

Benard, W., Kahn, H., Heuer, A., & Huff, M. (1998). Thin-film shape memory alloy actuated micropumps. *Journal of Microelectromechanical Systems*, *7*(2), 245–251. doi:10.1109/84.679390

Berg, A. V. D., & Lammerink, T. S. J. (1998). Micro total analysis systems: Microfluidic aspects, integration concept and applications in Microsystem Technology in Chemistry and Life Science. *Topics in Current Chemistry*, *194*, 21–49. doi:10.1007/3-540-69544-3_2

Berthier J, Brakke KA, Berthier E (2016-08-01). *Open Microfluidics*. Wiley. doi:10.1002/9781118720936

Beshana, S., Hussen, A., Leta, S., & Kaneta, T. (2022). Dispersive Liquid–Liquid Microextraction Coupled with Microfluidic Paper-Based Analytical Device for the Determination of Organophosphate and Carbamate Pesticides in the Water Sample. *Analytical Sciences*, *38*(10), 1359–1367. doi:10.100744211-022-00167-7 PMID:35908131

Bharti, A., Turchet, A., & Marmiroli, B. (2022). X-ray lithography for nanofabrication: Is there a future? *Frontiers in Nanotechnology*, *4*, 835701. doi:10.3389/fnano.2022.835701

Bhattacharya, S., Kumar, S., & Agarwal, A. (2019). *Paper Microfluidics- Theory and Applications*. Academic Press.

Bhattacharyya, A., & Klapperich, C. M. (2007). Design and testing of a disposable microfluidic chemiluminescent immunoassay for disease biomarkers in human serum samples. *Biomedical Microdevices*, *9*(2), 245–251. doi:10.100710544-006-9026-2 PMID:17165125

Bhattacharyya, R., Bhatt, G., Basu, A., & Basu, A. K. (2021). *Application of Polymers in Biomems Biomedical Devices and Related Challenges*. AIP Publishing. doi:10.1063/9780735423954_004

Bigdeli, A., Ghasemi, F., Abbasi-Moayed, S., Shahrajabian, M., Fahimi-Kashani, N., Jafarinejad, S., Farahmand Nejad, M. A., & Hormozi-Nezhad, M. R. (2019). Ratiometric Fluorescent Nanoprobes for Visual Detection: Design Principles and Recent Advances - A Review. *Analytica Chimica Acta, 1079*, 30–58. doi:10.1016/j.aca.2019.06.035 PMID:31387719

Bohr, A., Colombo, S., & Jensen, H. (2019). Future of microfluidics in research and in the market. In Micro and Nano Technologies, Microfluidics for Pharmaceutical Applications. William Andrew Publishing. doi:10.1016/B978-0-12-812659-2.00016-8

Bojang, A. A., & Wu, H. S. (2020). Design, fundamental principles of fabrication and applications of microreactors. *Processes (Basel, Switzerland), 8*(8), 891. Advance online publication. doi:10.3390/pr8080891

Boken, J., Soni, S. K., & Kumar, D. (2016). Microfluidic Synthesis of Nanoparticles and their Biosensing Applications. *Critical Reviews in Analytical Chemistry, 46*(6), 538–561. doi:10.1080/10408347.2016.1169912 PMID:27070948

Bonk, S., Stubbe, M., Buehler, S., Tautorat, C., Baumann, W., Klinkenberg, E.-D., & Gimsa, J. (2015). Design and characterization of a sensorized microfluidic cell-culture system with electrothermal micro-pumps and sensors for cell adhesion, oxygen, and pH on a glass chip. *Biosensors (Basel), 5*(3), 513–536. doi:10.3390/bios5030513 PMID:26263849

Boonkaew, S., Yakoh, A., Chuaypen, N., Tangkijvanich, P., Rengpipat, S., & Chailapakul, O. (2021). An automated fast-flow/delayed paper-based platform for the simultaneous electrochemical detection of hepatitis B virus and hepatitis C virus core antigen. *Biosensors & Bioelectronics, 193*, 113543. doi:10.1016/j.bios.2021.113543 PMID:34416431

Bouaidat, S., Hansen, O., Bruus, H., Berendsen, C., Bau-Madsen, N.K., & Thomsen, P. (2005). Surface-directed capillary system; theory, experiments and applications. *Lab on a Chip, 5* (8), 827–836. . doi:10.1039/b502207j

Bouchaala, A., Jaber, N., Yassine, O., Shekhah, O., Chernikova, V., Eddaoudi, M., & Younis, M. (2016). Nonlinear-based MEMS sensors and active switches for gas detection. *Sensors (Basel), 16*(6), 758. doi:10.339016060758 PMID:27231914

Bragheri, F., Martinez Vazquez, R. & Osellame, R. (2016). Microfluidics. *Three-Dimensional Microfabrication Using Two-Photon Polymerization: Fundamentals, Technology, and Applications,* 310–334. William Andrew. . doi:10.1016/B978-0-323-35321-2.00016-9

Braun, D. (2005). *Polymer Synthesis: Theory and Practice—Fundamentals, Methods, Experiments* (4th ed.). Springer.

Breslauer, D. N., Lee, P. J., & Lee, L. P. (2006). Microfluidics-based systems biology. *Molecular BioSystems, 2*(2), 97–112. doi:10.1039/b515632g PMID:16880927

Bressan, L. P., Robles-Najar, J., Adamo, C. B., Quero, R. F., Costa, B. M. C., de Jesus, D. P., & da Silva, J. A. F. (2019). 3D-printed microfluidic device for the synthesis of silver and gold nanoparticles. *Microchemical Journal, 146*, 1083–1089. doi:10.1016/j.microc.2019.02.043

Bresser, D., Mueller, F., Fiedler, M., Krueger, S., Kloepsch, R., Baither, D., Winter, M., Paillard, E., & Passerini, S. (2013). Transition-Metal-Doped Zinc Oxide Nanoparticles as a New Lithium-Ion Anode Material. *Chemistry of Materials, 25*(24), 4977–4985. doi:10.1021/cm403443t

Brling, K. (2011). UV-Induced Fluorescence Spectra and Lifetime Determination for Detection of Leaf Rust (Puccinia Triticina) in Susceptible and Resistant Wheat (Triticum Aestivum) Cultivars. *Functional Plant Biology, 38*(4), 337–345. doi:10.1071/FP10171 PMID:32480889

Brown, W. L., Venkatesan, T., & Wagner, A. (1981). Ion beam lithography. *Nuclear Instruments and Methods in Physics Research, 191*(1-3), 157–168. doi:10.1016/0029-554X(81)90999-X

Bruck, E. (2017). *Handbook of Magnetic Materials*. Elsevier.

Bußmann, A. B., Durasiewicz, C. P., Sebastian, H. A. K., & Wald, C. K. (2021). Piezoelectric titanium based microfluidic pump and valves for implantable medical applications. *Sensors and Actuators A: Physical, 323*. doi:10.1016/j.sna.2021.112649

Cai, G., Xue, L., Zhang, H., & Lin, J. (2017). A Review on Micromixers. *Micromachines*, *8*(9), 274. doi:10.3390/mi8090274 PMID:30400464

Cameron, N. S., Roberge, H., Veres, T., Jakeway, S. C., & Crabtree, H. J. (2006). High fidelity, high yield production of microfluidic devices by hot embossing lithography: Rheology and stiction. *Lab on a Chip*, *6*(7), 936–941. doi:10.1039/b600584e PMID:16804599

Canhete de Moraes, N., Marques Petroni, J., de Lima, F., Souza Ferreira, V., & Gabriel Lucca, B. (2022). Paper-based electrochemical platform modified with graphene nanoribbons: A new and affordable approach for analysis of 5-hydroxy-l-tryptophan. *Microchemical Journal*, *183*, 108030. doi:10.1016/j.microc.2022.108030

Cao, L., Fang, C., Zeng, R., Zhao, X., Jiang, Y., & Chen, Z. (2017). Paper-based microfluidic devices for electrochemical immunofiltration analysis of human chorionic gonadotropin. *Biosensors & Bioelectronics*, *92*, 87–94. doi:10.1016/j.bios.2017.02.002 PMID:28189070

Capretto, L., Cheng, W., Hill, M., & Zhang, X. (2011). *Micromixing Within Microfluidic Devices*. doi:10.1007/128_2011_150

Capretto, L., Mazzitelli, S., Brognara, E., Lampronti, I., Carugo, D., Hill, M., Zhang, X., Gambari, R., & Nastruzzi, C. (2012). Mithramycin encapsulated in polymeric micelles by microfuidic technology as novel therapeutic protocol for beta-thalassemia. *International Journal of Nanomedicine*, *7*, 307–324. PMID:22287841

Carroll, N. J., Rathod, S. B., Derbins, E., Mendez, S., Weitz, D. A., & Petsev, D. N. (2008). Droplet-based microfluidics for emulsion and solvent evaporation synthesis of monodisperse mesoporous silica microspheres. *Langmuir*, *24*(3), 658–661. doi:10.1021/la7032516 PMID:18171093

Carter, S. B. (1967). Haptotactic islands: A method of confining single cells to study individual cell reactions and clone formation. *Experimental Cell Research*, *48*(1), 189–193. doi:10.1016/0014-4827(67)90298-4 PMID:4862713

Carvalho, B. G., Ceccato, B. T., Michelon, M., Han, S. W., & de la Torre, L. G. (2022). Advanced Microfluidic Technologies for Lipid Nano-Microsystems from Synthesis to Biological Application. *Pharmaceutics*, *14*(1), 141. doi:10.3390/pharmaceutics14010141 PMID:35057037

Cate, D. M., Adkins, J. A., Mettakoonpitak, J., & Henry, C. S. (2015). Recent Developments in Paper-Based Microfluidic Devices. *Analytical Chemistry*, *87*(1), 19–41. doi:10.1021/ac503968p PMID:25375292

Chacón-Torres, J. C., Reinoso, C., Navas-León, D. G., Briceño, S., & González, G. (2020). Optimized and scalable synthesis of magnetic nanoparticles for RNA extraction in response to developing countries' needs in the detection and control of SARS-CoV-2. *Scientific Reports*, *10*(1), 19004. doi:10.103841598-020-75798-9 PMID:33149153

Chaibun, T., Puenpa, J., Ngamdee, T., Boonapatcharoen, N., Athamanolap, P., O'Mullane, A. P., Vongpunsawad, S., Poovorawan, Y., Lee, S. Y., & Lertanantawong, B. (2021). Rapid electrochemical detection of coronavirus SARS-CoV-2. *Nature Communications*, *12*(1), 802. doi:10.103841467-021-21121-7 PMID:33547323

Chai, Y., Wang, Y., Li, B., Qi, W., Su, R., & He, Z. (2021). Microfluidic Synthesis of Lignin/Chitosan Nanoparticles for the pH-Responsive Delivery of Anticancer Drugs. *Langmuir*, *37*(23), 7219–7226. doi:10.1021/acs.langmuir.1c00778 PMID:34078082

Chaiyo, S., Mehmeti, E., Siangproh, W., Hoang, T. L., Nguyen, H. P., Chailapakul, O., & Kalcher, K. (2018). Non-enzymatic electrochemical detection of glucose with a disposable paper-based sensor using a cobalt phthalocyanine-ionic liquid-graphene composite. *Biosensors & Bioelectronics*, *102*, 113–120. doi:10.1016/j.bios.2017.11.015 PMID:29128713

Chakraborty, S., Gupta, A. K., Roy, D., & Basumallick, A. (2019). Studies on nano-metal dispersed Cu-Cr matrix composite. *Materials Letters*, *257*, 126739. Advance online publication. doi:10.1016/j.matlet.2019.126739

Chakraborty, S., Kumar, A., & Sen, P. (2018). Special Issue on Microfluidics: Theory and Applications. *Journal of the Indian Institute of Science*, *98*(2), 83–84. doi:10.100741745-018-0078-0

Chandra, S., Bano, D., Sahoo, K., Kumar, D., Kumar, V., Kumar Yadav, P., & Hadi Hasan, S. (2022). Synthesis of Fluorescent Carbon Quantum Dots from Jatropha Fruits and Their Application in Fluorometric Sensor for the Detection of Chlorpyrifos. *Microchemical Journal*, *172*, 106953. doi:10.1016/j.microc.2021.106953

Chandrasekhar, S. (1981). *Hydrodynamic and Hydromagnetic Stability Dover.*

Chang, H. C., & Yeo, L. (2009). *Electrokinetically Driven Microfluidics and Nanofluidics*. Cambridge University Press.

Chang, Y., Hasan, D., Dong, B., Wei, J., Ma, Y., Zhou, G., Ang, K., & Lee, C. (2018). All-dielectric surface-enhanced infrared absorption-based gas sensor using guided resonance. *ACS Applied Materials & Interfaces*, *10*(44), 38272–38279. doi:10.1021/acsami.8b16623 PMID:30360088

Chan, J. F. W. (2020). Simulation of the Clinical and Pathological Manifestations of Coronavirus Disease 2019 (COVID-19) in a Golden Syrian Hamster Model: Implications for Disease Pathogenesis and Transmissibility. *Clinical Infectious Diseases*, *71*, 2428–2446. doi:10.1093/cid/ciaa644 PMID:32215622

Chappel, E., & Dumont-Fillon, D. (2021). Micropumps for drug delivery. In *Developments in Biomedical Engineering and Bioelectronics. Drug Delivery Devices and Therapeutic Systems.* Academic Press. doi:10.1016/B978-0-12-819838-4.00015-8

Chee, P. S., Minjal, M. N., Leow, P. L., & Ali, M. S. M. (2015). Wireless powered thermo-pneumatic micropump using frequency-controlled heater. *Sensors and Actuators. A, Physical*, *233*, 1–8. doi:10.1016/j.sna.2015.06.017

Chee, P. S., Nafea, M., Leow, P. L., & Ali, M. S. M. (2016). Thermal analysis of wirelessly powered thermo-pneumatic micropump based on planar LC circuit. *Journal of Mechanical Science and Technology*, *30*(6), 2659–2665. doi:10.100712206-016-0527-5

Cheheltani, R., Ezzibdeh, R. M., Chhour, P., Pulaparthi, K., Kim, J., Jurcova, M., Hsu, J. C., Blundell, C., Litt, H. I., Ferrari, V. A., Allcock, H. R., Sehgal, C. M., & Cormode, D. P. (2016). Tunable, biodegradable gold nanoparticles as contrast agents for computed tomography and photoacoustic imaging. *Biomaterials*, *102*, 87–97. Advance online publication. doi:10.1016/j.biomaterials.2016.06.015 PMID:27322961

Chen, C. H., & Che, H. L. (2017). Mass Spectrometry Pesticide Screening with Paper-Based Microfluidic Cassette for 2D Paper Chromatography and Electrospray Ionization. *TRANSDUCERS 2017 - 19th International Conference on Solid-State Sensors, Actuators and Microsystems*, 1483–86. 10.1109/TRANSDUCERS.2017.7994339

Chen, C. H., & Santiago, J. (2002). A Planar Electroosmotic Micropump. *Journal of Microelectromechanical Systems*, *11*(6), 672–683. doi:10.1109/JMEMS.2002.805055

Chen, G., Zhu, X., Xing, C., Wang, Y., Xu, X., Bao, J., Huang, J., Zhao, Y., Wang, X., Zhou, X., Du, X., & Wang, X. (2023). Machine Learning-Assisted Microfluidic Synthesis of Perovskite Quantum Dots. *Advanced Photonics Research*, *4*(1), 2200230. Advance online publication. doi:10.1002/adpr.202200230

Cheng, X., Liu, Y. S., Irimia, D., Demirci, U., Yang, L., Zamir, L., Rodríguez, W. R., Toner, M., & Bashir, R. (2007). Cell detection and counting through cell lysate impedance spectroscopy in microfluidic devices. *Lab on a Chip*, *7*(6), 746–755. doi:10.1039/B705082H PMID:17538717

Chen, H.-Y., Elkasabi, Y., & Lahann, J. (2006). Surface modification of confined microgeometries via vapor-deposited polymer coatings. *Journal of the American Chemical Society*, *128*(1), 374–380. doi:10.1021/ja057082h PMID:16390168

Chen, J. M., Huang, P.-C., & Lin, M.-G. (2007). Analysis and experiment of capillary valves for microfluidics on a rotating disk. *Microfluidics and Nanofluidics*, *4*(5), 427–437. doi:10.100710404-007-0196-x

Chen, Q., & Fung, Y. (2010). Capillary Electrophoresis with Immobilized Quantum Dot Fluorescence Detection for Rapid Determination of Organophosphorus Pesticides in Vegetables. *Electrophoresis*, *31*(18), 3107–3114. doi:10.1002/elps.201000260 PMID:20872612

Chen, T., Yao, Q., Nasaruddin, R. R., & Xie, J. (2019). Electrospray Ionization Mass Spectrometry: A Powerful Platform for Noble-Metal Nanocluster Analysis. *Angewandte Chemie International Edition*, *58*(35), 11967–11977. doi:10.1002/anie.201901970 PMID:30924248

Chen, T., Zhao, M., Shi, Q., Yang, Z., Liu, H., Sum, L., Ouyang, J., & Lee, C. (2018). Novel augmented reality interface using a self-powered triboelectric based virtual reality 3D-control sensor. *Nano Energy*, *51*, 162–172. doi:10.1016/j.nanoen.2018.06.022

Chen, W., Yan, M., Yang, L., Ding, B., He, B., Wang, Y., Liu, X., Liu, C., Zhu, H., You, B., Huang, S., Zhang, J., Mu, F., Xiang, Z., Feng, X., Wen, J., Fang, J., Yu, J., Yang, H., & Wang, J. (2005). Coronavirus Transmitted from Human to Pig. *Emerging Infectious Diseases*, *11*(3), 2–4. doi:10.3201/eid1103.040824 PMID:15757562

Chen, Z., Zhang, Z., Zhai, X., Li, Y., Lin, L., Zhao, H., Bian, L., Li, P., Yu, L., Wu, Y., & Lin, G. (2020). Rapid and Sensitive Detection of anti-SARS-CoV-2 IgG, Using Lanthanide-Doped Nanoparticles-Based Lateral Flow Immunoassay. *Analytical Chemistry*, *92*(10), 7226–7231. doi:10.1021/acs.analchem.0c00784 PMID:32323974

Cheong, H. R., Lai, K. C., & Chee, P. S. (2018). A wireless powered electroactive polymer using magnetic resonant coupling. *IOP Conf Ser Mater Sci Eng, 409*, 1-6. 10.1088/1757-899X/409/1/012002

Chikkaveeraiah, B. V., Mani, V., Patel, V., Gutkind, J. S., & Rusling, J. F. (2011). Microfluidic electrochemical immunoarray for ultrasensitive detection of two cancer biomarker proteins in serum. *Biosensors & Bioelectronics*, *26*(11), 4477–4483. doi:10.1016/j.bios.2011.05.005 PMID:21632234

Chin, C. D., Laksanasopin, T., Cheung, Y. K., Steinmiller, D., Linder, V., Parsa, H., Wang, J., Moore, H., Rouse, R., Umviligihozo, G., Karita, E., Mwambarangwe, L., Braunstein, S. L., van de Wijgert, J., Sahabo, R., Justman, J. E., El-Sadr, W., & Sia, S. K. (2011). Microfluidics-based diagnostics of infectious diseases in the developing world. *Nature Medicine*, *17*(8), 1015–1019. doi:10.1038/nm.2408 PMID:21804541

Chircov, C., Bîrcă, A. C., Vasile, B. S., Oprea, O. C., Huang, K. S., & Grumezescu, A. M. (2022). Microfluidic Synthesis of -NH2- and -COOH-Functionalized Magnetite Nanoparticles. *Nanomaterials (Basel, Switzerland)*, *12*(18), 3160. Advance online publication. doi:10.3390/nano12183160 PMID:36144948

Cho, M., Zhu, J., Kang, K., & Park, I. (2019). Half-pipe palladium nanotube network hydrogen sensor based on electrospun nanofiber scaffolds. *Proceedings of the 2019 20th International Conference on Solid-State Sensors, Actuators and Microsystems & Eurosensors*, 1183–1186. 10.1109/TRANSDUCERS.2019.8808214

Cho, I., Kang, K., Yang, D., Yun, J., & Park, I. (2017). Localized liquid-phase synthesis of porous SnO_2 nanotubes on MEMS platform for low-power, high performance gas sensors. *ACS Applied Materials & Interfaces*, *9*(32), 27111–27119. doi:10.1021/acsami.7b04850 PMID:28714311

Cho, I., & Yoon, E. (2010). Design and fabrication of a single membrane push-pull SPDT RF MEMS switch operated by electromagnetic actuation and electrostatic hold. *Journal of Micromechanics and Microengineering*, *20*(3), 035028. doi:10.1088/0960-1317/20/3/035028

Chokkalingam, V., Weidenhof, B., Krämer, M., Maier, W. F., Herminghaus, S., & Seemann, R. (2010, July). Optimizeddroplet-basedmicrofluidicsschemeforol-gelreactions. *Lab on a Chip*, *10*(13), 17001705. doi:10.1039/b926976b PMID:20405061

Chorsi, M. T., Curry, E. J., Chorsi, H. T., Das, R., Baroody, J., Purohit, P. K., Ilies, H., & Nguyen, T. D. (2019). Piezoelectric Biomaterials for Sensors and Actuators. *Advanced Materials*, *31*(1), e1802084. doi:10.1002/adma.201802084 PMID:30294947

Cho, S. K., Zhao, Y., & Kim, C. J. (2007). Concentration and binary separation of micro particles for droplet-based digital microfluidics. *Lab on a Chip*, *7*(4), 490–498. doi:10.1039/b615665g PMID:17389966

Chou, S. Y., Krauss, P. R. & Renstrom, P. J. (1996). Imprint lithography with 25-nanometer resolution. *Science, 272*, 85–87.

Chowdhury, S. R., Mukherjee, P., & Bhattachrya, S. (2016). Palladium and palladium–copper alloy nano particles as superior catalyst for electrochemical oxidation of methanol for fuel cell applications. *International Journal of Hydrogen Energy*, *41*(38), 17072–17083. doi:10.1016/j.ijhydene.2016.05.239

Chung, B. G., Manbachi, A., Saadi, W., Lin, F., Jeon, N. L., & Khademhosseini, A. (2007). A gradient-generating microfluidic device for cell biology. *Journal of Visualized Experiments*, *7*(7), 271. doi:10.3791/271 PMID:18989442

Chung, C. K., Shih, T. R., Chang, C. K., Lai, C. W., & Wu, B. H. (2011). Design and experiments of a short-mixing-length baffled microreactor and its application to microfluidic synthesis of nanoparticles. *Chemical Engineering Journal*, *168*(2), 790–798. doi:10.1016/j.cej.2010.12.035

Chungchai, W., Amatatongchai, M., Meelapsom, R., Seebunrueng, K., Suparsorn, S., & Jarujamrus, P. (2020). Development of a Novel Three-Dimensional Microfluidic Paper-Based Analytical Device (3D-MPAD) for Chlorpyrifos Detection Using Graphene Quantum-Dot Capped Gold Nanocomposite for Colorimetric Assay. *International Journal of Environmental Analytical Chemistry*, *100*(10), 1160–1178. doi:10.1080/03067319.2019.1650921

Cinti, S., Cusenza, R., Moscone, D., & Arduini, F. (2018). Paper-based synthesis of Prussian Blue Nanoparticles for the development of whole blood glucose electrochemical biosensor. *Talanta*, *187*, 59–64. doi:10.1016/j.talanta.2018.05.015 PMID:29853066

Cinti, S., Minotti, C., Moscone, D., Palleschi, G., & Arduini, F. (2017). Fully Integrated Ready-to-Use Paper-Based Electrochemical Biosensor to Detect Nerve Agents. *Biosensors & Bioelectronics*, *93*, 46–51. doi:10.1016/j.bios.2016.10.091 PMID:27839735

Cioffi, A., Mancini, M., Gioia, V., & Cinti, S. (2021). Office Paper-Based Electrochemical Strips for Organophosphorus Pesticide Monitoring in Agricultural Soil. *Environmental Science & Technology*, *55*(13), 8859–8865. doi:10.1021/acs.est.1c01931 PMID:34165948

Clark, J., Kaufman, M., & Fodor, P. (2018). Mixing Enhancement in Serpentine Micromixers with a Non-Rectangular Cross-Section. *Micromachines*, *9*(3), 107. doi:10.3390/mi9030107 PMID:30424041

Colozza, N., Tazzioli, S., Sassolini, A., Agosta, L., di Monte, M. G., Hermansson, K., & Arduini, F. (2021). Vertical-Flow Paper Sensor for On-Site and Prompt Evaluation of Chloride Contamination in Concrete Structures. *Analytical Chemistry*, *93*(43), 14369–14374. doi:10.1021/acs.analchem.1c03363 PMID:34669396

Cotta, M. A. (2020). Quantum Dots and Their Applications: What Lies Ahead? *ACS Applied Nano Materials*, *3*(6), 4920–4924. doi:10.1021/acsanm.0c01386

Crits-Christoph, A., Kantor, R. S., Olm, M. R., Whitney, O. N., Al-Shayeb, B., Lou, Y. C., Flamholz, A., Kennedy, L. C., Greenwald, H., Hinkle, A., Hetzel, J., Spitzer, S., Koble, J., Tan, A., Hyde, F., Schroth, G., Kuersten, S., Banfield, J. F., & Nelson, K. L. (2021). Genome Sequencing of Sewage Detects Regionally Prevalent SARS-CoV-2 Variants. *mBio*, *12*(1), e02703–e02720. doi:10.1128/mBio.02703-20 PMID:33468686

Cros, F. (2021). BioMEMS. *Biomedical Materials (Bristol, England)*, 581–620.

Cuchiara, M. P., Allen, A. C. B., Chen, T. M., Miller, J. S., & West, J. L. (2010). Multilayer microfluidic PEGDA hydrogels. *Biomaterials*, *31*(21), 5491–5497. doi:10.1016/j.biomaterials.2010.03.031 PMID:20447685

Cunningham, J. C., Brenes, N. J., & Crooks, R. M. (2014). Paper electrochemical device for detection of DNA and thrombin by target-induced conformational switching. *Analytical Chemistry*, *86*(12), 6166–6170. doi:10.1021/ac501438y PMID:24871788

Dabbagh, S. R., Becher, E., Ghaderinezhad, F., Havlucu, H., Ozcan, O., Ozkan, M., Yetisen, A. K., & Tasoglu, S. (2021). Increasing the Packing Density of Assays in Paper-Based Microfluidic Devices. *Biomicrofluidics*, *15*(1), 011502. doi:10.1063/5.0042816 PMID:33569089

Dai, R., Li, G., Xiao, L., Li, Y., Cui, Z., Jia, L., Zhou, M., Song, Y., Yang, Y., Cai, Y., Chen, C., & Yin, K. (2022). A droplet-driven micro-surfboard with dual gradients for programmable motion. *Chemical Engineering Journal*, *446*, 136874. doi:10.1016/j.cej.2022.136874

Daoudi, K., Gaidi, M., & Columbus, S. (2017). Biointerface Research in Applied Chemistry. *Ultrasound*, *69*, 70. doi:10.3390/mi13020164

Das, J., Ivanov, I., Sargent, E. H., & Kelley, S. O. (2016). DNA Clutch Probes for Circulating Tumor DNA Analysis. *Journal of the American Chemical Society*, *138*(34), 11009–11016. doi:10.1021/jacs.6b05679 PMID:27513828

Das, S., Gagandeep, & Bhatia, R. (2022). Paper-based microfluidic devices: Fabrication, detection, and significant applications in various fields. *Reviews in Analytical Chemistry*, *41*(1), 112–136. doi:10.1515/revac-2022-0037

Davis. (2018). Pneumatic Actuators. *Actuators*, *7*, 62.

Dayanandan, N., Kapoor, A., & Sivaraman, P. (2021). Studies on membrane distillation towards mitigating thermal pollution. *Chemical Papers*, *75*(6), 2819–2833. doi:10.100711696-021-01525-x

DebiotechS. A. (n.d.). https://www.debiotech.com/

Dejace, L., Laubeuf, N., Furfaro, I., & Lacour, S. (2019). Gallium-based thin films for wearable human Motion sensors. *Advanced Intelligent Systems*, *1*(5), 1900079. doi:10.1002/aisy.201900079

Demierre, N., Braschler, T., Muller, R., & Renaud, P. (2008). Focusing and continuous separation of cells in a microfluidic device using lateral dielectrophoresis. *Sensors and Actuators. B, Chemical*, *132*(2), 388–396. doi:10.1016/j.snb.2007.09.078

Demirci, U., & Montesano, G. (2007). Single cell epitaxy by acoustic picolitre droplets. *Lab on a Chip*, *7*(9), 1139–1145. doi:10.1039/b704965j PMID:17713612

Dempsey, E., Diamond, D., Smyth, M. R., Urban, G., Jobst, G., Moser, I., Verpoorte, E. M., Manz, A., Widmer, H. M., & Rabenstein, K. (1997). Design and development of a miniaturised total chemical analysis system for on-line lactate and glucose monitoring in biological samples. *Analytica Chimica Acta*, *346*(3), 341–349. doi:10.1016/S0003-2670(97)90075-1

Dentinger, P. M., Clift, W. M., & Goods, S. H. (2002). Removal of SU-8 photoresist for thick film applications. *Microelectronic Engineering, 61*, 993–1000. doi:10.1016/S0167-9317(02)00490-2

Desai, S. P., Taff, B. M., & Voldman, J. (2007). A photopatternable silicone for BioMEMS applications. Proc. Micro Total Anal. Syst.(-TAS).

Deshmukh, S. P., Patil, S. M., Mullani, S. B., & Delekar, S. D. (2019). Silver nanoparticles as an effective disinfectant: A review. *Materials Science and Engineering C, 97*, 954–965. doi:10.1016/j.msec.2018.12.102 PMID:30678983

Dewangan, L., Korram, J., Karbhal, I., Nagwanshi, R., Ghosh, K. K., Pervez, S., & Satnami, M. L. (2022). Alkaline Phosphatase Immobilized CdTe/ZnS Quantum Dots for Dual-Purpose Fluorescent and Electrochemical Detection of Methyl Paraoxon. *Industrial & Engineering Chemistry Research, 61*(10), 3636–3646. doi:10.1021/acs.iecr.1c05065

Dhakar, L., Gudla, S., Shan, X., Wang, Z., Tay, F. E. H., Heng, C.-H., & Lee, C. (2016). Large scale triboelectric nanogenerator and self-powered pressure sensor array using low cost roll-to-roll UV embossing. *Scientific Reports, 6*(1), 22253. doi:10.1038rep22253 PMID:26905285

Dhouha, J., Marcoccio, E., Moscone, D., Dridi, C., & Arduini, F. (2020). Highly sensitive paper-based electrochemical sensor for reagent free detection of bisphenol A. *Talanta, 216*, 120924. doi:10.1016/j.talanta.2020.120924 PMID:32456933

Diercks, A. H., Ozinsky, A., Hansen, C. L., Spotts, J. M., Rodriguez, D. J., & Aderem, A. (2009). A microfluidic devicefor multiplexed protein detection in nano-liter volumes. *Analytical Biochemistry, 386*(1), 30–35. doi:10.1016/j.ab.2008.12.012 PMID:19133224

Dincer, C., Bruch, R., Kling, A., Dittrich, P. S., & Urban, G. A. (2017). Multiplexed Point-of-Care Testing – xPOCT. *Trends in Biotechnology, 35*(8). doi:10.1016/j.tibtech.2017.03.013

Ding, J., Li, B., Chen, L., & Qin, W. (2016). A Three-Dimensional Origami Paper-Based Device for Potentiometric Biosensing. *Angewandte Chemie International Edition, 55*(42), 13033–13037. doi:10.1002/anie.201606268 PMID:27634584

Din, M. I., & Rehan, R. (2017). Synthesis, Characterization, and Applications of Copper Nanoparticles. *Analytical Letters, 50*(1), 50–62. doi:10.1080/00032719.2016.1172081

Dinnes, J., Deeks, J. J., Berhane, S., Taylor, M., Adriano, A., Davenport, C., Dittrich, S., Emperador, D., Takwoingi, Y., Cunningham, J., Beese, S., Domen, J., Dretzke, J., Ferrante di Ruffano, L., Harris, I. M., Price, M. J., Taylor-Phillips, S., Hooft, L., Leeflang, M. M., ... Van den Bruel, A.Cochrane COVID-19 Diagnostic Test Accuracy Group. (2021). Rapid, point-of-care antigen and molecular-based tests for diagnosis of SARS-CoV-2 infection. *Cochrane Database of Systematic Reviews, 3*(3), CD013705. Advance online publication. doi:10.1002/14651858.CD013705.pub2 PMID:33760236

Dobhal, A., Kulkarni, A., Dandekar, P., & Jain, R. (2017). A microreactor-based continuous process for controlled synthesis of poly-methyl-methacrylate-methacrylic acid (PMMA) nanoparticles. *Journal of Materials Chemistry. B, Materials for Biology and Medicine, 5*(18), 3404–3417. doi:10.1039/C7TB00560A PMID:32264406

Dong, J. S., Liu, R. G., Liu, W. S., Chen, Q. Q., Yang, Y., Wu, Y., Yang, Z. G., & Lin, B. S. (2017). Design of a piezoelectric pump with dual vibrators. *Sensors and Actuators. A, Physical, 257*, 165–172. doi:10.1016/j.sna.2017.02.001

Dressaire, E., & Sauret, A. (2017). Clogging of microfluidic systems. *Soft Matter, 13*(1), 37–48. doi:10.1039/C6SM01879C PMID:27801463

Dukle, A., Nathanael, A. J., Panchapakesan, B., & Oh, T. H. (2022). Role of Paper-Based Sensors in Fight against Cancer for the Developing World. *Biosensors (Basel), 12*(9), 737. doi:10.3390/bios12090737 PMID:36140122

Dumont-Fillon, D., Lamaison, D., & Chappel, E. (2020). Design and characterization of 3-stack MEMS-based passive flow regulators for implantable and ambulatory infusion pumps. *Journal of Microelectromechanical Systems*, *29*(2), 1–12. doi:10.1109/JMEMS.2019.2962107

Dungchai, W., Chailapakul, O., & Henry, C. S. (2009). Electrochemical Detection for Paper-Based Microfluidics. *Analytical Chemistry*, *81*(14), 5821–5826. doi:10.1021/ac9007573 PMID:19485415

Dutse, S. W., & Yusof, N. A. (2011). Microfluidics-based lab-on-chip systems in DNA-based biosensing: An overview. *Sensors (Basel)*, *11*(6), 5754–5768. doi:10.3390110605754 PMID:22163925

Duty & Sandborn. (2011). *Wet Isotropic and Anisotropic Etching of Silicon Silicon wafer etching, a process used in the fabrication of microelectronic devices, is the removal*. Academic Press.

Duwairi, H., & Abdullah, M. (2006). Thermal and flow analysis of a magneto-hydrodynamic micropump. *Microsystem Technologies*, *13*(1), 33–39. doi:10.100700542-006-0258-0

Eduati F, Utharala R, Madhavan D, Neumann UP, Longerich T, Cramer T, et al. (June 2018). A microfluidics platform for combinatorial drug screening on cancer biopsies. *Nature Communications*, *9*(1), 2434. . doi:10.1038/s41467-018-04919-w

Ehrfeld, W., Hessel, V., & Lehr, H. (1998). *Microsystem Technology in Chemistry and Life Science* (Vol. 194). Springer-Verlag. doi:10.1007/3-540-69544-3_10

Ehsani & Nejat. (2017). Conceptual design and performance analysis of a novel flexible-valve micropump using magnetofluid–solid interaction. *Smart Mater Struct, 26.*

Ellerbee, A. K., Phillips, S. T., Siegel, A. C., Mirica, K. A., Martinez, A. W., Striehl, P., Jain, N., Prentiss, M., & Whitesides, G. M. (2009). Quantifying colorimetric assays in paper-based microfluidic devices by measuring the transmission of light through paper. *Analytical Chemistry*, *81*(20), 8447–8452. doi:10.1021/ac901307q PMID:19722495

Engel, J. M., Chen, J., Bullen, D., & Liu, C. (2005). *18th IEEE International Conference on Micro Electro Mechanical Systems, MEMS*. IEEE.

Engelking, L. R. (2015). Enzyme Kinetics. In *Textbook of Veterinary Physiological Chemistry* (3rd ed.). Academic Press. https://www.sciencedirect.com/science/article/pii/B9780123919090500062

Erickson, D., & Li, D. (2004). Integrated microfluidic devices. *Analytica Chimica Acta*, *507*(1), 11–26. doi:10.1016/j.aca.2003.09.019

Eriksson, E., Sott, K., Lundqvist, F., Sveningsson, M., Scrimgeour, J., & Hanstorp, D. (2010). *A microfluidic device for reversible environmental changes around single cells using optical tweezers for cell selection and positioning*. doi:10.1039/B913587A

Exposito, A. J., Barrie, P. J., & Torrente-Murciano, L. (2020). Fast Synthesis of CeO2Nanoparticles in a Continuous Microreactor Using Deep Eutectic Reline As Solvent. *ACS Sustainable Chemistry & Engineering*, *8*(49), 18297–18302. doi:10.1021/acssuschemeng.0c06949

Fabiani, L., Saroglia, M., Galatà, G., De Santis, R., Fillo, S., Luca, V., Faggioni, G., D'Amore, N., Regalbuto, E., Salvatori, P., Terova, G., Moscone, D., Lista, F., & Arduini, F. (2021). Magnetic beads combined with carbon black-based screen-printed electrodes for COVID-19: A reliable and miniaturized electrochemical immunosensor for SARS-CoV-2 detection in saliva. *Biosensors & Bioelectronics*, *171*, 112686. doi:10.1016/j.bios.2020.112686 PMID:33086175

Fabozzi, A., Della Sala, F., di Gennaro, M., Barretta, M., Longobardo, G., Solimando, N., Pagliuca, M., & Borzacchiello, A. (2023). Design of functional nanoparticles by microfluidic platforms as advanced drug delivery systems for cancer therapy. In Lab on a Chip (Vol. 23, Issue 5). doi:10.1039/D2LC00933A

Farooq, M. A., Aquib, M., Khan, D. H., Ghayas, S., Ahsan, A., Ijaz, M., Banerjee, P., Khan, M. A., Ahmad, M. M., & Wang, B. (2019). Nanocarrier-mediated co-delivery systems for lung cancer therapy: Recent developments and prospects. *Environmental Chemistry Letters*, *17*(4), 1565–1583. doi:10.100710311-019-00897-7

Fava, E. L., Silva, T. A., Prado, T. M., Moraes, F. C., Faria, R. C., & Fatibello-Filho, O. (2019). Electrochemical Paper-Based Microfluidic Device for High Throughput Multiplexed Analysis. *Talanta*, *203*(May), 280–286. doi:10.1016/j.talanta.2019.05.081 PMID:31202339

Fedder, G. K. (1994). *Simulation of microelectromechanical systems*. University of California.

Flader, I. B., Chen, Y., Yang, Y., Ng, E. J., Shin, D. D., Heinz, D. B., Ortiz, L. C., Alter, A. L., Park, W., Goodson, K. E., & … . (2019). Micro-tethering for fabrication of encapsulated inertial sensors with high sensitivity. *Journal of Microelectromechanical Systems*, *28*(3), 372–381. doi:10.1109/JMEMS.2019.2900761

Focke, M., Kosse, D., Müller, C., Reinecke, H., Zengerle, R., & von Stetten, F. (2010). Lab-on-a-foil: Microfluidics on thin and flexible films. *Lab on a Chip*, *10*(11), 1365–1386. doi:10.1039/c001195a PMID:20369211

Folch, A. (2016). *Introduction to bioMEMS*. CRC Press. doi:10.1201/b12263

Fong, J., Xiao, Z., & Takahata, K. (2015). Wireless implantable chip with integrated nitinol-based pump for radio-controlled local drug delivery. *Lab on a Chip*, *15*(4), 1050–1058. doi:10.1039/C4LC01290A PMID:25473933

Friend, J., & Yeo, L. (2010). Fabrication of microfluidic devices using polydimethylsiloxane. *Biomicrofluidics*, *4*(2), 026502-1, 026502–026505. doi:10.1063/1.3259624 PMID:20697575

Frisk, T., Rönnholm, D., van der Wijngaart, W., Stemme, G. (2006). A micromachined interface for airborne sample-to-liquid transfer and its application in a biosensor system. *Lab on a Chip*, *6*(12), 1504 - 1509. . PMID 17203153. doi:10.1039/B612526N

Fu, L. M., & Wang, Y. N. (2018). Detection Methods and Applications of Microfluidic Paper-Based Analytical Devices. *Trends in Analytical Chemistry*, *107*, 107. doi:10.1016/j.trac.2018.08.018

Funano, S.-i., Ota, N., & Tanaka, Y. (2021). A simple and reversible glass–glass bonding method to construct a microfluidic device and its application for cell recovery. *Lab on a Chip*, *21*(11), 2244–2254. doi:10.1039/D1LC00058F PMID:33908537

Fuyal, M., & Giri, B. (2020). A Combined System of Paper Device and Portable Spectrometer for the Detection of Pesticide Residues. *Food Analytical Methods*, *13*(7), 1492–1502. doi:10.100712161-020-01770-y

Gabbarini, V., Rossi, R., Ciparisse, J.-F., Malizia, A., Divizia, A., De Filippis, P., Anselmi, M., Carestia, M., Palombi, L., Divizia, M., & Gaudio, P. (2019). Laser-Induced Fluorescence (LIF) as a Smart Method for Fast Environmental Virological Analyses: Validation on Picornaviruses. *Scientific Reports*, *9*(1), 3–9. doi:10.103841598-019-49005-3 PMID:31467322

Gad-el-Hak, M. (2002). Flow Physics. In M. Gad-el-Hak (Ed.), *The MEMS Handbook*. CRC Press.

Gagliardi, A., Giuliano, E., Venkateswararao, E., Fresta, M., Bulotta, S., Awasthi, V., & Cosco, D. (2021). Biodegradable Polymeric Nanoparticles for Drug Delivery to Solid Tumors. *Frontiers in Pharmacology*, *12*(February), 1–24. doi:10.3389/fphar.2021.601626 PMID:33613290

Galdames, A., Ruiz-Rubio, L., Orueta, M., Sánchez-Arzalluz, M., & Vilas-Vilela, J. L. (2020). Zero-valent iron nanoparticles for soil and groundwater remediation. *International Journal of Environmental Research and Public Health, 17*(16), 5817. Advance online publication. doi:10.3390/ijerph17165817 PMID:32796749

Gambhire, G., & Massai, D. (2011). A Review on Different Micromixers and its Micromixing within Microchannel. *International Journal of Current Engineering and Technology*. doi:10.14741/Ijcet/22774106/spl.4.2016.83

Gao, Y., Pinho, B., & Torrente-Murciano, L. (2022). Tailoring the size of silver nanoparticles by controlling mixing in microreactors. *Chemical Engineering Journal, 432*, 134112. doi:10.1016/j.cej.2021.134112

Gao, Y., Wu, M., Lin, Y., & Xu, J. (2020). Trapping and control of bubbles in various microfluidic applications. *Lab on a Chip, 20*(24), 4512–4527. doi:10.1039/D0LC00906G PMID:33232419

García-López, E. I., & Marcì, G. (2021). Preparation of photocatalysts by physical methodologies. In *Materials Science in Photocatalysis* (pp. 37–62). Elsevier. doi:10.1016/B978-0-12-821859-4.00007-6

Gdowski, A., Johnson, K., Shah, S., Gryczynski, I., Vishwanatha, J., & Ranjan, A. (2018). Optimization and scale up of microfluidic nanolipomer production method for preclinical and potential clinical trials. *Journal of Nanobiotechnology, 16*(1), 12. Advance online publication. doi:10.118612951-018-0339-0 PMID:29433518

Génot, V., Desportes, S., Croushore, C., Lefèvre, J.-P., Pansu, R. B., Delaire, J. A., & von Rohr, P. R. (2010). Synthesis of organic nanoparticles in a 3D flow focusing microreactor. *Chemical Engineering Journal, 161*(1–2), 234–239. doi:10.1016/j.cej.2010.04.029

Gervais, L., Hitzbleck, M., & Delamarche, E. (2011). Capillary-driven multiparametric microfluidic chips for one-step immunoassays. *Biosensors & Bioelectronics, 27*(1), 64–70. doi:10.1016/j.bios.2011.06.016 PMID:21752632

Getahun, Y. W., Gardea-Torresdey, J., Manciu, F. S., Li, X., & El-Gendy, A. A. (2022). Green synthesized superparamagnetic iron oxide nanoparticles for water treatment with alternative recyclability. *Journal of Molecular Liquids, 356*, 118983. Advance online publication. doi:10.1016/j.molliq.2022.118983

Ghazali, F. A. M., Mah, C. K., & Abuzaiter, A. (2017). Soft dielectric elastomer actuator micropump. *Sensors and Actuators. A, Physical, 263*, 276–284. doi:10.1016/j.sna.2017.06.018

Ghubade, A., Mandal, S., Chaudhury, R., Singh, R. K., & Bhattacharya, S. (2009). Dielectrophoresis assisted concentration of micro-particles and their rapid quantitation based on optical means. *Biomedical Microdevices, 11*(5), 987–995. doi:10.100710544-009-9316-6 PMID:19415499

Giner, J., Maeda, D., Ono, K., Shkel, A., & Sekiguchi, T. (2019). MEMS gyroscope with concentrated springs suspensions demonstrating single digit frequency split and temperature robustness. *Journal of Microelectromechanical Systems, 28*(1), 25–35. doi:10.1109/JMEMS.2018.2881209

Gioria, E., Signorini, C., Wisniewski, F., & Gutierrez, L. (2020). Green synthesis of time-stable palladium nanoparticles using microfluidic devices. *Journal of Environmental Chemical Engineering, 8*(5), 104096. Advance online publication. doi:10.1016/j.jece.2020.104096

Gnyawali, V., Strohm, E.M., Wang, J.Z., Tsai, S.S., Kolios, M.C. (2019). Simultaneous acoustic and photoacoustic microfluidic flow cytometry for label-free analysis. *Scientific Reports, 9*(1), 1585. . doi:10.1038/s41598-018-37771-5

Gomez, F. A. (Ed.). (2008). *Biological Applications of Microfluidics* (1st ed.). Wiley-Interscience.

Gong, M. M., & Sinton, D. (2017). Turning the Page: Advancing Paper-Based Microfluidics for Broad Diagnostic Application. *Chemical Reviews, 117*(12), 8447–8480. doi:10.1021/acs.chemrev.7b00024 PMID:28627178

Gonidec, M., & Puigmartí-Luis, J. (2018). Continuous- versus Segmented-Flow Microfluidic Synthesis in Materials Science. *Crystals*, *9*(1), 12. doi:10.3390/cryst9010012

Gosálvez, M. A., Zubel, I., & Viinikka, E. (2010). Wet etching of silicon. In *Handbook of silicon based MEMS materials and technologies* (pp. 447–480). Elsevier. doi:10.1016/B978-0-12-817786-0.00017-7

Goswami, S., & Manna, K. (2021). Organochlorine Pesticide Residues and Other Toxic Substances in Salted Tenualosa Ilisha L.: Northeastern Part of India. *Foods and Raw Materials*, *9*(2), 201–206. doi:10.21603/2308-4057-2021-2-201-206

Gowri, A., Kumar, N. A., & Anand, B. S. (2021). Recent advances in nanomaterials based biosensors for point of care (PoC) diagnosis of COVID-19–a minireview. *Trends in Analytical Chemistry*, *137*, 116205. doi:10.1016/j.trac.2021.116205 PMID:33531721

Gravesen, P., Branebjerg, J., & Jensen, O. S. (1993). Microfluidics-a review. *Journal of Micromechanics and Microengineering*, *3*(4), 168–182. doi:10.1088/0960-1317/3/4/002

Greener, J., Tumarkin, E., Debono, M., Dicks, A.P., Kumacheva, E. (2012). Education: a microfluidic platform for university-level analytical chemistry laboratories. *Lab on a Chip, 12*(4), 696–701. . doi:10.1039/c2lc20951a

Greener, J., Tumarkin, E., Debono, M., Kwan, C.H., Abolhasani, M., Guenther, A., Kumacheva, E. (2012). Development and applications of a microfluidic reactor with multiple analytical probes. *The Analyst*, *137*(2) 444–450. . doi:10.1039/C1AN15940B

Guan, L., Peng, K. W., Yang, Y. L., Qiu, X. H., & Wang, C. (2009). *Nanotechnology*, 20. Institute of Physics.

Gubbi, J., Buyya, R., Marusic, S., & Palaniswami, M. (2013). Internet of Things (IoT): A vision, architectural elements, and future directions. *Future Generation Computer Systems*, *29*(7), 1645–1660. doi:10.1016/j.future.2013.01.010

Guerine, A., Merzouki, T., Hami, A. E., & Zineb, T. B. (2018). Uncertainty analysis of an actuator for a shape memory alloy micro-pump with uncertain parameters. *Advances in Engineering Software*, *122*, 22–30. doi:10.1016/j.advengsoft.2018.02.011

Gu, L., & Livermore, C. (2010). Passive self-tuning energy harvester for extracting energy from rotational motion. *Applied Physics Letters*, *97*(8), 081904. doi:10.1063/1.3481689

Guo, G., Wu, X., Liu, D., Liao, L., Zhang, D., Zhang, Y., Mao, T., He, Y., Huang, P., & Wang, W. (2021). A Self-Regulated Microfluidic Device with Thermal Bubble Micropumps. *Micromachines, 13.*

Guo, X., Zong, L., Jiao, Y., Han, Y., Zhang, X., Xu, J., Li, L., Zhang, C. W., Liu, Z., Ju, Q., Liu, J., Xu, Z., Yu, H. D., & Huang, W. (2019). Signal-Enhanced Detection of Multiplexed Cardiac Biomarkers by a Paper-Based Fluorogenic Immunodevice Integrated with Zinc Oxide Nanowires. *Analytical Chemistry*, *91*(14), 9300–9307. doi:10.1021/acs.analchem.9b02557 PMID:31241314

Gupta, A., & Pal, P. (2019). 8 - Micro-electro-mechanical system–based drug delivery devices, In Woodhead Publishing Series in Electronic and Optical Materials. Bioelectronics and Medical Devices, 183-210.

Gupta, R. K., Shi, Q., Dhakar, L., Wang, T., Heng, C. H., & Lee, C. (2017). Broadband energy harvester using non-linear polymer spring and electromagnetic/triboelectric hybrid mechanism. *Scientific Reports*, *7*(1), 41396. doi:10.1038rep41396 PMID:28120924

H. Bruus, Theoretical Microfluidics. (2008). OUP.

Hajji, I., Serra, M., Geremie, L., Ferrante, I., Renault, R., Viovy, J.L., Descroix, S., & Ferraro, D. (2020). Droplet microfluidic platform for fast and continuous-flow RT-qPCR analysis devoted to cancer diagnosis application. *Sensors and Actuators B: Chemical, 303*, 127171. . S2CID 208705450 doi:10.1016/j.snb.2019.127171

Haldkar, R. K., Khalatkar, A., Gupta, V. K., & Sheorey, T. (2021). New piezoelectric actuator design for enhance the micropump flow. *Materials Today: Proceedings*, *44*(1), 776-781. doi:10.1016/j.matpr.2020.10.702

Haldkar, R.K., Gupta, V.K., Sheorey, T., & Parinov, I.A., (2021). Design, Modeling, and Analysis of Piezoelectric-Actuated Device for Blood Sampling. *Appl. Sci., 11.*

Halfmann, P. J., Hatta, M., Chiba, S., Maemura, T., Fan, S., Takeda, M., Kinoshita, N., Hattori, S., Sakai-Tagawa, Y., Iwatsuki-Horimoto, K., Imai, M., & Kawaoka, Y. (2020). Transmission of SARS-CoV-2 in Domestic Cats. *The New England Journal of Medicine*, *383*(6), 592–594. doi:10.1056/NEJMc2013400 PMID:32402157

Hamid, N. A., Majlis, B. Y., Yunas, J., Syafeeza, A. R., Wong, Y. C., & Ibrahim, M. (2016). A stack bonded thermopneumatic micro-pump utilizing polyimide-based actuator membrane for biomedical applications. *Microsystem Technologies*, *23*(9), 4037–4043. doi:10.100700542-016-2951-y

Hammoudi, N.-E.-H., Sobhi, W., Attoui, A., Lemaoui, T., Erto, A., & Benguerba, Y. (2021). In Silico Drug Discovery of Acetylcholinesterase and Butyrylcholinesterase Enzymes Inhibitors Based on Quantitative Structure-Activity Relationship (QSAR) and Drug-Likeness Evaluation. *Journal of Molecular Structure*, *1229*, 129845. doi:10.1016/j.molstruc.2020.129845

Hao, N., Xu, Z., Nie, Y., Jin, C., Closson, A. B., Zhang, M., & Zhang, J. X. J. (2019). Microfluidics-enabled rational design of ZnO micro-/nanoparticles with enhanced photocatalysis, cytotoxicity, and piezoelectric properties. *Chemical Engineering Journal*, *378*(July), 122222. doi:10.1016/j.cej.2019.122222 PMID:32831625

Hao, Y., Zhang, S., Fang, B., Sun, F., Liu, H., & Li, H. (2022). A Review of Smart Materials for the Boost of Soft Actuators, Soft Sensors, and Robotics Applications. *Chinese Journal of Mechanical Engineering*, *35*(1), 37. doi:10.118610033-022-00707-2

Hara, T. O., & Singh, B. (2021). Electrochemical Biosensors for Detection of Pesticides and Heavy Metal Toxicants in Water: Recent Trends and Progress. *ACS ES&T Water*, *1*(3), 462–478. doi:10.1021/acsestwater.0c00125

Hardouin, J., Duchateau, M., Joubert-Caron, R., Caron, M. (2006). Usefulness of an integrated microfluidic device (HPLC-Chip-MS) to enhance confidence in protein identification by proteomics. *Rapid Communications in Mass Spectrometry, 20*(21), 3236244. . doi:10.1002/rcm.2725

Harmon, J. B., Gray, H. K., Young, C. C., & Schwab, K. J. (2020). Microfluidic droplet application for bacterial surveillance in fresh-cut produce wash waters. *PLoS One*, *15*(6), e0233239. doi:10.1371/journal.pone.0233239 PMID:32516315

Harries, N., Burns, J. R., Barrow, D. A., & Ramshaw, C. (2003). A numerical model for segmented flow in a microreactor. *International Journal of Heat and Mass Transfer*, *46*(17), 3313–3322. doi:10.1016/S0017-9310(03)00120-0

Hasan, D., & Lee, C. (2018). Hybrid metamaterial absorber platform for sensing of CO_2 gas at mid-IR. *Advancement of Science*, *5*(5), 1700581. doi:10.1002/advs.201700581 PMID:29876204

Hasani-Sadrabadi, M. M., Taranejoo, S., Dashtimoghadam, E., Bahlakeh, G., Majedi, F. S., VanDersarl, J. J., Janmaleki, M., Sharifi, F., Bertsch, A., Hourigan, K., Tayebi, L., Renaud, P., & Jacob, K. I. (2016). Microfluidic Manipulation of Core/Shell Nanoparticles for Oral Delivery of Chemotherapeutics: A New Treatment Approach for Colorectal Cancer. *Advanced Materials*, *28*(21), 4134–4141. doi:10.1002/adma.201502697 PMID:27001745

Hasan, M. I., Ali, A. J. F., & Tufah, R. S. (2017). Numerical study of the effect of channel geometry on the performance of Magnetohydrodynamic micro pump. *Eng Sci Technol Int J*, *20*(3), 982–989. doi:10.1016/j.jestch.2017.01.008

Ha, Y., Ko, S., Kim, I., Huang, Y., Mohanty, K., Huh, C., & Maynard, J. A. (2018). Recent Advances Incorporating Superparamagnetic Nanoparticles into Immunoassays. *ACS Applied Nano Materials*, *1*(2), 512–521. doi:10.1021/acsanm.7b00025 PMID:29911680

Heckele, M., & Schomburg, W. K. (2004). Review on micro molding of thermoplastic polymers. *Journal of Micromechanics and Microengineering*, *14*(3), R1–R14. doi:10.1088/0960-1317/14/3/R01

Heo, H. Y., Chung, S., Kim, Y. T., Kim, D. H., & Seo, T. S. (2016). A valveless rotary microfluidic device for multiplex point mutation identification based on ligation-rolling circle amplification. *Biosensors & Bioelectronics*, *78*, 140–146. doi:10.1016/j.bios.2015.11.039 PMID:26606304

Herr, A. E., Hatch, A. V., Throckmorton, D. J., Tran, H. M., Brennan, J. S., Giannobile, W. V., & Singh, A. K. (2007). Microfluidic immunoassays as rapid saliva-based clinical diagnostics. *Proceedings of the National Academy of Sciences of the United States of America*, *104*(13), 5268–5273. doi:10.1073/pnas.0607254104 PMID:17374724

He, T., Sun, Z., Shi, Q., Zhu, M., Anaya, D. V., Chen, T., Yuce, M. R., Thean, A. V.-Y., & Lee, C. (2019). Self-powered glove-based intuitive interface for diversified control applications in real/cyber space. *Nano Energy*, *58*, 641–651. doi:10.1016/j.nanoen.2019.01.091

Higgins, D. C., Meza, D., & Chen, Z. (2010). Nitrogen-Doped Carbon Nanotubes as Platinum Catalyst Supports for Oxygen Reduction Reaction in Proton Exchange Membrane Fuel Cells. *The Journal of Physical Chemistry C*, *114*(50), 21982–21988. doi:10.1021/jp106814j

Hillson, N. J., Rosengarten, R. D., & Keasling, J. D. (2012). J.D. Keasling js DNA assembly design automation software. *ACS Synthetic Biology*, *1*(1), 14–21. doi:10.1021b2000116 PMID:23651006

Ho, C. M., & Tai, Y. C. (1998). Micro-electro-mechanical-systems (MEMS) and fluid flows. *Annual Review of Fluid Mechanics*, *30*(1), 579–612. doi:10.1146/annurev.fluid.30.1.579

Hochstetter, A. (2019, December). Presegmentation Procedure Generates Smooth-Sided Microfluidic Devices: Unlocking Multiangle Imaging for Everyone? *ACS Omega*, *4*(25), 20972–20977. doi:10.1021/acsomega.9b02139 PMID:31867488

Hoehr, L., Bavendiek, J., Sackmann, J., & Schomburg, W. K. (2019). Ultrasonic Welding of Polymer Micro Fluidic Devices by Inserting Metal Parts. *Microsystem Technologies*, *25*(2), 673–681. doi:10.100700542-018-3995-y

Homola, J., Yee, S. S., & Gauglitz, G. (1999). Surface plasmon resonance sensors. *Sensors and Actuators. B, Chemical*, *54*(1-2), 3–15. doi:10.1016/S0925-4005(98)00321-9

Hong, T. F., Ju, W. J., Wu, M. C., Tai, C. H., Tsai, C. H., & Fu, L. M. (2010). Rapid prototyping of PMMA microfluidic chips utilizing a CO2 laser. *Microfluidics and Nanofluidics*, *9*(6), 1125–1133. doi:10.100710404-010-0633-0

Horst, J., de Andreade, P., Duvoisin, C. A., & Vieira, R. (2020). Fabrication of conductive filaments for 3D-printing: Polymer nanocomposites. *Biointerface Research in Applied Chemistry*, *10*(6), 6577–6586. doi:10.33263/BRIAC106.65776586

Hosseini, H. R., Nikookar, H., Yesiloz, G., Naseh, M., & Mohammadi, M. (2021). An overview on micropumps, micromixers, and their applications in bioprocess. In Advances in Bioenergy and Microfluidic Applications. Elsevier. doi:10.1016/B978-0-12-821601-9.00015-7

Hosseini, S., Espinoza-Hernandez, M., Garcia-Ramirez, R., Cerda-Kipper, A. S., Reveles-Huizar, S., & Acosta-Soto, L. (2019). *BioMEMS*. Springer.

Hsieh, C. L., Goldsmith, J. A., Schaub, J. M., DiVenere, A. M., Kuo, H. C., Javanmardi, K., Le, K. C., Wrapp, D., Lee, A. G., Liu, Y., Chou, C. W., Byrne, P. O., Hjorth, C. K., Johnson, N. V., Ludes-Meyers, J., Nguyen, A. W., Park, J., Wang, N., Amengor, D., ... McLellan, J. S. (2020). Structure-based design of prefusion-stabilized SARS-CoV-2 spikes. *Science*, *369*(6510), 1501–1505. doi:10.1126cience.abd0826 PMID:32703906

Huang, E., Huang, D., Wang, Y., Cai, D., Luo, Y., Zhong, Z., & Liu, D. (2022). Active droplet-array microfluidics-based chemiluminescence immunoassay for point-of-care detection of procalcitonin. *Biosensors & Bioelectronics*, *195*, 113684. doi:10.1016/j.bios.2021.113684 PMID:34607116

Huang, H., du Toit, H., Besenhard, M. O., Ben-Jaber, S., Dobson, P., Parkin, I., & Gavriilidis, A. (2018). Continuous flow synthesis of ultrasmall gold nanoparticles in a microreactor using trisodium citrate and their SERS performance. *Chemical Engineering Science*, *189*, 422–430. doi:10.1016/j.ces.2018.06.050

Huang, Y., Shi, Y., Yang, H. Y., & Ai, Y. (2015). A novel single-layered MoS2 nanosheet based microfluidic biosensor for ultrasensitive detection of DNA. *Nanoscale*, *7*(6), 2245–2249. Advance online publication. doi:10.1039/C4NR07162J PMID:25567642

Hui, Y., Liu, Tang, Song, Madou, Xia, & Wu. (2019). Determination of Mercury(II) on a Centrifugal Microfluidic Device Using Ionic Liquid Dispersive Liquid-Liquid Microextraction. *Micromachines*, *10*(8), 523. doi:10.3390/mi10080523 PMID:31398936

Hung, P. J., Lee, P. J., Sabounchi, P., Aghdam, N., Lin, R., & Lee, L. P. (2005). A novel high aspect ratio microfluidic design to provide a stable and uniform microenvironment for cell growth in a high throughput mammalian cell culture array. *Lab on a Chip*, *5*(1), 44–48. doi:10.1039/b410743h PMID:15616739

Hupert, M. L., Guy, W. J., Llopis, S. D., Shadpour, H., Rani, S., Nikitopoulos, D. E., & Soper, S. A. (2007). Evaluation of micromilled metal mold masters for the replication of microchip electrophoresis devices. *Microfluidics and Nanofluidics*, *3*(1), 1–11. doi:10.100710404-006-0091-x

Hu, R., Liu, C., Xuan, J., Xu, Y., Li, T., Liu, B.-F., Li, Y., & Yang, Y. (2019). 3D hydrodynamic flow focusing-based micromixer enables high-resolution imaging for studying the early folding kinetics of G-quadruplex. *Sensors and Actuators. B, Chemical*, *293*, 312–320. doi:10.1016/j.snb.2019.05.026

Hu, S., Zeng, S., Zhang, B., Yang, C., Song, P., Hang Danny, T. J., Lin, G., Wang, Y., Anderson, T., Coquet, P., Liu, L., Zhang, X., & Yong, K. T. (2014). Preparation of biofunctionalized quantum dots using microfluidic chips for bioimaging. *Analyst (London)*, *139*(18), 4681–4690. Advance online publication. doi:10.1039/C4AN00773E PMID:25054471

Hu, X., Zhang, Y., Ding, T., Liu, J., & Zhao, H. (2020). Multifunctional Gold Nanoparticles: A Novel Nanomaterial for Various Medical Applications and Biological Activities. *Frontiers in Bioengineering and Biotechnology*, *8*, 990. Advance online publication. doi:10.3389/fbioe.2020.00990 PMID:32903562

Hwang, C. M., Sim, W. Y., Lee, S. H., Foudeh, A. M., Bae, H., Lee, S.-H., & Khademhosseini, A. (2010). Benchtop fabrication of PDMS microstructures by an unconventional photolithographic method. *Biofabrication*, *2*(4), 045001. doi:10.1088/1758-5082/2/4/045001 PMID:21076185

Hwang, J., Cho, Y. H., Park, M. S., & Kim, B. H. (2019). Microchannel Fabrication on Glass Materials for Microfluidic Devices. *International Journal of Precision Engineering and Manufacturing*, *20*(3), 479–495. doi:10.100712541-019-00103-2

Iannacci, J. (2017). Introduction to MEMS and RF-MEMS: From the early days of microsystems to modern RF-MEMS passives. In *Rf-Mems Technology for High-Performance Passives* (pp. 2053–2563). IOP Publishing Ltd.

il Kim, Y. (2020). Infection and Rapid Transmission of SARS-CoV-2 in Ferrets. *Cell Host & Microbe, 27*(5), 704–709. e2. doi:10.1016/j.chom.2020.03.023 PMID:32259477

Iliescu, C., & Tresset, G. (2015). Microfluidics-Driven Strategy for Size-Controlled DNA Compaction by Slow Diffusion through Water Stream. *Chemistry of Materials, 27*(24), 8193–8197. doi:10.1021/acs.chemmater.5b04129

Imre, A., Ocola, L. E., Rich, L., & Klingfus, J. (2010). Microelectronics and Nanometer Structures : Processing, Measurement, and Phenomena. *Journal of Vacuum Science & Technology. B: An Official Journal of the American Vacuum Society, 28*, 304–309.

Irshad, M. A., Nawaz, R., Rehman, M. Z., Adrees, M., Rizwan, M., Ali, S., Ahmad, S., & Tasleem, S. (2021). Synthesis, characterization and advanced sustainable applications of titanium dioxide nanoparticles: A review. *Ecotoxicology and Environmental Safety, 212*, 111978. doi:10.1016/j.ecoenv.2021.111978 PMID:33561774

Ishfaque, A., Rahaman, A., & Kim, B. (2019). Bioinspired low noise circular-shaped MEMS directional microphone. *Journal of Micro/Nanolithography, MEMS, and MOEMS : JM3, 18*(1), 1. doi:10.1117/1.JMM.18.1.010501

Ismail, R. A., Mohsin, M. H., Ali, A. K., Hassoon, K. I., & Erten-Ela, S. (2020). Preparation and characterization of carbon nanotubes by pulsed laser ablation in water for optoelectronic application. *Physica E, Low-Dimensional Systems and Nanostructures, 119*, 113997. Advance online publication. doi:10.1016/j.physe.2020.113997

IWA23. (2016). *Interoperability of Microfluidic Devices— Guidelines for Pitch Spacing Dimensions and Initial Device Classification.* IWA 23.

Jafry, A. T., Lee, H., Tenggara, A. P., Lim, H., Moon, Y., Kim, S.-H., Lee, Y., Kim, S.-M., Park, S., Byun, D., & Lee, J. (2019). Double-Sided Electrohydrodynamic Jet Printing of Two-Dimensional Electrode Array in Paper-Based Digital Microfluidics. *Sensors and Actuators. B, Chemical, 282*, 831–837. doi:10.1016/j.snb.2018.11.135

James, M., Revia, R. A., Stephen, Z., & Zhang, M. (2020). Microfluidic Synthesis of Iron Oxide Nanoparticles. *Nanomaterials (Basel, Switzerland), 10*(11), 2113. doi:10.3390/nano10112113 PMID:33114204

Jang, W. I., Choi, C. A., Jun, C. H., Kim, Y. T., & Esashi, M. (2004). Surface micromachined thermally driven micropump. *Sensors and Actuators A: Physical, 115*(1), 151-158. doi:10.1016/j.sna.2004.04.024

Jang, I., Carrão, D. B., Menger, R. F., Moraes de Oliveira, A. R., & Henry, C. S. (2020). Pump-Free Microfluidic Rapid Mixer Combined with a Paper-Based Channel. *ACS Sensors, 5*(7), 2230–2238. doi:10.1021/acssensors.0c00937 PMID:32583663

Javadi, A., Habibi, M., & Taheri, F.S. (2013). Effect of wetting on capillary pumping in microchannels. *Sci Rep., 3.*

Jeong, G. S., Chung, S., Kim, C. B., & Lee, S.-H. (2010). Applications of micromixing technology. *Analyst (London), 135*(3), 460–473. doi:10.1039/b921430e PMID:20174696

Jeong, H.-H., Issadore, D., & Lee, D. (2016). Recent developments in scale-up of microfluidic emulsion generation via parallelization. *Korean Journal of Chemical Engineering, 33*(6), 1757–1766. doi:10.100711814-016-0041-6

Jeon, J. S., Chung, S., Kamm, R. D., & Charest, J. L. (2011, April). Hot embossing for fabrication of a microfluidic 3D cell culture platform. *Biomedical Microdevices, 13*(2), 325–333. doi:10.100710544-010-9496-0 PMID:21113663

Jeon, N. L., Chiu, D. T., Wargo, C. J., Wu, H. K., Choi, I. S., Anderson, J. R., & Whitesides, G. M. (2002). *Biomedical Microdevices, 4*(2), 117–121. doi:10.1023/A:1014683114796

Jerman, J. H., & Terry, S. C. (1981). A miniature gas chromatograph for atmospheric monitoring. *Environment International, 5*(2), 77–83. doi:10.1016/0160-4120(81)90126-4

Jia, M., Zhang, Z., Li, J., Shao, H., Chen, L., & Yang, X. (2017). A Molecular Imprinting Fluorescence Sensor Based on Quantum Dots and a Mesoporous Structure for Selective and Sensitive Detection of 2,4-Dichlorophenoxyacetic Acid. *Sensors and Actuators. B, Chemical, 252*, 934–943. doi:10.1016/j.snb.2017.06.090

Jing, G., Polaczyk, A., Oerther, D. B., & Papautsky, I. (2007). Development of a microfluidic biosensor for detection of environmental mycobacteria. *Sensors and Actuators. B, Chemical, 123*(1), 614–621. doi:10.1016/j.snb.2006.07.029

Jin, J. H., Kim, J. H., Lee, S. K., Choi, S. J., Park, C. W., & Min, N. K. (2018). A Fully Integrated Paper-Microfluidic Electrochemical Device for Simultaneous Analysis of Physiologic Blood Ions. *Sensors (Basel), 18*(1), 104. doi:10.339018010104 PMID:29301270

Jin, L., Hao, Z., Zheng, Q., Chen, H., Zhu, L., Wang, C., Liu, X., & Lu, C. (2020). A Facile Microfluidic Paper-Based Analytical Device for Acetylcholinesterase Inhibition Assay Utilizing Organic Solvent Extraction in Rapid Detection of Pesticide Residues in Food. *Analytica Chimica Acta, 1100*, 215–224. doi:10.1016/j.aca.2019.11.067 PMID:31987143

Ji, T., Liu, Z., Wang, G., Guo, X., Akbar Khan, S., Lai, C., Chen, H., Huang, S., Xia, S., Chen, B., Jia, H., Chen, Y., & Zhou, Q. (2020). Detection of COVID-19: A review of the current literature and future perspectives. *Biosensors & Bioelectronics, 166*, 112455. doi:10.1016/j.bios.2020.112455 PMID:32739797

Johnson & Borkholder. (2016). Towards an Implantable, Low Flow Micropump That Uses No Power in the Blocked-Flow State. *Micromachines, 7.*

Joshi-Imre, A., & Bauerdick, S. (2014). Direct-write ion beam lithography. *Journal of Nanotechnology, 2014*, 1–26. doi:10.1155/2014/170415

Kaigala, G. V., Lovchik, R. D., & Delamarche, E. (2012, November). Microfluidics in the "open space" for performing localized chemistry on biological interfaces. *Angewandte Chemie, 51*(45), 11224–11240. doi:10.1002/anie.201201798 PMID:23111955

Kakinuma, Y., Yasuda, N., & Aoyama, T. (2008). Micromachining of Soft Polymer Material applying Cryogenic Cooling. *Journal of Advanced Mechanical Design, Systems and Manufacturing*, 2(4), 560–569. doi:10.1299/jamdsm.2.560

Kakran, M., Li, L., & Müller, R. H. (2012). Overcoming the challenge of poor drug solubility. *Pharmaceutical Engineering, 32*(4), 82–89.

Kamat, V., Marathe, I., Ghormade, V., Bodas, D., & Paknikar, K. (2015). Synthesis of Monodisperse Chitosan Nanoparticles and in Situ Drug Loading Using Active Microreactor. *ACS Applied Materials & Interfaces*, 7(41), 22839–22847. doi:10.1021/acsami.5b05100 PMID:26448128

Kane. (2020). Passive Controlled Flow for Parkinson's Disease Neuronal Cell Culture in 3D Microfluidic Devices. *Organs-on-a-Chip, 2.* https://www.sciencedirect.com/science/article/pii/S2666102020300057

Karabacak, N. M., Spuhler, P. S., Fachin, F., Lim, E. J., Pai, V., Ozkumur, E., Martel, J. M., Kojic, N., Smith, K., Chen, P., Yang, J., Hwang, H., Morgan, B., Trautwein, J., Barber, T. A., Stott, S. L., Maheswaran, S., Kapur, R., Haber, D. A., & Toner, M. (2014, March). Microfluidic, marker-free isolation of circulating tumor cells from blood samples. *Nature Protocols, 9*(3), 694–710. doi:10.1038/nprot.2014.044 PMID:24577360

Karniadakis, G. M., Beskok, A., & Aluru, N. (2005). *Microflows and Nanoflows. Springer Verlag. Bruus H (2007). Theoretical Microfluidics.* Oxford University Press.

Karthikeyan, K., & Sujatha, L. (2018d). Study of Permissible Flow Rate and Mixing Efficiency of the Micromixer Devices. *International Journal of Chemical Reactor Engineering, 17*(1).

Karthikeyan, K., Sujatha, L., & Sudharsan, N. M. (2018e). Numerical Modeling and Parametric Optimization of Micromixer for Low Diffusivity Fluids. *International Journal of Chemical Reactor Engineering, 16*(3).

Karthikeyan, K. (2023). Numerical simulation and parameter optimization of micromixer device using fuzzy logic technique. *RSC Advances, 13*(7), 4504–4522. doi:10.1039/D2RA07992E PMID:36760289

Karthikeyan, K., Kishor, A., Karthikeyan, P., & Karthikeyan, S. (2022). Design and Numerical Simulation of RF Antenna for Bio-Medical Application. *International Conference on Applied Artificial Intelligence and Computing (ICAAIC),* 1222-1227.

Karthikeyan, K., & Sujatha, L. (2017) Design and fabrication of microfluidic device for mercury ions detection in water. *International Conference on Nextgen Electronic Technologies: Silicon to Software (ICNETS2)*, 218-221. 10.1109/ICNETS2.2017.8067934

Karthikeyan, K., & Sujatha, L. (2018a). Detection of Mercury Ion in a Microfluidic Device Using Colorimetric Method. *International Journal of Advances in Science, Engineering and Technology, 6*(2), 67–70.

Karthikeyan, K., & Sujatha, L. (2018c). Fluorometric Sensor for Mercury Ion Detection in a Fluidic MEMS Device. *IEEE Sensors Journal, 18*(13), 5225–5231. doi:10.1109/JSEN.2018.2840331

Karthikeyan, K., Sujatha, L., & Sundar, R. (2021). A Low-Cost Fabrication and Numerical Simulation of a MEMS Acoustic Transducer Using Polyimide Membrane on FR4 Substrate. *Journal of Electronic Materials, 50*(11), 6489–6503. doi:10.100711664-021-09184-0

Karthikeyan, K., Sujatha, L., Sundar, R., & Sharma, S. K. (2018b). Dimension Tolerances in Fabrication of Polymer Microfluidic Devices. *Journal of Semiconductor Technology and Science-IEIE, 18*(2), 262–269. doi:10.5573/JSTS.2018.18.2.262

Kasoju, A., Shahdeo, D., Khan, A. A., Shrikrishna, N. S., Mahari, S., Alanazi, A. M., Bhat, M. A., Giri, J., & Gandhi, S. (2020). Author Correction: Fabrication of Microfluidic Device for Aflatoxin M1 Detection in Milk Samples with Specific Aptamers. *Scientific Reports, 10*(1), 9222. doi:10.103841598-020-64241-8 PMID:32494019

Kašpar, O., Koyuncu, A. H., Hubatová-Vacková, A., Balouch, M., & Tokárová, V. (2020). Influence of channel height on mixing efficiency and synthesis of iron oxide nanoparticles using droplet-based microfluidics. *RSC Advances, 10*(26), 15179–15189. Advance online publication. doi:10.1039/D0RA02470H PMID:35495462

Kašpar, O., Koyuncu, A. H., Pittermannová, A., Ulbrich, P., & Tokárová, V. (2019). Governing factors for preparation of silver nanoparticles using droplet-based microfluidic device. *Biomedical Microdevices, 21*(4), 88. doi:10.100710544-019-0435-4 PMID:31571009

Katoch, V., Sharma, N., Sharma, M., Baghoria, M., Panda, J. J., Singh, M., & Prakash, B. (2021). Microflow synthesis and enhanced photocatalytic dye degradation performance of antibacterial Bi2O3 nanoparticles. *Environmental Science and Pollution Research International, 28*(15), 19155–19165. doi:10.100711356-020-11711-1 PMID:33398764

Katuri, P., Maralla, Y., Sharada, S., & Narasaiah Tumma, B. (2022). Synthesis of perbutyric acid with amberlyst catalyst in a continuous microreactor. *Materials Today: Proceedings*. Advance online publication. doi:10.1016/j.matpr.2022.06.202

Kaur, H. (2021). Effect of Synthesis Methods and Conditions on Properties and Applications of Carbon Dots for the Detection of Potential Water Contaminants: A Review. *Critical Reviews in Analytical Chemistry*, 1–24. https://www.tandfonline.com/doi/abs/10.1080/10408347.2021.1977608 PMID:34605318

Kaushik, A., Tiwari, S., Dev Jayant, R., Marty, A., & Nair, M. (2016). Towards detection and diagnosis of Ebola virus disease at point-of-care. *Biosensors & Bioelectronics, 75*, 254–272. doi:10.1016/j.bios.2015.08.040 PMID:26319169

Kawun, P., Leahy, S., & Lai, Y. (2016). A thin PDMS nozzle/diffuser micropump for biomedical applications. *Sensors and Actuators. A, Physical, 249*, 149–154. doi:10.1016/j.sna.2016.08.032

Keller, S., Blagoi, G., Lillemose, M., Haefliger, D., & Boisen, A. (2008). Processing of thin SU-8 films. *Journal of Micromechanics and Microengineering, 18*(12), 125020. doi:10.1088/0960-1317/18/12/125020

Kenawy, S., & Khalil, A. (2020). Advanced ceramics and relevant polymers for environmental and biomedical applications. *Biointerface Research in Applied Chemistry, 10*(4), 5747–5754. doi:10.33263/BRIAC104.747754

Kentsch, J., Breisch, S., & Stelzle, M. (2006). Low temperature adhesion bonding for BioMEMS. *Journal of Micromechanics and Microengineering, 16*(4), 802–807. doi:10.1088/0960-1317/16/4/017

KERN. (2002). *Riweta Material Selector*. KERN. www.kern.de.

Khalil, A. T., Ayaz, M., Ovais, M., Wadood, A., Ali, M., Shinwari, Z. K., & Maaza, M. (2018). In Vitro Cholinesterase Enzymes Inhibitory Potential and in Silico Molecular Docking Studies of Biogenic Metal Oxides Nanoparticles. *Inorganic and Nano-Metal Chemistry, 48*(9), 441–448. doi:10.1080/24701556.2019.1569686

Khan Malek, C., Robert, L., Boy, J.-J., & Blind, P. (2007). Deep microstructuring in glass for microfluidic applications. *Microsystem Technologies, 13*(5), 447–453. doi:10.100700542-006-0185-0

Khanna, P., Kaur, A., & Goyal, D. (2019). Algae-based metallic nanoparticles: Synthesis, characterization and applications. *Journal of Microbiological Methods, 163*, 105656. doi:10.1016/j.mimet.2019.105656 PMID:31220512

Khan, S. A., Jain, M., Pandey, A., Pant, K. K., Ziora, Z. M., Blaskovich, M. A. T., Shetti, N. P., & Aminabhavi, T. M. (2022). Leveraging the potential of silver nanoparticles-based materials towards sustainable water treatment. *Journal of Environmental Management, 319*, 115675. doi:10.1016/j.jenvman.2022.115675 PMID:35834856

Kharisov, B. I., Kharissova, O. V., & Ortiz-Mendez, U. (2016). *CRC concise encyclopedia of nanotechnology*. CRC Press. doi:10.1201/b19457

Khawwaf, J., Zheng, J., Lu, R., Al-Ghanimi, A., Kazem, B. I., & Man, Z. (2017). Robust tracking control of an IPMC actuator using nonsingular terminal sliding mode. *Smart Materials and Structures, 26*(9), 095042. doi:10.1088/1361-665X/aa7d69

Khoshroo, A., Fattahi, A., & Hosseinzadeh, L. (2022). Development of paper-based aptasensor for circulating tumor cells detection in the breast cancer. *Journal of Electroanalytical Chemistry (Lausanne, Switzerland), 910*, 116182. doi:10.1016/j.jelechem.2022.116182

Khunkitchai, N., Nuchtavorn, N., Rypar, T., Vlcnovska, M., Nejdl, L., Vaculovicova, M., & Macka, M. (2022). UV-Light-Actuated in-Situ Preparation of Paper@ZnCd Quantum Dots for Paper-Based Enzymatic Nanoreactors. *Chemical Engineering Journal, 428*, 132508. doi:10.1016/j.cej.2021.132508

Kilic, T., Erdem, A., Ozsoz, M., & Carrara, S. (2018). microRNA biosensors: Opportunities and challenges among conventional and commercially available techniques. *Biosensors & Bioelectronics, 99*, 525–546. doi:10.1016/j.bios.2017.08.007 PMID:28823978

Kim, H., Guzman, A., Thapa, H., Devarenne, T., & Han, A. (2016). A droplet-microfluidics platform for rapid microalgal growth and oil production analysis. *Biotechnolm Bioeng, 8,* 1691-1701.

Kim, H., Hwang, H., Baek, S., & Kim, D. (2018). Design, fabrication and performance evaluation of a printed-circuit-board microfluidic electrolytic pump for lab-on-a-chip devices. *Sensors and Actuators A: Physical, 277*, 73-84. doi:10.1016/j.sna.2018.04.042

Kim, H., Kim, K., & Lee, S.J. (2016). Compact and Thermosensitive Natureinspired Micropump. *Scientific Reports, 6.*

Kim, C., Yoo, Y. K., Lee, N. E., Lee, J., Kim, K. H., Lee, S., Kim, J., Park, S. J., Lee, D., Lee, S. W., Hwang, K. S., Han, S. I., Lee, D., Yoon, D. S., & Lee, J. H. (2022). Nanoelectrokinetic-assisted lateral flow assay for COVID-19 antibody test. *Biosensors & Bioelectronics, 212*, 114385. doi:10.1016/j.bios.2022.114385 PMID:35623254

Kim, D., Raj, A., Zhu, L., Masel, R. I., & Shannon, M. A. (2008). Non-Equilibrium Electrokinetic Nanofluidic Mixers. *Electrochemistry (Tokyo, Japan)*, 1–3.

Kim, H. J., Kim, Y., Park, S. J., Kwon, C., & Noh, H. (2018). Development of Colorimetric Paper Sensor for Pesticide Detection Using Competitive-Inhibiting Reaction. *Biochip Journal, 12*(4), 326–331. doi:10.100713206-018-2404-z

Kim, S., Wang, H., Yan, L., Zhang, X., & Cheng, Y. (2020). Continuous preparation of itraconazole nanoparticles using droplet-based microreactor. *Chemical Engineering Journal, 393*(March), 124721. doi:10.1016/j.cej.2020.124721

Kirby, B. J. (2010). Micro- and Nanoscale Fluid Mechanics: Transport in Microfluidic Devices. Cambridge University Press.

Kirby, B. J. (2010). *Micro- and Nanoscale Fluid Mechanics: Transport in Microfluidic Devices*. Cambridge University Press.

Kirby, B. J. (2010). *Micro-and Nanoscale Fluid Mechanics: Transport in Microfluidic Devices*. Cambridge University Press.

Kitchen, L. C., Berman, M., Halper, J., & Chazot, P. (2022). Rationale for 1068 nm Photobiomodulation Therapy (PBMT) as a Novel, Non-Invasive Treatment for COVID-19 and Other Coronaviruses: Roles of NO and Hsp70. *International Journal of Molecular Sciences, 23*(9), 5221. doi:10.3390/ijms23095221 PMID:35563611

Kjeang, E., Djilali, N., & Sinton, D. (2009). Microfluidic fuel cells, a review. *Journal of Power Sources, 186*(2), 353–369. doi:10.1016/j.jpowsour.2008.10.011

Ko, C.-H., Liu, C.-C., Chen, K.-H., Sheu, F., Fu, L.-M., & Chen, S.-J. (2021). Microfluidic colorimetric analysis system for sodium benzoate detection in foods. *Food Chemistry, 345*, 128773. . doi:10.1016/j.foodchem.2020.128773

Köhler, J. M., Li, S., & Knauer, A. (2013). Why is Micro Segmented Flow Particularly Promising for the Synthesis of Nanomaterials? *Chemical Engineering & Technology, 36*(6), 887–899. doi:10.1002/ceat.201200695

Konda, A., & Morin, S. A. (2017, June). Flow-directed synthesis of spatially variant arrays of branched zinc oxide mesostructures. *Nanoscale, 9*(24), 8393–8400. doi:10.1039/C7NR02655B PMID:28604901

Konermann, L., Metwally, H., Duez, Q., & Peters, I. (2019). Charging and Supercharging of Proteins for Mass Spectrometry: Recent Insights into the Mechanisms of Electrospray Ionization. *Analyst (London), 144*(21), 6157–6171. doi:10.1039/C9AN01201J PMID:31560020

Kotnala, R., & Shah, J. (2015). Ferrite materials: nano to spintronics regime. In *Handbook of magnetic materials* (Vol. 23, pp. 291–379). Elsevier.

Kou, S., Cheng, D., Sun, F., & Hsing, I. M. (2016). Microfluidics and microbial engineering. Lab on a Chip. doi:10.1039/C5LC01039J

Kovarik, M. L., Gach, P. C., Ornoff, D. M., Wang, Y., Balowski, J., Farrag, L., & Allbritton, N. L. (2012). Micro Total Analysis Systems for Cell Biology and Biochemical Assays. *Analytical Chemistry, 84*(2), 516–540. doi:10.1021/ac202611x PMID:21967743

Krishnan, J., & Shukla, S. (2019). The behaviour of soil stabilised with nanoparticles: an extensive review of the present status and its applications. In Arabian Journal of Geosciences (Vol. 12, Issue 14). doi:10.100712517-019-4595-6

Kulkarni, M. B., & Goel, S. (2020). Microfluidic devices for synthesizing nanomaterials—A review. *Nano Express, 1*(3), 032004. doi:10.1088/2632-959X/abcca6

Kulkarni, S. A., Mhaisalkar, S. G., Mathews, N., & Boix, P. P. (2019). Perovskite Nanoparticles: Synthesis, Properties, and Novel Applications in Photovoltaics and LEDs. *Small Methods, 3*(1), 1800231. doi:10.1002mtd.201800231

Kumar, A., & Whitesides, G. M. (1993). Features of gold having micrometer to centimeter dimensions can be formed through a combination of stamping with an elastomeric stamp and an alkanethiol "ink"followed by chemical etching. *Applied Physics Letters, 63*(14), 2002–2004. doi:10.1063/1.110628

Kumar, G. N. S., Mitra, S. K., & Bhattacharjee, S. (2009). *Dielectrophoretic Mixing With Novel Electrode Geometry* (Vol. 2). Fora. doi:10.1115/FEDSM2009-78260

Kuswandi, B., Hidayat, M. A., & Noviana, E. (2022). Paper-Based Electrochemical Biosensors for Food Safety Analysis. *Biosensors (Basel), 12*(12), 1088. doi:10.3390/bios12121088 PMID:36551055

Kwak, C. H., Park, J. P., Lee, S. S., Muruganantham, R., Kwon, S., Roh, C., Kim, S.-W., & Huh, Y. S. (2017). Droplet-Based Microfluidic Reactor for Synthesis of Size-Controlled CdSe Quantum Dots. *Journal of Nanoscience and Nanotechnology, 18*(2), 1339–1342. doi:10.1166/jnn.2018.14918 PMID:29448587

Lababidi, N., Sigal, V., Koenneke, A., Schwarzkopf, K., Manz, A., & Schneider, M. (2019). Microfluidics as tool to prepare size-tunable PLGA nanoparticles with high curcumin encapsulation for efficient mucus penetration. *Beilstein Journal of Nanotechnology, 10*, 2280–2293. doi:10.3762/bjnano.10.220 PMID:31807413

Labouta, H. I., Hooshmand, N., Upreti, T., & El-Sayed, M. A. (2021). Localized Plasmonic Photothermal Therapy as a Life-saving Treatment Paradigm for Hospitalized COVID-19 Patients. *Plasmonics, 16*(4), 1029–1033. doi:10.100711468-020-01353-x PMID:33495692

Lakhera, P., Chaudhary, V., Bhardwaj, B., Kumar, P., & Kumar, S. (2022). Development and recent advancement in microfluidics for point of care biosensor applications: A review. *Biosensors and Bioelectronics, 11*. doi:10.1016/j.biosx.2022.100218

Lammertyn, J., Verboven, P., Veraverbeke, E. A., Vermeir, S., Irudayaraj, J., & Nicolaï, B. M. (2006). Analysis of fluid flow and reaction kinetics in a flow injection analysis biosensor. *Sensors and Actuators. B, Chemical, 114*(2), 728–736. doi:10.1016/j.snb.2005.06.050

Landari, H., Dussault, M.-A., Ruel, J., Begin-Drolet, A., & Miled, A. (2018). Biocompatible compact micropump with integrated unidirectional microvalves for low pressure microfluidic applications. *Sensors and Actuators A: Physical, 276*, 246-258. doi:10.1016/j.sna.2018.03.046

Lawton, M. (2015). *The symbiosis of digital and physical security*. Available: https://futurelab.assaabloy.com/en/thesymbiosis-of-digital-and-physical-security/

Lebon, G. S. B., Tzanakis, I., Pericleous, K., & Eskin, D. (2018). Experimental and numerical investigation of acoustic pressures in different liquids. *Ultrasonics Sonochemistry, 42*, 411–421. doi:10.1016/j.ultsonch.2017.12.002 PMID:29429686

Lecault, V., White, A. K., Singhal, A., & Hansen, C. L. (2012). Microfluidic single cell analysis: From promise to practice. *Current Opinion in Chemical Biology, 16*(3-4), 381–390. doi:10.1016/j.cbpa.2012.03.022 PMID:22525493

Lee, D., & Yang, S. (2012). Surface modification of PDMS by atmospheric-pressure plasma-enhanced chemical vapor deposition and analysis of long-lasting surface hydrophilicity. *Sensors and Actuators. B, Chemical, 162*(1), 425–434. doi:10.1016/j.snb.2011.12.017

Lee, S., & Jun, B.-H. (2019). Silver Nanoparticles: Synthesis and Application for Nanomedicine. *International Journal of Molecular Sciences*, *20*(4), 865. doi:10.3390/ijms20040865 PMID:30781560

Lee, S., Park, J., & Park, J. K. (2018). Foldable Paper-Based Analytical Device for the Detection of an Acetylcholinesterase Inhibitor Using an Angle-Based Readout. *Sensors and Actuators. B, Chemical*, *273*, 322–327. doi:10.1016/j.snb.2018.06.059

Lee, Y. S., Bhattacharjee, N., & Folch, A. (2018). 3D-printed Quake-style microvalves and micropumps. *Lab on a Chip*, *18*(8), 1207–1214. doi:10.1039/C8LC00001H PMID:29553156

Leidinger, M., Rieger, M., Sauerwald, T., Alépée, C., & Schütze, A. (2016). Integrated pre-concentrator gas sensor microsystem for ppb level benzene detection. *Sensors and Actuators. B, Chemical*, *236*, 988–996. doi:10.1016/j.snb.2016.04.064

Lei, K. F. (2012). Microfluidic systems for diagnostic applications: A review. *JALA*, *17*(5), 330–347. doi:10.1177/2211068212454853 PMID:22893635

Lei, K. F., & Li, W. J. (2008). A novel in-plane microfluidic mixer using vortex pumps for fluidic discretization. *Journal of the Association for Laboratory Automation*, *13*(4), 227–236. doi:10.1016/j.jala.2008.03.008

Leinders, S. M., Westerveld, W. J., Pozo, J., Van Neer, P. L. M. J., Snyder, B., O'Brien, P., Urbach, H. P., De Jong, N., & Verweij, M. D. (2015). A sensitive optical micro-machined ultrasound sensor (OMUS) based on a silicon photonic ring resonator on an acoustical membrane. *Scientific Reports*, *5*(1), 1–10. doi:10.1038rep14328 PMID:26392386

Le, S., & Hegab, H. (2017). Investigation of a multistage micro gas compressor cascaded in series for increase pressure rise. *Sensors and Actuators. A, Physical*, *256*, 66–76. doi:10.1016/j.sna.2017.01.016

Leung, M. H. M., & Shen, A. Q. (2018). Microfluidic Assisted Nanoprecipitation of PLGA Nanoparticles for Curcumin Delivery to Leukemia Jurkat Cells. *Langmuir*, *34*(13), 3961–3970. doi:10.1021/acs.langmuir.7b04335 PMID:29544247

Lewpiriyawong, N., Kandaswamy, K., Yang, C., Ivanov, V., & Stocker, R. (2011, December). Microfluidic characterization and continuous separation of cells and particles using conducting poly(dimethyl siloxane) electrode induced alternating current-dielectrophoresis. *Analytical Chemistry*, *83*(24), 9579–9585. doi:10.1021/ac202137y PMID:22035423

Li, C., Boban, M., & Tuteja, A. (2017). Open-channel, water-in-oil emulsification in paper-based microfluidic devices. *Lab on a Chip, 17*(8), 1436–1441. . doi:10.1039/C7LC00114B

Li, J., Gershow, M., Stein, D., Brandin, E., & Golovchenko, J.A. (2003). DNA molecules and configurations in a solid-state nanopore microscope. *Nature Materials, 2*(9), 611–615. . doi:10.1038/nmat965

Li, A., Wang, Y., Deng, L., Zhao, X., Yan, Q., Cai, Y., Lin, J., Bai, Y., Liu, S., & Zhang, Y. (2013). Use of nitrocellulose membranes as a scaffold in cell culture. *Cytotechnology*, *65*(1), 71–81. doi:10.100710616-012-9458-x PMID:22717658

Liang, L.G., Kong, M.Q., Zhou, S., Sheng, Y.F., Wang, P., & Yu, T. (2017). An integrated double-filtration microfluidic device for isolation, enrichment and quantification of urinary extracellular vesicles for detection of bladder cancer. *ScientificReports, 7*(1), 46224. . doi:10.1038/srep46224

Liang, S., Wang, C., Li, Y., Xu, M., & Jia, H. (2021). Microfluidic Fabrication of ZrO2 Microspheres Using Improved External Gelation Aided by Polyacrylamide Networks. *Ceramics International*, *47*(15), 21576–21581. doi:10.1016/j.ceramint.2021.04.169

Liao, X., Song, W., Zhang, X., Zhan, H., Liu, Y., Wang, Y., & Zheng, Y. (2019). Hetero-contact microstructure to program discerning tactile interactions for virtual reality. *Nano Energy*, *60*, 127–136. doi:10.1016/j.nanoen.2019.03.048

Li, D. (2004). *Electrokinetics in microfluidics*. Elsevier.

Li, F., Guo, L., Hu, Y., Li, Z., Liu, J., He, J., & Cui, H. (2020). Multiplexed Chemiluminescence Determination of Three Acute Myocardial Infarction Biomarkers Based on Microfluidic Paper-Based Immunodevice Dual Amplified by Multifunctionalized Gold Nanoparticles. *Talanta*, *207*, 120346. doi:10.1016/j.talanta.2019.120346 PMID:31594588

Li, L.-L., Li, X., & Wang, H. (2017). Microfluidic Synthesis of Nanomaterials for Biomedical Applications. *Small Methods*, *1*(8), 1700140. doi:10.1002mtd.201700140

Li, L., Wang, X., Pu, Q., & Liu, S. (2019). Advancement of electroosmotic pump in microflow analysis: A Review. *Analytica Chimica Acta*, *1060*, 1–16. doi:10.1016/j.aca.2019.02.004 PMID:30902323

Lim, W. Y., Thevarajah, T. M., Goh, B. T., & Khor, S. M. (2019). Corrigendum to: 'Paper Microfluidic Device for Early Diagnosis and Prognosis of Acute Myocardial Infarction via Quantitative Multiplex Cardiac Biomarker Detection' [Biosensors and Bioelectronics 128 (2019) 176-185]. *Biosensors & Bioelectronics*, *129*, 299. doi:10.1016/j.bios.2019.01.036 PMID:30736959

Ling, F. W. M., Abdulbari, H. A., & Chin, S. Y. (2021). Microfluidic Chips for Formulation of Silica Nanoparticles and Enzyme Immobilization. *Chemical Engineering & Technology*, *44*(8), 1423–1431. doi:10.1002/ceat.202100098

Lin, P., & Smith, S. (2018). A tactile feedback glove for reproducing realistic surface roughness and continual lateral stroking perception. In D. Prattichizzo, H. Shinoda, H. Tan, E. Ruffaldi, & A. Frisoli (Eds.), *Haptic Sensing and Touch Enabled Computer Applications* (Vol. 10894, pp. 169–179). Springer. doi:10.1007/978-3-319-93399-3_16

Lin, S.-C., Lu, J.-C., Sung, Y.-L., Lin, C.-T., & Tung, Y.-C. (2013). A low sample volume particle separation device with electrokinetic pumping based on circular travelling-wave electroosmosis. *Lab on a Chip*, *13*(15), 3082–3089. doi:10.1039/c3lc50343g PMID:23753015

Lin, S.-L., Lin, T.-Y., & Fuh, M.-R. (2014). Microfluidic chip-based liquid chromatography coupled to mass spectrometry for determination of small molecules in bioanalytical applications: An update. *Electrophoresis*, *35*(9), 1275–1284. doi:10.1002/elps.201300415 PMID:24165927

Li, S., Liu, Y., Guo, P., & Wang, C. (2017). Self-Climbed Amorphous Carbon Nanotubes Filled with Transition Metal Oxide Nanoparticles for Large Rate and Long Lifespan Anode Materials in Lithium Ion Batteries. *ACS Applied Materials & Interfaces*, *9*(32), 26818–26825. doi:10.1021/acsami.7b06394 PMID:28745487

Li, S., Pang, C., Ma, X., Zhang, Y., Xu, Z., Li, J., Zhang, M., & Wang, M. (2021). Microfluidic Paper-Based Chip for Parathion-Methyl Detection Based on a Double Catalytic Amplification Strategy. *Mikrochimica Acta*, *188*(12), 438. doi:10.100700604-021-05084-6 PMID:34839414

Liu Z, Lee Y, Jang JH, Li Y, Han X, Yokoi K, et al. (2015). Microfluidic cytometric analysis of cancer cell transportability and invasiveness. *Scientific Reports, 5*(1), 14272. . doi:10.1038/srep14272

Liu, B., Sun, J., Li, D., Zhe, J., & Oh, K. W. (2016). A high flow rate thermal bubble-driven micropump with induction heating. *Microfluidics and Nanofluidics*, *20*(11), 155. doi:10.100710404-016-1822-2

Liu, C. (2012). *Foundations of MEMS* (2nd ed.). Pearson Education, Inc.

Liu, C. C., Ko, C. H., Wang, Y. N., Fu, L. M., & Lee, S. Z. (2021). Rapid detection of artificial sweeteners in food using microfluidic chromatography detection system. *Chemical Engineering Journal*, *425*, 131528. doi:10.1016/j.cej.2021.131528

Liu, C., Xu, T., Cheng, G., & Zhang, X. (2021). Target-Triggered Regioselective Assembly of Nanoprobes for Raman Imaging of Dual Cancer Biomarkers in Living Cells. *Sensors and Actuators. B, Chemical*, *330*, 129319. doi:10.1016/j.snb.2020.129319

Liu, H., Huang, J., Sun, D., Lin, L., Lin, W., Li, J., Jiang, X., Wu, W., & Li, Q. (2012). Microfluidic biosynthesis of silver nanoparticles: Effect of process parameters on size distribution. *Chemical Engineering Journal*, *209*, 568–576. doi:10.1016/j.cej.2012.08.049

Liu, H., Soon, B., Wang, N., Tay, C., Quan, C., & Lee, C. (2012). Feasibility study of a 3D vibration-driven electromagnetic MEMS energy harvester with multiple vibration modes. *Journal of Micromechanics and Microengineering*, *22*(12), 125020. doi:10.1088/0960-1317/22/12/125020

Liu, H., Zhang, S., Kathiresan, R., Kobayashi, T., & Lee, C. (2012). Development of piezoelectric microcantilever flow sensor with wind-driven energy harvesting capability. *Applied Physics Letters*, *100*(22), 223905. doi:10.1063/1.4723846

Liu, H., Zhao, P., Xiu, W., Zhang, L., Zhu, P., Ge, S., & Yu, J. (2022). SERS Paper Slip Based on 3D Dendritic Gold Nanomaterials Coupling with Urchin-like Nanoparticles for Rapid Detection of Thiram. *Sensors and Actuators. B, Chemical*, *355*, 131264. doi:10.1016/j.snb.2021.131264

Liu, J., Fang, H., Xu, Z., Mao, X., Shen, X., Chen, D., Liao, H., & Cai, B. (2008). A MEMS-based piezoelectric power generator array for vibration energy harvesting. *Microelectronics*, *39*(5), 802–806. doi:10.1016/j.mejo.2007.12.017

Liu, L. Y., Zhao, Y., Zhang, N., Wang, K.-N., Tian, M., Pan, Q., & Lin, W. (2021). Ratiometric Fluorescence Imaging for the Distribution of Nucleic Acid Content in Living Cells and Human Tissue Sections. *Analytical Chemistry*, *93*(3), 1612–1619. doi:10.1021/acs.analchem.0c04064 PMID:33381958

Liu, X., & Li, S. (2020). Fabrication of a Three-Layer PDMS Pneumatic Microfluidic Chip for Micro Liquid Sample Operation. *SLAS Technology*, *25*(2), 151–161. doi:10.1177/2472630319870126 PMID:31425005

Liu, X., Zhang, D., Liu, J., Wang, L., & Li, S. (2021). RGB Color Model Analysis for a Simple Structured Polydimethylsiloxane Pneumatic Micromixer. *SLAS Technology*, *26*(5), 510–518. doi:10.1177/24726303211008868 PMID:34384310

Liu, Y., Sun, L., Zhang, H., Shang, L., & Zhao, Y. (2021). Microfluidics for Drug Development: From Synthesis to Evaluation. *Chemical Reviews*, *121*(13), 7468–7529. doi:10.1021/acs.chemrev.0c01289 PMID:34024093

Liu, Y., Yang, G., Hui, Y., Ranaweera, S., & Zhao, C. X. (2022). Microfluidic Nanoparticles for Drug Delivery. *Small*, *2106580*(36), 1–33. doi:10.1002mll.202106580 PMID:35396770

Li, W., & Zhao, D. (2013). An overview of the synthesis of ordered mesoporous materials. *Chemical Communications*, *49*(10), 943–946. Advance online publication. doi:10.1039/C2CC36964H PMID:23249963

Li, X., & Liu, X. (2016). A Microfluidic Paper-Based Origami Nanobiosensor for Label-Free, Ultrasensitive Immunoassays. *Advanced Healthcare Materials*, *5*(11), 1326–1335. doi:10.1002/adhm.201501038 PMID:27122227

Li, Y., Yao, C., Wang, C., Yang, P., Wu, R., Fei, L., Zhang, Y., & Jiang, Y. (2021). An approach through steam to form sulfur nanoparticles for lithium sulfur batteries. *Electrochemistry Communications*, *125*, 107010. doi:10.1016/j.elecom.2021.107010

Li, Z., Li, F., Hu, J., Wee, W. H., Han, Y. L., Pingguan-Murphy, B., Lu, T. J., & Xu, F. (2015). Direct writing electrodes using a ball pen for paper-based point-of-care testing. *Analyst (London)*, *140*(16), 5526–5535. doi:10.1039/C5AN00620A PMID:26079757

Lo, S. C., Lai, W. C., Chang, C. I., Lo, Y. Y., Wang, C., Bai, M. R., & Fang, W. (2015). Development of a no-back-plate SOI MEMS condenser microphone. *Proceedings of the Transducers—2015 18th International Conference on Solid-State Sensors, Actuators and Microsystems (TRANSDUCERS)*, 1085–1088. 10.1109/TRANSDUCERS.2015.7181115

Lorenz, H., Despont, M., Fahrni, N., LaBianca, N., Renaud, P., & Vettiger, P. (1997). SU-8: A low-cost negative resist for MEMS. *Journal of Micromechanics and Microengineering*, *7*(3), 121–124. doi:10.1088/0960-1317/7/3/010

Luan, R. S., Wang, X., Sun, X., Chen, X. S., Zhou, T., Liu, Q. H., Lü, X., Wu, X. P., Gu, D. Q., Tang, M. S., Cui, H. J., Shan, X. F., Ouyang, J., Zhang, B., & Zhang, W., & Sichuan University Covid-, E. (2020). Sichuan da xue xue bao. Yi xue ban =. *Sichuan Da Xue Xue Bao. Yi Xue Ban = Journal of Sichuan University. Medical Science Edition*, *51*(2), 131–138. doi:10.12182/20200360505 PMID:32220177

Luo, R., Feng, Z., Shen, G., Xiu, Y., Zhou, Y., Niu, X., & Wang, H. (2018). Acetylcholinesterase Biosensor Based On Mesoporous Hollow Carbon Spheres/Core-Shell Magnetic Nanoparticles-Modified Electrode for the Detection of Organophosphorus Pesticides. *Sensors (Basel)*, *18*(12), 4429. doi:10.339018124429 PMID:30558201

Luo, T., Fan, L., Zhu, R., & Sun, D. (2019). Microfluidic Single-Cell Manipulation and Analysis: Methods and Applications. *Micromachines*, *10*(2), 104. doi:10.3390/mi10020104 PMID:30717128

Lu, Q., Su, T., Shang, Z., Jin, D., Shu, Y., Xu, Q., & Hu, X. (2021). Flexible paper-based Ni-MOF composite/AuNPs/CNTs film electrode for HIV DNA detection. *Biosensors & Bioelectronics*, *184*, 113229. doi:10.1016/j.bios.2021.113229 PMID:33894427

Lu, Y., O'Riordan, E., Cottone, F., Boisseau, S., Galayko, D., Blokhina, E., Marty, F., & Basset, P. (2016). A batch-fabricated electret-biased wideband MEMS vibration energy harvester with frequency-up conversion behavior powering a UHF wireless sensor node. *Journal of Micromechanics and Microengineering*, *26*(12), 124004. doi:10.1088/0960-1317/26/12/124004

MahariS.RobertsA.ShahdeoD.GandhiS. eCovSens-ultrasensitive novel in-house built printed circuit board based electrochemical device for rapid detection of nCovid-19. bioRxiv 2020. doi:10.1101/2020.04.24.059204

Mahdavi, Z., Rezvani, H., & Keshavarz Moraveji, M. (2020). Core–shell nanoparticles used in drug delivery-microfluidics: A review. *RSC Advances*, *10*(31), 18280–18295. doi:10.1039/D0RA01032D PMID:35517190

Maluf, N., & Williams, K. (2004). An Introduction to Microelectromechanical Systems Engineering. Artech House, Inc.

Maluf, N. (2000). *An Introduction to Microelectromechanical Systems Engineering*. Artech House.

Maluf, N., & Williams, K. (2004). *An introduction to microelectromechanical systems engineering*. Artech House.

Mandal, S., Kumar Gupta, A., Echeverria, E., McIlroy, D. N., Poplawsky, J. D., & Sachan, R. (2022). Laser-assisted nanofabrication of multielement complex oxide core–shell nanoparticles. *Materials & Design*, *220*, 110882. Advance online publication. doi:10.1016/j.matdes.2022.110882

Manz, A., Graber, N., & Widmer, H. M. (1990). Miniaturized total chemical analysis systems: A novel concept for chemical sensing. *Sensors and Actuators. B, Chemical*, *1*(1-6), 244–248. doi:10.1016/0925-4005(90)80209-I

Manzari, M. T., Shamay, Y., Kiguchi, H., Rosen, N., Scaltriti, M., & Heller, D. A. (2021). Targeted drug delivery strategies for precision medicines. *Nature Reviews. Materials*, *6*(4), 351–370. doi:10.103841578-020-00269-6 PMID:34950512

Mao, Z., Asai, Y., Wiranata, A., Kong, D., & Man, J. (2022). Eccentric actuator driven by stacked electrohydrodynamic pumps. *Journal of Zhejiang University. Science A*, *23*(4), 329–334. doi:10.1631/jzus.A2100468

Market and Market. (2020a). Microfluidics Market. *Market and Market.* https://www.marketsandmarkets.com/Market-Reports/microfluidics-market-1305.html

Market and Market. (2022b). *Microfluidics Market by Product (Devices, Components (Chip, Sensor, Pump, Valve)), Application (IVD (POC, Clinical, Veterinary), Research, Manufacturing, Therapeutics), End User (Hospital, Diagnostic Center, Academic Institutes) - Global Forecasts (2022 -).* Markets and Markets Research Private Ltd. https://www.marketsandmarkets.com/Market-Reports/microfluidics-market-1305.html

Markets and Markets. (2023). Microelectromechanical Systems (MEMS) Size. *Markets and Markets.* https://www.marketsandmarkets.com/Market-Reports/mems-market-13689179.html?gclid=EAIaIQobChMIrdXB9ve3_gIVhNCWCh3o2g70EAAYASAAEgIi9fD_BwE

MarketsandMarkets. (n.d.). *About.* Markets and Markets. https://www.marketsandmarkets.com

Martinez, A. W., Phillips, S. T., Butte, M. J., & Whitesides, G. M. (2007). Patterned paper as a platform for inexpensive, low-volume, portable bioassays. *Angewandte Chemie International Edition, 46*(8), 1318–1320. doi:10.1002/anie.200603817 PMID:17211899

Martinez, A. W., Phillips, S. T., Whitesides, G. M., & Carrilho, E. (2010). Diagnostics for the developing world: Microfluidic paper-based analytical devices. *Analytical Chemistry, 82*(1), 3–10. doi:10.1021/ac9013989 PMID:20000334

Martinez, J., Garcia, A., Oliver, M., Molina, J., & Gonzalez, P. (2016). Identifying virtual 3D geometric shapes with a vibrotactile glove. *IEEE Computer Graphics and Applications, 36*(1), 42–51. doi:10.1109/MCG.2014.81 PMID:25137722

Mastroianni, L., Vajglová, Z., Eränen, K., Peurla, M., Di Serio, M. Yu., Murzin, D., Russo, V., & Salmi, T. (2022). Microreactor technology in experimental and modelling study of alcohol oxidation on nanogold. *Chemical Engineering Science, 260,* 117920. doi:10.1016/j.ces.2022.117920

Mata, A., Fleischman, A. J., & Roy, S. (2006). Fabrication of multi-layer SU-8 microstructures. *Journal of Micromechanics and Microengineering, 16*(2), 276–284. doi:10.1088/0960-1317/16/2/012

Mathur, A., Nayak, H. C., Rajput, S., Roy, S., Nagabooshanam, S., Wadhwa, S., & Kumar, R. (2021). An Enzymatic Multiplexed Impedimetric Sensor Based on α-MnO_2/GQD Nano-Composite for the Detection of Diabetes and Diabetic Foot Ulcer Using Micro-Fluidic Platform. *Chemosensors (Basel, Switzerland), 9*(12), 339. doi:10.3390/chemosensors9120339

Mathur, A., & Roy, S. (2022). 10-Microfluidics and lab-on-a-chip. In G. Maruccio & J. Narang (Eds.), *Electrochemical Sensors* (pp. 261–287). Woodhead Publishing. doi:10.1016/B978-0-12-823148-7.00010-6

Matteucci, M., Pérennès, F., Marmiroli, B., Miotti, P., Vaccari, L., Gosparini, A., Turchet, A., & Di Fabrizio, E. (2006). Compact micropumping system based on LIGA fabricated microparts. *Microelectronic Engineering, 83*(4–9). doi:10.1016/j.mee.2006.01.259

Mauriz, E. (2020). Recent Progress in Plasmonic Biosensing Schemes for Virus Detection. *Sensors (Basel), 20*(17), 4745. doi:10.339020174745 PMID:32842601

Maurya, R. A., Min, K. I., & Kim, D. P. (2014). Continuous flow synthesis of toxic ethyl diazoacetate for utilization in an integrated microfluidic system. *Green Chemistry, 16*(1), 116–120. doi:10.1039/C3GC41226A

Ma, X., Özliseli, E., Zhang, Y., Pan, G., Wang, D., & Zhang, H. (2019). Fabrication of redox-responsive doxorubicin and paclitaxel prodrug nanoparticles with microfluidics for selective cancer therapy. *Biomaterials Science, 7*(2), 634–644. doi:10.1039/C8BM01333K PMID:30534690

McDonnell, E. E., Han, S., Hilty, C., Pierce, K. L., & Pines, A. (2005). NMR analysis on microfluidic devices by remote detection. *Analytical Chemistry, 77*(24), 8109–8114. doi:10.1021/ac051320+ PMID:16351162

McKeen, L. W. (2016). *Permeability Properties of Plastics and Elastomers* (4th ed.).

Mclaren, A., Valdes-Solis, T., Li, G., & Tsang, S. C. (2009). Shape and Size Effects of ZnO Nanocrystals on Photocatalytic Activity. *Journal of the American Chemical Society, 131*(35), 12540–12541. doi:10.1021/ja9052703 PMID:19685892

Meftaul, M. (2020). Pesticides in the Urban Environment: A Potential Threat That Knocks at the Door. *Science of The Total Environment, 711*, 134612. https://www.sciencedirect.com/science/article/pii/S0048969719346030

Melin, J., van der Wijngaart, W., & Stemme, G. (2005, June). Behaviour and design considerations for continuous flow closed-open-closed liquid microchannels. *Lab on a Chip*, *5*(6), 682–686. doi:10.1039/b501781e PMID:15915262

Meng, A. H., Nguyen, N.-T., & White, R. M. (2000). Focused flow micropump using ultrasonic flexural plate waves. *Biomedical Microdevices*, *2*(3), 169–174. doi:10.1023/A:1009905926740

Meredith, N. A., Quinn, C., Cate, D. M., Reilly, T. H. III, Volckens, J., & Henry, C. S. (2016). Paper-based analytical devices for environmental analysis. *Analyst (London)*, *141*(6), 1874–1887. doi:10.1039/C5AN02572A PMID:26901771

Miller, H. A., Vizza, F., Marelli, M., Zadick, A., Dubau, L., Chatenet, M., Geiger, S., Cherevko, S., Doan, H., Pavlicek, R. K., Mukerjee, S., & Dekel, D. R. (2017). Highly active nanostructured palladium-ceria electrocatalysts for the hydrogen oxidation reaction in alkaline medium. *Nano Energy*, *33*, 293–305. doi:10.1016/j.nanoen.2017.01.051

Mills, J. K., & Needham, D. (1999). Targeted drug delivery. *Expert Opinion on Therapeutic Patents*, *9*(11), 1499–1513. doi:10.1517/13543776.9.11.1499

Ming, T., Wang, Y., Luo, J., Liu, J., Sun, S., Xing, Y., Xiao, G., Jin, H., & Cai, X. (2019). Folding Paper-Based Aptasensor Platform Coated with Novel Nanoassemblies for Instant and Highly Sensitive Detection of 17β-Estradiol. *ACS Sensors*, *4*(12), 3186–3194. doi:10.1021/acssensors.9b01633 PMID:31775503

Min, K. I., Im, D. J., Lee, H. J., & Kim, D. P. (2014). Three-dimensional flash flow microreactor for scale-up production of monodisperse PEG-PLGA nanoparticles. *Lab on a Chip*, *14*(20), 3987–3992. doi:10.1039/C4LC00700J PMID:25133684

Minnella, W. (2019). Microfluidics applications: a short review - Elveflow. *Elveflow* https://www.elveflow.com/microfluidic-reviews/general-microfluidics/microfluidics-applications-a-short-review/

Mishra, S., & Yadava, V. (2015). Laser Beam MicroMachining (LBMM). A review. *Optics and Lasers in Engineering, 73*, 89-122. doi:10.1016/j.optlaseng.2015.03.017

Mishra, N., Dhwaj, A., Verma, D., & Prabhakar, A. (2022). Cost-effective microabsorbance detection based nanoparticle immobilized microfluidic system for potential investigation of diverse chemical contaminants present in drinking water. *Analytica Chimica Acta*, *1205*, 339734. Advance online publication. doi:10.1016/j.aca.2022.339734 PMID:35414395

Mittal, M., Sardar, S., & Jana, A. (2021). Nanofabrication techniques for semiconductor chemical sensors. In *Handbook of Nanomaterials for Sensing Applications* (pp. 119–137). Elsevier. doi:10.1016/B978-0-12-820783-3.00023-3

Mohan, J. M., Amreen, K., Kulkarni, M. B., Javed, A., Dubey, S. K., & Goel, S. (2021). Optimized ink jetted paper device for electroanalytical detection of picric acid. *Colloids and Surfaces. B, Biointerfaces*, *208*, 112056. doi:10.1016/j.colsurfb.2021.112056 PMID:34425529

Mohith, S., Navin Karanth, P., & Kulkarni, S. M. (2019). Recent trends in mechanical micropumps and their applications: A review. *Mechatronics, 60*, 34-55. doi:10.1016/j.mechatronics.2019.04.009

Mora, M. F., Stockton, A. M., & Willis, P. A. (2015). Analysis of thiols by microchip capillary electrophoresis for in situ planetary investigations. In Microchip Capillary Electrophoresis Protocols. Methods in Molecular Biology (Vol. 1274, pp. 43–52). HumanaPress., , PMID 25673481. doi:10.1007/978-1-4939-2353-3_4

Mourzina, Y., Steffen, A., & Offenhäusser, A. (2005). The evaporated metal masks for chemical glass etching for BioMEMS. *Microsystem Technologies*, *11*(2), 135–140. doi:10.100700542-004-0430-3

Mund, D., & Leib, J. (2004). Novel microstructuring technology for glass on silicon and glass-substrates. *2004 Proceedings 54th Electronic Components and Technology Conference.*

Mun, S., & Choi, S. J. (2015). Detection of Salmonella typhimurium by antibody/enzyme-conjugated magnetic nanoparticles. *Biochip Journal, 9*(1), 10–15. doi:10.100713206-014-9102-2

Nahavandi, S., Baratchi, S., Soffe, R., Tang, S.-Y., Nahavandi, S., Mitchell, A., & Khoshmanesh, K. (2014). Microfluidic platforms for biomarker analysis. *Lab on a Chip, 14*(9), 1496–1514. doi:10.1039/C3LC51124C PMID:24663505

Nak, H., Gurkan, I., Cebeci, H., & Ergenc, A. F. (2020). Fabrication and Characterization of a Microfluidic Device with Vertically Aligned Multi Walled Carbon Nanotube Channels. *IFAC-PapersOnLine, 53*(2), 11761–11766. doi:10.1016/j.ifacol.2020.12.682

Nam, D. N. C., & Ahn, K. K. (2012). Design of an IPMC diaphragm for micropump application. *Sensors and Actuators. A, Physical, 187*, 174–182. doi:10.1016/j.sna.2012.08.027

Narang, J., Singhal, C., Mathur, A., Sharma, S., Singla, V., & Pundir, C. S. (2018). Portable bioactive paper based genosensor incorporated with Zn-Ag nanoblooms for herpes detection at the point-of-care. *International Journal of Biological Macromolecules, 107*(Pt B), 2559-2565.

Nasseri, B., Soleimani, N., Rabiee, N., Kalbasi, A., Karimi, M., & Hamblin, M. R. (2018). Point-of-care microfluidic devices for pathogen detection. *Biosensors & Bioelectronics, 117*, 112–128. doi:10.1016/j.bios.2018.05.050 PMID:29890393

Nathanson, H. C., Newell, W. E., Wickstrom, R. A., & Davis, J. R. (1967, March). The resonant gate transistor. *IEEE Transactions on Electron Devices, 14*(3), 117–133. doi:10.1109/T-ED.1967.15912

Nemani, K. V., Moodie, K. L., Brennick, J. B., Su, A., & Gimi, B. (2013). In vitro and in vivo evaluation of SU-8 biocompatibility. *Materials Science and Engineering C, 33*(7), 4453–4459. doi:10.1016/j.msec.2013.07.001 PMID:23910365

Nguyen, N.-T. (2012a). Fundamentals of mass transport in the microscale. Micromixers. doi:10.1016/B978-1-4377-3520-8.00002-4

Nguyen, N.-T. (2012b). Micromixers based on chaotic advection. Micromixers. doi:10.1016/B978-1-4377-3520-8.00006-1

Nguyen, N.-T. (2012c). Micromixers based on molecular diffusion. Micromixers. doi:10.1016/B978-1-4377-3520-8.00005-X

Nguyen, N.-T., & Wereley, S. T. (2006). Fundamentals and Applications of Microfluidics. Artech House, Inc.

Nguyen, H. (2020). Editorial of Special Issue "Nanostructured Light-Emitters". *Micromachines, 11*(6), 601. doi:10.3390/mi11060601 PMID:32575847

Nguyen, N. I. T., & Wereley, S. (2006). *Fundamentals and Applications of Microfluidics*. Artech House.

Nguyen, N. T., & Wu, Z. G. (2005). Micromixers—A review. *Journal of Micromechanics and Microengineering, 15*(2), R1–R16. doi:10.1088/0960-1317/15/2/R01

Nguyen, N.-T., Huang, X., & Chuan, T. K. (2002). MEMS-micropumps; A Review. *Journal of Fluids Engineering, 124*(2), 384–392. doi:10.1115/1.1459075

Niculescu, A. G., Chircov, C., Bîrcă, A. C., & Grumezescu, A. M. (2021). Fabrication and Applications of Microfluidic Devices: A Review. *International Journal of Molecular Sciences, 22*(4), 1–26. doi:10.3390/ijms22042011 PMID:33670545

Niculescu, A.-G., Mihaiescu, D. E., & Grumezescu, A. M. (2022). A Review of Microfluidic Experimental Designs for Nanoparticle Synthesis. *International Journal of Molecular Sciences, 23*(15), 8293. doi:10.3390/ijms23158293 PMID:35955420

Nie, C., Frijns, A.J.H., Mandamparambil, R., & Toonder, J.M.J.D. (2015). A microfluidic device based on an evaporation-driven micropump. *Biomed Microdevices, 17*(47).

Nie, Y., Liu, Y., Su, X., & Ma, Q. (2019). Nitrogen-Rich Quantum Dots-Based Fluorescence Molecularly Imprinted Paper Strip for p-Nitroaniline Detection. *Microchemical Journal, 148*, 162–168. doi:10.1016/j.microc.2019.04.080

Nikolova, M. P., & Chavali, M. S. (2020). Metal Oxide Nanoparticles as Biomedical Materials. *Biomimetics, 5*(2), 27. doi:10.3390/biomimetics5020027 PMID:32521669

Nilghaz, A., Mousavi, S. M., Li, M., Tian, J., Cao, R., & Wang, X. (2021). Paper-based microfluidics for food safety and quality analysis. *Trends in Food Science & Technology, 118*, 273–284. doi:10.1016/j.tifs.2021.08.029

Nisar, A., Afzulpurkar, N., Mahaisavariya, B., & Tuantranont, A. (2008). MEMS-based micropumps in drug delivery and biomedical applications. *Sensors and Actuators B: Chemical, 130*(2), 917-942. doi:10.1016/j.snb.2007.10.064

Nishat, S., Jafry, A., Martinez, A., & Awan, F. (2021). Paper-based microfluidics: Simplified fabrication and assay methods. *Sensors and Actuators. B, Chemical, 336*, 129681. doi:10.1016/j.snb.2021.129681

Nitzan, S., Zega, V., Li, M., Ahn, C., Corigliano, A., Kenny, T., & Horsley, D. (2015). Self-induced parametric amplification arising from nonlinear elastic coupling in a micromechanical resonating disk gyroscope. *Scientific Reports, 5*(1), 9036. doi:10.1038rep09036 PMID:25762243

Niu, X., Gulati, S., Edel, J. B., & deMello, A. J. (2008, November). Pillar-induced droplet merging in microfluidic circuits. *Lab on a Chip, 8*(11), 1837–1841. doi:10.1039/b813325e PMID:18941682

Noiphung, J., Songjaroen, T., Dungchai, W., Henry, C. S., Chailapakul, O., & Laiwattanapaisal, W. (2013). Electrochemical detection of glucose from whole blood using paper-based microfluidic devices. *Analytica Chimica Acta, 788*, 39–45. doi:10.1016/j.aca.2013.06.021 PMID:23845479

Nontawong, N., Ngaosri, P., Chunta, S., Jarujamrus, P., Nacapricha, D., Lieberzeit, P. A., & Amatatongchai, M. (2022). Smart sensor for assessment of oxidative/nitrative stress biomarkers using a dual-imprinted electrochemical paper-based analytical device. *Analytica Chimica Acta, 1191*, 339363. doi:10.1016/j.aca.2021.339363 PMID:35033235

Nordström, M., Keller, S., Lillemose, M., Johansson, A., Dohn, S., Haefliger, D., Blagoi, G., Havsteen-Jakobsen, M., & Boisen, A. (2008). SU-8 cantilevers for bio/chemical sensing; fabrication, characterisation and development of novel read-out methods. *Sensors (Basel), 8*(3), 1595–1612. doi:10.33908031595 PMID:27879783

Nunes, P. S., Ohlsson, P. D., Ordeig, O., & Kutter, J. P. (2010). Cyclic olefin polymers: Emerging materials for lab-on-a-chip applications. *Microfluidics and Nanofluidics, 9*(2-3), 145–161. doi:10.100710404-010-0605-4

Oddy, M. H., Santiago, J. G., & Mikkelsen, J. C. (2001). Electrokinetic instability micromixing. *Analytical Chemistry, 73*(24), 5822–5832. doi:10.1021/ac0155411 PMID:11791550

Ogawa, M., Higashi, K., & Miki, N. (2015). Development of hydrogel microtubes for microbe culture in open environment. *2015 37th Annual International Conference of the IEEE Engineering in Medicine and Biology Society (EMBC)*. (pp. 5896–5899). IEEE.

Oh, K. W., & Ahn, C. H. (2006). A review of microvalves. *Journal of Micromechanics and Microengineering, 16*(5), R13–R39. doi:10.1088/0960-1317/16/5/R01

Okamoto, Y., Ryoson, H., Fujimoto, K., Ohba, T., & Mita, Y. (2020). On-Chip CMOS-MEMS-Based Electroosmotic Flow Micropump Integrated With High-Voltage Generator. *Journal of Microelectromechanical Systems*, *29*(1), 1–10. doi:10.1109/JMEMS.2019.2953290

Olszynko-Gryn, J. (2017). Thin blue lines: Product placement and the drama of pregnancy testing in British cinema and television. *British Journal for the History of Science*, *50*(3), 495–520. doi:10.1017/S0007087417000619 PMID:28923127

Oreshkova, N., Molenaar, R. J., Vreman, S., Harders, F., Oude Munnink, B. B., Hakze-van der Honing, R. W., Gerhards, N., Tolsma, P., Bouwstra, R., Sikkema, R. S., Tacken, M. G. J., de Rooij, M. M. T., Weesendorp, E., Engelsma, M. Y., Bruschke, C. J. M., Smit, L. A. M., Koopmans, M., van der Poel, W. H. M., & Stegeman, A. (2020). SARS-CoV-2 infection in farmed minks, the Netherlands, April and May 2020. *Eurosurveillance*, *25*(23), 1. doi:10.2807/1560-7917.ES.2020.25.23.2001005 PMID:32553059

Otzen, D. E., Buell, A. K., & Jensen, H. (2021). Microfluidics and the quantification of biomolecular interactions. *Current Opinion in Structural Biology*, *70*, 8–15. doi:10.1016/j.sbi.2021.02.006 PMID:33831785

Oung, E.W., Berthier, E., Guckenberger, D.J., Sackmann, E., Lamers, C., Meyvantsson, I. (2011). Rapid prototyping of arrayed microfluidic systems in polystyrene for cell-based assays. *Analytical Chemistry, 83* (4), 1408–1417. . PMC 3052265. PMID 21261280. doi:10.1021/ac102897h

Pabst, O., Ho¨lzer, S., Beckert, E., Perelaer, J., Schubert, U. S., Eberhardt, R., & Tünnermann, A. (2014). Inkjet printed micropump actuator based on piezoelectric polymers, Device performance and morphology studies. *Organic Electronics*, *15*(11), 3306–3315. doi:10.1016/j.orgel.2014.09.007

Pagaduan, J. V., Sahore, V., & Woolley, A. T. (2015). Applications of microfluidics and microchip electrophoresis for potential clinical biomarker analysis. *Analytical and Bioanalytical Chemistry*, *407*(23), 6911–6922. doi:10.100700216-015-8622-5 PMID:25855148

Park, C., Lee, J., Lee, D., & Jang, J. (2021). Paper-based electrochemical peptide sensor for label-free and rapid detection of airborne Bacillus anthracis simulant spores. *Sensors and Actuators. B, Chemical*, *355*, 131321. doi:10.1016/j.snb.2021.131321

Park, J. H., & Sudarshan, T. S. (Eds.). (2001). *Chemical vapor deposition* (Vol. 2). ASM international.

Park, S., Zhang, Y., Lin, S., Wang, T.-H., & Yang, S. (2011). Advances in microfluidic PCR for point-of-care infectious disease diagnostics. *Biotechnology Advances*, *29*(6), 830–839. doi:10.1016/j.biotechadv.2011.06.017 PMID:21741465

Partridge, A., Reynolds, J., Chui, B., Chow, E., Fitzgerald, A., Zhang, L., Maluf, N., & Kenny, T. (2000). A high-performance planar piezoresistive accelerometer. *Journal of Microelectromechanical Systems*, *9*(1), 58–66. doi:10.1109/84.825778

Parvizian, F., Rahimi, M., Azimi, N., & Alsairafi, A. A. (2014). CFD Modeling of Micromixing and Velocity Distribution in a 1.7-MHz Tubular Sonoreactor. *Chemical Engineering & Technology*, *37*(1), 113–122. doi:10.1002/ceat.201300231

Passaro, V., Dell'Olio, F., Casamassima, B., & De Leonardis, F. (2007). Guided-Wave Optical Biosensors. *Sensors (Basel)*, *7*(4), 508–536. doi:10.33907040508

Patel, D., Mishra, A., & Nabeel, M. (2022). Heat Transfer Characteristics of Nanofluids of Silicon Oxides (Sio2) with Conventional Fluid, 2022 2nd International Conference on Innovative Practices in Technology and Management (ICIPTM), (pp. 420-423). IEEE. 10.1109/ICIPTM54933.2022.9754181

Patel, D., Sharma, A., & Mishra, A. (2021). Study of Convective Heat Transfer Characteristics of Nano Fluids in Circular Tube 2021 International Conference on Technological Advancements and Innovations (ICTAI), (pp. 264-267). IEEE., 10.1109/ICTAI53825.2021.9673432

Patel, D., Singh, R. P., Rajput, R. S., & Tiwari, P. (2022). *Thermophyscical Properties And Applications Nanofluids – On Review.* 2022 5th International Conference on Contemporary Computing and Informatics (IC3I), Uttar Pradesh, India. 10.1109/IC3I56241.2022.10072842

Patil, J., Rajput, R., Patil, P., Mujumdar, A., & Naik, J. (2020). Generation of sustained release chitosan nanoparticles for delivery of ketorolac tromethamine: A tubular microreactor approach. *International Journal of Polymeric Materials, 69*(8), 516–524. doi:10.1080/00914037.2019.1581201

Pawell, R. S., Inglis, D. W., Barber, T. J., & Taylor, R. A. (2013). Manufacturing and wetting low-cost microfluidic cell separation devices. *Biomicrofluidics, 7*(5), 56501. doi:10.1063/1.4821315 PMID:24404077

Pecoraro, V., Negro, A., Pirotti, T., & Trenti, T. (2022). Estimate false-negative RT-PCR rates for SARS-CoV-2. A systematic review and meta-analysis. *European Journal of Clinical Investigation, 52*(2), e13706. doi:10.1111/eci.13706 PMID:34741305

Pekkari, A., Say, Z., Susarrey-Arce, A., Langhammer, C., Härelind, H., Sebastian, V., & Moth-Poulsen, K. (2019). Continuous Microfluidic Synthesis of Pd Nanocubes and PdPt Core–Shell Nanoparticles and Their Catalysis of NO $_2$ Reduction. *ACS Applied Materials & Interfaces, 11*(39), 36196–36204. doi:10.1021/acsami.9b09701 PMID:31418548

Peng, X., Jiang, M., Wang, X., Li, D., Sun, S., Shao, Y., & Zheng, Y. (2021). Photocatalytic purification of contaminated air in intensive care units by ZnSn(OH)6 nanoparticles. *Environmental Science and Pollution Research International, 28*(24), 31770–31777. Advance online publication. doi:10.100711356-021-12970-2 PMID:33611731

Pennemann, H., Watts, P., Haswell, S. J., Hessel, V., & Löwe, H. (2004). Benchmarking of microreactor applications. *Organic Process Research & Development, 8*(3), 422–439. doi:10.1021/op0341770

Peoples, M. C., & Karnes, H. T. (2008). Microfluidic immunoaffinity separations for bioanalysis. *Journal of Chromatography. B, Analytical Technologies in the Biomedical and Life Sciences, 866*(1–2), 14–25. doi:10.1016/j.jchromb.2007.08.030 PMID:17869593

Petersen, N., Jensen, H., Hansen, S., Foss, S., Snakenborg, D., & Pedersen-Bjergaard, S. (2010). On-chip electro membrane extraction. *Microfluidics and Nanofluidics, 9*(4-5), 881–888. doi:10.100710404-010-0603-6

Pfohl, T., Mugele, F., Seemann, R., & Herminghaus, S. (2003, December). Trends in microfluidics with complex fluids. *ChemPhysChem, 4*(12), 1291–1298. doi:10.1002/cphc.200300847 PMID:14714376

Pindyurin, V., Goldenberg, B., Petrova, E., Ancharova, U., Eliseev, V., Korolkov, V., & Nasyrov, R. (2008). Dynamic X-ray lithography for blazed diffractive optics fabrication. *Proceedings of SPIE - The International Society for Optical Engineering, 7102*. 10.1117/12.797728

Placer, L., Lavilla, I., Pena-Pereira, F., & Bendicho, C. (2023). A 3D Microfluidic Paper-Based Analytical Device with Smartphone-Assisted Colorimetric Detection for Iodine Speciation in Seaweed Samples. *Sensors and Actuators. B, Chemical, 377*, 133109. doi:10.1016/j.snb.2022.133109

Pokpas, K., Jahed, N., Bezuidenhout, P., Smith, S., Land, K., & Iwuoha, E. (2022). Nickel contamination analysis at cost-effective silver printed paper-based electrodes based on carbon black dimethylglyoxime ink as electrode modifier: Original scientific paper. *Journal of Electrochemical Science and Engineering, 12*(1), 153–164. doi:10.5599/jese.1173

Pollack, M.G., & Fair, R.B. (2000). Electrowetting-based actuation of liquid droplets for microfluidic applications. Appl. Phys. Lett., 77, 1725–1726.

Poolakkandy, R. R., & Menamparambath, M. M. (2020). Soft-template-assisted synthesis: A promising approach for the fabrication of transition metal oxides. *Nanoscale Advances, 2*(11), 5015–5045. doi:10.1039/D0NA00599A PMID:36132034

Postek, W., & Garstecki, P. (2022). Droplet Microfluidics for High-Throughput Analysis of Antibiotic Susceptibility in Bacterial Cells and Populations. *Accounts of Chemical Research, 55*(5), 605–615. doi:10.1021/acs.accounts.1c00729 PMID:35119826

Pourmand, A., Seyed, A. M. S., Ghavifekr, H. B., Aghdam, E. N., Dokmeci, M. R., Khademhosseini, A., & Zhang, Y. S. (2018). Fabrication of whole-thermoplastic normally closed microvalve, micro check valve, and micropump. *Sensors and Actuators B: Chemical, 262*, 625-636. doi:10.1016/j.snb.2017.12.132

Prakash, M., & Gershenfeld, N. (2007). Microfluidic bubble logic. *Science, 315* . doi:10.1126/science.1136907

Premraj, A., Aleyas, A. G., Nautiyal, B., & Rasool, T. J. (2020). Nucleic Acid and Immunological Diagnostics for SARS-CoV-2: Processes, Platforms and Pitfalls. *Diagnostics (Basel), 10*(11), 866. doi:10.3390/diagnostics10110866 PMID:33114057

Proskurina, O. V., Abiev, R. S., Danilovich, D. P., Panchuk, V. V., Semenov, V. G., Nevedomsky, V. N., & Gusarov, V. V. (2019). Formation of nanocrystalline BiFeO3 during heat treatment of hydroxides co-precipitated in an impinging-jets microreactor. *Chemical Engineering and Processing, 143*(July), 107598. doi:10.1016/j.cep.2019.107598

Qian, S., Duval, J. F. L., & Mahi, F. T. (2016). Mixers. In Reference Module in Materials Science and Materials Engineering (pp. 1–51). Elsevier. doi:10.1016/B978-0-12-803581-8.00547-6

Qi, J., Li, B., Zhou, N., Wang, X., Deng, D., Luo, L., & Chen, L. (2019). The strategy of antibody-free biomarker analysis by in-situ synthesized molecularly imprinted polymers on movable valve paper-based device. *Biosensors & Bioelectronics, 142*, 111533. doi:10.1016/j.bios.2019.111533 PMID:31377573

Qiu, G., Gai, Z., Tao, Y., Schmitt, J., Kullak-Ublick, G. A., & Wang, J. (2020). Dual-Functional Plasmonic Photothermal Biosensors for Highly Accurate Severe Acute Respiratory Syndrome Coronavirus 2 Detection. *ACS Nano, 14*(5), 5268–5277. doi:10.1021/acsnano.0c02439 PMID:32281785

Qu, Y., Zhou, J., & Wu, W. (2018). Theoretical and Experimental Research on Bubble Actuated Micro-Pumps. *Micromachines, 9*(225), 1–11. doi:10.3390/mi9050225 PMID:30424158

Rabarot, M., Bablet, J., Ruty, M., Kipp, M., Chartier, I., & Dubarry, C. (2003). Thick SU-8 photolithography for BioMEMS. *Micromachining and Microfabrication Process Technology Viii, 4979*, 382–393. doi:10.1117/12.478244

Rahaman, A., Ishfaque, A., Jung, H., & Kim, B. (2019). Bio-inspired rectangular shaped piezoelectric MEMS directional microphone. *IEEE Sensors Journal, 19*(1), 88–96. doi:10.1109/JSEN.2018.2873781

Rajaram, V., Qian, Z., Kang, S., Calisgan, S., McGruer, N., & Rinaldi, M. (2018). Zero-power electrically tunable micromechanical photoswitches. *IEEE Sensors Journal, 18*(19), 7833–7841. doi:10.1109/JSEN.2018.2850898

Ramezani, S., Zahedi, P., Bahrami, S. H., & Nemati, Y. (2019). Microfluidic Fabrication of Nanoparticles Based on Ethyl Acrylate-Functionalized Chitosan for Adsorption of Methylene Blue from Aqueous Solutions. *Journal of Polymers and the Environment, 27*(8), 1653–1665. Advance online publication. doi:10.100710924-019-01463-6

Ran, G., Fu, Q., & Xu, W. (2015). Microfluidic-based controllable synthesis of Pt nanocatalysts supported on carbon for fuel cells. *RSC Advances, 5*(19), 14740–14746. Advance online publication. doi:10.1039/C4RA12145G

Rani, M., Keshu, & Shanker, U. (2021). Sunlight-induced photocatalytic degradation of organic pollutants by biosynthesized hetrometallic oxides nanoparticles. *Environmental Science and Pollution Research International, 28*(43), 61760–61780. doi:10.100711356-021-15003-0 PMID:34189687

Rathi, N., & Gaikar, V. G. (2017). Optimization of Continuous Synthesis of Cross-Linked Chitosan Nanoparticles Using Microreactors. *Chemical Engineering & Technology, 40*(3), 506–513. doi:10.1002/ceat.201600339

Rattanarat, P., Dungchai, W., Siangproh, W., Chailapakul, O., & Henry, C. S. (2012). Sodium dodecyl sulfate-modified electrochemical paper-based analytical device for determination of dopamine levels in biological samples. *Analytica Chimica Acta, 744*, 1–7. doi:10.1016/j.aca.2012.07.003 PMID:22935367

Raza, W., Hossain, S., & Kim, K. Y. (2020). A review of passive micromixers with a comparative analysis. *Micromachines, 11*(5), 455. Advance online publication. doi:10.3390/mi11050455 PMID:32349452

Ren, K., Dai, W., Zhou, J., Su, J., & Wu, H. (2011). Whole-teflon microfluidic chips. *Proceedings of the National Academy of Sciences of the United States of America, 108*(20), 8162–8166. doi:10.1073/pnas.1100356108 PMID:21536918

Ren, K., Zhou, J., & Wu, H. (2013). Materials for microfluidic chip fabrication. *Accounts of Chemical Research, 46*(11), 2396–2406. doi:10.1021/ar300314s PMID:24245999

Reseach, G. V. (2021). Injection Molded Plastics Market Size Report, 2021-2028. *Grand View Research.* https://www.grandviewresearch.com/industry-analysis/injection-molded-plastics-market

Research and Markets. (n.d.). *Home.* Research and Markets. https://www.researchandmarkets.com

Rezai, P., Wu, W., & Selvaganapathy, P. (2012). Microfabrication of polymers for bioMEMS. In *MEMS for Biomedical Applications* (pp. 3–45). Elsevier. doi:10.1533/9780857096272.1.3

Riahi, R., Tamayol, A., Shaegh, S. A. M., Ghaemmaghami, A. M., Dokmeci, M. R., & Khademhosseini, A. (2015). Microfluidics for advanced drug delivery systems. *Current Opinion in Chemical Engineering, 7*, 101–112. doi:10.1016/j.coche.2014.12.001 PMID:31692947

Richardson, J. R., Fitsanakis, V., Westerink, R. H. S., & Kanthasamy, A. G. (2019). Neurotoxicity of Pesticides. *Acta Neuropathologica, 138*(3), 343–362. doi:10.100700401-019-02033-9 PMID:31197504

Rizkin, B. A., Shkolnik, A. S., Ferraro, N. J., & Hartman, R. L. (2020). Combining automated microfluidic experimentation with machine learning for efficient polymerization design. *Nature Machine Intelligence, 2*(4), 200–209. doi:10.103842256-020-0166-5

Robertson, J. M., Rodriguez, R. X., Holmes, L. R. Jr, Mather, P. T., & Wetzel, E. D. (2016). Thermally driven microfluidic pumping via reversible shape memory polymers. *Smart Materials and Structures, 25*(8), 085043. doi:10.1088/0964-1726/25/8/085043

Rockx, B. (2020). Comparative pathogenesis of COVID-19, MERS, and SARS in a nonhuman primate model. *Science, 368*, 1012–1015.

Romasanta, L. J., D'Alençon, L., Kirchner, S., Pradère, C., & Leng, J. (2019). Thin Coatings of Cerium Oxide Nanoparticles with Anti-Reflective Properties. *Applied Sciences (Basel, Switzerland), 9*(18), 3886. doi:10.3390/app9183886

Rotake, D. R., Darji, A. D., & Kale, N. S. (2019). Design of microfluidic experimental setup for the detection of heavy metal ions using piezoresistive biomems sensor. *Microelectronics International, 33*(1), 10–28. doi:10.1108/MI-05-2019-0025

Roy, D., Chakraborty, S., Gupta, A. K., BasuMallick, A., Scattergood, R. O., & Koch, C. C. (2021). Synergistic effect of Nb and Zr additions on the structure-property relationships of nanocrystalline Cu processed by mechanical alloying and hot pressing. *Journal of Alloys and Compounds, 854*, 157174. Advance online publication. doi:10.1016/j.jallcom.2020.157174

Rusli, M., Chee, P. S., Arsat, R., Lau, K. X., & Leow, P. L. (2018). Electromagnetic actuation dual-chamber bidirectional flow micropump. *Sensors and Actuators. A, Physical, 282*, 17–27. doi:10.1016/j.sna.2018.08.047

Russel, M., Selvaganapathy, P., & Ching, C. (2016). Ion drag electrohydrodynamic (EHD) micro-pumps under a pulsed voltage. *Journal of Electrostatics*, *82*, 48–54. doi:10.1016/j.elstat.2016.05.003

Saadati, A., Kholafazad Kordasht, H., Ehsani, M., Hasanzadeh, M., Seidi, F., & Shadjou, N. (2021). An innovative flexible and portable DNA based biodevice towards sensitive identification of Haemophilus influenzae bacterial genome: A new platform for the rapid and low cost recognition of pathogenic bacteria using point of care (POC) analysis. *Microchemical Journal*, *169*, 106610. doi:10.1016/j.microc.2021.106610

Sackmann, E. K., Fulton, A. L., & Beebe, D. J. (2014). The present and future role of microfluidics in biomedical research. *Nature*, *507*(7491), 181–189. doi:10.1038/nature13118 PMID:24622198

Sadler, D. J., Changrani, R., Roberts, P., Chou, C.-F., & Zenhausern, F. (2002). Thermal management of BioMEMS. *ITherm 2002 Eighth Intersociety Conference on Thermal and Thermomechanical Phenomena in Electronic Systems.*

Sadler, D. J., Changrani, R., Roberts, P., Chou, C.-F., & Zenhausern, F. (2003). Thermal management of BioMEMS: Temperature control for ceramic-based PCR and DNA detection devices. *IEEE Transactions on Components and Packaging Technologies*, *26*(2), 309–316. doi:10.1109/TCAPT.2003.815093

Said, M.M., Yunas, J., & Bais, B. (2017). The design, fabrication, and testing of an electromagnetic micropump with a matrix-patterned magnetic polymer composite actuator membrane. *Micromachines, 9*(13).

Saliterman, S. (2006). *Fundamentals of BioMEMS and medical microdevices* (Vol. 153). SPIE press.

Salve, M., Mandal, A., Amreen, K., Pattnaik, P. K., & Goel, S. (2020). Greenly synthesized silver nanoparticles for supercapacitor and electrochemical sensing applications in a 3D printed microfluidic platform. *Microchemical Journal*, *157*, 104973. Advance online publication. doi:10.1016/j.microc.2020.104973

Sánchez-Cano, A., Ruiz-Vega, G., Vicente-Gómez, S., de la Serna, E., Sulleiro, E., Molina, I., Sánchez-Montalvá, A., & Baldrich, E. (2021). Development of a Fast Chemiluminescent Magneto-Immunoassay for Sensitive *Plasmodium falciparum* Detection in Whole Blood. *Analytical Chemistry*, *93*(37), 12793–12800. doi:10.1021/acs.analchem.1c03242 PMID:34496566

Sang, S. (2010). *An approach to the design of surface stress-based PDMS micro-membrane biosensors-concept, numerical simulations and prototypes*. Academic Press.

Sanjay, S. T., Fu, G., Dou, M., Xu, F., Liu, R., Qi, H., & Li, X. J. (2015). Biomarker detection for disease diagnosis using cost-effective microfluidic platforms. *Analyst (London)*, *140*(21), 7062–7081. doi:10.1039/C5AN00780A PMID:26171467

Sankar, K., Lenisha, D., Janaki, G., Juliana, J., Kumar, R. S., Selvi, M. C., & Srinivasan, G. (2020). Digital Image-Based Quantification of Chlorpyrifos in Water Samples Using a Lipase Embedded Paper Based Device. *Talanta*, *208*, 120408. doi:10.1016/j.talanta.2019.120408 PMID:31816733

Sasaki, K., Miyake, K., Uchida, Y., & Nishiyama, N. (2022). Mechanochemical Synthesis of Dispersible Platinum Nanosheets for Enhanced Catalysis in a Microreactor. *ACS Applied Nano Materials*, *5*(4), 4998–5005. doi:10.1021/acsanm.1c04583

Sato, T., Yamanishi, Y., Cacucciolo, V., Kuwajima, Y., Shigemune, H., Cianchetti, M., Laschi, C., & Maeda, S. (2017). Electrohydrodynamic conduction pump with asymmetrical electrode structures in the microchannels. *Chemistry Letters*, *46*(7), 950–952. doi:10.1246/cl.170217

Savjani, K. T., Gajjar, A. K., & Savjani, J. K. (2012). Drug Solubility: Importance and Enhancement Techniques. *ISRN Pharmaceutics*, *2012*, 1–10. doi:10.5402/2012/195727 PMID:22830056

Sawetwong, P., Chairam, S., Jarujamrus, P., & Amatatongchai, M. (2020, December). 2021. "Enhanced Selectivity and Sensitivity for Colorimetric Determination of Glyphosate Using Mn–ZnS Quantum Dot Embedded Molecularly Imprinted Polymers Combined with a 3D-Microfluidic Paper-Based Analytical Device." *Talanta, 225*, 122077. doi:10.1016/j.talanta.2020.122077 PMID:33592801

Sayar, E., & Farouk, B. (2013). Electroosmotic Augmentation in Flexural Plate Wave Micropumps. *Journal of Microelectromechanical Systems*, *22*(2), 372–385. doi:10.1109/JMEMS.2012.2226930

Schaller, R. R. (1997, June). Moore's law: Past, present and future. *IEEE Spectrum*, *34*(6), 52–59. doi:10.1109/6.591665

Scholten, K., & Meng, E. (2016). Electron-beam lithography for polymer bioMEMS with submicron features. *Microsystems & Nanoengineering*, *2*(1), 1–7. doi:10.1038/micronano.2016.53 PMID:31057839

Schrott, W., Svoboda, M., Slouka, Z., & Snita, D. (2009). Metal electrodes in plastic microfluidic systems. *Microelectronic Engineering*, *86*(4-6), 1340–1342. doi:10.1016/j.mee.2009.01.001

Scordo, G., Moscone, D., Palleschi, G., & Arduini, F. (2018). A Reagent-Free Paper-Based Sensor Embedded in a 3D Printing Device for Cholinesterase Activity Measurement in Serum. *Sensors and Actuators. B, Chemical*, *258*, 1015–1021. doi:10.1016/j.snb.2017.11.134

Scullion, M. G., Di Falco, A., & Krauss, T. F. (2011). Slotted photonic crystal cavities with integrated microfluidics for biosensing applications. *Biosensors & Bioelectronics*, *27*(1), 101–105. doi:10.1016/j.bios.2011.06.023 PMID:21764290

Sebastian, V., Arruebo, M., & Santamaria, J. (2014). Reaction Engineering Strategies for the Production of Inorganic Nanomaterials. *Small, 10*(5), 835–853. doi:10.1002mll.201301641 PMID:24123934

Sen, Y.H., & Karnik, R. (2009). Investigating the translocation of lambda-DNA molecules through PDMS nanopores. *Analytical and Bioanalytical Chemistry, 394* (2), 437–446. . doi:10.1007/s00216-008-2529-3

Sen, N., Shaikh, T., Singh, K. K., Sirsam, R., & Shenoy, K. T. (2020). Synthesis of polyacrylamide (PAM) beads in microreactors. *Chemical Engineering and Processing*, *157*(April), 108105. doi:10.1016/j.cep.2020.108105

Seo, G., Lee, G., Kim, M. J., Baek, S. H., Choi, M., Ku, K. B., Lee, C. S., Jun, S., Park, D., Kim, H. G., Kim, S. J., Lee, J. O., Kim, B. T., Park, E. C., & Kim, S. I. (2020). Rapid Detection of COVID-19 Causative Virus (SARS-CoV-2) in Human Nasopharyngeal Swab Specimens Using Field-Effect Transistor-Based Biosensor. *ACS Nano*, *14*(4), 5135–5142. doi:10.1021/acsnano.0c02823 PMID:32293168

Shaegh, S. A. M., Wang, Z., Ng, S. H., Wu, R., Nguyen, H. T., Chan, L. C. Z., Toh, A. G. G., & Wang, Z. (2015). Plug-and-play microvalve and micropump for rapid integration with microfluidic chips. *Microfluidics and Nanofluidics*, *19*(3), 557–564. doi:10.100710404-015-1582-4

Shah, A., Aftab, S., Nisar, J., Ashiq, M. N., & Iftikhar, F. J. (2021). Nanocarriers for targeted drug delivery. *Journal of Drug Delivery Science and Technology*, *62*(February), 102426. doi:10.1016/j.jddst.2021.102426

Sharif, N., Alzahrani, K. J., Ahmed, S. N., & Dey, S. K. (2021). Efficacy, Immunogenicity and Safety of COVID-19 Vaccines: A Systematic Review and Meta-Analysis. *Frontiers in Immunology*, *12*, 714170. doi:10.3389/fimmu.2021.714170 PMID:34707602

Sharma, S., & Madou, M. (2012). Micro and nano patterning of carbon electrodes for bioMEMS. *Bioinspired. Biomimetic and Nanobiomaterials*, *1*(4), 252–265. doi:10.1680/bbn.12.00010

Shemesh, J., Bransky, A., Khoury, M., & Levenberg, S. (2010). Advanced microfluidic droplet manipulation based on piezoelectric actuation. *Biomedical Microdevices, 12* (5), 907–914. . doi:10.1007/s10544-010-9445-y

Shen, L.-L., Zhang, G.-R., & Etzold, B. J. M. (2020). Paper-Based Microfluidics for Electrochemical Applications. *ChemElectroChem*, *7*(1), 10–30. doi:10.1002/celc.201901495 PMID:32025468

Shestopalov, I., Tice, J.D., & Ismagilov, R.F. (2004). Multi-step synthesis of nanoparticles performed on millisecond time scale in a microfluidic droplet-based system. *Lab on a Chip, 4*(4), 316–321. doi: 15269797. doi:10.1039/b403378g.PMID

Shi, J. (2020). Susceptibility of ferrets, cats, dogs, and other domesticated animals to SARS-coronavirus 2. *Science, 368*, 1016–1020.

Shih, S. C. C., Goyal, G., Kim, P. W., Koutsoubelis, N., Keasling, J. D., Adams, P. D., Hillson, N. J., & Singh, A. K. (2015). A versatile microfluidic device for automating synthetic biology. *ACS Synthetic Biology*, *10*(10), 1151–1164. doi:10.1021/acssynbio.5b00062 PMID:26075958

Shilton, R. J., Travagliati, M., Beltram, F., & Cecchini, M. (2014, August). Nanoliter-droplet acoustic streaming via ultra high frequency surface acoustic waves. *Advanced Materials*, *26*(29), 4941–4946. doi:10.1002/adma.201400091 PMID:24677370

Shin, D., Ahn, C., Chen, Y., Christensen, D., Flader, I., & Kenny, T. Environmentally robust differential resonant accelerometer in a wafer-scale encapsulation process. *Proceedings of the IEEE 30th International Conference on Micro Electro Mechanical Systems (MEMS),* 17–20. 10.1109/MEMSYS.2017.7863328

Shiri, M. S. Z., Henderson, W., & Mucalo, M. R. (2019). A review of the lesser-studied microemulsion-based synthesis methodologies used for preparing nanoparticle systems of the noble metals, Os, Re, Ir and Rh. In Materials (Vol. 12, Issue 12). doi:10.3390/ma12121896

Shkolnikov, V. (2019). *Principles of Microfluidics.*

Shoji, S., Sato, H., & Zengerle, R. (2008). Liquid Micropumps. In Comprehensive Microsystems. Elsevier. doi:10.1016/B978-044452190-3.00041-0

Shoji, E. (2016). Fabrication of a diaphragm micropump system utilizing the ionomer-based polymer actuator. *Sensors and Actuators. B, Chemical*, *237*, 660–665. doi:10.1016/j.snb.2016.06.153

Shrimal, P., Jadeja, G., Naik, J., & Patel, S. (2019). Continuous microchannel precipitation to enhance the solubility of telmisartan with poloxamer 407 using Box-Behnken design approach. *Journal of Drug Delivery Science and Technology*, *53*(August), 101225. doi:10.1016/j.jddst.2019.101225

Siavashy, S., Soltani, M., Ghorbani-Bidkorbeh, F., Fallah, N., Farnam, G., Mortazavi, S. A., Shirazi, F. H., Tehrani, M. H. H., & Hamedi, M. H. (2021). Microfluidic platform for synthesis and optimization of chitosan-coated magnetic nanoparticles in cisplatin delivery. *Carbohydrate Polymers*, *265*, 118027. doi:10.1016/j.carbpol.2021.118027 PMID:33966822

Silva, M. K. L., Sousa, G. S., Simoes, R. P., & Cesarino, I. (2022). Fabrication of paper-based analytical devices using a PLA 3D-printed stencil for electrochemical determination of chloroquine and escitalopram. *Journal of Solid State Electrochemistry*, *26*(2), 581–586. doi:10.100710008-021-05075-w PMID:34751209

Sima, A.H., Salari, A., & Shafii, M.B. (2015). Low-cost reciprocating electromagnetic-based micropump for high-flow rate applications. *J Micro/Nanolithogr MEMS MOEMS, 14.*

Simmons, M. D., Jones, N., Evans, D. J., Wiles, C., Watts, P., Salamon, S., Escobar Castillo, M., Wende, H., Lupascu, D. C., & Francesconi, M. G. (2015). Doping of inorganic materials in microreactors-preparation of Zn doped Fe3O4 nanoparticles. *Lab on a Chip*, *15*(15), 3154–3162. doi:10.1039/C5LC00287G PMID:26099495

Simonnet, C., & Groisman, A. (2005). Chaotic mixing in a steady flow in a microchannel. *Physical Review Letters*, *94*(13), 8–11. doi:10.1103/PhysRevLett.94.134501 PMID:15903994

Singh, S., Kumar, N., George, D., & Sen, A. K. (2015). Analytical modeling, simulations and experimental studies of a PZT actuated planar valveless PDMS micropump. *Sensors and Actuators A: Physical, 225*, 81-94. doi:10.1016/j.sna.2015.02.012

Singh, A. K., & Patel, D. (2020). Optimization of Air Flow over a Car by Wind Tunnel. *2020 International Conference on Computation, Automation and Knowledge Management (ICCAKM)*, (pp. 1-4). IEEE. 10.1109/ICCAKM46823.2020.9051457

Singh, A. P., Balayan, S., Hooda, V., Sarin, R. K., & Chauhan, N. (2020). Nano-Interface Driven Electrochemical Sensor for Pesticides Detection Based on the Acetylcholinesterase Enzyme Inhibition. *International Journal of Biological Macromolecules, 164*, 3943–3952. doi:10.1016/j.ijbiomac.2020.08.215 PMID:32882280

Singhal, T. (2020). A Review of Coronavirus Disease-2019 (COVID-19). *Indian Journal of Pediatrics, 87*(4), 281–286. doi:10.100712098-020-03263-6 PMID:32166607

Singh, R. K., Kumar, A., Kant, R., Gupta, A., Suresh, E., & Bhattacharya, S. (2014). Design and fabrication of 3-dimensional helical structures in polydimethylsiloxane for flow control applications. *Microsystem Technologies, 20*(1), 101–111. doi:10.100700542-013-1738-7

Singh, R., & Lillard, J. W. Jr. (2009). Nanoparticle-based targeted drug delivery. *Experimental and Molecular Pathology, 86*(3), 215–223. doi:10.1016/j.yexmp.2008.12.004 PMID:19186176

Sinha, A., Basu, M., & Chandna, P. (2022). Paper based microfluidics: A forecast toward the most affordable and rapid point-of-care devices. *Progress in Molecular Biology and Translational Science, 186*(1), 109–158. doi:10.1016/bs.pmbts.2021.07.010 PMID:35033281

Sinha, N., Wabiszewski, G., Mahameed, R., Felmetsger, V., Tanner, S., Carpick, R., & Piazza, G. (2009). Piezoelectric aluminum nitride nanoelectromechanical actuators. *Applied Physics Letters, 95*(5), 4–7. doi:10.1063/1.3194148

Soon, B. W., Singh, N., Tsai, J. M., & Lee, C. (2013). A Bistable silicon nanofin: An ideal device for nonvolatile memory applications. *IEEE Nanotechnology Magazine, 7*(2), 24–28. doi:10.1109/MNANO.2013.2260461

Spieth, S., Schumacher, A., & Kallenbach, C. (2012). The Neuro Medicator a micropump integrated with silicon microprobes for drug delivery in neural research. *J Micromech Microeng, 22.*

Spille, E., & Feder, R. (1977). X-ray lithography. *X-Ray Optics*, 35-92.

Squires, T. M., Messinger, R. J. & Manalis, S. R. (2008). Making it stick: convection, reaction and diffusion in surface-based biosensors. *Nature Biotechnology, 26,* 417–426.

Squires, M., & Quake Microfluidics, S. R. (2005). fluid physics at the nanoliter scale. *Reviews of Modern Physics, 3*(3), 977–1026. doi:10.1103/RevModPhys.77.977

Sri Sruthi, P., Balasubramanian, S., Senthil Kumar, P., Kapoor, A., Ponnuchamy, M., Mariam Jacob, M., & Prabhakar, S. (2021). Eco-friendly pH detecting paper-based analytical device: Towards process intensification. *Analytica Chimica Acta, 1182*, 338953. doi:10.1016/j.aca.2021.338953 PMID:34602199

Sridhar, A., Kapoor, A., Kumar, P. S., Ponnuchamy, M., Sivasamy, B., & Vo, D.-V. N. (2022). Lab-on-a-chip technologies for food safety, processing, and packaging applications: A review. *Environmental Chemistry Letters, 20*(1), 901–927. doi:10.100710311-021-01342-4 PMID:34803553

Stevanus, J., Villanueva, A. H. R., Paliath, N., & Rifat, R. A. (2022). *JMN455–SSI Design Lab II: Manufacturing Processes for Smart Systems Lab Report*. Academic Press.

Stolzenburg, P., Lorenz, T., Dietzel, A., & Garnweitner, G. (2018). Microfluidic synthesis of metal oxide nanoparticles via the nonaqueous method. *Chemical Engineering Science*, *191*, 500–510. Advance online publication. doi:10.1016/j.ces.2018.07.007

Sudarsan, A. P., & Ugaz, V. M. (2006a). Fluid mixing in planar spiral microchannels. *Lab on a Chip*, *6*(1), 74–82. doi:10.1039/B511524H PMID:16372072

Sudarsan, A. P., & Ugaz, V. M. (2006b). Multivortex micromixing. *Proceedings of the National Academy of Sciences of the United States of America*, *103*(19), 7228–7233. doi:10.1073/pnas.0507976103 PMID:16645036

Sudhakara Prasad, K., Cao, X., Gao, N., Jin, Q., Sanjay, S. T., Henao-Pabon, G., & Li, X. (2020). A Low-Cost Nanomaterial-based Electrochemical Immunosensor on Paper for High-Sensitivity Early Detection of Pancreatic Cancer. *Sensors and Actuators. B, Chemical*, 305. PMID:32863588

Sugioka, K., Cheng, Y., Masuda, M., & Midorikawa, K. (2004). In P. R. Herman, J. Fieret, A. Pique, T. Okada, F. G. Bachmann, W. Hoving, K. Washio, X. Xu, J. J. Dubowski, D. B. Geohegan, & F. Trager (Eds.), *Photon Processing in Microelectronics and Photonics Iii* (Vol. 5339, pp. 205–213). Spie-Int Soc Optical Engineering. doi:10.1117/12.531630

Sunaina, Kaur, H., Kumari, N., Sharma, A., Sachdeva, M., & Mutreja, V. (2022). Optical and Electrochemical Microfluidic Sensors for Water Contaminants: A Short Review. *Materials Today: Proceedings*, *48*, 1673–1679. doi:10.1016/j.matpr.2021.09.534

Sundaram, N., & Tafti, D. K. (2004). Evaluation of microchamber geometries and surface conditions for electrokinetic driven mixing. *Analytical Chemistry*, *76*(13), 3785–3793. doi:10.1021/ac0499591 PMID:15228355

Sung, W., Sung, S., Lee, J., Kang, T., Lee, Y., & Lee, J. (2008). Development of a lateral velocity-controlled MEMS vibratory gyroscope and its performance test. *Journal of Micromechanics and Microengineering*, *18*(5), 055028. doi:10.1088/0960-1317/18/5/055028

Sun, T., Park, W.-T., Tsang, J. W. M., Yee, T. B., & Je, M. (2013). Cytocompatibility assessment of Si, plasma enhanced chemical vapor deposition-formed SiO2 and Si3N4 used for neural prosthesis: A comparative study. *Nanoscience and Nanotechnology Letters*, *5*(8), 916–920. doi:10.1166/nnl.2013.1629

Swerdlow, H., & Gesteland, R. (1990). Capillary gel electrophoresis for rapid, high resolution DNA sequencing. *Nucleic Acids Research*, *18*(6), 1415–1419. https://sci-hub.do/https://academic.oup.com/nar/article-abstract/18/6/1415/1325051. doi:10.1093/nar/18.6.1415 PMID:2326186

Szmigiel, D., Pamula, E., Sobczak, J. W., & Grabiec, P. (2007). Surface properties of silicone rubber for sensor applications. *SMCBS'2007 International Workshop*.

Tabeling, P. (2005). *Introduction to Microfluidics*. Oxford University Press.

Tahir, N., Madni, A., Li, W., Correia, A., Khan, M. M., Rahim, M. A., & Santos, H. A. (2020). Microfluidic fabrication and characterization of Sorafenib-loaded lipid-polymer hybrid nanoparticles for controlled drug delivery. *International Journal of Pharmaceutics*, *581*, 119275. Advance online publication. doi:10.1016/j.ijpharm.2020.119275 PMID:32229283

Tahmasebi, S., Jerkiewicz, G., Baranton, S., Coutanceau, C., Furuya, Y., & Ohma, A. (2018). How Stable Are Spherical Platinum Nanoparticles Applied to Fuel Cells? *The Journal of Physical Chemistry C*, *122*(22), 11765–11776. Advance online publication. doi:10.1021/acs.jpcc.7b10617

Takashima, A., Kojima, K., & Suzuki, H. (2010). Autonomous Microfluidic Control by Chemically Actuated Micropumps and Its Application to Chemical Analyses. *Analytical Chemistry*, *82*(16), 6870–6876. doi:10.1021/ac1009657 PMID:20669895

Tanaka, Y., Noguchi, Y., Yalikun, Y., & Kamamichi, N. (2017). Earthworm muscle driven bio-micropump. *Sensors and Actuators. B, Chemical*, *242*, 1186–1192. doi:10.1016/j.snb.2016.09.123

Tang, M. Y. H., & Shum, H. C. (2016). One-step immunoassay of C-reactive protein using droplet microfluidics. *Lab on a Chip*, *16*(22), 4359–4365. doi:10.1039/C6LC01121G PMID:27738692

Tang, Z., Kim, W. S., & Yu, T. (2019). Studies on morphology changes of copper sulfide nanoparticles in a continuous Couette-Taylor reactor. *Chemical Engineering Journal*, *359*, 1436–1441. doi:10.1016/j.cej.2018.11.034

Tao, K., Lye, S., Miao, J., Tang, L., & Hu, X. (2015). Out-of-plane electret-based MEMS energy harvester with the combined nonlinear effect from electrostatic force and a mechanical elastic stopper. *Journal of Micromechanics and Microengineering*, *25*(10), 104014. doi:10.1088/0960-1317/25/10/104014

Tay, A., Pavesi, A., Yazdi, S. R., Lim, C. T., & Warkiani, M. E. (2016). Advances in microfluidics in combating infectious diseases. *Biotechnology Advances*, *34*(4), 404–421. doi:10.1016/j.biotechadv.2016.02.002 PMID:26854743

Tayyab, M., Sami, M. A., Raji, H., Mushnoori, S., & Javanmard, M. (2020). Potential microfluidic devices for COVID-19 antibody detection at point-of-care (POC): A review. *IEEE Sensors Journal*, *21*(4), 4007–4017. doi:10.1109/JSEN.2020.3034892

Tenje, M., Fornell, A., Ohlin, M., & Nilsson, J. (2018, February). Particle Manipulation Methods in Droplet Microfluidics. *Analytical Chemistry*, *90*(3), 1434–1443. doi:10.1021/acs.analchem.7b01333 PMID:29188994

Terry, S. C., Jerman, J. H., & Angell, J. B. (1979). A gas chromatographic air analyzer fabricated on a silicon wafer. *IEEE Transactions on Electron Devices*, *26*(12), 1880–1886. doi:10.1109/T-ED.1979.19791

Thu, V. T., Mai, A. N., Van Trung, H., Thu, P. T., Tien, B. Q., Thuat, N. T., & Lam, T. D. (2016). Fabrication of PDMS-Based Microfluidic Devices: Application for Synthesis of Magnetic Nanoparticles. *Journal of Electronic Materials*, *45*(5), 2576–2581. Advance online publication. doi:10.100711664-016-4424-6

Tian, T., Bi, Y., Xu, X., Zhu, Z., & Yang, C. (2018). Integrated Paper-Based Microfluidic Devices for Point-of-Care Testing. *Analytical Methods*, *10*(29), 3567–3581. doi:10.1039/C8AY00864G

Tian, Z., Ge, X., Wang, Y., & Xu, J. (2018). Nanoparticles and nanocomposites with microfluidic technology. In *Polymer-Based Multifunctional Nanocomposites and Their Applications*. Elsevier Inc. doi:10.1016/B978-0-12-815067-2.00001-9

Tiwari, G., Tiwari, R., Bannerjee, S., Bhati, L., Pandey, S., Pandey, P., & Sriwastawa, B. (2012). Drug delivery systems: An updated review. *International Journal of Pharmaceutical Investigation*, *2*(1), 2. doi:10.4103/2230-973X.96920 PMID:23071954

Tofighi, G., Lichtenberg, H., Gaur, A., Wang, W., Wild, S., Herrera Delgado, K., Pitter, S., Dittmeyer, R., Grunwaldt, J. D., & Doronkin, D. E. (2022). Continuous synthesis of Cu/ZnO/Al2O3nanoparticles in a co-precipitation reaction using a silicon based microfluidic reactor. *Reaction Chemistry & Engineering*, *7*(3), 730–740. Advance online publication. doi:10.1039/D1RE00499A

Tokachichu, D. R., & Bhushan, B. (2006). Bioadhesion of polymers for BioMEMS. *IEEE Transactions on Nanotechnology*, *5*(3), 228–231. doi:10.1109/TNANO.2006.874047

Tomei, M. R., Cinti, S., Interino, N., Manovella, V., Moscone, D., & Arduini, F. (2019). Paper-based electroanalytical strip for user-friendly blood glutathione detection. *Sensors and Actuators. B, Chemical*, *294*, 294. doi:10.1016/j.snb.2019.02.082

Tong, X., Lin, X., Duan, N., Wang, Z., & Wu, S. (2022). Laser-Printed Paper-Based Microfluidic Chip Based on a Multicolor Fluorescence Carbon Dot Biosensor for Visual Determination of Multiantibiotics in Aquatic Products. *ACS Sensors*, *7*(12), 3947–3955. Advance online publication. doi:10.1021/acssensors.2c02008 PMID:36454704

Toprak, A., & Tigli, O. (2015). MEMS scale PVDF-TrFE-based piezoelectric energy harvesters. *Journal of Microelectromechanical Systems, 24*(6), 1989–1997. doi:10.1109/JMEMS.2015.2457782

Toropov, N., & Vartanyan, T. (2019). 1.03 Noble Metal Nanoparticles: Synthesis and Optical Properties. *Comprehensive Nanoscience and Nanotechnology*, 61.

Torul, H., Yarali, E., Eksin, E., Ganguly, A., Benson, J., Tamer, U., Papakonstantinou, P., & Erdem, A. (2021). Paper-Based Electrochemical Biosensors for Voltammetric Detection of miRNA Biomarkers Using Reduced Graphene Oxide or MoS_2 Nanosheets Decorated with Gold Nanoparticle Electrodes. *Biosensors (Basel), 11*(7), 236. doi:10.3390/bios11070236 PMID:34356708

Trache, D., Tarchoun, A. F., Derradji, M., Hamidon, T. S., Masruchin, N., Brosse, N., & Hussin, M. H. (2020). Nanocellulose: From Fundamentals to Advanced Applications. *Frontiers in Chemistry, 8*, 392. doi:10.3389/fchem.2020.00392 PMID:32435633

Tripathi, D., Sharma, A., & Be'g, O. A. (2017). Electrothermal transport of nanofluids via peristaltic pumping in a finite micro-channel, effects of Joule heating and Helmholtz-Smoluchowski velocity. *International Journal of Heat and Mass Transfer, 111*, 138–149. doi:10.1016/j.ijheatmasstransfer.2017.03.089

Trofimchuk, E., Nilghaz, A., Sun, S., & Lu, X. (2020). Determination of norfloxacin residues in foods by exploiting the coffee-ring effect and paper-based microfluidics device coupling with smartphone-based detection. *Journal of Food Science. 85* (3), 736–743. . doi:10.1111/1750-3841.15039

Truckenmüller, R., Rummler, Z., Schaller, T., & Schomburg, W. K. (2002). Low-cost thermoforming of micro fluidic analysis chips. *Journal of Micromechanics andMicroengineering. 12* (4), 375–379. . doi:10.1088/0960-1317/12/4/304

Tsagkaris, A. S., Migliorelli, D., Uttl, L., Filippini, D., Pulkrabova, J., & Hajslova, J. (2021). A Microfluidic Paper-Based Analytical Device (MPAD) with Smartphone Readout for Chlorpyrifos-Oxon Screening in Human Serum. *Talanta, 222*, 121535. doi:10.1016/j.talanta.2020.121535 PMID:33167243

Tsai, M., Liu, Y., & Fang, W. (2012). A three-axis CMOS-MEMS accelerometer structure with vertically integrated fully differential sensing electrodes. *Journal of Microelectromechanical Systems, 21*(6), 1329–1337. doi:10.1109/JMEMS.2012.2205904

Tsai, Y., Jhu, W., Chen, C., Kao, C., & Chen, C. (2019). Unity game engine: Interactive software design using digital glove for virtual reality baseball pitch training. *Microsystem Technologies, 9*, 1–17.

Tsang, H. F., Chan, L., Cho, W., Yu, A., Yim, A., Chan, A., Ng, L., Wong, Y., Pei, X. M., Li, M., & Wong, S. C. (2021). An update on COVID-19 pandemic: The epidemiology, pathogenesis, prevention and treatment strategies. *Expert Review of Anti-Infective Therapy, 19*(7), 877–888. doi:10.1080/14787210.2021.1863146 PMID:33306423

Tymm, C., Zhou, J., Tadimety, A., Burklund, A., & Zhang, J. (2020). Scalable COVID-19 Detection Enabled by Lab-on-Chip Biosensors. *Cellular and Molecular Bioengineering, 13*(4), 313–329. doi:10.100712195-020-00642-z PMID:32837587

Udaya Kumar, A., Sai Ganesh, D., Vamsi Krishna, T., Sashank, B., & Satyanarayana, T. (2022). Modeling and investigation on mixing characteristics of T & Y-shaped micromixers for microfluidic devices. *Materials Today: Proceedings, 59*, 501–505. doi:10.1016/j.matpr.2021.11.474

Uhlig, Gaudet, Langa, Schimmanz, Conrad, Kaiser, & Schenk. (2018). Electrostatically Driven In-Plane Silicon Micropump for Modular Configuration. *Micromachines, 9*(4).

Uhlig, S., Gaudet, M., & Langa, S. (2018). Electrostatically driven inplane silicon micropump for modular configuration. *Micromachines, 9*.

Ulkir, O., Girit, O., & Ertugrul, I. (2021). Design and Analysis of a Laminar Diffusion-Based Micromixer with Microfluidic Chip. *Journal of Nanomaterials*, *2021*, 1–10. Advance online publication. doi:10.1155/2021/6684068

Umapathi, U., Shin, P., Nakagaki, K., Leithinger, D. & Ishii, H. (2018). Programmable Droplets for Interaction. 1–1. . doi:10.1145/3170427.3186607

Unger, M. A., Chou, H. P., Thorsen, T., Scherer, A., & Quake, S. R. (2000). Monolithic microfabricated valves and pumps by multilayer soft lithography. *Science*, *288*(5463), 113–116. doi:10.1126cience.288.5463.113 PMID:10753110

Uram, J.D., Ke, K., Hunt, A.J., & Mayer, M. (2006). Label-free affinity assays by rapid detection of immune complexes in submicrometer pores. *Angewandte Chemie*. *45* (14), 2281–2285. . hdl:2027.42/50668. PMID 16506296. doi:10.1002/anie.200502862

Uson, L., Arruebo, M., Sebastian, V., & Santamaria, J. (2018). Single phase microreactor for the continuous, high-temperature synthesis of <4 nm superparamagnetic iron oxide nanoparticles. *Chemical Engineering Journal*, *340*, 66–72. doi:10.1016/j.cej.2017.12.024

Uvarov, I.V., Lemekhov, S.S., & Melenev, A.E. (2016). A simple electrochemical micropump, design and fabrication. *J Phys Conf Ser, 741*.

Uvarov, I. V., Lemekhov, S. S., Melenev, A. E., & Svetovoy, V. B. (2017). Exploding microbubbles driving a simple electrochemical micropump. *Journal of Micromechanics and Microengineering*, *27*(10), 105009. doi:10.1088/1361-6439/aa8914

Uvarov, I. V., Shlepakov, P. S., Melenev, A. E., Ma, K., Svetovoy, V. B., & Krijnen, G. J. M. (2021). A Peristaltic Micropump Based on the Fast-Electrochemical Actuator, Design, Fabrication, and Preliminary Testing. *Actuators*, *10*(3), 62. doi:10.3390/act10030062

Uygun, M., Tasdelen, M. A., & Yagci, Y. (2010). Influence of Type of Initiation on Thiol-Ene "Click" Chemistry. *Macromolecular Chemistry and Physics*, *211*(1), 103–110. doi:10.1002/macp.200900442

Vaishampayan, V., Kapoor, A., & Gumfekar, S. P. (2023). Enhancement in the limit of detection of lab-on-chip microfluidic devices using functional nanomaterials. *Canadian Journal of Chemical Engineering*, cjce.24915. Advance online publication. doi:10.1002/cjce.24915

van Deemter, J. J., Zuiderweg, F. J., & Klinkenberg, A. (1995). Longitudinal diffusion and resistance to mass transfer as causes of nonideality in chromatography. *Chemical Engineering Science*, *50*(24), 3869–3882. doi:10.1016/0009-2509(96)81813-6

van Lintel, H. T. G., van De Pol, F. C. M., & Bouwstra, S. (1988). A piezoelectric micropump based on micromachining of silicon. *Sensors and Actuators*, *15*(2), 153–167. doi:10.1016/0250-6874(88)87005-7

Vargason, A. M., Anselmo, A. C., & Mitragotri, S. (2021). The evolution of commercial drug delivery technologies. *Nature Biomedical Engineering*, *5*(9), 951–967. doi:10.103841551-021-00698-w PMID:33795852

Vasudev, A., Kaushik, A., Jones, K., & Bhansali, S. (2012). Prospects of low temperature co-fired ceramic (LTCC) based microfluidic systems for point-of-care biosensing and environmental sensing. *Microfluidics and Nanofluidics*, *14*(3–4), 683–702.

Vasudev, M. C., Anderson, K. D., Bunning, T. J., Tsukruk, V. V., & Naik, R. R. (2013). Exploration of plasma-enhanced chemical vapor deposition as a method for thin-film fabrication with biological applications. *ACS Applied Materials & Interfaces*, *5*(10), 3983–3994. doi:10.1021/am302989x PMID:23668863

Verma, K., Tarafdar, A., & Badgujar, P. C. (2021). Microfluidics assisted tragacanth gum based sub-micron curcumin suspension and its characterization. *LWT, 135*. . ISSN 0023-6438. S2CID 224875232 doi:10.1016/j.lwt.2020.110269

Vermisoglou, E., Panáček, D., Jayaramulu, K., Pykal, M., Frébort, I., Kolář, M., Hajdúch, M., Zbořil, R., & Otyepka, M. (2020). Human virus detection with graphene-based materials. *Biosensors & Bioelectronics, 166*, 112436. doi:10.1016/j.bios.2020.112436 PMID:32750677

Vodova, M., Nejdl, L., Pavelicova, K., Zemankova, K., Rrypar, T., Skopalova Sterbova, D., Bezdekova, J., Nuchtavorn, N., Macka, M., Adam, V., & Vaculovicova, M. (2022). Detection of Pesticides in Food Products Using Paper-Based Devices by UV-Induced Fluorescence Spectroscopy Combined with Molecularly Imprinted Polymers. *Food Chemistry, 380*, 132141. https://www.sciencedirect.com/science/article/pii/S0308814622001029. doi:10.1016/j.foodchem.2022.132141 PMID:35101791

Volpatti, L. R., & Yetisen, A. K. (2014, July). Commercialization of microfluidic devices. *Trends in Biotechnology, 32*(7), 347–350. doi:10.1016/j.tibtech.2014.04.010 PMID:24954000

Wagh, P., Mujumdar, A., & Naik, J. B. (2019). Preparation and characterization of ketorolac tromethamine-loaded ethyl cellulose micro-/nanospheres using different techniques. *Particulate Science and Technology, 37*(3), 347–357. doi:10.1080/02726351.2017.1383330

Walgama, C., Pollok, N., Boatner, L., & Crooks, R. (2021). A Hybrid Paper and 3D Printed Microfluidic Platform for Electrochemical Detection of Ag-Labeled Magnetic Particles. *ECS Meeting Abstracts, MA2021-02*(57), 1858. 10.1149/MA2021-02571858mtgabs

Wang, F. & Burns, M. (2009). A Performance of nanotechnology. *Microdevices, 11*, 1071–1080.

Wang, Y.-N. & Fu, L. -M. (2018). Micropumps and biomedical applications: A review. *Microelectronic Engineering, 195*, 121-138. doi:10.1016/j.mee.2018.04.008

Wang, C., Mu, X., Huo, J., Zhang, B., & Zhang, K. (2021). Highly-Efficient SERS Detection for E. Coli Using a Microfluidic Chip with Integrated NaYF4:Yb,Er@SiO2@Au under near-Infrared Laser Excitation. *Microsystem Technologies, 27*(9), 3285–3291. doi:10.100700542-020-05087-y

Wang, J., Liu, Y., Shen, Y., Chen, S., & Yang, Z. (2016). A resonant piezoelectric diaphragm pump transferring gas with compact structure. *Micromachines, 7*(12), 219. doi:10.3390/mi7120219 PMID:30404390

Wang, J., & Song, Y. (2017). Microfluidic Synthesis of Nanohybrids. *Small, 13*(18), 1604084. doi:10.1002mll.201604084 PMID:28256806

Wang, J., Zhao, K., Shen, X., Zhang, W., Ji, S., Song, Y., Zhang, X., Rong, R., & Wang, X. (2015). Microfluidic synthesis of ultra-small magnetic nanohybrids for enhanced magnetic resonance imaging. *Journal of Materials Chemistry. C, Materials for Optical and Electronic Devices, 3*(48), 12418–12429. doi:10.1039/C5TC02279G

Wang, Q., Yin, Q., Fan, Y., Zhang, L., Xu, Y., Hu, O., Guo, X., Shi, Q., Fu, H., & She, Y. (2019). Double Quantum Dots-Nanoporphyrin Fluorescence-Visualized Paper-Based Sensors for Detecting Organophosphorus Pesticides. *Talanta, 199*(February), 46–53. doi:10.1016/j.talanta.2019.02.023 PMID:30952284

Wang, T., Chen, J., Zhou, T., & Song, L. (2018). Fabricating Microstructures on Glass for Microfluidic Chips by Glass Molding Process. *Micromachines, 9*(6), 269. doi:10.3390/mi9060269 PMID:30424202

Wang, Y., Luo, J., Liu, J., Li, X., Kong, Z., Jin, H., & Cai, X. (2018). Electrochemical integrated paper-based immunosensor modified with multi-walled carbon nanotubes nanocomposites for point-of-care testing of 17β-estradiol. *Biosensors & Bioelectronics, 107*, 47–53. doi:10.1016/j.bios.2018.02.012 PMID:29428366

Wang, Y., Luo, J., Liu, J., Sun, S., Xiong, Y., Ma, Y., Yan, S., Yang, Y., Yin, H., & Cai, X. (2019). Label-free microfluidic paper-based electrochemical aptasensor for ultrasensitive and simultaneous multiplexed detection of cancer biomarkers. *Biosensors & Bioelectronics*, *136*, 84–90. doi:10.1016/j.bios.2019.04.032 PMID:31039491

Wang, Y., Zang, D., Ge, S., Ge, L., Yu, J., & Yan, M. (2013). A Novel Microfluidic Origami Photoelectrochemical Sensor Based on CdTe Quantum Dots Modified Molecularly Imprinted Polymer and Its Highly Selective Detection of S-Fenvalerate. *Electrochimica Acta*, *107*, 147–154. doi:10.1016/j.electacta.2013.05.154

Watson, J. D. (1990). The human genome project: Past, present, and future. *Science*, *248*(4951), 44–49. doi:10.1126cience.2181665 PMID:2181665

Watt, F., Bettiol, A., Van Kan, J., Teo, E. J., & Breese, M. (2005). Ion beam lithography and nanofabrication: A review. *International Journal of Nanoscience*, *4*(03), 269–286. doi:10.1142/S0219581X05003139

Wei, B., Mao, K., Liu, N., Zhang, M., & Yang, Z. (2018). Graphene nanocomposites modified electrochemical aptamer sensor for rapid and highly sensitive detection of prostate specific antigen. *Biosensors & Bioelectronics*, *121*, 41–46. doi:10.1016/j.bios.2018.08.067 PMID:30196046

Weingart, H. M., & … . (2004). Susceptibility of Pigs and Chickens to SARS Coronavirus. *Emerging Infectious Diseases*, *10*(2), 179–184. doi:10.3201/eid1002.030677 PMID:15030680

Weng, X., & Neethirajan, S. (2017). *Paper based microfluidic aptasensor for food safety*. Academic Press.

Weng, X., Gaur, G., & Neethirajan, S. (2016). Rapid detection of food allergens by microfluidics ELISA-based optical sensor. *Biosensors (Basel)*, *6*(2), 24. doi:10.3390/bios6020024 PMID:27338488

West, J., Becker, M., Tombrink, S., & Manz, A. (2008). J, Becker M, Tombrink S, Manz A. Micro total analysis systems: Latest achievements. *Analytical Chemistry*, *80*(12), 4403–4419. doi:10.1021/ac800680j PMID:18498178

White, F. M. (2016). *Fluid Mechanics*. McGraw-Hill Education.

Whitesides, G. M. (2006). The origins and the future of microfluidics. *Nature, 442*(7101), 368-373.

Whitesides, G. M. (2002). The origins and the future of microfluidics. Nature. 2006;442(7101):368–373.[66] Thorsen T, Maerkl SJ, Quake SR. Microfluidic large-scale integration. *Science*, *298*(5593), 580–584. PMID:12351675

Whitesides, G. M. (2006). The origins and the future of microfluidics. *Nature*, *442*(7101), 368–373. doi:10.1038/nature05058 PMID:16871203

Wikramaratna, P. S., Paton, R. S., Ghafari, M., & Lourenço, J. (2020). Estimating the false-negative test probability of SARS-CoV-2 by RT-PCR. Euro surveillance: bulletin Europeen sur les maladies transmissibles = European communicable disease bulletin, 25(50), 2000568. doi:10.2807/1560-7917.ES.2020.25.50.2000568

Windhorn, K., Meixner, L., Drost, S., Schroder, H., Kilgus, E., Scheel, W., Reichl, H., Ebling, F., Kuchler, M., & Nester, J. (2003). Polymer BioMEMs with integrated optical and fluidic functionality. *Proc. Micro Syst. Technol*, *2003*, 416–423.

Wlodarczyk, K. L., Hand, D. P., & Maroto-Valer, M. M. (2019). Maskless, rapid manufacturing of glass microfluidic devices using a picosecond pulsed laser. *Scientific Reports*, *9*(1), 1–13. doi:10.103841598-019-56711-5 PMID:31882878

Wong, S., Ward, M., & Wharton, C. (2004). Micro T-mixer as a rapid mixing micromixer. *Sensors and Actuators. B, Chemical*, *100*(3), 359–379. doi:10.1016/j.snb.2004.02.008

Woolley, A. T., & Mathies, R. A. (1994). Ultra-high-speed DNA fragment separations using microfabricated capillary array electrophoresis chips. *Proceedings of the National Academy of Sciences of the United States of America*, *91*(24), 11348–11352. doi:10.1073/pnas.91.24.11348 PMID:7972062

World Health Organization. (2021). WHO Coronavirus Disease (COVID-19) Dashboard With Vaccination Data. *World Health Organization.* https://covid19.who.int/%0Ahttps://covid19.who.int/%0Ahttps://covid19.who.int/region/searo/country/bd.

Wu, H., Chen, J., Yang, Y., Yu, W., Chen, Y., Lin, P., & Liang, K. (2022). Smartphone-Coupled Three-Layered Paper-Based Microfluidic Chips Demonstrating Stereoscopic Capillary-Driven Fluid Transport towards Colorimetric Detection of Pesticides. *Analytical and Bioanalytical Chemistry, 414*(5), 1759–1772. doi:10.100700216-021-03839-x PMID:35059790

Wu, H., Odom, T. W., Chiu, D. T., & Whitesides, G. M. (2003). Fabrication of complex three-dimensional microchannel systems in PDMS. *Journal of the American Chemical Society, 125*(2), 554–559. doi:10.1021/ja021045y PMID:12517171

Wu, X., Sun, S., Yu, X., Zhu, X., Xu, M., Wu, G., & Xu, J. (2022). Review on Microfluidic Construction of Advanced Nanomaterials for High-Performance Energy Storage Applications. *Energy & Fuels, 36*(9), 4708–4727. doi:10.1021/acs.energyfuels.2c00576

Wu, Y., Yu, C., Wu, F., Li, C., Zhou, J., Gong, Y., Rao, Y., & Chen, Y. (2017). A highly sensitive fiber-optic microphone based on graphene oxide membrane. *Journal of Lightwave Technology, 35*(19), 4344–4349. doi:10.1109/JLT.2017.2737639

Xavier, M. S., & Charbel, D. (2022). Soft Pneumatic Actuators, A Review of Design, Fabrication, Modeling, Sensing, Control and Applications. *IEEE Access : Practical Innovations, Open Solutions, 10*, 59442–59485. doi:10.1109/ACCESS.2022.3179589

Xia, N., Hunt, T.P., Mayers, B.T., Alsberg, E., Whitesides, G.M., Westervelt, R.M., Ingber, D.E. (2006). Combined microfluidic-micromagnetic separation of living cells in continuous flow. *Biomedical Microdevices, 8* (4), 299–308. . PMID 17003962. S2CID 14534776. doi:10.1007/s10544-006-0033-0

Xu, C., Akakuru, O. U., Zheng, J., & Wu, A. (2019). Applications of Iron Oxide-Based Magnetic Nanoparticles in the Diagnosis and Treatment of Bacterial Infections. *Frontiers in Bioengineering and Biotechnology, 7*, 141. doi:10.3389/fbioe.2019.00141 PMID:31275930

Xu, J., Hu, X., Khan, H., Tian, M., & Yang, L. (2019). Converting Solution Viscosity to Distance-Readout on Paper Substrates Based on Enzyme-Mediated Alginate Hydrogelation: Quantitative Determination of Organophosphorus Pesticides. *Analytica Chimica Acta, 1071*, 1–7. doi:10.1016/j.aca.2019.04.017 PMID:31128750

Xu, J., Zhang, S., MacHado, A., Lecommandoux, S., Sandre, O., Gu, F., & Colin, A. (2017). Controllable Microfluidic Production of Drug-Loaded PLGA Nanoparticles Using Partially Water-Miscible Mixed Solvent Microdroplets as a Precursor. *Scientific Reports, 7*(1), 1–12. doi:10.103841598-017-05184-5 PMID:28684775

Xu, L., Peng, J., Srinivasakannan, C., Chen, G., & Shen, A. Q. (2015). Synthesis of copper nanocolloids using a continuous flow based microreactor. *Applied Surface Science, 355*, 1–6. doi:10.1016/j.apsusc.2015.07.070

Xu, L., Srinivasakannan, C., Peng, J., Zhang, D., & Chen, G. (2015). Synthesis of nickel nanoparticles by aqueous reduction in continuous flow microreactor. *Chemical Engineering and Processing, 93*, 44–49. doi:10.1016/j.cep.2015.04.010

Xu, L., Srinivasakannan, C., Peng, J., Zhang, L., & Zhang, D. (2017). Synthesis of Cu-CuO nanocomposite in microreactor and its application to photocatalytic degradation. *Journal of Alloys and Compounds, 695*, 263–269. doi:10.1016/j.jallcom.2016.10.195

Yadav, A. K., Banerjee, S., Kumar, R., Kar, K. K., Ramkumar, J., & Dasgupta, K. (2018). Mechanical Analysis of Nickel Particle-Coated Carbon Fiber-Reinforced Epoxy Composites for Advanced Structural Applications. *ACS Applied Nano Materials, 1*(8), 4332–4339. Advance online publication. doi:10.1021/acsanm.8b01193

Yakoh, A., Pimpitak, U., Rengpipat, S., Hirankarn, N., Chailapakul, O., & Chaiyo, S. (2021). Paper-based electrochemical biosensor for diagnosing COVID-19: Detection of SARS-CoV-2 antibodies and antigen. *Biosensors & Bioelectronics, 176*, 112912. doi:10.1016/j.bios.2020.112912 PMID:33358057

Yamada, M., Kobayashi, J., Yamato, M., Seki, M., & Okano, T. (2008). Millisecond treatment of cells using microfluidic devices via two-step carrier-medium exchange. [PubMed: 18432348]. *Lab on a Chip, 8*(5), 772–778. doi:10.1039/b718281c PMID:18432348

Yang, J., & Li, X. (2021). Research and development of a novel pneumaticelectric hybrid actuator. *Journal of Mechanical Science and Technology, 35*(6).

Yang, Z., Dong, L., Wang, M., Liu, G., Li, X., & Li, Y. (2022). A wearable insulin delivery system based on a piezoelectric micropump. *Sensors and Actuators A: Physical, 347*, 113909. doi:10.1016/j.sna.2022.113909

Yang, B., & Lee, C. (2010). Non-resonant electromagnetic wideband energy harvesting mechanism for low frequency vibrations. *Microsystem Technologies, 16*(6), 961–966. doi:10.100700542-010-1059-z

Yang, H., Wei, S. X., Chen, H., Chen, L., Au, C. T., Xie, T. L., & Yin, S. F. (2022). A high-throughput chaotic advection microreactor for preparation of uniform and aggregated barium sulfate nanoparticles. *AIChE Journal. American Institute of Chemical Engineers, 68*(April), 1–10. doi:10.1002/aic.17810

Yang, J., Zhang, M., Si, C., Han, G., Ning, J., Yang, F., & Wang, X. (2019). A T-shape aluminum nitride thin-film piezoelectric MEMS resonant accelerometer. *Journal of Microelectromechanical Systems, 28*(5), 776–781. doi:10.1109/JMEMS.2019.2924956

Yang, K., Peretz-Soroka, H., Liu, Y., & Lin, F. (2016). Novel Developments in Mobile Sensing Based on the Integration of Microfluidic Devices and Smartphones. *Lab on a Chip, 16*(6), 943–958. doi:10.1039/C5LC01524C PMID:26899264

Yang, N., Shaheen, N., Xie, L., Yu, J., Ahmad, H., & Mao, H. (2019). Pesticide residues identification by optical spectrum in the time-sequence of enzyme inhibitors performed on microfluidic paper-based analytical devices (μPADs). *Molecules (Basel, Switzerland), 24*(13), 2428. doi:10.3390/molecules24132428 PMID:31269660

Yang, N., Shen, K., Guo, J., Tao, X., Xu, P., & Mao, H. (2017). Error Analysis for Pesticide Detection Performed on Paper-Based Microfluidic Chip Devices. *Modern Physics Letters B, 31*(19–21), 1–6. doi:10.1142/S0217984917400243

Yang, N., Zhou, X., Yu, D., Jiao, S., Han, X., Zhang, S., Yin, H., & Mao, H. (2020). Pesticide residues identification by impedance time-sequence spectrum of enzyme inhibition on multilayer paper-based microfluidic chip. *Journal of Food Process Engineering, 43*(12), e13544. doi:10.1111/jfpe.13544

Yang, W., Zhou, X., Zhao, J., & Xu, W. (2018). A Cascade Amplification Strategy of Catalytic Hairpin Assembly and Hybridization Chain Reaction for the Sensitive Fluorescent Assay of the Model Protein Carcinoembryonic Antigen. *Mikrochimica Acta, 185*(2), 100. doi:10.100700604-017-2620-6 PMID:29594400

Yang, Z., Dong, L., Wang, M., Jia, Y., Wang, C., Li, P., & Liu, G. (2022). Sensors and Actuators : A. Physical Controllable synthesis of silver nanoparticles using a multi-stage microfluidic reactor driven by two valveless piezoelectric pumps. *Sensors and Actuators. A, Physical, 346*(September), 113871. doi:10.1016/j.sna.2022.113871

Yang, Z., Matsumoto, S., Goto, H., Matsumoto, M., & Maeda, R. (2001). Ultrasonic micromixer for microfluidic systems. *Sensors and Actuators. A, Physical, 93*(3), 266–272. doi:10.1016/S0924-4247(01)00654-9

Yan, J., Xu, X., Zhou, J., Liu, C., Zhang, L., Wang, D., Yang, F., & Zhang, H. (2020). Fabrication of a pH/Redox-Triggered Mesoporous Silica-Based Nanoparticle with Microfluidics for Anticancer Drugs Doxorubicin and Paclitaxel Codelivery. *ACS Applied Bio Materials, 3*(2), 1216–1225. Advance online publication. doi:10.1021/acsabm.9b01111 PMID:35019322

Yazdi, S. A. F. F., Corigliano, A., & Ardito, R. (2019). 3-D design and simulation of a piezoelectric micropump. *Micromachines, 10*(4), 259. doi:10.3390/mi10040259 PMID:31003481

Yeo, L. Y., Chang, H.-C., Chan, P. P. Y., & Friend, J. R. (2011). Microfluidic devices for bioapplications. *Small, 7*(1), 12–48. doi:10.1002mll.201000946 PMID:21072867

Yildirim, Y. A., Toprak, A., & Tigli, O. (2018). Piezoelectric membrane actuators for micropump applications using PVDF-TrFE. *Journal of Microelectromechanical Systems, 27*(1), 86–94. doi:10.1109/JMEMS.2017.2773090

Yildiz, M. Z., & Asadi Dereshgi, H. (2019). Design of PZT micropumps for biomedical applications, glaucoma treatment. *Journal of Engineering Research, 7*, 226–241.

Yin, H. Y., Li, Y. T., Tsai, W. C., Dai, H. Y., & Wen, H. W. (2022). An immunochromatographic assay utilizing magnetic nanoparticles to detect major peanut allergen Ara h 1 in processed foods. *Food Chemistry, 375*, 131844. doi:10.1016/j.foodchem.2021.131844 PMID:34952385

Yue, J., Falke, F. H., Schouten, J. C., & Nijhuis, T. A. (2013). Microreactors with integrated UV/Vis spectroscopic detection for online process analysis under segmented flow. *Lab on a Chip, 13*(24), 4855. Advance online publication. doi:10.1039/c3lc50876e PMID:24178763

Yu, H., Wang, M., Zhou, J., Yuan, B., Luo, J., Wu, W., Chen, Z., & Yu, R. (2020). Microreactor-assisted synthesis of α-alumina nanoparticles. *Ceramics International, 46*(9), 13272–13281. doi:10.1016/j.ceramint.2020.02.105

Yus, C., Arruebo, M., Irusta, S., & Sebastián, V. (2020). Microflow nanoprecipitation of positively charged gastroresistant polymer nanoparticles of Eudragit® RS100: A study of fluid dynamics and chemical parameters. *Materials (Basel), 13*(13), 2925. Advance online publication. doi:10.3390/ma13132925 PMID:32629799

Zainal, M. A., Sahlan, S., & Mohamed, S. M. A. (2015). Micromachined Shape-Memory-Alloy Microactuators and Their Application in Biomedical Devices. *Micromachines, 7*(7), 879–901. doi:10.3390/mi6070879

Zang, D., Ge, L., Yan, M., Song, X., & Yu, J. (2012). Electrochemical immunoassay on a 3D microfluidic paper-based device. *Chemical Communications, 48*(39), 4683–4685. doi:10.1039/c2cc16958d PMID:22301713

Zanoli, L. M., & Spoto, G. (2012). Isothermal amplification methods for the detection of nucleic acids in microfluidic devices. *Biosensors (Basel), 3*(1), 18–43. doi:10.3390/bios3010018 PMID:25587397

Zeng, L., & Palmer, J. (2005). Enhancement of Micromixing Tees Using Ultrasound Energy. In Microreactor Technology and Process Intensification (pp. 322–333). doi:10.1021/bk-2005-0914.ch020

Zengerle, R., Ulrich, J., Kluge, S., Richter, M., & Richter, A. (1995). A bidirectional silicon micropump. *Sensors and Actuators. A, Physical, 50*(1-2), 81–86. doi:10.1016/0924-4247(96)80088-4

Zhang, C., Xing, D., & Li, Y. (2007). Micropumps, microvalves, and micromixers within PCR microfluidic chips: Advances and trends. *Biotechnology Advances, 25*(5), 483-514. doi:10.1016/j.biotechadv.2007.05.003

Zhang, Y. & Nguyen, N.T. (2017). Magnetic digital microfluidics - a review. *Lab on a Chip, 17*(6), 994–1008. . hdl:10072/344389. PMID 28220916. S2CID 5013542. doi:10.1039/C7LC00025A

Zhang, Z., Demir, K.G., & Gu, G.X. (2019). Developments in 4D-printing: a review on current smart materials, technologies, and applications. *IJSNM*, 205-224. doi:10.1080/19475411.2019.1591541

Zhang, D., Ye, K., Yao, Y., Liang, F., Qu, T., Ma, W., Yang, B., Dai, Y., & Watanabe, T. (2019). Controllable synthesis of carbon nanomaterials by direct current arc discharge from the inner wall of the chamber. *Carbon, 142*, 278–284. Advance online publication. doi:10.1016/j.carbon.2018.10.062

Zhang, H., Wei, X., Ding, Y., Jiang, Z., & Ren, J. (2019). A low noise capacitive MEMS accelerometer with anti-spring structure. *Sensors and Actuators. A, Physical*, *296*, 79–86. doi:10.1016/j.sna.2019.06.051

Zhang, J., Yang, L., Xu, H., Zhou, J., Sang, Y., Cui, Z., Liu, C., Liu, J., Guo, T., Wang, X., Wang, L., Chen, G., & Chen, X. (2022). Dip-Coating Self-Assembly Fabrication and Polarization Sensitive Photoresponse of Aligned Single-Walled Carbon Nanotube Film. *Sensors (Basel)*, *22*(2), 490. doi:10.339022020490 PMID:35062451

Zhang, J., Zhu, L., Jia, H., Wei, K., & Wen, L. (2022). Microreactor facilitated preparation and Ni-doping of MnO2 nanoparticles for supercapacitors. *Journal of Alloys and Compounds*, *889*, 161772. doi:10.1016/j.jallcom.2021.161772

Zhang, L., Feng, Q., Wang, J., Zhang, S., Ding, B., Wei, Y., Dong, M., Ryu, J.-Y., Yoon, T.-Y., Shi, X., Sun, J., & Jiang, X. (2015). Microfluidic Synthesis of Hybrid Nanoparticles with Controlled Lipid Layers: Understanding Flexibility-Regulated Cell–Nanoparticle Interaction. *ACS Nano*, *9*(10), 9912–9921. doi:10.1021/acsnano.5b05792 PMID:26448362

Zhang, Q., Zhang, Z., Xu, S., Da, L., Lin, D., & Jiang, C. (2022). Enzyme-Free and Rapid Visual Quantitative Detection for Pesticide Residues Utilizing Portable Smartphone Integrated Paper Sensor. *Journal of Hazardous Materials*, *436*, 129320. doi:10.1016/j.jhazmat.2022.129320 PMID:35739808

Zhang, S., Qiu, G., Ting, Y. P., & Chung, T. S. (2013). Silver-PEGylated dendrimer nanocomposite coating for anti-fouling thin film composite membranes for water treatment. *Colloids and Surfaces. A, Physicochemical and Engineering Aspects*, *436*, 207–214. Advance online publication. doi:10.1016/j.colsurfa.2013.06.027

Zhang, S., Wang, T., Lou, L., Tsang, W., Sawada, R., Kwong, D., & Lee, C. (2014). Annularly grooved diaphragm pressure sensor with embedded silicon nanowires for low pressure application. *Journal of Microelectromechanical Systems*, *23*(6), 1396–1407. doi:10.1109/JMEMS.2014.2313635

Zhang, Y., Bauer, R., Whitmer, W. M., Jackson, J. C., Windmill, J. F. C., & Uttamchandani, D. (2018). A MEMS microphone inspired by ormia for spatial sound detection. *Proceedings of the 2018 IEEE Micro Electro Mechanical Systems (MEMS)*, 253–256. 10.1109/MEMSYS.2018.8346532

Zhang, Y., Chan, H. F., & Leong, K. W. (2013). Advanced materials and processing for drug delivery: The past and the future. *Advanced Drug Delivery Reviews*, *65*(1), 104–120. doi:10.1016/j.addr.2012.10.003 PMID:23088863

Zhang, Y., & Jiang, H. R. (2016). A review on continuous-flow microfluidic PCR in droplets: Advances, challenges and future. *Analytica Chimica Acta*, *914*, 7–16. doi:10.1016/j.aca.2016.02.006 PMID:26965323

Zhang, Y., & Ozdemir, P. (2009). Microfluidic DNA amplification – a review. *Analytica Chimica Acta*, *638*(2), 115–125. doi:10.1016/j.aca.2009.02.038 PMID:19327449

Zhang, Y., Wang, T., Luo, A., Hu, Y., Li, X., & Wang, F. (2018). Micro electrostatic energy harvester with both broad bandwidth and high normalized power density. *Applied Energy*, *212*, 362–371. doi:10.1016/j.apenergy.2017.12.053

Zhang, Z., Kan, J., Cheng, G., Wang, H., & Jiang, Y. (2013). A piezoelectric micropump with an integrated sensor based on space-division multiplexing. *Sensors and Actuators. A, Physical*, *203*, 29–36. doi:10.1016/j.sna.2013.08.027

Zhang, Z., Ma, X., Li, B., Zhao, J., Qi, J., Hao, G., Jianhui, R., & Yang, X. (2020). Fluorescence Detection of 2,4-Dichlorophenoxyacetic Acid by Ratiometric Fluorescence Imaging on Paper-Based Microfluidic Chips. *Analyst (London)*, *145*(3), 963–974. doi:10.1039/C9AN01798D PMID:31828260

Zhao, B., MacMinn, C. W., & Juanes, R. (2016). Wettability control on multiphase flow in patterned microfluidics. *Proceedings of the National Academy of Sciences of the United States of America*, *113*(37), 10251–10256. doi:10.1073/pnas.1603387113 PMID:27559089

Zhao, C. X., He, L., Qiao, S. Z., & Middelberg, A. P. J. (2011). Nanoparticle synthesis in microreactors. *Chemical Engineering Science*, *66*(7), 1463–1479. doi:10.1016/j.ces.2010.08.039

Zhao, H., Wang, J. X., Wang, Q. A., Chen, J. F., & Yun, J. (2007). Controlled liquid antisolvent precipitation of hydrophobic pharmaceutical nanoparticles in a MicroChannel reactor. *Industrial & Engineering Chemistry Research*, *46*(24), 8229–8235. doi:10.1021/ie070498e

Zhao, T., Li, L., Chen, R., Wu, H., Zhang, X., Chen, S., Xie, M., Wu, F., Lu, J., & Amine, K. (2015). Design of surface protective layer of LiF/FeF3 nanoparticles in Li-rich cathode for high-capacity Li-ion batteries. *Nano Energy*, *15*, 164–176. Advance online publication. doi:10.1016/j.nanoen.2015.04.013

Zhao, Y., Cheng, Y., Shang, L., Wang, J., Xie, Z., & Gu, Z. (2015). Microfluidic Synthesis of Barcode Particles for Multiplex Assays. *Small*, *11*(2), 151–174. doi:10.1002mll.201401600 PMID:25331055

ZhaoZ.. A simple magnetic nanoparticles-based viral RNA extraction method for efficient detection of SARS-CoV-2. bioRxiv 2020.02.22.961268 (2020). doi:10.1101/2020.02.22.961268

Zheng, X., Li, L., Cui, K., Zhang, Y., Zhang, L., Ge, S., & Yu, J. (2018). Ultrasensitive Enzyme-free Biosensor by Coupling Cyclodextrin Functionalized Au Nanoparticles and High-Performance Au-Paper Electrode. *ACS Applied Materials & Interfaces*, *10*(4), 3333–3340. doi:10.1021/acsami.7b17037 PMID:29318878

Zhou, J., Li, B., Qi, A., Shi, Y., Qi, J., Xu, H., & Chen, L. (2020). ZnSe Quantum Dot Based Ion Imprinting Technology for Fluorescence Detecting Cadmium and Lead Ions on a Three-Dimensional Rotary Paper-Based Microfluidic Chip. *Sensors and Actuators. B, Chemical*, *305*, 305. doi:10.1016/j.snb.2019.127462

Zhou, L., Chen, Y., Fang, X., Liu, Y., Du, M., Lu, X., Li, Q., Sun, Y., Ma, J., & Lan, T. (2020). Microfluidic-RT-LAMP chip for the point-of-care detection of emerging and re-emerging enteric coronaviruses in swine. *Analytica Chimica Acta*, *1125*, 57–65. doi:10.1016/j.aca.2020.05.034 PMID:32674781

Zhou, X., Gao, M., & Gui, L. (2017). A Liquid-Metal Based Spiral Magnetohydrodynamic Micropump. *Micromachines*, *8*(365), 1–10. PMID:30400555

Zhou, X., Poenar, D. P., Liu, K. Y., Li, W., Tse, M. S., Chen, H., Heng, C.-K., & Tan, S. N. (2007). Glass-based BioMEMS devices for optically excited cell impedance measurement. *Sensors and Actuators. A, Physical*, *133*(2), 301–310. doi:10.1016/j.sna.2006.06.037

Zhou, Y., Wu, Y., Ding, L., Huang, X., & Xiong, Y. (2021). Point-of-care COVID-19 diagnostics powered by lateral flow assay. *Trends in analytical chemistry. Trends in Analytical Chemistry*, *145*, 116452. doi:10.1016/j.trac.2021.116452 PMID:34629572

Zhu, K.Y., Leung, K.W., Ting, A.K., Wong, Z.C., Ng, W.Y., Choi, R.C. (March 2012). Microfluidic chip based nano liquid chromatography coupled to tandem mass spectrometry for the determination of abused drugs and metabolites in human hair. *Analytical and Bioanalytical Chemistry*, *402* (9).. PMID 22281681. S2CID 7748546 doi:10.1007/s00216-012-5711-6

Zhu, X., Wang, K., Yan, H., Liu, C., Zhu, X., & Chen, B. (2022). Microfluidics as an Emerging Platform for Exploring Soil Environmental Processes: A Critical Review. In Environmental Science and Technology (Vol. 56, Issue 2). doi:10.1021/acs.est.1c03899

Zhuang, S., Lee, E. S., Lei, L., Nunna, B. B., Kuang, L., & Zhang, W. (2016). Synthesis of nitrogen-doped graphene catalyst by high-energy wet ball milling for electrochemical systems. *International Journal of Energy Research*, *40*(15), 2136–2149. Advance online publication. doi:10.1002/er.3595

Zhu, J., Chen, Q., Kutsanedzie, F. Y. H., Yang, M., Ouyang, Q., & Jiang, H. (2017). Highly Sensitive and Label-Free Determination of Thiram Residue Using Surface-Enhanced Raman Spectroscopy (SERS) Coupled with Paper-Based Microfluidics. *Analytical Methods*, *9*(43), 6186–6193. doi:10.1039/C7AY01637A

Zhu, J., Guo, X., & Huang, R. (2018). Asymmetric disappearance and periodic asymmetric phenomena of rocking dynamics in micro dual-capacitive energy harvester. *AIP Conference Proceedings*, *1971*. doi:10.1063/1.5041144

Zhu, J., Wang, A., Hu, H., & Zhu, H. (2017). Hybrid electromagnetic and triboelectric nanogenerators with multi-impact for wideband frequency energy harvesting. *Energies*, *10*(12), 2024. doi:10.3390/en10122024

Zhu, L., Mei, X., Peng, Z., Yang, J., & Li, Y. (2022). A Paper-Based Microfluidic Sensor Array Combining Molecular Imprinting Technology and Carbon Quantum Dots for the Discrimination of Nitrophenol Isomers. *Journal of Hazardous Materials*, *435*, 129012. doi:10.1016/j.jhazmat.2022.129012 PMID:35504132

Zhu, S., Li, H., Yang, M., & Pang, S. W. (2018). Label-free detection of live cancer cells and DNA hybridization using 3D multilayered plasmonic biosensor. *Nanotechnology*, *29*(36), 365503. doi:10.1088/1361-6528/aac8fb PMID:29848789

Zhu, X., Sarwar, M., Zhu, J.-J., Zhang, C., Kaushik, A., & Li, C.-Z. (2018). Using a glucose meter to quantitatively detect disease biomarkers through a universal nanozyme integrated lateral fluidic sensing platform. *Biosensors & Bioelectronics*, 126. PMID:30544082

Ziaie, B., Baldi, A., Lei, M., Gu, Y., & Siegel, R. A. (2004). Hard and soft micromachining for BioMEMS: Review of techniques and examples of applications in microfluidics and drug delivery. *Advanced Drug Delivery Reviews*, *56*(2), 145–172. doi:10.1016/j.addr.2003.09.001 PMID:14741113

Zou, S., Zhao, X., Ouyang, W., & Xu, S. (2022). Microfluidic Synthesis, Doping Strategy, and Optoelectronic Applications of Nanostructured Halide Perovskite Materials. *Micromachines*, *13*(10), 1647. doi:10.3390/mi13101647 PMID:36296000

About the Contributors

Rajeev Kumar Singh is working as Associate Professor (Mechanical Engineering) in Amity University Uttar Pradesh Noida UP since 2017. He had worked as Associate Professor (Mechanical Engineering) in Glocal University, Saharanpur from 2014 to 2017. Prior to this, he was working as Assistant Professor (Sr. Grade) in PSG Institute of Advanced Studies, Coimbatore TN from 2013 to 2014. Dr. Rajeev Kumar Singh has worked in faculty position in Northern India Engineering College, Lucknow from 2000 to 2007. Apart from this, he has also worked in Nagpur Alloy Casting Limited, Nagpur MS and Crown Brass Pvt. Ltd., Bahadurgarh Distt. Jhajjar HR. Dr. Rajeev Kumar Singh has completed Ph.D. in Mechanical Engineering from Indian Institute of Technology, Kanpur in 2014; M. Tech. in Mechanical Engineering (Industrial Systems Engineering) from Kamla Nehru Institute of Technology, Sultanpur UP in 2007 and B.E. (Mechanical Engineering) from Yeshwant Rao Chavan College of Engineering, Nagpur in 1992.

Rakesh Kumar Phanden is B.Tech. Mechanical Engineering, MTech. Integrated Product Design and Manufacturing from the Department of Mechanical Engineering, G.J.U. of Science and Technology, Hisar, India. He completed his PhD in Mechanical Engineering from the National Institute of Technology, Kurukshetra, India in 2012. He is currently working as Associate Professor in the Department of Mechanical Engineering at the Amity University, Uttar Pradesh, India. He has chaired many technical sessions in conferences and reviewed articles for many SCIs and Scopus indexed international journals of repute. He has organized various international conferences namely ICNDME 2014, FLAME 2020, FLUTE 2021, FLUTE 2023 and FLAME 2022 in the capacity of organizing secretary and convener. He is the sponsor of RSP Scholarship Program in which Rs. 50K is awarded to 5th standard students of his village every year on 15th August (Indian Independence Day). He has published 4 books and edited 5 special issue for a Scopus indexed journal. He is serving as an editorial board member for many journals. Currently, he is working as Guest Editor in 4 Special Issues on Industry 4.0, Sustainability Manufacturing, Lean Six Sigma to be published with Scopus indexed International Journals of Repute. He has 16 years of teaching experience at private and government institutes and universities in India and abroad. He has contributed more than 60 papers at the national/international levels including SCI journals namely International Journal of Computer Integrated Manufacturing, International Journal of Advanced Manufacturing Technology, Journal of Enhanced Heat Transfer, Journal of Manufacturing Technology and many more. His current areas of interest include thermal management of electronic devices, MEMS and Microfluidic Systems, Cyber-physical production systems, energy-aware modelling of systems, manufacturing systems, tool path optimization, production scheduling, integration of process planning and scheduling and product design and manufacturing. He has received International Travel Grant from DST to visit the USA and visited London, Singapore, Turkey, Dubai, and many African countries for various research work in collaboration with Amity University.

Basant Singh Sikarwar is a Postdoc from Iowa State University USA, Ph.D. from Indian Institute of Technology Kanpur, India, and M.Tech from Indian Institute of Technology Roorkee, India. His research experience is quite diversified, and he has been involved in experimental and simulation in many aspects of thermo-fluids research with an emphasis on phase change phenomena and the formation of interfaces. The subject is poised for growth because of several engineering applications such as the optimization of cooling techniques for thermal management of engineering systems, cooling of electronic equipment, heat pipes fabrication, water harvesting from moist air, and thermal energy storage. During his Ph.D., he developed a Finite Volume-based CFD solver and many students of his research groups at Amity university have been using this CFD solver for their research in thermal and fluid engineering. He published more than 40 SCI/Scopus-indexed journals and more than 100 Scopus-indexed conference publications. He completed one SERB-funded project and three funded projects from various funding agencies (BRNS and DST) are going on. He chaired many conferences. He is a member of the editorial board of many journals and book series, and he is a guest editor in the journal Enhanced Heat Transfer (Begell House Corp.). He is a consultant in many industries, and he has guided four Ph.D. Currently, four students are carrying out his research under his supervision. He has been heading the department of Mechanical engineering, Amity University Uttar Pradesh, since 2018. In addition, he is an Assistant Director of the Amity directorate of Engineering, Technology, and Innovation (ADETI).

J. Paulo Davim received his Ph.D. degree in Mechanical Engineering in 1997, M.Sc. degree in Mechanical Engineering (materials and manufacturing processes) in 1991, Mechanical Engineering degree (5 years) in 1986, from the University of Porto (FEUP), the Aggregate title (Full Habilitation) from the University of Coimbra in 2005 and the D.Sc. from London Metropolitan University in 2013. He is Eur Ing by FEANI-Brussels and Senior Chartered Engineer by the Portuguese Institution of Engineers with an MBA and Specialist title in Engineering and Industrial Management. Currently, he is Professor at the Department of Mechanical Engineering of the University of Aveiro, Portugal. He has more than 30 years of teaching and research experience in Manufacturing, Materials, Mechanical and Industrial Engineering, with special emphasis in Machining & Tribology. He has also interest in Management, Engineering Education and Higher Education for Sustainability. He has guided large numbers of postdoc, Ph.D. and master's students as well as coordinated & participated in several financed research projects. He has received several scientific awards. He has worked as evaluator of projects for international research agencies as well as examiner of Ph.D. thesis for many universities. He is the Editor in Chief of several international journals, Guest Editor of journals, books Editor, book Series Editor and Scientific Advisory for many international journals and conferences. Presently, he is an Editorial Board member of 25 international journals and acts as reviewer for more than 80 prestigious Web of Science journals. In addition, he has also published as editor (and co-editor) more than 100 books and as author (and co-author) more than 10 books, 80 book chapters and 400 articles in journals and conferences (more than 200 articles in journals indexed in Web of Science core collection/h-index 45+/6000+ citations and SCOPUS/h-index 53+/8500+ citations).

* * *

Anjali Awasthi is working as Assistant Professor in Chemical Engineering Department, at HBTU Kanpur. She did her M.tech from HBTI Kanpur and Ph.D. from Malaviya National Institute of Technology, Jaipur.

Soma Banerjee is working as an Assistant Professor in the department of plastic technology, HBTU Kanpur. PhD from IIT KANPUR. BTECH and MTECH from Univ. of Calcutta.

Nitin Chauhan has completed B. Tech. 2010 in Mechanical Engineering from Jawahar Lal Nehru Govt Engineering College, Sundernagar, H.P. He has done post gradution in M. Tech. in 2012 Mechanical Engineering - CAD-CAM from NIT Hamirpur. He has published more than 10 research articles in reputed journals. His area of interest is in tribology, machine design, microfluidics, MEMS.

Anil Ghubade was born in India, in 1985. He has received BE in Mechanical from the University of Pune in 2006 and M. Tech in Manufacturing Science from IIT Kanpur in 2009. He is currently pursuing Ph.D. from IIT Ropar. From 2009 to 2021, he is working as an Assistant Professor in the School of Mechanical Engineering, at Lovely Professional University, Punjab, India. His area of interest is in Manufacturing processes, machining, non-traditional machining, Bio-MEMS, Product design, and development, etc. He has published more than 10 research papers.

Aseem Grover received his degree of MPhil (Zoology) in 2021, PG Dipolma in Cyber Law in 2020, and MSc (Forensic Science) in 2019. Now pursuing PhD in Forensic Science from Amity University Haryana. He has research experience of 2 years and published many peer review articles in international journals.

Ankur Gupta is presently an assistant professor at the Indian Institute of Technology, Jodhpur, India. He received his PhD degree in mechanical engineering from the Indian Institute of Technology, Kanpur. He received the "ISEES Young Scientist Award-2017" and the "IEI Young Engineer Award 2019" from the Institution of Engineers (India) in the Production division, He has also participated in the "2nd BRICS Young Scientist Conclave," held at Zhejiang University, China, in 2017. He has also received the "Indian National Young Academy of Sciences (INYAS) membership 2023" in the "Engineering Sciences" category by INYAS, a recognized scientific body under the Indian National Science Academy (INSA), along with the "SERB International Research Experience (SIRE) Fellowship Award (2022)" for a research visit at the Karlsruhe Institute of Technology, Germany.

Karthikeyan K. is working as an Associate Professor at M. Kumarasamy College of Engineering, Karur. He has more than 2.3 years academic experience, 4 years electronics core industrial experience and 5 years full time research and development experience. He completed his full time Ph.D in the field of MEMS & Microfluidics at Rajalakshmi Engineering College under Anna University Chennai. His research interests are MEMS and Microfluidic Device for Lab-On-Chip Applications.

Ashish Kapoor is a faculty in Department of Chemical Engineering at Harcourt Butler Technical University, Kanpur, India. His research interests include microfluidics, transport phenomena, water treatment, lab-on-a-chip sensors, modeling and simulation.

Mohammad Mohsin Khan is an accomplished Assistant Professor in the Mechanical Engineering Department at the esteemed National Institute of Technology, Srinagar. He has garnered a reputation as a prolific researcher and educator with extensive experience in teaching and research. Dr. Khan's research contributions are highly regarded in the academic community. He has published over 40 papers in national and international journals of repute and has presented his research findings at various international conferences. As a member of several prestigious research societies, including the Material Research Society of India, Tribology Society of India, and the Institution of Engineers, Dr. Khan is dedicated to advancing the field of engineering. In addition to his research pursuits, Dr. Khan has also distinguished himself as an exceptional teacher and project supervisor. He teaches at both the undergraduate and postgraduate levels and has guided over 25 projects for B.Tech and M.Tech students. Dr. Khan's dedication to education and research has been recognized with numerous awards and accolades. Dr. Khan's research interests are focused on ultra-high-temperature ceramic composites, light metal alloys, and bio-composites. His work in these areas has the potential to make significant contributions to the field of materials science and engineering.

Sujatha L. is heading the Centre of Excellence in MEMS & Microfluidics at Rajalakshmi Engineering College, Chennai, India. She completed her PhD and Post-doctoral Programs at Indian Institute of Technology Madras. She has received funding for 15 projects from various funding agencies. She has 32 years teaching and research experience. Her research interests are microfabrication of chemical and biosensors.

Jegatha Nambi Krishnan received her B.Tech in the major of Chemical and Electrochemical Engineering from Central Electrochemical Research Institute (CECRI), Karaikudi. She has qualified both GRE and TOEFL from ETS, USA during her B. Tech final year. She received distinguished Korea Science and Engineering Foundation (KOSEF) award to pursue her higher education in South Korea. She pursued her Integrative Ph.D. in Micro / Nano Systems Engineering from Korea Institute of Science and Technology (KIST), Seoul, South Korea in 2011. She received the KIST-IRDA Academic Excellence Award for her research accomplishments. Since 2011, she served as Scientist I in Bio Electronics Program at the Institute of Microelectronics (IME), A*STAR, Singapore. From Sep. 2013, she is a faculty member of Department of Chemical Engineering, BITS Pilani K K Birla Goa Campus. She has published several research papers in both reputed journals and international conferences. In particular, she has patented three of her findings internationally (1 Korean and 2 US patents). While six of her findings have been patented in India. Many of her works have been selected for extremely reputed conference proceedings. She has received the OPERA (Outstanding Potential for Excellence in Academics and Research) award from BITS Pilani to continue her research career at Goa Campus. She has involved and completed both government and industrial projects with the worth of several crores successfully. She has guided and supervised nine M.E students' dissertation work. Presently, one PhD student has successfully submitted his thesis. Two Ph. D scholars are working under her supervision. Her research areas at BITS Pilani Goa Campus include the development of Advanced materials for Electrochemical Systems, Bio-MEMS and Sensors. She is an active member of InDA, HAI, IIChe, IIM, IMS, ISAMPE, ISS, MRSI, SPSI and Biochip Society.

Harsh Nigam completed his Bachelor of Engineering (Chemical Engineering) from BITS Pilani, K.K. Birla Goa Campus, India in 2023. Harsh has worked on a number of projects in the fields of MEMS,

Membrane Condensers, Carbon Capture, Utilization and Storage (CCUS) and Machine Learning both for academia and industry. His research interests lie in the field of sustainable engineering.

Yogesh M. Nimdeo received his Ph.D. degree in Chemical Engineering in 2017 from the Indian Institute of Technology Kanpur, India. M. Chem. Engg. degree in Chemical Engineering in 2009 from the Institute of Chemical Technology Mumbai, India. Chemical Engineering degree in 2007, from the University of Amravati. Just after his Ph.D., he worked as an Institute Post-doctoral fellow at the Indian Institute of Technology Bombay, Mumbai, India (3 years). Currently, he is a Professor at the Department of Chemical Engineering, Indian Institute of Technology Jammu, India. He is working as head of Chemical Engineering. He has more than 10 years of teaching and research experience in Colloids and suspensions, Interfacial dynamics of Complex Fluids, Chemical Engineering, and Industrial Engineering. He is guiding several Ph.D. and master's students as well as coordinating & participated in several financed research projects. He has published several journal papers, Book chapters and received scientific awards.

Dhirendra Patel is an Assistant Professor in the Department of Mechanical Engineering at Amity University Greater Noida. He received his M.Tech from NIT Kurukshetra Haryana in 2009 respectively, and his BTech from Govt. Engineering College Rewa MP, in 2006 and Ph.D. Pursuing. Research areas are micro-and nano-fluidics, heat transfer.

Ankit Patil is a research scholar in the Department of Chemical Engineering at the Indian Institute of Technology, Jammu, India. He pursued his M. Tech degree in Chemical Engineering in 2021 from the University Institute of Chemical Technology, KBC North Maharashtra University, Jalgaon, MS, India. And B. E. in Chemical Engineering in 2018 from Savitribai Phule Pune University, India. Currently, he is working on the application of Pickering Emulsion for Sustainable Drug Delivery as his research work.

Parveen Sharma has completed Ph.D. from the Department of mechanical engineering at the National Institute of Technology (NIT), Kurukshetra, India. The area of research includes Facility Layout Planning for manufacturing industries. He has published more than 10 research papers in national and International Journals.

Ayushi Thakur is working as Assistant Professor in Department of Mechanical Engineering at Amity University Greater Noida. She is also a part time PhD scholar at Amity University, Noida. She has gained the master's degree in mechanical engineering from UPTU Lucknow. She is having almost 10 years of academic experience. Her research interest includes the additive manufacturing of smart materials such as shape memory polymers or alloys, bio mimic properties of smart materials and characterization of smart materials. Applications of smart materials, 4D Printing.

Vijay Vaishampayan is currently pursuing his doctorate degree in Chemical Engineering at Indian Institute of Technology Ropar (IIT Ropar). He holds B.E. in chemical engineering from Savitribai Phule Pune University and obtained his M.Tech in chemical engineering from SRM Institute of Science and Technology, Kattankulathur, Tamil Nadu, India. He has experience in developing frugal technologies, foldscope, composite materials and microfluidics. His research interests include the development of frugal technologies, printed electronics, microfluidic lab-on-a-chip devices, energy storage devices, and electrochemistry. sensors.

Index

I

L

M

N

O

P

R

S

W

www.ingramcontent.com/pod-product-compliance
Lightning Source LLC
LaVergne TN
LVHW062139300926
849161LV00010B/2573